Deformation Theory and Symplectic Geometry

MATHEMATICAL PHYSICS STUDIES

Series Editor:

M. FLATO, *Université de Bourgogne, Dijon, France*

VOLUME 20

Deformation Theory and Symplectic Geometry

Proceedings of the Ascona Meeting, June 1996

edited by

Daniel Sternheimer

C.N.R.S. and
Université de Bourgogne,
Dijon, France

John Rawnsley

Mathematics Institute,
University of Warwick,
Coventry, England, U.K.

and

Simone Gutt

Université Libre de Bruxelles, Belgium
and
Université de Metz, France

KLUWER ACADEMIC PUBLISHERS

DORDRECHT / BOSTON / LONDON

A C.I.P. Catalogue record for this book is available from the Library of Congress

ISBN 0-7923-4525-8

Published by Kluwer Academic Publishers,
P.O. Box 17, 3300 AA Dordrecht, The Netherlands.

Sold and distributed in the U.S.A. and Canada
by Kluwer Academic Publishers,
101 Philip Drive, Norwell, MA 02061, U.S.A.

In all other countries, sold and distributed
by Kluwer Academic Publishers,
P.O. Box 322, 3300 AH Dordrecht, The Netherlands.

Printed on acid-free paper

Printed in the Netherlands

TABLE OF CONTENTS

Preface vii

Alain Connes 1
Noncommutative Differential Geometry and the Structure of Space Time

Ludwig Faddeev and Alexander Volkov 35
Shift operator for the discrete $sl(2)$ current algebra

Moshé Flato, Giuseppe Dito and Daniel Sternheimer 43
Nambu mechanics, n-ary operations and their quantization

Jürg Fröhlich 67
The Geometrization Principle

Christian Frønsdal 69
Generalization and deformations of quantum groups

Murray Gerstenhaber 85
Quantized intersection cardinalities

Simone Gutt and John Rawnsley 103
Deformation quantization and Mp^c structures

Michio Jimbo, Hitoshi Konno and Tetsuji Miwa 117
Massless XXZ model and degeneration of the elliptic algebra $\mathcal{A}_{q,p}(\widehat{sl}_2)$

Maxim Kontsevich 139
Formality conjecture

Bertram Kostant 157
Quantum cohomology of the flag manifold as an algebra of rational functions on a unipotent algebraic group

Olivier Mathieu 177
Homologies associated with Poisson structures

Anthony Bloch, Mohamed Ould El Hadrami, Hermann Flaschka and Tudor S. Ratiu 201
Maximal tori of some symplectomorphism groups and applications to convexity

Michael Polyak and Nicolai Reshetikhin 223
On 2D Yang-Mills theory and invariants of links

Marc Rosso 249
Some applications of quantum shuffles

Wilfried Schmid 259
Character formulas and localization of integrals

Jim Stasheff 271
Deformation theory and the Batalin–Vilkovisky Master Equation

Ryszard Nest and Boris Tsygan 285
The Fukaya type categories for associative algebras

Alan Weinstein 301
Tangential deformation quantization and polarized symplectic groupoids

Martin Bordemann and Stefan Waldmann 315
Formal GNS construction and WKB expansion in deformation quantization

Jens Hoppe 321
On the deformation of time harmonic flows

André Lichnerowicz 327
Distinguished essential extensions of infinite dimensional classical Lie algebras

Hideki Omori, Yoshiaki Maeda, Naoya Miyazaki and Akira Yoshioka 333
Noncommutative contact algebras

Shahn Majid 339
Braided geometry and the inductive construction of Lie algebras and quantum groups

Carlos Moreno and Luis Valero 345
Star products in the triangular case

Giuseppe Dito 353
Aspetti fisici della teoria delle deformazioni

Scientific programme 355

List of participants 358

PREFACE

A short history of time and money.

In early 1994, Michel Cahen had contacted Moshé Flato with the idea of organizing in 1996 a meeting on deformation quantization in Belgium. Discussions on the feasibility of such a project had started when suddenly one midsummer night in July 1994, our friend Dr. G. (a.k.a. Murray Gerstenhaber) had a dream: one should have in mid 1996 a conference to check what happened to his intellectual child, deformation theory, in the past decade. Dr. G. then wrote an e-mail message to Moshé Flato, suggesting that such a conference be organized, possibly "on a shoestring".

The two ideas were put together and shortly afterwards, at the Paris IAMP Congress, Konrad Osterwalder (then dean of Sciences at ETH) mentioned to one of us (D.S.) the new facility which he had created, Centro Stefano Franscini (CSF) in Ascona; there, in addition to very nice accommodation for some 60 participants, approved meetings benefited from a "dowry" coming (via ETH) from the Swiss National Fund and equivalent to about half of the local expenses. He suggested that we apply for a one week slot, which we did: Jürg Fröhlich agreed to be the "main applicant" to CSF and the other 3 applicants to CSF follow him in the list of organizers (in the "Swiss order"); the last two organizers were added shortly afterwards. In July 1995 the Scientific Council of CSF approved the application and the organizers started more actively to contact again possible speakers and participants and to look seriously for additional funding.

Since a large number of participants would be coming from the European Union, Moshé Flato thought of requesting EU support (for them), to complement Swiss and participants' support. The rule of thumb in scientific relations with Switzerland (not an EU member) being that each part covers its own side, a positive preliminary reaction was received in September 1995 and (after formal application) he was informally told in March 1996 that the request would be approved. Formal approval should have been given a month later but bureaucratic complications (including the mad cow crisis backlash on the EU) were such that it came ... at the end of July 1996, at which time we learned that this EU support (EC-DGXII Grant PSS*0981) would by rule be limited to 43% of the total budget. Fortunately this was the case and in December 1996 the EU share of the stay (representing the bulk of the above Grant) could be paid to CSF for our June Conference.

After the EU direct participation, support for the Conference came with a *most generous Swiss participation* from Centro Stefano Franscini and Fondazione Monte Verità. This support took the form both of formal funding (the abovementioned "dowry") and of a very reasonable charge for use of the superb and recently renovated facilities. In addition CSF had gracefully extended credit even though before July there was no legally binding document to show.

Among other *de facto* contributors we should mention the EU Network Contract ERB CHRX-CT94-0701 (coordinated by Flato), which covered the travel to Ascona of 19 participants, the Erwin Schrödinger Institute in Vienna and Université de Bourgogne (which "lent" resp. 4 and 3 participants among their visitors), Japanese grants which supported travel for the 3 Japanese participants, US grants which contributed in part to the presence of many US participants, and others. Last but not least we must mention Kluwer Academic Publishers (and ourselves) who (in lieu of royalties for the Editors) agreed to send to every participant a free copy of these Proceedings. To all, a very grateful *danke merci* as people say in Zürich!

Confident that a reasonable level of funding would eventually be available, the organizers put most of their efforts into the scientific side of the meeting. The existence of two other meetings (in the US and Europe) on related topics of interest to some of the first invited participants and held almost at the same time made the attendance fluid until rather late in the Spring but eventually, as can be seen from the list of participants, the efforts of the organizers and coworkers were very successful. Those who came are happy to have participated in a first class Conference.

A brief history of space.

The Centre where the conference took place has a very interesting history. Stefano Franscini (1796-1857) was a major political figure in Ticino since (at least) 1830, a member of the advanced liberal party and advocating a higher level of education. In 1848 he became Federal minister for education and interior, the first Ticinese to be member of the Federal Council. In this capacity he had a major rôle in the creation, and location in the main city of Zürich (and not in the Federal capital, Bern) of the Swiss Federal Polytechnic (Eidgenössische Technische Hochschule, founded 1855).

In 1869 the Russian anarchist Michail Bakunin settled in Locarno and lived from 1873 to 1874 in Minusio (close to Ascona on Lago di Locarno) where he developed his concept of Utopia, a society in which no one rules over anyone else – an essential, if not always respected, factor in the advancement of Science. In 1889 a member of the Swiss Parliament and theosophist from Locarno, Alfredo Pioda, planned the foundation of a theosophical cloister on what is now called Monte Verità. Starting in 1904, following Raphael Friederberg, M.D., many anarchists came again to Ascona, dreamed to have there "a republic of the homeless, the exiled and the destitute" and planned a university which would lead to a return to the communist paradise (by the way, quite a few Russian scientists came to our meeting). An anarchist-communist agricultural community was founded in Minusio in 1919 by a former secretary of the federation of Swiss labour unions, and lasted until 1964; another community existed between 1924 and 1938 in Ronco sopra Ascona.

In 1900-1920 the champions of an alternative to capitalism and communism founded a cooperative based initially on primitive socialism, later on individualistic vegetarian principles and in a third phase constructed the Monte Verità sanatorium. Many artists, especially dancers, writers and intellectual émigrés (during WW I) visited Ascona in those days. Rudolf von Laban established his "School of Art" on Monte Verità in 1913-1918; the New Dance was born there and Isadora Duncan visited it. In 1918 artists flocked to Ascona in great numbers and that flow continued mainly during the Thirties. The founders of Monte Verità emigrated to Spain and then Brazil in 1920 and in 1923 it became a hotel business run by three artists, bought in 1926 by Baron Eduard von der Heydt (a German banker and famous art collector) and discovered by the artists of the Bauhaus school in 1927. That year the present main building of the Monte Verità hotel was built in the Bauhaus style by Fahrenkamp. After the death of the Baron, Monte Verità became the property of the Republic and Canton Ticino in 1964, and the site of cultural and university meetings (since 1968, mostly centered around psychoanalysis) next to which 3 museums were established.

In 1989 ownership was transferred to a trust and a contract was signed with the ETH Zürich giving it access to the complex for 15-20 weeks a year for seminars, organized by its Centro Stefano Franscini. Renovation of the main building and construction of the main Lecture Hall were done in 1990-1992 together with a number of workshops; the nearby Villa Semiramis was renovated in 1994-1995. Presently the Monte Verità Seminar Centre, supported by a private foundation made up of the Canton Ticino, the town council of Ascona and the Federal Polytechnics of Lausanne and Zürich, makes almost constant use of these superb surroundings by about 20 "Seminari del Monte Verità" conceived by public and private organizations in Ticino, and 20-25 scientific meetings of Centro Stefano Franscini. In keeping with the spirit of the place, the organizers are requested to schedule on one evening a general public lecture in Italian aimed at presenting to the local population the field to which the meeting is dedicated; the abstract of the lecture that was prepared is reproduced in these Proceedings after the scientific contributions. The volume ends with a faithful reproduction of the posters (in Italian and English) issued by CSF to advertise the meeting, with a picture of the main building as a background to the text.

The heart of the matter.

The volume is organized as follows. In the first part we have grouped the contributions written by the main Lecturers in Ascona, possibly with coauthors, in the alphabetic order of family names of the *participants* who delivered the lecture. The analysis below will not completely follow that order because we performed some modest topical regroupings. The second part (following immediately the first part with no blank page) contains the other contributions (again in the same order except for the abstract in Italian which is the last contribution; for the Editors we chose the first names order). The title may differ somewhat from that of the lecture actually given in Ascona, as one can check from the Programme which is reproduced at the end of this volume together with the list (with affiliations) of the participants, but the content is in almost all cases quite close and (for the few exceptions) include new developments to which the informal discussions held in Ascona were an important factor.

As a matter of fact – and this was one of the aims of the Organizers – there were numerous and fruitful informal contacts between participants, favoured by the inspiring beauty of Nature on Monte Verità, the superb accommodation, the (generally fine) weather and (last but not least) the right size of the meeting and the quality of its attendance. Its impact on the development of Science will continue to be felt in the years to come.

The volume starts with its longest contribution, by *Alain Connes.* Noncommutative geometry is known to be closely related to deformation quantization as can be seen e.g. in Connes' book with the former title (his Ref. 7). A few years ago Connes cast the standard model for elementary particles into noncommutative geometry. In the present paper he presents his way of going one step further, giving a consistent framework in which gravity is coupled with the standard model by developing the metric aspect of noncommutative geometry. This is not the first attempt to couple gravity with matter and certainly not the last, but it is an important step in that direction and we are happy to have it here.

Our Swiss organizer *Fröhlich* spoke at Ascona about supersymmetric quantum theory and geometry, which became part of about 450 pages of Lecture Notes (completed in March 1997): by honesty and modesty he limited his contribution to an extended abstract, with references.

At the opposite end of space-time approaches *Faddeev*, once a strong advocate of physics in "realistic" 3+1 space-time dimensions, followed the path of painting in the last century and after doing seminal work in continuum theories in 1+1 dimensions turned "abstract" and presents an instructive development in 1+1 dimensional discrete space-time, a kind of lattice deformation of the Sugawara construction.

Reshetikhin's contribution (with Polyak) deals with an entirely different aspect of modern symplectic geometry, Yang–Mills theory in 2 dimensions and its relation with invariants of knots in circle bundles over surfaces and the quantization of the algebra of functions on the moduli space flat bundles. In this preface we cannot elaborate more in details on this original contribution but the reader is invited to have a closer look at it.

Jim *Stasheff* translates the Batalin–Vilkovisky (BV) approach of master equations, which are integrability conditions for deformations of algebras and differential algebras (in the classical and quantum cases, resp.), into the language of deformation theory. He first reviews the jet bundle approach to Lagrangian field theory and interprets the BV equation in terms of his strong homotopy Lie algebras. Then he discusses some examples where free Lagrangians are deformed to interactive Lagrangians.

Our Japanese speakers *Jimbo* and *Miwa* gave an inspiring development of the connection between the bosonization of the massless XXZ model and the degeneration of some elliptic algebras.

In his contribution, *Frønsdal* goes beyond the universally admitted classes of quantum groups using a universal R-matrix approach and deforming these R-matrices. His approach, starting from basics and questioning conventional thinking in the domain, is original and gives in particular new elliptic quantum groups.

Our expert in quantum shuffles, *Rosso*, makes a superb use of that theory (to which he contributed greatly in its establishment) to the difficult problem of finding some intrinsic definition of quantized enveloping algebras (QUE). The dual algebras (of functions, with star products) were, of course, intrinsically defined but to express the deformed coproduct in enveloping algebras one had to use generators and relations. The general framework here is that of the cotensor Hopf algebras on a Hopf bimodule which generalizes the classical shuffle algebra and in which the QUE fit naturally. Simple but interesting examples are given.

A group of 6 papers deals directly with various aspects of deformation quantization. *Flato* presented a new type of deformation quantization developed in the past couple of years, aimed at providing a quantization of n-ary operations of the type introduced by Nambu in 1973 (Jacobians). No satisfactory quantization of such structures had been found and the solution is original in many respects (generalized deformations, decomposition of polynomials into irreducibles, methods of second quantization etc.).

Gerstenhaber, the "father" of deformation theory of algebras comes with a new development – and a big surprise. He achieves an inductive construction of formulas for quantization of functions of the cardinality of the intersection of two sets (e.g. gravitational force between equal masses placed at the vertices of a cube in n-space), avoiding obstructions. Just read it!

Kontsevich proposes a reasonable conjecture of homological nature (the formality conjecture) which would among others give a positive answer to a longstanding open problem, to prove that there exist star products on *any* Poisson manifold, regular or not. Dealing with the next obvious question, how unique is this construction, he finds a strange flow on the space of Poisson manifolds. His contribution, though not the final answer, is certainly seminal, and is a follow-up of discussions held in Ascona.

Mathieu has a different shot at the quantization of general Poisson manifolds X and in particular describes a test which could give a negative answer, and a counterexample to Kontsevich's formality conjecture; this test is however not easy to verify. The bulk of his paper is devoted to the investigation of various homologies attached to Poisson and symplectic manifolds. His main tool is a generalization of the Koszul differential δ_∞ to a one parameter family δ_τ of differentials. This enables him to define a canonical map Θ^0 between homologies of functions on X and homologies $H^0_*(X)[1]$ of δ_0 (for X Poisson) which (for X symplectic) can be extended to maps Θ^n into $H^0_*(X)[2n+1]$ (similar to the maps S^n in Connes' cyclic cohomology).

Two of us (*S.G.* and *J.R.*) give a proof of an integrality condition for the existence of an Mp^c connection with a prescribed central curvature. We show how these Mp^c connections arise in the geometry of deformation quantization when one tries to build approximate representations (à la Fedosov) of a star product.

Weinstein's paper is a direct development of the Ascona meeting. An important example of a nonregular Poisson manifold is given by the Lie–Poisson structure on the dual $\mathfrak{g}^*$ of a Lie algebra. The question of the compatibility of a differential deformation quantization of $\mathfrak{g}^*$ with its coadjoint orbit decomposition was shown by 4 participants in Ascona to require a very strong algebraic condition. Weinstein gives analogues of that result using a new analogy between deformation quantization and quantization by symplectic groupoids ("geometric deformation quantization").

In his contribution *Kostant* shows that the quantum cohomology of the flag manifold (of a complex semisimple Lie group) naturally injects into the field of rational functions on a unipotent subgroup with generators explicitly given in terms of coefficients of a representation. Quantum cohomology was introduced (heuristically) by Witten in 1988 using a deformation of the usual cup-product of cohomology classes, in connection with supersymmetric nonlinear sigma models with target X, then proved to exist mathematically for a large class of Kähler manifolds X, but its computation remained a very difficult problem. Kostant's construction is particularly elegant both in the end result and the methods used which include Newton's equations for the Toda lattice where the quantum parameters appear as forces.

Nest and *Tsygan* start from Feigin's idea that the cohomology of deformed algebras may be related to Lagrangian intersections and Fukaya's study of the latter which lead him to construct a category of symplectomorphisms. They show (with examples) a Fukaya-type construction, based upon an idea of Manin, in terms of noncommutative geometry for any associative ring, which resembles closely Fukaya's construction when the latter is a result of deformation quantization.

Ratiu's joint contribution is an attempt to study the largely unexplored world of symplectomorphism groups by drawing analogies from the classical theory of compact Lie groups. They summarize their approach in a "conversion dictionary" in which they propose the infinite-dimensional analogues of a number of classical notions, a dictionary which is supported by a number of recent results (of theirs and others), some of which are described here, but which remains largely heuristic so far. In deriving their analogies they are partly guided by Toeplitz quantization, a complex (Berezin style) variant of geometric (and deformation) quantization.

The virtuoso of noncommutative harmonic analysis, *Schmid*, presents briefly but clearly a series of results and problems around the central notion of character formulas for representations of semi-simple groups. More specifically he describes the Weyl (summation over the Weyl group) and Kirillov (integration over a flag variety) character formulas for compact groups and their relation via the Atiyah–Bott fixed point and Berline–Vergne localization formulas, then turns to their noncompact versions, with dominant contributions by Kashiwara, Rossmann (and Schmid), and suggests the search for a localization formula for Hamiltonian actions.

The contributions by other participants were (by lack of space) limited to the size of a short report. In the spirit of noncommutative geometry, Bordemann and Waldmann show how they have cast the GNS construction and WKB expansion of operator algebras in the deformation quantization formalism. Hoppe shows how to deform time-harmonic flows while keeping integrability. Lichnerowicz shows that the Chevalley nontrivial 2-cocycle by which star brackets start can be traced to the cohomologies of Lie algebras of vector fields and gives a distinguished essential extension of the latter. Maeda (together with his coworkers) generalizes the formalism of deformation quantization to contact algebras. Majid gives a brief account of an application of his braided geometry to a semiclassical version of the inductive construction of double-bosonisation for quantum groups. Finally Moreno and Valero give a complete proof of Drinfeld's classification of invariant star products on a simply connected Lie group.

DANIEL STERNHEIMER, JOHN RAWNSLEY & SIMONE GUTT

NONCOMMUTATIVE DIFFERENTIAL GEOMETRY AND THE STRUCTURE OF SPACE TIME

ALAIN CONNES
Collège de France, 3, rue d'Ulm, 75005 PARIS
and
I.H.E.S., 35, route de Chartres, 91440 BURES-sur-YVETTE

Notes by Olga Kravchenko

Abstract. We describe the metric aspect of non commutative geometry and apply it to interpret the Lagrangian of gravity coupled with the stantard model as a simple spectral function of the line element.

1. Generalities

The basic data of Riemannian geometry consists of a manifold M whose points are locally labeled by a finite number of real coordinates $\{x^\mu\}$ and a *metric*, which is given by the infinitesimal line element:

$$ds^2 = g_{\mu\nu}\, dx^\mu\, dx^\nu\,. \tag{1.1}$$

The distance between two points $x, y \in M$ is given by

$$d(x,y) = \text{Inf}\{\text{Length}\ \gamma \,|\, \gamma \text{ is a path between } x \text{ and } y\} \tag{1.2}$$

where

$$\text{Length}\ \gamma = \int_\gamma ds\,. \tag{1.3}$$

Riemannian geometry is flexible enough to give a good description of space–time in general relativity (up to a sign change). The essential point here is that the differential and integral calculus allows to go from the local to the global, while simple notions of Euclidean geometry continue to make sense.

D. Sternheimer et al. (eds.), Deformation Theory and Symplectic Geometry, 1–33.

For instance the idea of a straight line gives rise to the notion of geodesic. The geodesic equation

$$\frac{d^2 x^\mu}{dt^2} = -\Gamma^\mu_{\nu\rho} \frac{dx^\nu}{dt} \frac{dx^\rho}{dt} \tag{1.4}$$

where $\Gamma^\mu_{\nu\rho} = \frac{1}{2} g^{\mu\alpha}(g_{\alpha\nu,\rho} + g_{\alpha\rho,\nu} - g_{\nu\rho,\alpha})$, gives the Newton equation of the motion of a particle in the Newtonian potential V provided one uses the metric $dx^2 + dy^2 + dz^2 - (1 + 2V(x,y,z))dt^2$ (cf. [38] for the more precise formulation). Recent experimental data on binary pulsars confirms through general relativity [14] that Riemannian geometry works well as a model for space–time on a sufficiently large scale. However, it is not clear [32] whether this geometry is adequate for the description of the small scale structure of space–time. The Planck length:

$$\ell_p = (G\hbar/c^3)^{1/2} \sim 10^{-33}\text{cm} \tag{1.5}$$

is considered as a natural lower limit for the precision at which coordinates of an event in the space–time make sense. (See for example [17] or [16] for a physical argument establishing this limit using quantum mechanics).

In these lectures we shall present a new notion of a geometric space where *points* do not play the central role, thus giving much more freedom for describing the small scale texture of space–time. The proposed framework is sufficiently general to treat discrete spaces, Riemannian manifolds, configuration spaces of quantum field theory, and the duals of discrete groups which are not necessarily commutative. The main problem is to show that the usual geometrical ideas and the tools of the infinitesimal calculus do adapt to this more general framework. It turns out that the operator formalism of quantum mechanics together with the analysis of logarithmic divergences of traces of operators give the generalization of the differential and integral calculus that we are looking for. Several direct applications of this approach are described in theorems 1, 2, and 4.

We consider a geometric space to be a *spectral triple:*

$$(\mathcal{A}, \mathcal{H}, D) \tag{1.6}$$

where $\mathcal{A}$ is an involutive algebra of operators in a Hilbert space $\mathcal{H}$ and D is a selfadjoint operator on $\mathcal{H}$. The involutive algebra $\mathcal{A}$ corresponds to a given space M like in the classical duality "Space $\leftrightarrow$ Algebra" in algebraic geometry. The operator $D^{-1} = ds$ corresponds to the infinitesimal line element in Riemannian geometry.

One can see the difference between this *spectral geometry* and Riemannian geometry in two ways. Firstly it is very important that one does not assume that the algebra $\mathcal{A}$ is commutative anymore. Secondly the infinitesimal ds in spectral geometry becomes an operator and does not commute with elements of $\mathcal{A}$ even if the algebra $\mathcal{A}$ is commutative itself.

As we will see, simple commutation relations between ds and elements of $\mathcal{A}$, together with Poincaré duality, characterize the spectral triples (1.6) which come from Riemannian manifolds (Theorem 6). When the algebra $\mathcal{A}$ is commutative the spectrum of its norm closure $\bar{\mathcal{A}}$ in bounded operators on $\mathcal{H}$ is a compact space M. A point of M is a character of $\bar{\mathcal{A}}$, i.e. a homomorphism from $\bar{\mathcal{A}}$ to $\mathbb{C}$,

$$\chi : \bar{\mathcal{A}} \to \mathbb{C}\,,\ \chi(a+b) = \chi(a) + \chi(b)\,,\ \chi(\lambda a) = \lambda\,\chi(a)\,,\ \chi(ab) = \chi(a)\,\chi(b)\,, \tag{1.7}$$
$$\forall\, a, b \in \bar{\mathcal{A}}\,,\ \forall\,\lambda \in \mathbb{C}.$$

As an example let us take $\mathcal{A}$ to be the group algebra $\mathbb{C}\Gamma$ for a discrete group Γ acting on the Hilbert space $\mathcal{H} = \ell^2(\Gamma)$ by the regular (left) representation of Γ. When the group Γ and hence the algebra $\mathcal{A}$ are commutative then the characters of $\bar{\mathcal{A}}$ are elements of the Pontryagin dual of Γ,

$$\hat{\Gamma} = \{\chi : \Gamma \to U(1)\ ;\ \chi(g_1\, g_2) = \chi(g_1)\,\chi(g_2) \qquad \forall\, g_1, g_2 \in \Gamma\}\,. \tag{1.8}$$

Elementary notions of differential geometry for the space $\hat{\Gamma}$ continue to make sense in the general case when Γ is no longer commutative. The right column in the following dictionary does not use the commutativity of the algebra $\mathcal{A}$:

Space X	**Algebra $\mathcal{A}$**
Vector bundle	Finite projective module
Differential form of degree k	Hochschild cycle of dimension k
De Rham current of dimension k	Hochschild cocycle of dimension k
De Rham homology	Cyclic cohomology of $\mathcal{A}$

The power of this generalization to the noncommutative case is demonstrated for example in the proof of Novikov conjecture [28] for hyperbolic groups Γ [10].

In the general case the notion of a point given by (1.7) is not of much interest; but the notion of probability measure keeps its full meaning. Such a measure φ, also called a state, is a normalized positive linear form on $\mathcal{A}$ such that $\varphi(1) = 1$,

$$\varphi : \bar{\mathcal{A}} \to \mathbb{C}\,,\ \varphi(a^* a) \geq 0\,, \quad \forall\, a \in \bar{\mathcal{A}}\,,\ \varphi(1) = 1\,. \tag{1.9}$$

Instead of measuring distances between points using the formula (1.2) we measure distances between states φ, ψ on $\bar{A}$ by a dual formula. This dual formula involves *sup* instead of *inf* and does not use paths in the space

$$d(\varphi,\psi) = \operatorname{Sup}\{|\varphi(a)-\psi(a)| \; ; \; a \in \mathcal{A} \, , \; \|[D,a]\| \leq 1\}. \tag{1.10}$$

Let us show that this formula indeed gives the geodesic distance in the Riemannian case. Let M be a Riemannian compact manifold with a K-orientation, i.e. a spin structure. The associated spectral triple $(\mathcal{A},\mathcal{H},D)$ is given by the representation

$$(f\,\xi)(x) = f(x)\,\xi(x) \qquad \forall\, x \in M \;,\; f \in \mathcal{A} \;,\; \xi \in \mathcal{H} \tag{1.11}$$

of the algebra $\mathcal{A}$ of functions on M in the Hilbert space

$$\mathcal{H} = L^2(M,S) \tag{1.12}$$

of the square integrable sections of the spinor bundle.

The operator D is the Dirac operator (cf. [25]). One can check that the commutator $[D,f]$, for $f \in \mathcal{A} = C^\infty(M)$ is an operator which is the Clifford multiplication by the gradient ∇f and that its operator norm is:

$$\|[D,f]\| = \operatorname{Sup}_{x\in M} \|\nabla f(x)\| = \text{Lipschitz norm } f. \tag{1.13}$$

Let $x,y \in M$ and φ,ψ be the corresponding characters: $\varphi(f) = f(x)$, $\psi(f) = f(y)$ for $\forall\; f \in \mathcal{A}$. Then formula (1.10) gives the same result as formula (1.2), i.e. it gives the geodesic distance between x and y.

Unlike the formula (1.2) its dual formula (1.10) makes sense in general, namely, for example for discrete spaces and even totally disconnected spaces.

The usual notion of *dimension* of a space is replaced by the *dimension spectrum* which is a subset of $\mathbb{C}$ with real part not bigger than some $\alpha > 0$ if

$$\lambda_n^{-1} = O(n^{-\alpha}) \tag{1.14}$$

where λ_n is the nth eigenvalue of $|D|$.

The relation between local and global data is given by the local index formula (Theorem 4) [11].

The characteristic property of *differentiable manifolds* which is carried over to the noncommutative case is *Poincaré duality*. Poincaré duality in ordinary homology is not sufficient to describe homotopy type of manifolds [27] but D. Sullivan [36] showed (in the simply connected PL case of dimension ≥ 5 ignoring 2–torsion) that it is sufficient to replace ordinary homology by KO–homology.

Moreover, K-homology (cf. for example [1]) thanks to the work of Brown, Douglas, Fillmore, Atiyah, and Kasparov admits a fairly simple definition in algebraic terms, given by

Space X	Algebra $\mathcal{A}$
$K_1(X)$	Stable homotopy class of the spectral triple $(\mathcal{A}, \mathcal{H}, D)$
$K_0(X)$	Stable homotopy class of $\mathbb{Z}/2$ graded spectral triple

(i.e. for K_0 we suppose that $\mathcal{H}$ is $\mathbb{Z}/2$–graded by γ, where $\gamma = \gamma^*$, $\gamma^2 = 1$ and $\gamma a = a\gamma \quad \forall\, a \in \mathcal{A}$, $\gamma D = -D\gamma$).

This description works for the complex K–homology which is 2–periodic.

The *fundamental class* of a noncommutative space is a class μ in the KR–homology of the algebra $\mathcal{A} \otimes \mathcal{A}^0$ equipped with the involution

$$\tau(x \otimes y^0) = y^* \otimes (x^*)^0 \qquad \forall\, x, y \in \mathcal{A} \tag{1.15}$$

where $\mathcal{A}^0$ denotes the algebra opposite to $\mathcal{A}$. The Kasparov intersection product [21] allows one to formulate Poincaré duality in terms of the invertibility of μ. In KR-homology besides the above data one has to give an anti-linear isometry J on $\mathcal{H}$ which implements the involution τ in as much as

$$JwJ^{-1} = \tau(w) \qquad \forall\, w \in \mathcal{A} \otimes \mathcal{A}^0, \tag{1.16}$$

KR-homology is periodic with period 8 and the dimension modulo 8 is specified by the following commutation rules. One has $J^2 = \varepsilon$, $JD = \varepsilon' DJ$, $J\gamma = \varepsilon'' \gamma J$ where $\varepsilon, \varepsilon', \varepsilon'' \in \{-1, 1\}$ and with n the dimension modulo 8,

n	0	1	2	3	4	5	6	7
ε	1	1	-1	-1	-1	-1	1	1
ε'	1	-1	1	1	1	-1	1	1
ε''	1		-1		1		-1	

The anti-linear isometry J is given in Riemannian geometry by the charge conjugation operator and in the noncommutative case by the Tomita operator [37]. When an operator algebra $\mathcal{A}$ admits a cyclic vector which is cyclic for the commutant $\mathcal{A}'$, the Tomita operator establishes an anti-isomorphism

$$\mathcal{A}'' \to \mathcal{A}' : a \mapsto Ja^*J^{-1}. \tag{1.17}$$

The class μ specifies only the stable homotopy class of the spectral triple $(\mathcal{A},\mathcal{H},D)$ equipped with the isometry J (and $\mathbb{Z}/2$–grading γ if n is even). The non-triviality of this homotopy class shows up in the intersection form

$$K(\mathcal{A})\times K(\mathcal{A})\to\mathbb{Z} \tag{1.18}$$

given by the Fredholm index of D with coefficients in $K(\mathcal{A}\otimes\mathcal{A}^0)$.

In order to compare different spectral triples in the same homotopy class defined by μ we shall use the following spectral functional

$$\mathrm{Trace}\,(\varpi(D))\,, \tag{1.19}$$

where $\varpi:\mathbb{R}\to\mathbb{R}_+$ is a suitable positive function.

Once the algebra $\mathcal{A}$ fixed, a spectral geometry is determined by the unitary equivalence class of a spectral triple $(\mathcal{A},\mathcal{H},D)$ together with the isometry J. Denote the representation of $\mathcal{A}$ in $\mathcal{H}$ by π. $(\pi_1,\mathcal{H}_1,D_1,J_1)$ is unitarily equivalent to $(\pi_2,\mathcal{H}_2,D_2,J_2)$ if there exists a unitary $U:\mathcal{H}_1\to\mathcal{H}_2$ such that

$$U\pi_1(a)U^*=\pi_2(a)\quad\forall\,a\in\mathcal{A}\,,\ UD_1U^*=D_2\,,\ UJ_1U^*=J_2 \tag{1.20}$$

(and $U\gamma_1U^*=\gamma_2$ for even n).

The automorphism group $\mathrm{Aut}(\mathcal{A})$ of the involutive algebra $\mathcal{A}$ acts on the variety of spectral geometries by the following composition:

$$\pi'(a)=\pi(\alpha^{-1}(a))\qquad\forall\,a\in\mathcal{A}\,,\ \alpha\in\mathrm{Aut}(\mathcal{A})\,. \tag{1.21}$$

Let $\mathrm{Aut}^+(\mathcal{A})$ be the subgroup of automorphisms which preserve the class μ. $\mathrm{Aut}^+(\mathcal{A})$ acts on the stable homotopy class determined by μ and preserves the action functional (1.19). In general this subgroup is not compact. For Riemannian manifolds it coincides with the group of diffeomorphisms preserving the K–orientation, $\mathrm{Diff}^+(M)$, i.e. the subgroup preserving the spin structure on M. On the other hand, the isotropy group of a given geometry is automatically *compact* (for a unital $\mathcal{A}$). This implies that the action functional (1.19) automatically produces the phenomenon of spontaneous symmetry breaking.

We will show that for the right choice of the algebra $\mathcal{A}$ the functional of action (1.19) added to $\langle\xi,D\xi\rangle$, $\xi\in\mathcal{H}$ gives the standard Glashow-Weinberg-Salam model coupled with gravity. The algebra $\mathcal{A}$ is the tensor product of the algebra of functions on a Riemannian space M and a finite dimensional noncommutative algebra whose spectral geometry is specified by the phenomenological data from particle physics.

2. Infinitesimal calculus

Here we shall give a precise meaning to the notion of infinitesimal variable using the operator formalism of quantum mechanics.

The notion of infinitesimal is supposed to have an obvious intuitive meaning. Let us take a particular example [5]. Let us consider darts thrown arbitrarily into a target Ω. Let $dp(x)$ be the probability for the dart to hit the point $x \in \Omega$. It is clear that $dp(x) < \varepsilon$ for any $\varepsilon > 0$ but nevertheless saying that $dp(x) = 0$ is unsatisfactory. The usual formalisms of measure theory or of the theory of differential forms avoid the problem by giving sense to the following expression

$$\int f(x)\, dp(x) \qquad f : \Omega \to \mathbb{C} \tag{2.1}$$

but it is insufficient to give any meaning to expressions like $e^{-\frac{1}{dp(x)}}$. The answer given by nonstandard analysis, a so called nonstandard real, is equally deceiving. From every nonstandard real number one can construct canonically a subset of the interval $[0,1]$, which is not Lebesgue measurable. No such set can be exhibited [34]. This implies that not a single nonstandard real number can actually be exhibited. The formalism which we propose below will give a substantial and computable answer to the above problem.

Our framework consists of a separable Hilbert space decomposed into two orthogonal infinite dimensional subspaces. This decomposition can be described by the linear operator F on $\mathcal{H}$ which is the identity, $F\xi = \xi$, on one subspace and minus identity, $F\xi = -\xi$, on the other; so one has

$$F = F^* \ , \ F^2 = 1 \, . \tag{2.2}$$

This framework is determined uniquely up to isomorphism. The following is the beginning of a long dictionary which translates classical notions into operator language:

Classical	**Quantum**
Complex variable	Operator in $\mathcal{H}$
Real variable	Selfadjoint operator
Infinitesimal	Compact operator
Infinitesimal of order α	Compact operator with eigenvalues μ_n satisfying $\mu_n = O(n^{-\alpha})$, $n \to \infty$

Differential of a real real or complex variable	$đf = [F, f] = Ff - fF$
Integral of an infinitesimal of order 1	$⨍ T =$ Coefficient of logarithmic divergence in the trace of T.

The first two lines of the dictionary are familiar from quantum mechanics. The range of a complex variable corresponds to the *spectrum* of an operator. The holomorphic functional calculus gives a meaning to $f(T)$ for all holomorphic functions f on the spectrum of T. It is only holomorphic functions which operate in this generality which reflects the difference between complex and real analysis. When $T = T^*$ is selfadjoint then $f(T)$ has a meaning for all Borel functions f. Notice that every usual random variable X on the probability space (Ω, P) can be trivially considered as a selfadjoint operator. One takes $\mathcal{H} = L^2(\Omega, P)$ and T to be the operator of multiplication by X:

$$(T\xi)(p) = X(p)\,\xi(p) \qquad \forall p \in \Omega\,,\ \xi \in \mathcal{H}\,. \tag{2.3}$$

Let us consider the third line of the dictionary. We look for a definition of "infinitesimal variables", i.e. operators T in $\mathcal{H}$ such that

$$\|T\| < \varepsilon \qquad \forall \varepsilon > 0\,, \tag{2.4}$$

where $\|T\| = \mathrm{Sup}\,\{\|T\xi\|\ ;\ \|\xi\| = 1\}$ is the operator norm. Clearly, if one takes (2.4) literally then one gets $\|T\| = 0$ and $T = 0$ as the unique solution, but one can weaken (2.4) in the following way:

$$\forall \varepsilon > 0\,,\ \exists \text{ finite dimensional space } E \subset \mathcal{H} \text{ such that } \|T|_{E^\perp}\| < \varepsilon\,. \tag{2.5}$$

Here $E^\perp$ denotes the orthogonal complement to E in the space $\mathcal{H}$:

$$E^\perp = \{\xi \in \mathcal{H}\ ;\ \langle \xi, \eta\rangle = 0 \quad \forall \eta \in E\} \tag{2.6}$$

which is a subspace of finite codimension in $\mathcal{H}$. $T|_{E^\perp}$ denotes the restriction of T to this subspace,

$$T|_{E^\perp} : E^\perp \to \mathcal{H}\,. \tag{2.7}$$

Operators which satisfy the condition (2.5) are the *compact operators*, i.e. those for which the image of the unit ball of $\mathcal{H}$ has compact closure. The operator T is compact if and only if $|T| = \sqrt{T^*T}$ is compact, and this means that the spectrum of $|T|$ is a sequence of eigenvalues $\mu_0 \geq \mu_1 \geq \mu_2 \ldots\,,\ \mu_n \downarrow 0$.

These eigenvalues are the characteristic values of T and one has

$$\mu_n(T) = \mathrm{Inf}\,\{\|T - R\|\ ;\ R \text{ is an operator of rank } \leq n\} \tag{2.8}$$

$$\mu_n(T) = \operatorname{Inf}\{\|T/E^{\perp}\| \; ; \; \dim E \leq n\}. \tag{2.9}$$

Compact operators form a two-sided ideal $\mathcal{K}$ in the algebra $\mathcal{L}(\mathcal{H})$ of bounded operators in $\mathcal{H}$.

Consider now the fourth entry in our dictionary. The size of the infinitesimal $T \in \mathcal{K}$ is governed by the order of decay of the sequence $\mu_n = \mu_n(T)$ as $n \to \infty$. In particular, for all real positive α the following condition defines infinitesimals of order α:

$$\mu_n(T) = O(n^{-\alpha}) \qquad \text{when } n \to \infty \tag{2.10}$$

(i.e. there exists $C > 0$ such that $\mu_n(T) \leq C n^{-\alpha} \quad \forall n \geq 1$). Infinitesimals of order α also form a two-sided ideal as it can be checked using (2.8), (cf. [7]) and moreover,

$$T_j \text{ of order } \alpha_j \Rightarrow T_1 T_2 \text{ of order } \alpha_1 + \alpha_2 \,. \tag{2.11}$$

(For $\alpha < 1$ this ideal is a normed ideal given by real interpolation between the ideal $\mathcal{L}^1$ of trace class operators and the ideal $\mathcal{K}$ [7].)

Hence, apart from commutativity, intuitive properties of the infinitesimal calculus are fulfilled.

Since the size of the infinitesimal is measured by the sequence $\mu_n \to 0$ it might seem that one does not need the operator formalism at all, that it would be enough to replace the ideal $\mathcal{K}$ in $\mathcal{L}(\mathcal{H})$ by the ideal $c_0(\mathbb{N})$ of sequences converging to zero in the algebra $\ell^{\infty}(\mathbb{N})$ of bounded sequences. A variable would be a bounded sequence, and an infinitesimal a sequence μ_n, $\mu_n \downarrow 0$. However, this commutative version does not allow for the existence of variables with range a continuum since all elements of $\ell^{\infty}(\mathbb{N})$ have a point spectrum and a discrete spectral measure. Only *noncommutativity* of $\mathcal{L}(\mathcal{H})$ makes it possible to include in one framework variables with Lebesgue spectrum together with infinitesimal variables.

Noncommutativity of $\mathcal{L}(\mathcal{H})$ is also crucial for the next line of the dictionary. The differential df of a real or complex variable

$$df = \Sigma \frac{\partial f}{\partial x^{\mu}} dx^{\mu} \tag{2.12}$$

is replaced by the commutator

$$đ f = [F, f] \,. \tag{2.13}$$

Going from (2.12) to (2.13) is similar to taking the commutator $[f, g] = fg - gf$ of quantum observables instead of the Poisson bracket $\{f, g\}$ of two functions f, g in classical mechanics.

For a given algebra $\mathcal{A}$ of operators in $\mathcal{H}$ the *dimension* of the corresponding space (in the sense of the dictionary on page 3) is governed by the size of the differentials $\not{d}f$, for $f \in \mathcal{A}$. In dimension p one has

$$\not{d}f \quad \text{of order } \frac{1}{p}, \quad \text{for any } f \in \mathcal{A}. \tag{2.14}$$

We shall see below (Theorem 1) concrete examples where p is the Hausdorff dimension of Julia sets. Simple algebraic manipulations on the functional

$$\tau(f^0, \ldots, f^n) = \text{Trace}\,(f^0\, \not{d}f^1 \ldots \not{d}f^n) \qquad n \text{ odd}, \ n > p \tag{2.15}$$

show that τ is a cyclic cocycle and moreover that the cocycle τ is *integral*, i.e. that

$$\langle \tau, K_1(\mathcal{A}) \rangle \subset \mathbb{Z}. \tag{2.16}$$

This integrality result is a very powerful tool, see [7].

If the dictionary would have ended here the calculus would lack a vital tool, the *locality*, namely the possibility of neglecting infinitesimals of order > 1. These are contained in the following two-sided ideal

$$\left\{ T \in \mathcal{K}\ ;\ \mu_n(T) = o\left(\frac{1}{n}\right) \right\}. \tag{2.17}$$

where little o has the usual meaning, namely, that $n\mu_n(T) \to 0$ when $n \to \infty$.

Hence if we use the trace like in (2.15) for integration we would encounter two problems:

1. infinitesimals of order 1 are not in the domain of the trace.
2. the trace of infinitesimals of order > 1 is not zero.

The natural domain of the trace is the two-sided ideal $\mathcal{L}^1(\mathcal{H})$ of trace-class operators

$$\mathcal{L}^1 = \left\{ T \in \mathcal{K}\ ;\ \sum_{o}^{\infty} \mu_n(T) < \infty \right\}. \tag{2.18}$$

The trace of an operator $T \in \mathcal{L}^1(\mathcal{H})$ is given by the sum

$$\text{Trace}\,(T) = \sum \langle T\xi_i, \xi_i \rangle \tag{2.19}$$

independent of the choice of the orthonormal basis (ξ_i) of $\mathcal{H}$. One has

$$\text{Trace}\,(T) = \sum_{o}^{\infty} \mu_n(T) \qquad \text{for } \forall T \geq 0. \tag{2.20}$$

Let $T \geq 0$ be an infinitesimal of order 1, the only condition on $\mu_n(T)$ is

$$\mu_n(T) = O\left(\frac{1}{n}\right) \tag{2.21}$$

which is not sufficient to insure the finiteness of the sum in (2.20). This shows the nature of both of the problems since the trace is not zero even on the smallest ideal in $\mathcal{L}(\mathcal{H})$, namely, the ideal of finite rank operators, $\mathcal{R}$. The solution is obtained by employing the Dixmier trace [15] , i.e. by the following analysis of the logarithmic divergence of the partial traces

$$\mathrm{Trace}_N(T) = \sum_{0}^{N-1} \mu_n(T) \ , \ T \geq 0 . \tag{2.22}$$

In fact, it is useful to define $\mathrm{Trace}_\Lambda(T)$ for any positive real $\Lambda > 0$ by the interpolation formula

$$\mathrm{Trace}_\Lambda(T) = \mathrm{Inf}\{\|A\|_1 + \Lambda\|B\| \ ; \ A + B = T\} \tag{2.23}$$

where $\|A\|_1$ is the $\mathcal{L}^1$ norm of A, $\|A\|_1 = \mathrm{Trace}\,|A|$, and $\|B\|$ is the operator norm of B. This formula coincides with (2.22) for integer Λ and gives its piecewise affine interpolation for noninteger Λ. Then also [7]

$$\mathrm{Trace}_\Lambda(T_1 + T_2) \leq \mathrm{Trace}_\Lambda(T_1) + \mathrm{Trace}_\Lambda(T_2) \qquad \forall\, \Lambda > 0 \tag{2.24}$$

$$\mathrm{Trace}_{\Lambda_1+\Lambda_2}(T_1 + T_2) \geq \mathrm{Trace}_{\Lambda_1}(T_1) + \mathrm{Trace}_{\Lambda_2}(T_2) \qquad \forall\, \Lambda_1, \Lambda_2 > 0 \tag{2.25}$$

where T_1, T_2 are positive in (2.25).
Define for all order 1 operators $T \geq 0$

$$\tau_\Lambda(T) = \frac{1}{\log \Lambda} \int_e^\Lambda \frac{\mathrm{Trace}_\mu(T)}{\log \mu} \frac{d\mu}{\mu} \tag{2.26}$$

which is the Cesaro mean of the function $\frac{\mathrm{Trace}_\mu(T)}{\log \mu}$ over the scaling group $\mathbb{R}_+^*$.
For $T \geq 0$, an infinitesimal of order 1, one has

$$\mathrm{Trace}_\Lambda(T) \leq C \log \Lambda \tag{2.27}$$

so that $\tau_\Lambda(T)$ is bounded. The essential property is the following *asymptotic additivity* of the coefficient $\tau_\Lambda(T)$ of the logarithmic divergence (2.27):

$$|\tau_\Lambda(T_1 + T_2) - \tau_\Lambda(T_1) - \tau_\Lambda(T_2)| \leq 3C \, \frac{\log(\log \Lambda)}{\log \Lambda} \tag{2.28}$$

for $T_j \geq 0$.

An easy consequence of (2.28) is that any limit point τ of the nonlinear functionals τ_Λ for $\Lambda \to \infty$ defines a positive and linear trace on the two–sided ideal of infinitesimals of order 1,

$$\begin{aligned} \tau(\lambda_1 T_1 + \lambda_2 T_2) = \lambda_1\, \tau(T_1) + \lambda_2\, \tau(T_2) \qquad & \forall\, \lambda_j \in \mathbb{C} \\ \tau(T) \geq 0 \qquad & \forall\, T \geq 0 \\ \tau(ST) = \tau(TS) \qquad & \text{for any bounded } S \\ \tau(T) = 0 \text{ if } \mu_n(T) = o\left(\frac{1}{n}\right). & \end{aligned} \tag{2.29}$$

In practice the choice of the limit point τ is irrelevant because in all important examples (in particular as a corollary of the axioms in the general framework cf. Section 4) T is a *measurable* operator, i.e.:

$$\tau_\Lambda(T) \text{ converges when } \Lambda \to \infty. \tag{2.30}$$

Thus the value $\tau(T)$ is independent of the choice of the limit point τ and is denoted

$$⨍ T. \tag{2.31}$$

The first interesting example is provided by pseudodifferential operators T on a differentiable manifold M. When T is of order 1 (in the sense of (2.21)) it is measurable and $⨍ T$ is the non-commutative residue of T [39, 22]. It has a local expression in terms of the distribution kernel $k(x,y)$, $x, y \in M$. For T of order 1 the kernel $k(x,y)$ diverges logarithmically near the diagonal,

$$k(x,y) = -a(x) \log|x-y| + 0(1) \ (\text{for } \ y \to x) \tag{2.32}$$

where $a(x)$ is a 1–density independent of the choice of Riemannian distance $|x-y|$. Then one has (up to normalization),

$$⨍ T = \int_M a(x). \tag{2.33}$$

The right hand side of this formula makes sense for all pseudodifferential operators (cf. [39]) since one can see that the kernel of such an operator is asymptotically of the form

$$k(x,y) = \sum a_k(x, x-y) + a(x) \log|x-y| + 0(1) \tag{2.34}$$

where $a_k(x,\xi)$ is homogeneous of degree $-k$ in ξ, and the 1–density $a(x)$ is defined intrinsically.

The same principle of extension of $⨍$ to infinitesimals of order < 1 works for hypoelliptic operators and more generally (cf. Theorem 4) for spectral triples whose dimension spectrum is simple.

After this description of the general framework let us discuss examples. The infinitesimal variable $dp(x)$, which is the probability in the darts game, is given by the operator

$$dp = \Delta^{-1} \tag{2.35}$$

where Δ is the Dirichlet Laplacian for the domain Ω, acting on the Hilbert space $L^2(\Omega)$. Also the algebra of functions $f(x_1, x_2)$, $f : \Omega \to \mathbb{C}$ acts by multiplication operators on $L^2(\Omega)$(cf. (2.3)). From the H. Weyl theorem on the asymptotic behavior of eigenvalues of Δ it follows that dp is of order 1, $f\, dp$ is measurable, and that

$$⨍ f\, dp = \int_\Omega f(x_1, x_2)\, dx_1 \wedge dx_2 \tag{2.36}$$

gives the usual expression for the probability. Note that $e^{-\frac{1}{dp(x)}}$ has now full meaning as the heat kernel.

Now let us show how this infinitesimal calculus can be used to make sense of some expressions like the area of for 4–dimensional manifolds, which do not exist in the usual calculus.

There is a canonical quantization of the infinitesimal calculus over $\mathbb{R}$ which is invariant under translations and dilations. It is given by the representation of the algebra of functions f on $\mathbb{R}$ as multiplication operators in $L^2(\mathbb{R})$ (cf. (2.3)), while the operator F in $\mathcal{H} = L^2(\mathbb{R})$ is the Hilbert transform [33]

$$(f\xi)(s) = f(s)\,\xi(s) \ \ \forall\, s \in \mathbb{R},\ \xi \in L^2(\mathbb{R})\,; \quad (F\xi)(t) = \frac{1}{\pi i} \int \frac{\xi(s)}{s-t}\, ds\,. \tag{2.37}$$

One has a unitary equivalent description for $S^1 = P_1(\mathbb{R})$ with $\mathcal{H} = L^2(S^1)$ and

$$F\, e_n = \mathrm{Sign}\,(n)\, e_n\,,\ e_n(\theta) = \exp\,(in\theta) \ \ \forall\, \theta \in S^1\,,\ (\mathrm{Sign}\,(0) = 1)\,. \tag{2.38}$$

The operator $đ f = [F, f]$, for $f \in L^\infty(\mathbb{R})$, is represented by the kernel $\frac{1}{\pi i}\, k(s,t)$, with

$$k(s,t) = \frac{f(s) - f(t)}{s-t}\,. \tag{2.39}$$

Since f and F are bounded operators, $đ f = [F, f]$ is also bounded for all bounded measurable f on S^1, and it makes sense to talk about $|đ f|^p$ for all $p > 0$. Let us give an example where one has to use such an expression for a non differentiable f. Let J be the Julia set associated to the iterations of the map ($c \in \mathbb{C}$)

$$\varphi(z) = z^2 + c \, , \quad J = \partial B \, , \quad B = \{z \in \mathbb{C} \, ; \ \sup_{n \in \mathbb{N}} |\varphi^n(z)| < \infty\} \, . \qquad (2.40)$$

For small c, J is a Jordan curve and B is the bounded component of its complement. The Riemann mapping theorem provides us with a conformal equivalence $D \sim B$ of the unit disc $D = \{z \in \mathbb{C}, \, |z| < 1\}$ with B. By a theorem of Caratheodory it extends continuously to a homeomorphism $Z : S^1 \to J$ where $S^1 = \partial D$. Since (by a result of D. Sullivan) the Hausdorff dimension p of J is strictly bigger than 1 (for $c \neq 0$) the function Z is nowhere of bounded variation on S^1 and $|Z'|$, the absolute value of the derivative of Z, does not make sense as a distribution. However $|đ Z|$ is well defined and one has:

Theorem 1 .

1. *$|đ Z|$ is an infinitesimal of order $\frac{1}{p}$.*
2. *For every continuous function h on J, the operator $h(Z) \, |đ Z|^p$ is measurable.*
3. *$\exists \, \lambda > 0$,*

$$⨍ h(Z) \, |đ Z|^p = \lambda \int_J h \, d\Lambda_p \qquad \forall \, h \in C(J)$$

where $d\Lambda_p$ denotes the Hausdorff measure on J.

The first statement of the theorem uses a result of V.V. Peller which characterizes functions f for which Trace $(|đ f|^\alpha) < \infty$. The constant λ is determined by the asymptotic expansion in $n \in \mathbb{N}$ for the distance in $L^\infty(S^1)$ between Z and restrictions to S^1 of rational fractions with at most n poles outside the unit disc. This constant is of order $\sqrt{p-1}$ and so it is zero for $p = 1$. This is related to a specific feature of dimension 1 manifolds, namely, the differential $đ f$ of a function $f \in C^\infty(S^1)$ is not just of order $(\dim S^1)^{-1} = 1$ but is even trace-class, with

$$\text{Trace}\,(f^0 \, đ \, f^1) = \frac{1}{\pi i} \int_{S^1} f^0 \, df^1 \qquad \forall \, f^0, f^1 \in C^\infty(S^1) \, . \qquad (2.41)$$

In fact, by a classical result of Kronecker $đ f$ is of finite rank if and only if f is a rational fraction (cf. [29]).

The quantized calculus can be used in the same manner to describe the projective space $P_1(K)$ over any local field K (i.e. non discrete locally compact). This calculus is invariant under projective transformations. The special cases of $K = \mathbb{C}$ and $K = \mathbb{H}$ (the field of quaternions) are examples of the calculus on oriented conformal even-dimensional compact manifolds, $M = M^{2n}$. The calculus is defined as follows:

$$\mathcal{H} = L^2(M, \Lambda^n T^*), \quad (f\xi)(p) = f(p)\xi(p), \quad F = 2P - 1\,, \tag{2.42}$$

$\forall f \in L^\infty(M)$, where the scalar product on the Hilbert space of differential forms of degree $n = \frac{1}{2}\dim M$ is given by $\langle \omega_1, \omega_2 \rangle = \int \omega_1 \wedge * \omega_2$ which only depends on the conformal structure on M. The operator P is the orthogonal projection on the subspace of exact forms.

Consider $n = 1$, i.e. M being a Riemann surface. An easy calculation shows that

$$⨍ đ f\, đ g = -\frac{1}{\pi} \int df \wedge * dg \qquad \forall f, g \in C^\infty(M)\,. \tag{2.43}$$

Let X be a smooth map from M to the space $\mathbb{R}^N$ equipped with a Riemannian metric $g_{\mu\nu}\, dx^\mu\, dx^\nu$. The components X^μ of the map X are functions on M. One has

$$⨍ g_{\mu\nu}\, đ X^\mu\, đ X^\nu = -\frac{1}{\pi} \int_M g_{\mu\nu}\, dX^\mu \wedge * dX^\nu \tag{2.44}$$

where the right hand side is the Polyakov action in string theory. However, the equality (2.44) does not hold for $n = 4$: the right hand side is not very interesting because it is not conformally invariant but the left hand side is still conformally invariant, because $đ X^\mu = [F, X^\mu]$ and F are conformally invariant. It defines the natural conformal analog of the Polyakov action in the 4-dimensional case. A calculation yields:

Theorem 2 . *Let X be a smooth map from M^4 to $(\mathbb{R}^N, g_{\mu\nu}\, dx^\mu\, dx^\nu)$,*

$$⨍ g_{\mu\nu}(X)\, đ X^\mu\, đ X^\nu = (16\pi^2)^{-1} \int_M g_{\mu\nu}(X)$$
$$\left\{ \tfrac{1}{3} r \langle dX^\mu, dX^\nu \rangle - \Delta \langle dX^\mu, dX^\nu \rangle + \langle \nabla dX^\mu, \nabla dX^\nu \rangle - \tfrac{1}{2} (\Delta X^\mu)(\Delta X^\nu) \right\} dv$$

where for the right hand side one uses a Riemannian structure η on M, compatible with the given conformal structure. The scalar curvature r, the Laplacian Δ, the Levi–Civita connection ∇, and the measure dv are defined by η but the result is independent of its choice.

Theorem 2 is related with the following formula expressing the Hilbert–Einstein action as the area of the four dimensional manifold (cf. [23, 20])

$$⨍ ds^2 = \frac{-1}{96\pi^2} \int_{M_4} r \sqrt{g}\, d^4x \tag{2.45}$$

($dv = \sqrt{g}\, d^4x$ is the volume form and $ds = D^{-1}$ the length element, i.e. the inverse of the Dirac operator).

When the metric $g_{\mu\nu}\, dx^\mu\, dx^\nu$ on $\mathbb{R}^N$ is invariant under translations the action functional of Theorem 2 is given by the Paneitz operator on M. It is a fourth order operator which plays the role of the Laplacian in conformal geometry [4]. Its conformal anomaly was computed by T. Branson [3].

Let us go back to the case $n = 2$ and modify the conformal structure on M by a Beltrami differential $\mu(z,\bar z)\, d\bar z/dz$, $|\mu(z,\bar z)| < 1$. Thus if z is a conformal local coordinate, we now measure angles at $z \in M$ by the identification

$$T_z(M) \to \mathbb{C} : \ X \mapsto \langle X, dz + \mu(z,\bar z)\, d\bar z\rangle \tag{2.46}$$

instead of $X \mapsto \langle X, dz\rangle$. The quantized calculus on M associated to the new conformal structure has the same $\mathcal{H}$, $\mathcal{A}$, and representation of $\mathcal{A}$ in $\mathcal{H}$ unchanged but the operator F is replaced by F' with

$$F' = (\alpha F + \beta)(\beta F + \alpha)^{-1}\ ,\ \alpha = (1-m^2)^{-1/2}\ ,\ \beta = m(1-m^2)^{-1/2}. \tag{2.47}$$

Here m is the operator in $\mathcal{H} = L^2(M, \Lambda^1 T^*)$ given by the endomorphism of the vector bundle $\Lambda^1 T^* = \Lambda^{(1,0)} \oplus \Lambda^{(0,1)}$ with matrix:

$$m(z,\bar z) = \begin{bmatrix} 0 & \bar\mu(z,\bar z)\, d\bar z/dz \\ \mu(z,\bar z)\, dz/d\bar z & 0 \end{bmatrix}. \tag{2.48}$$

The crucial properties of the operator $m \in \mathcal{L}(\mathcal{H})$ are as follows:

$$\|m\| < 1\ ,\ m = m^*\ ,\ m\, f = f\, m \qquad \forall f \in \mathcal{A} = C^\infty(M) \tag{2.49}$$

and the deformation (2.47) is a particular case of the

Proposition 3 . *Let $\mathcal{A}$ be an involutive algebra of operators in $\mathcal{H}$ and*

$$N = \mathcal{A}' = \{T \in \mathcal{L}(\mathcal{H})\,;\, Ta = aT \quad \forall a \in \mathcal{A}\}$$

the von Neumann algebra commutant of $\mathcal{A}$.

1. *The following formula defines an action of the group $G = GL_1(N)$ of invertible elements of N on the operators F, that satisfy $F = F^*$, $F^2 = 1$*

$$g(F) = (\alpha F + \beta)\,(\beta F + \alpha)^{-1} \qquad \forall g \in G$$

where $\alpha = \frac{1}{2}(g - (g^{-1})^*)$, $\beta = \frac{1}{2}(g + (g^{-1})^*)$.

2. *One has*

$$[g(F), a] = Y[F,a]\,Y^* \quad \forall a \in \mathcal{A}, \quad \textit{where} \quad Y = (\beta F + \alpha)^{*-1}.$$

The last equality shows that the transformation $F \to g(F)$ preserves the condition

$$[F, a] \in J \tag{2.50}$$

for all two-sided ideals $J \subset \mathcal{L}(\mathcal{H})$. Only *measurability* of the Beltrami differential μ is needed for m to satisfy (2.50), and similarly one only needs that the conformal structure on M is measurable in order to define the associated quantized calculus. More to that, the second equality in proposition 3 shows that the regularity condition on $a \in L^\infty(M)$ imposed by (2.50) only depends on the quasi-conformal structure on the manifold M [13]. A local homeomorphism φ of $\mathbb{R}^n$ is *quasi-conformal* if and only if there exists $K < \infty$ such that, $\forall x \in \text{Domain}\, \varphi$,

$$H_\varphi(x) = \text{Lim sup}_{r \to 0} \frac{\max\{|\varphi(x) - \varphi(y)|\,;\, |x-y| = r\}}{\min\{|\varphi(x) - \varphi(y)|\,;\, |x-y| = r\}} \leq K\,. \tag{2.51}$$

A quasi-conformal structure on a topological manifold M^n is given by a quasi-conformal atlas. The discussion above applies to the general case (n even) [13] and shows that the quantized calculus is well defined on all quasi-conformal manifolds. The result of D. Sullivan [35] based on [24] shows that all topological manifolds M^n, $n \neq 4$ admit a quasi-conformal structure. Using the quantized calculus and cyclic cohomology instead of the differential calculus and Chern–Weil theory one gets [13] a local formula for the topological Pontryagin classes of M^n.

3. Local index formula and the transverse fundamental class

In this section we show how the infinitesimal calculus allows to go from local to global in the general framework of spectral triples $(\mathcal{A}, \mathcal{H}, D)$. We will apply the general result to the cross product of a manifold by a group of diffeomorphisms.

Let us make the following regularity hypothesis on $(\mathcal{A}, \mathcal{H}, D)$

$$a \text{ and } [D, a] \in \cap \,\text{Dom}\, \delta^k, \ \forall\, a \in \mathcal{A} \tag{3.1}$$

where δ is the derivation $\delta(T) = [|D|, T]$ for any operator T.

Let $\mathcal{B}$ denote the algebra generated by $\delta^k(a)$, $\delta^k([D,a])$. The *dimension* of a spectral triple is bounded above by $p > 0$ if and only if $a(D+i)^{-1}$ is an infinitesimal of order $\frac{1}{p}$ for any $a \in \mathcal{A}$. When $\mathcal{A}$ is unital it depends only on the spectrum of D.

The precise notion of dimension is given by the subset $\Sigma \subset \mathbb{C}$ of singularities of the analytic functions

$$\zeta_b(z) = \text{Trace}\,(b|D|^{-z}) \qquad \text{Re}\, z > p\,,\ b \in \mathcal{B}\,. \tag{3.2}$$

We assume that Σ is discrete and simple, i.e. that ζ_b can be extended to $\mathbb{C}/\Sigma$ with simple poles in Σ.

We refer to [11] for the case of a spectrum with multiplicities.

The Fredholm index of the operator D determines an additive map $K_1(\mathcal{A}) \xrightarrow{\varphi} \mathbb{Z}$ given by the equality

$$\varphi([u]) = \text{Index}\,(PuP)\ ,\ u \in GL_1(\mathcal{A}) \tag{3.3}$$

where P is the projector $P = \frac{1+F}{2}$, $F = \text{Sign}\,(D)$.

This map is calculated by the pairing of $K_1(\mathcal{A})$ with the following cyclic cocycle

$$\tau(a^0, \ldots, a^n) = \text{Trace}\,(a^0[F, a^1] \ldots [F, a^n]) \qquad \forall\, a^j \in \mathcal{A} \tag{3.4}$$

where $F = \text{Sign}\ D$ and $n \geq p$ is an odd integer.

It is difficult to compute τ in general because the formula (3.4) employs the ordinary trace instead of the local trace $⨍$.

This problem is solved by the following general formula:

Theorem 4 . *[11].*

Let $(\mathcal{A}, \mathcal{H}, D)$ be a spectral triple satisfying the hypothesis (3.1) and (3.2). Then

1. *The equality*

$$⨍ P = \text{Res}_{z=0}\ \text{Trace}\,(P|D|^{-z})$$

defines a trace on the algebra generated by $\mathcal{A}$, $[D, \mathcal{A}]$ and $|D|^z$, where $z \in \mathbb{C}$.

2. *There is only a finite number of non-zero terms in the following formula. It defines the odd components $(\varphi_n)_{n=1,3,\ldots}$ of a cocycle in the bicomplex (b, B) of $\mathcal{A}$,*

$$\varphi_n(a^0, \ldots, a^n) = \sum_k c_{n,k} ⨍ a^0[D, a^1]^{(k_1)} \ldots [D, a^n]^{(k_n)}\, |D|^{-n-2|k|},$$

$\forall\, a^j \in \mathcal{A}$, where the following notations are used: $T^{(k)} = \nabla^k(T)$ and $\nabla(T) = D^2T - TD^2$, k is a multi-index, $|k| = k_1 + \ldots + k_n$,

$$c_{n,k} = \frac{(-1)^{|k|}\,\sqrt{2i}\ \Gamma\left(|k| + \frac{n}{2}\right)}{k_1! \ldots k_n!\,(k_1+1) \ldots (k_1 + k_2 + \ldots + k_n + n)}.$$

3. *The pairing of the cyclic cohomology class $(\varphi_n) \in HC^*(\mathcal{A})$ with $K_1(\mathcal{A})$ gives the Fredholm index of D with coefficients in $K_1(\mathcal{A})$.*

Let us recall that the bicomplex (b, B) is given by the following operators acting on multi-linear forms on $\mathcal{A}$,

$$(b\varphi)(a^0, \ldots, a^{n+1}) = \sum_0^n (-1)^j \varphi(a^0, \ldots, a^j a^{j+1}, \ldots, a^{n+1})$$
$$+(-1)^{n+1} \varphi(a^{n+1}a^0, a^1, \ldots, a^n), \quad (3.5)$$

$$B = AB_0, \quad \text{where}$$
$$(B_0\varphi)(a^0, \ldots, a^{n-1}) = \varphi(1, a^0, \ldots, a^{n-1})$$
$$-(-1)^n \varphi(a^0, \ldots, a^{n-1}, 1),$$
$$(A\psi)(a^0, \ldots, a^{n-1}) = \sum_0^{n-1} (-1)^{(n-1)j} \psi(a^j, a^{j+1}, \ldots, a^{j-1}). \quad (3.6)$$

For the normalization of the pairing between HC^* and $K(\mathcal{A})$ see [7].

Remarks.

(a) The statement of Theorem 4 remains valid when D is replaced by $D|D|^\alpha$, $\alpha \geq 0$.

(b) In the even case, i.e. when $\mathcal{H}$ is $\mathbb{Z}/2$ graded by γ,

$$\gamma = \gamma^*, \quad \gamma^2 = 1, \quad \gamma a = a\gamma \quad \forall\, a \in \mathcal{A}, \; \gamma D = -D\gamma,$$

there is an analogous formula for a cocycle (φ_n), n even, which gives the Fredholm index of D with coefficients in K_0. However, φ_0 is not expressed in terms of the residue $⨍$ because it is not local for a finite dimensional $\mathcal{H}$ (cf. [11]).

(c) There exists an analogous formula for the case when the dimension spectrum Σ has multiplicities. There are some correction terms, their number is finite and bounded independently of the multiplicity (cf. [11]).

The dimension spectrum of a manifold M is the set $\{0, 1, \ldots, n\}$, $n = \dim M$; it is simple. Multiplicities appear for singular manifolds. Cantor sets provide examples of complex points $z \notin \mathbb{R}$ in the dimension spectrum.

Starting from a manifold M we are going to perform a general geometrical construction which will yield a spectral triple satisfying the above hypothesis (3.1) and (3.2) and give the fundamental class in K-homology of a K-oriented manifold M without breaking the symmetry of $\mathrm{Diff}^+(M)$, the group of diffeomorphisms of M which preserve K-orientation. More precisely, we are looking for a spectral triple, $(C^\infty(M), \mathcal{H}, D)$ in the same K-homology class as the Dirac operator associated to a Riemannian metric (cf. (1.12) and (1.13)) but which is equivariant under the action of the group $\mathrm{Diff}^+(M)$ in the sense of [21]. This means that one has a unitary representation $\varphi \to U(\varphi)$ of $\mathrm{Diff}^+(M)$ in $\mathcal{H}$ such that

$$U(\varphi)\, f\, U(\varphi)^{-1} = f \circ \varphi^{-1} \qquad \forall f \in C^\infty(M)\,,\ \varphi \in \mathrm{Diff}^+(M) \tag{3.7}$$

and

$$U(\varphi)\, D\, U(\varphi)^{-1} - D \text{ is bounded for all } \varphi \in \mathrm{Diff}^+(M)\,. \tag{3.8}$$

When D is the Dirac operator associated to a Riemannian structure, the principal symbol of D in turn determines the Riemannian metric and hence diffeomorphisms which satisfy (3.8) are isometries.

A solution to this problem is essential in order to define the transverse geometry of foliations. It is obtained in two steps. The first step consists of using the negative curvature metric on $GL(n)/O(n)$ and the "dual Dirac" operator of Miscenko and Kasparov [8] to reduce the problem to the action of $\mathrm{Diff}^+(M)$ on the total space P of the bundle of metrics over M. The second step, following the idea of Hilsum and Skandalis [19], is to use hypoelliptic operators to construct D on P.

Although the equivariant geometry obtained on P is finite dimensional and satisfies hypothesis (3.1) and (3.2), the geometry obtained for M by using the intersection product with the "dual Dirac" is infinite dimensional and θ-summable,

$$\mathrm{Trace}\,(e^{-\beta D^2}) < \infty \qquad \forall \beta > 0\,. \tag{3.9}$$

By construction the fiber of $P \xrightarrow{\pi} M$ is the quotient $F/O(n)$ of the $GL(n)$–principal bundle F of frames on M by the action of the orthogonal group $O(n) \subset GL(n)$. The space P admits a canonical foliation: the vertical foliation $V \subset TP$, $V = \mathrm{Ker}\,\pi_*$ and on the fibers V and on $N = (TP)/V$ the following Euclidean structures. A choice of $GL(n)$–invariant Riemannian metric on $GL(n)/O(n)$ determines a metric on V. The metric on N is defined tautologically: for every $p \in P$ one has a metric on $T_{\pi(p)}(M)$ which is isomorphic to N_p by π_*.

This construction is functorial for diffeomorphisms on M.

The hypoelliptic calculus adapted to this structure is a particular case of the pseudodifferential calculus on Heisenberg manifolds [2]. One simply modifies the notion of homogeneity of symbols $\sigma(p,\xi)$ by using the following homotheties:

$$\lambda \cdot \xi = (\lambda \xi_v, \lambda^2 \xi_n)\,,\ \forall \lambda \in \mathbb{R}_+^* \tag{3.10}$$

where ξ_v, ξ_n are vertical and perpendicular to vertical components of the covector ξ. The formula (3.10) depends on local coordinates (x_v, x_n) adapted to the vertical foliation, but the corresponding pseudodifferential calculus does not depend upon this choice. The principal symbol of a hypoelliptic operator of order k is a function, homogeneous of degree k in the sense of (3.10) on the fiber $V^* \oplus N^*$. The distribution kernel $k(x,y)$ of a pseudodifferential operator T in this hypoelliptic calculus has the following behavior near the diagonal

$$k(x,y) \sim \sum a_j(x, x-y) - a(x) \log |x-y|' + 0(1) \tag{3.11}$$

where a_j is homogeneous of degree $(-j)$ in $(x-y)$ in the sense of (3.10) and where the metric $|x-y|'$ is locally of the form

$$|x-y|' = ((x_v - y_v)^4 + (x_n - y_n)^2)^{1/4}. \tag{3.12}$$

Like in the ordinary pseudodifferential calculus the residue is extended to operators of all degrees and is given by the equality

$$⨍ T = \frac{1}{v+2m} \int a(x) \tag{3.13}$$

where the 1–density $a(x)$ does not depend on the choice of metric $|\ |'$ and where $v = \dim V$, $m = \dim N$ and $v + 2m$ is the Hausdorff dimension of the metric space $(P, |\ |')$.

The operator D is defined by the equation $D|D| = Q$ where Q is a differential hypoelliptic operator of degree 2. For v even, Q is obtained by combining the signature operator $d_V d_V^* - d_V^* d_V$ with the transverse Dirac operator, where d_V is the vertical differential. (We use the metaplectic cover $M\ell(n)$ of $GL(n)$ to define the spin structure on M.) The explicit formula for Q uses an affine connection on M. The choice of this connection does not affect the *principal hypoelliptic symbol* of Q and therefore of D which ensures that D is invariant as in (3.8) under diffeomorphisms of M.

Let us give the explicit formula for Q in the case $n = 1$, i.e. for $M = S^1$. We replace P by the suspension $SP = \mathbb{R} \times P$ in order to consider the case where the vertical dimension is even. A point of $SP = \mathbb{R} \times P$ is parameterized by three coordinates $\alpha \in \mathbb{R}$ and $p = (s, \theta)$ where $\theta \in S^1$ and $s \in \mathbb{R}$ defines the metric $e^{2s}(d\theta)^2$ for $\theta \in S^1$.

We endow SP with the measure $\nu = d\alpha\, ds\, d\theta$ and represent the algebra $C_c^\infty(SP)$ by multiplication operators on $\mathcal{H} = L^2(SP, \nu) \otimes \mathbb{C}^2$. Functoriality of the construction above gives the following unitary representation of the group $\mathrm{Diff}^+(S^1)$,

$$(U(\varphi)^{-1}\xi)(\alpha, s, \theta) = \varphi'(\theta)^{1/2}\, \xi(\alpha, s - \log \varphi'(\theta), \varphi(\theta)). \tag{3.14}$$

The operator Q is given by the formula

$$Q = -2\partial_\alpha \partial_s \sigma_1 + \frac{1}{i} e^{-s} \partial_\theta \sigma_2 + \left(\partial_s^2 - \partial_\alpha^2 - \frac{1}{4}\right) \sigma_3 \tag{3.15}$$

where $\sigma_1, \sigma_2, \sigma_3 \in M_2(\mathbb{C})$ are three Pauli matrices.

In the hypoelliptic calculus the operator ∂_θ has *degree 2* which shows the hypoellipticity of Q.

Long calculation gives the following result [12]:

Theorem 5 . *Let $\mathcal{A}$ be the crossed product of $C_c^\infty(SP)$ by* $\mathrm{Diff}^+(S^1)$.

1. *The spectral triple $(\mathcal{A}, \mathcal{H}, D)$ (where $\mathcal{A}$ acts on $\mathcal{H}$ by (3.14) and $D|D| = Q$) satisfies the hypotheses (3.1) and (3.2); its dimension spectrum is $\Sigma = \{0,1,2,3,4\}$.*
2. *The only nonzero term of the associated cocycle (Theorem 4) is φ_3, which is cohomological to 2ψ where ψ is the 3 cyclic cocycle of the transversal fundamental class of the crossed product.*

From Theorem 4 follows the *integrality* of 2ψ, i.e. that the pairing $\langle 2\psi, K_1(\mathcal{A})\rangle$ is an integer. The 3 cocycle ψ is given by the following formula (cf. [7])

$$\psi(f^0U(\varphi_0), f^1U(\varphi_1), f^2U(\varphi_2), f^3U(\varphi_3)) =$$
$$\begin{cases} 0 & \text{if } \varphi_0\varphi_1\varphi_2\varphi_3 \neq 1 \\ \int h^0\, dh^1 \wedge dh^2 \wedge dh^3 & \text{if } \varphi_0\varphi_1\varphi_2\varphi_3 = 1 \end{cases} \tag{3.16}$$

where $h^0 = f^0$, $h^1 = (f^1)^{\varphi_0}$, $h^2 = (f^2)^{\varphi_0\varphi_1}$, $h^3 = (f^3)^{\varphi_0\varphi_1\varphi_2}$.

The homology between φ_3 and 2ψ involves the action of the Hopf algebra generated by the following linear transformations of the algebra $\mathcal{A}$ (for the relation of δ_3 to Godbillon Vey invariant see [7]

$$\delta_1(fU(\varphi)) = (\partial_\alpha f)\, U(\varphi),\ \delta_2(fU(\varphi)) = (\partial_s f)\, U(\varphi), \tag{3.17}$$
$$\delta_3(fU(\varphi)) = f\, e^{-s}\, \partial_\theta \log(\varphi^{-1})'\, U(\varphi),\ X(fU(\varphi)) = e^{-s}(\partial_\theta f)\, U(\varphi).$$

The compatibility with the multiplication in $\mathcal{A}$ is given by the coproduct rules

$$\Delta\delta_j = \delta_j \otimes 1 + 1 \otimes \delta_j \qquad j = 1,2,3 \tag{3.18}$$

(i.e. δ_j are derivations in $\mathcal{A}$)

$$\Delta X = X \otimes 1 + 1 \otimes X - \delta_3 \otimes \delta_2\ . \tag{3.19}$$

The last equation shows that X is of degree 2 not only from its degree as a pseudodifferential operator in the hypoelliptic calculus but also in an algebraic manner since ΔX involves a tensor product of two derivations.

4. The notion of manifold and the axioms of geometry

Let us first characterize the spectral triples corresponding to ordinary Riemannian geometry (Theorem 6 below). Let the dimension n be given and $(\mathcal{A}, \mathcal{H}, D)$ be a spectral triple with $\mathbb{Z}/2$–grading γ when n is even.

The axioms for *commutative* geometry are the following:

(1) (Dimension) $ds = D^{-1}$ is an infinitesimal of order $\frac{1}{n}$.
(2) (Order one) $[[D, f], g] = 0 \quad \forall f, g \in \mathcal{A}$.
(3) (Smoothness) For all $f \in \mathcal{A}$, both f and $[D, f]$ belong to $\cap_k$ Domain δ^k, where δ is the derivation $\delta(T) = [|D|, T]$.
(4) (Orientability) There exists a Hochschild cocycle $c \in Z_n(\mathcal{A}, \mathcal{A})$ such that $\pi(c) = 1$ (n odd) or $\pi(c) = \gamma$ (n even), where π: $\mathcal{A}^{\otimes(n+1)} \to \mathcal{L}(\mathcal{H})$ is the unique linear map such that

$$\pi(a^0 \otimes a^1 \otimes \cdots \otimes a^n) = a^0[D, a^1] \ldots [D, a^n] \quad \forall a^j \in \mathcal{A}.$$

(5) (Finiteness) The $\mathcal{A}$-module $\mathcal{E} = \cap_k$ Domain D^k is finite projective. The following identity defines a Hermitian structure on $\mathcal{E}$,

$$\langle a\xi, \eta \rangle = \int\!\!\!\!\!\!- \, a(\xi, \eta)\, ds^n \qquad \forall \xi, \eta \in \mathcal{E}\ ,\ a \in \mathcal{A}.$$

(6) (Poincaré duality) The intersection form $K_*(\mathcal{A}) \times K_*(\mathcal{A}) \to \mathbb{Z}$ given by the composition of the Fredholm index of D with the diagonal,

$$m_* : K_*(\mathcal{A}) \times K_*(\mathcal{A}) \to K_*(\mathcal{A} \otimes \mathcal{A}) \to K_*(\mathcal{A}),$$

is *invertible.*
(7) (Reality) There exists an anti-linear isometry J on $\mathcal{H}$ such that

$$Ja^*J^{-1} = a \quad \forall a \in \mathcal{A} \ \text{ and } \ J^2 = \varepsilon, JD = \varepsilon' DJ, J\gamma = \varepsilon'' \gamma J$$

where the values of $\varepsilon, \varepsilon', \varepsilon'' \in \{-1, 1\}$ are given by the table (1.16) as functions of n modulo 8.

Axioms (2) and (4) describe the presentation of the abstract algebra denoted by $(\mathcal{A}, ds)$ generated by $\mathcal{A}$ and $ds = D^{-1}$.

Theorem 6 *Let $\mathcal{A} = C^\infty(M)$, where M is a compact smooth manifold.*

1. *Let π be unitary representation of $(\mathcal{A}, ds)$ satisfying Axioms (1) to (7). Then there exists a unique Riemannian structure g on M such that the geodesic distance is given by*

$$d(x, y) = \operatorname{Sup} \{|a(x) - a(y)|\ ;\ a \in \mathcal{A}\ ,\ \|[D, a]\| \leq 1\}.$$

2. *The metric $g = g(\pi)$ depends only on the unitary equivalence class of π. Fibers of the map {unitary equivalence classes} $\to g(\pi)$ are a finite union of affine spaces $\mathcal{A}_\sigma$ parameterized by the spin structures σ on M.*
3. *The functional $⨍ ds^{n-2}$ is a positive quadratic form on each $\mathcal{A}_\sigma$ with a unique minimum π_σ.*
4. *π_σ is the representation of $(\mathcal{A}, ds)$ in $L^2(M, S_\sigma)$ given by multiplication operators and the Dirac operator associated to the Levi–Civita connection of the metric g.*
5. *The value of $⨍ ds^{n-2}$ the π_σ is the Hilbert–Einstein action of the metric g,*

$$⨍ ds^{n-2} = -c_n \int r \sqrt{g}\, d^n x \ , \ c_n = \tfrac{n-2}{12} (4\pi)^{-n/2} 2^{[n/2]} \Gamma\left(\tfrac{n}{2}+1\right)^{-1} .$$

To understand the meaning of this theorem consider the most simple example, namely, the verification that the geometry of the circle S^1 of length 2π is completely specified by the presentation:

$$U^{-1}[D, U] = 1 \ , \ \text{where } UU^* = U^*U = 1 . \tag{4.1}$$

Then the algebra $\mathcal{A}$ is the algebra of smooth functions of the single element U. One has $S^1 =$ Spectrum $(\mathcal{A})$ and the equality 4.1 is the simplest case of Axiom (4).

Remarks.

(a) We should not have to assume in the statement of Theorem 6 that the algebra $\mathcal{A}$ is equal to the algebra of smooth functions on a manifold. For a commutative $\mathcal{A}$ it should in fact follow from the axioms that the spectrum of $\mathcal{A}$ is a smooth manifold M and that $\mathcal{A} = C^\infty(M)$. From Axioms (3) and (5) it follows that if $\mathcal{A}''$ is the von Neumann algebra generated by $\mathcal{A}$ ($\mathcal{A}''$ is the weak closure and the bicommutant of $\mathcal{A}$ in $\mathcal{H}$) one has

$$\mathcal{A} = \left\{ T \in \mathcal{A}'' \ ; \ T \in \bigcap_{k>0} \operatorname{Dom} \delta^k \right\} \tag{4.2}$$

$\mathcal{A}$ is uniquely specified inside $\mathcal{A}''$ by fixing D (i.e. the geometry ($\mathcal{A}$, $\mathcal{H}, D$) is determined by $(\mathcal{A}'', \mathcal{H}, D)$). This also implies that $\mathcal{A}$ is stable under the smooth functional calculus in its norm closure $\bar{\mathcal{A}} = A$ and in particular

$$\operatorname{Spectrum} \mathcal{A} = \operatorname{Spectrum} A . \tag{4.3}$$

Let $X = \operatorname{Spectrum} A$. It is a compact space. One should deduce from the axioms that the map from X to $\mathbb{R}^N$ given by $a_i^j \in \mathcal{A}$ (the components of the Hochschild cocycle c given by Axiom (4)) is an embedding of X as a smooth submanifold of $\mathbb{R}^N$ (cf. [7, Proposition 15, p.312]).

(b) Let us recall that a Hochschild cycle $c \in Z_n(\mathcal{A}, \mathcal{A})$ is an element of $\mathcal{A}^{\otimes(n+1)}$, $c = \sum a_i^0 \otimes a_i^1 \ldots \otimes a_i^n$ such that $bc = 0$, where b is the linear map $b : \mathcal{A}^{\otimes n+1} \to \mathcal{A}^{\otimes n}$ (cf. (3.5)). The class of the Hochschild cycle c determines the *volume form*.

(c) We use the convention that the scalar curvature r is positive for the sphere S^n, in particular, the sign of the action $⨍ ds^{n-2}$ is the correct one for the Euclidean formulation of gravity. For example for $n = 4$ the Hilbert–Einstein action

$$-\frac{1}{16\pi G} \int r \sqrt{g}\, d^4x$$

coincides with the area $\frac{1}{\ell_p^2} ⨍ ds^2$ in Planck units.

(d) When M is a spin manifold the map $\pi \to g(\pi)$ from Theorem 6 is surjective and if one fixes the cycle $c \in Z_n(\mathcal{A}, \mathcal{A})$ its image is the set of metrics whose volume form (Remark **(b)**) is given by the class of c.

(e) If one omits Axiom (7), one gets a result analogous to Theorem 6 replacing spin structures by spin^c–structures [25], but then there will be no more uniqueness in Theorem 6 *3.*. because of the choice of spin connection.

(f) It follows from Axiom (1) (see [7, Theorem 8, p.309]) that the operators $a\, ds^n$, $a \in \mathcal{A}$ are automatically measurable so that $⨍$ is well defined in Axiom (5).

Now let us consider the general noncommutative case. Given an involutive algebra of operators $\mathcal{A}$ on the Hilbert space $\mathcal{H}$, Tomita's theory associates to all vectors $\xi \in \mathcal{H}$, cyclic for $\mathcal{A}$ and for its commutant $\mathcal{A}'$

$$\overline{\mathcal{A}\xi} = \mathcal{H}\ ,\ \overline{\mathcal{A}'\xi} = \mathcal{H} \tag{4.4}$$

an anti-linear isometric involution $J : \mathcal{H} \to \mathcal{H}$ obtained from the polar decomposition of the operator

$$S\, a\xi = a^*\xi \qquad \forall\, a \in \mathcal{A}\,. \tag{4.5}$$

It satisfies the following commutation relation:

$$J\mathcal{A}'' J^{-1} = \mathcal{A}'\,. \tag{4.6}$$

In particular $[a, b^0] = 0 \quad \forall\, a, b \in \mathcal{A}$ where

$$b^0 = Jb^*J^{-1} \qquad \forall b \in \mathcal{A} \tag{4.7}$$

so $\mathcal{H}$ becomes an $\mathcal{A}$-bimodule using the representation of the opposite algebra $\mathcal{A}^0$ given by (4.7). There is no difference between module and bimodule structures in the commutative case because one has $a^0 = a \quad \forall\, a \in \mathcal{A}$.

Tomita's theorem is the key ingredient which guarantees the substance of the axioms in the general case. The axioms (1), (3), and (5) are left untouched, but in the axiom of reality (7) the equality $Ja^*J^{-1} = a \quad \forall a \in \mathcal{A}$ is replaced by

(7′) $\qquad [a, b^0] = 0 \qquad \forall\, a, b \in \mathcal{A}$ where $b^0 = Jb^*J^{-1}$

also Axiom (2) (order one) becomes

(2′) $\qquad [[D, a], b^0] = 0 \qquad \forall\, a, b \in \mathcal{A}$.

(Notice that since a and b^0 commute (2′) is equivalent to $[[D, a^0], b] = 0 \quad \forall\, a, b \in \mathcal{A}$.)

The Hilbert space $\mathcal{H}$ becomes an $\mathcal{A}$–bimodule by Axiom (7′) and gives a class μ of KR^n–homology for the algebra $\mathcal{A} \otimes \mathcal{A}^0$ equipped with the anti-linear automorphism τ,

$$\tau(x \otimes y^0) = y^* \otimes x^{*0}.$$

The Kasparov intersection product [21] allows to formulate the Poincaré duality in terms of the invertibility of μ,

(6′) $\exists\, \beta \in KR_n(\mathcal{A}^0 \otimes \mathcal{A})\ ,\ \beta \otimes_{\mathcal{A}} \mu = \mathrm{id}_{\mathcal{A}^0}\ ,\ \mu \otimes_{\mathcal{A}^0} \beta = \mathrm{id}_{\mathcal{A}}$.

It implies the isomorphism $K_*(\mathcal{A}) \xrightarrow{\cap\mu} K^*(\mathcal{A})$. The intersection form

$$K_*(\mathcal{A}) \times K_*(\mathcal{A}) \to \mathbb{Z}$$

is obtained from the Fredholm index of D with coefficients in $K_*(\mathcal{A} \otimes \mathcal{A}^0)$. Note that it is defined without using the diagonal map $m : \mathcal{A} \otimes \mathcal{A} \to \mathcal{A}$, which is not a homomorphism in the noncommutative case. This form is quadratic or symplectic according to the value of n modulo 8.

The Hochschild homology with coefficients in a bimodule makes perfect sense in the general case and Axiom (4) takes the following form:

(4′) There exists a Hochschild cycle $c \in Z_n(\mathcal{A}, \mathcal{A} \otimes \mathcal{A}^0)$ such that $\pi(c) = 1$ (n odd), or $\pi(c) = \gamma$ (n even).

(Where $\mathcal{A} \otimes \mathcal{A}^0$ is the $\mathcal{A}$–bimodule obtained by restriction of the structure of $\mathcal{A} \otimes \mathcal{A}^0$– bimodule of $\mathcal{A} \otimes \mathcal{A}^0$ to the subalgebra $\mathcal{A} \otimes 1 \subset \mathcal{A} \otimes \mathcal{A}^0$, i.e.

$$a(b \otimes c^0)d = abd \otimes c^0 \qquad \forall\, a, b, c, d \in \mathcal{A}\,.)$$

Axioms (1), (3) et (5) are unchanged in the noncommutative setting. The proof showing that the operators $a(ds)^n$, $(a \in \mathcal{A})$ are measurable stays valid in general.

We adopt Axioms (1), (2′), (3), (4′), (5), (6′) et (7′) in the general case as a definition of a *spectral manifold* of dimension n. Once fixing the algebra

$\mathcal{A}$ one can talk about the spectral geometry of $\mathcal{A}$ like in (1.20) and (1.21). One can show that the von Neumann algebra $\mathcal{A}''$ generated by $\mathcal{A}$ in $\mathcal{H}$ is automatically finite and hyperfinite and there is a complete list of such algebras up to isomorphism [7]. The algebra $\mathcal{A}$ is stable under smooth functional calculus in its norm closure $A = \bar{\mathcal{A}}$ so that $K_j(\mathcal{A}) \simeq K_j(A)$, i.e. $K_j(\mathcal{A})$ depends only on the underlying topology (defined by the C^* algebra A). The integer $\chi = \langle \mu, \beta \rangle \in \mathbb{Z}$ gives the Euler characteristic in the form

$$\chi = \text{Rang}\, K_0(\mathcal{A}) - \text{Rang}\, K_1(\mathcal{A})$$

and Theorem 4 gives a local formula for it.

The group of automorphisms of the involutive algebra $\mathcal{A}$, $\text{Aut}(\mathcal{A})$, in general plays the role of the group of diffeomorphisms, $\text{Diff}(M)$, of a manifold M. (There is a canonical isomorphism $\text{Diff}(M) \xrightarrow{\alpha} \text{Aut}(C^\infty(M))$ given by

$$\alpha_\varphi(f) = f \circ \varphi^{-1} \qquad \forall f \in C^\infty(M)\ ,\ \varphi \in \text{Diff}(M)\,.)$$

In the general noncommutative case, parallel to the normal subgroup $\text{Int}\,\mathcal{A} \subset \text{Aut}\,\mathcal{A}$ of inner automorphisms of $\mathcal{A}$,

$$\alpha(f) = ufu^* \qquad \forall f \in \mathcal{A} \tag{4.8}$$

where u is a unitary element of $\mathcal{A}$ (i.e. $uu^* = u^*u = 1$), there exists a natural foliation of the space of spectral geometries on $\mathcal{A}$ by equivalence classes of *inner deformations* of a given geometry. Such a deformation is obtained by the following formula without modifying neither the representation of $\mathcal{A}$ in $\mathcal{H}$ nor the anti-linear isometry J

$$D \to D + A + JAJ^{-1} \tag{4.9}$$

where $A = A^*$ is an arbitrary selfadjoint operator of the form

$$A = \Sigma\, a_i[D, b_i]\ ,\ a_i, b_i \in \mathcal{A}\,. \tag{4.10}$$

The newly obtained spectral triple also satisfies Axioms (1) through (7′). The action of the group $\text{Int}(\mathcal{A})$ on the spectral geometries (cf. (1.21)) is simply the following gauge transformation of A

$$\gamma_u(A) = u[D, u^*] + uAu^*\,. \tag{4.11}$$

The required unitary equivalence is implemented by the following representation of the unitary group of $\mathcal{A}$ in $\mathcal{H}$,

$$u \to uJuJ^{-1} = u(u^*)^0\,. \tag{4.12}$$

The transformation (4.9) is the identity operator for the usual Riemannian case. To get a nontrivial example it suffices to consider the product of a Riemannian triple by the unique spectral geometry on the finite-dimensional algebra $\mathcal{A}_F = M_N(\mathbb{C})$ of $N \times N$ matrices on $\mathbb{C}$, $N \geq 2$. One then has $\mathcal{A} = C^\infty(M) \otimes \mathcal{A}_F$, $\text{Int}(\mathcal{A}) = C^\infty(M, PSU(N))$ and inner deformations of the geometry are parameterized by the gauge potentials for the gauge theory of the group $SU(N)$. The space of pure states of the algebra $\mathcal{A}$, $P(\mathcal{A})$, is the product $P = M \times P_{N-1}(\mathbb{C})$ and the metric on $P(\mathcal{A})$ determined by the formula (1.10) depends on the gauge potential A. It coincide with the Carnot metric [18] on P defined by the horizontal distribution given by the connection associated to A (cf. [9]). The group $\text{Aut}(\mathcal{A})$ of automorphisms of $\mathcal{A}$ is the following semi-direct product

$$\text{Aut}(\mathcal{A}) = \mathcal{U} \rtimes \text{Diff}(M) \tag{4.13}$$

of the local gauge transformation group $\text{Int}(\mathcal{A})$ by the group of diffeomorphisms. In dimension $n = 4$, the Hilbert–Einstein action functional for the Riemannian metric and the Yang–Mills action for the vector potential A appear in the asymptotic expansion in $\frac{1}{\Lambda}$ of the number $N(\Lambda)$ of eigenvalues of D which are $\leq \Lambda$. One regularizes this expression replacing it by

$$\text{Trace}\ \varpi\left(\frac{D}{\Lambda}\right) \tag{4.14}$$

where $\varpi \in C_c^\infty(\mathbb{R})$ is an even function which is 1 on $[-1, 1]$, (cf. [6]). Other nonzero terms in the asymptotic expansion are cosmological, Weyl gravity and topological terms.

A more sophisticated example of a spectral manifold is provided by the noncommutative torus $\mathbb{T}_\theta^2$. The parameter $\theta \in \mathbb{R}/\mathbb{Z}$ defines the following deformation of the algebra of smooth functions on the torus $\mathbb{T}^2$, with generators U, V. The relations

$$VU = \exp 2\pi i\theta\ UV \quad \text{and} \quad UU^* = U^*U = 1\ ,\ VV^* = V^*V = 1 \tag{4.15}$$

define the presentation of the involutive algebra $\mathcal{A}_\theta = \{\Sigma\, a_{n,m} U^n V^n\ ;\ a = (a_{n,m}) \in \mathcal{S}(\mathbb{Z}^2)\}$ where $\mathcal{S}(\mathbb{Z}^2)$ is the Schwartz space of sequences with rapid decay. Geometries on $\mathbb{T}_\theta^2$ are parameterized by complex numbers τ with positive imaginary part like in the case of elliptic curves. Up to isometry the geometry depends only on the orbit of τ under the action of $PSL(2, \mathbb{Z})$ [7]. However, a new phenomenon appears in the noncommutative case, namely, the *Morita equivalence* which relates the algebras $\mathcal{A}_{\theta_1}$ and $\mathcal{A}_{\theta_2}$ if θ_1 and θ_2 are in the same orbit of the $PSL(2, \mathbb{Z})$ action on $\mathbb{R}$ [30, 31].

Given a spectral manifold $(\mathcal{A}, \mathcal{H}, D)$ and the Morita equivalence between $\mathcal{A}$ and an algebra $\mathcal{B}$ where

$$\mathcal{B} = \text{End}_{\mathcal{A}}(\mathcal{E}) \tag{4.16}$$

where $\mathcal{E}$ is a finite, projective, Hermitian right $\mathcal{A}$–module, one gets a spectral geometry on $\mathcal{B}$ by the choice of a *Hermitian connection* on $\mathcal{E}$. Such a connection ∇ is a linear map $\nabla : \mathcal{E} \to \mathcal{E} \otimes_{\mathcal{A}} \Omega^1_D$ satisfying the rules [7]

$$\nabla(\xi a) = (\nabla\xi)a + \xi \otimes da \qquad \forall \xi \in \mathcal{E}\ ,\ a \in \mathcal{A} \tag{4.17}$$

$$(\xi, \nabla\eta) - (\nabla\xi, \eta) = d(\xi, \eta) \qquad \forall \xi, \eta \in \mathcal{E} \tag{4.18}$$

where $da = [D, a]$ and where $\Omega^1_D \subset \mathcal{L}(\mathcal{H})$ is the $\mathcal{A}$–bimodule generated by operators of the form 4.10.

Any algebra $\mathcal{A}$ is Morita equivalent to itself (with $\mathcal{E} = \mathcal{A}$) and when one applies the above construction one gets exactly the inner deformations of the spectral geometry.

5. The spectral geometry of space–time

The experimental and theoretical data which one has about the structure of space–time is summarized in the following action functional:

$$\mathcal{L} = \mathcal{L}_E + \mathcal{L}_G + \mathcal{L}_{G\varphi} + \mathcal{L}_\varphi + \mathcal{L}_{\varphi f} + \mathcal{L}_f \tag{5.1}$$

where

$$\mathcal{L}_E = -\frac{1}{16\pi G}\int r\,\sqrt{g}\, d^4x$$

is the Hilbert–Einstein action while the five other terms constitute the standard model of particle physics with minimal coupling to gravity. Besides the metric $g_{\mu\nu}$ this Lagrangian involves several bosonic and fermionic fields. Spin 1 bosons are the photon γ, the intermediate bosons $W^\pm$ and Z, and the eight gluons. The zero spin bosons are the Higgs fields φ which are introduced in order to provide masses to various particles through the mechanism of spontaneous symmetry breaking without contradicting the renormalizability of nonabelian gauge fields. All fermions have spin $\frac{1}{2}$ and form three families of quarks and leptons.

The fields involved in the standard model have a priori a completely different status from the one of the metric $g_{\mu\nu}$. The symmetry group of these fields, namely, the group of local gauge transformations:

$$\mathcal{U} = C^\infty(M, U(1) \times SU(2) \times SU(3)) \tag{5.2}$$

is a priori quite different from the group Diff(M) of symmetries of the total Lagrangian $\mathcal{L}_E$. The natural group of symmetries of $\mathcal{L}$ is the semi–direct product $\mathcal{U} >\!\!\lhd \text{Diff}(M) = G$ of $\mathcal{U}$ by the natural action of Diff(M) by automorphisms of $\mathcal{U}$. The first requirement if one wants to obtain a pure geometrical understanding of $\mathcal{L}$ unifying gauge theory with gravity is to find a geometric space X such that $G = \text{Diff}(X)$. This determines the algebra $\mathcal{A}$ [1]

$$\mathcal{A} = C^\infty(M) \otimes \mathcal{A}_F \ , \ \mathcal{A}_F = \mathbb{C} \oplus \mathbb{H} \oplus M_3(\mathbb{C}) \tag{5.3}$$

where the involutive algebra $\mathcal{A}_F$ is the direct sum of the algebras $\mathbb{C}$, $\mathbb{H}$ (the quaternions), and $M_3(\mathbb{C})$ of 3×3 complex matrices.

The algebra $\mathcal{A}_F$ corresponds to a *finite* space where the standard model fermions and the Yukawa parameters (masses of fermions and mixing matrix of Kobayashi Maskawa) determine the spectral geometry in the following manner. The Hilbert space is finite-dimensional and admits the set of elementary fermions as a basis. For example for the first generation of leptons this set is

$$e_L, e_R, \nu_L, \bar{e}_L, \bar{e}_R, \bar{\nu}_L \,. \tag{5.4}$$

The algebra $\mathcal{A}_F$ admits a natural representation in $\mathcal{H}_F$ (see [9]). Denote by J_F the unique anti-linear involution which exchanges f and $\bar{f}$ for all vectors in the base. One checks the commutation relation

$$[a, Jb^* J^{-1}] = 0 \qquad \forall \, a, b \in \mathcal{A}_F \,, \tag{5.5}$$

which shows that Axiom (2) holds. The operator D_F is given simply by the matrix $\begin{bmatrix} Y & 0 \\ 0 & \bar{Y} \end{bmatrix}$ where Y is the Yukawa coupling matrix. The detailed structure of Y (and in particular the fact that color is not broken) allows to check the the following relation

$$[[D_F, a], b^0] = 0 \qquad \forall \, a, b \in \mathcal{A}_F \,. \tag{5.6}$$

The natural $\mathbb{Z}/2$-grading of $\mathcal{H}_F$ gives 1 for left-handed fermions $(e_L, \nu_L \ldots)$ and -1 for right-handed fermions; one has

$$\gamma_F = \varepsilon \, \varepsilon^0 \text{ where } \varepsilon = (1, -1, 1) \in \mathcal{A}_F \,. \tag{5.7}$$

We refer to [9] for the verification of the axioms (1) through (7′). The only drawback of this construction is that the number of families introduces a multiplicity in the intersection form, $K_0(\mathcal{A}) \times K_0(\mathcal{A}) \to \mathbb{Z}$, given by an integer multiple of the matrix

$$\begin{bmatrix} -1 & 1 & -1 \\ 1 & 0 & 1 \\ -1 & 1 & 0 \end{bmatrix} . \tag{5.8}$$

[1] taking into account the lifting of diffeomorphisms to the spinors

We will dwell on the significance of the specific spectral geometry $(\mathcal{A}_F, \mathcal{H}_F, D_F)$ at the end of this exposition.

The next step consists of the computation of internal deformations (formula (4.9)) of the product geometry $M \times F$ where M is a 4–dimensional Riemannian spin manifold. The computation gives the standard model gauge bosons $\gamma, W^{\pm}, Z$, the eight gluons and the Higgs fields φ with accurate quantum numbers. It also shows that

$$\mathcal{L}_{\varphi f} + \mathcal{L}_f = \langle \psi, D\psi \rangle \tag{5.9}$$

where $D = D_0 + A + JAJ^{-1}$ is the inner deformation of the product geometry (given by the operator $D_0 = \partial\!\!\!/ \otimes 1 + \gamma_5 \otimes D_F$).

The product structure of $M \times F$ gives a bigrading of Ω_D^1 and a decomposition $A = A^{(1,0)} + A^{(0,1)}$ of A which corresponds to the decomposition (5.9). The term $A^{(1,0)}$ is the sum of the vector potentials of all spin 1 bosons, the term $A^{(0,1)}$ is the Higgs boson which appears from the finite space $(\mathcal{A}_F, \mathcal{H}_F, D_F)$ as finite difference terms. This bigrading exists also on Ω_D^2, the analogue of 2–forms [7], and decomposes the curvature $\theta = dA + A^2$ in three terms $\theta = \theta^{(2,0)} + \theta^{(1,1)} + \theta^{(0,2)}$, which are pairwise orthogonal with respect to the scalar product

$$\langle \omega_1, \omega_2 \rangle = ⨍ \omega_1 \, \omega_2^* \, ds^4 \,. \tag{5.10}$$

Hence, the Yang-Mills action, $\langle \theta, \theta \rangle = ⨍ \theta^2 \, ds^4$, also decomposes into the sum of three terms. One can show that these terms are $\mathcal{L}_G$, $\mathcal{L}_{G\varphi}$ and $\mathcal{L}_\varphi$ respectively [7].

The Yang-Mills action $⨍ \theta^2 \, ds^4$ uses the decomposition $D = D_0 + A + JAJ^{-1}$ and hence depends on more than just the geometry fixed by D. Since we want to unify matter with gravity by an action which is purely geometric we need a better formula that only involves D alone. In the simplest case, as it was shown in formula (4.14), the sum $\mathcal{L}_E + \mathcal{L}_G$ appears directly in the asymptotic expansion of the number of eigenvalues of D which are smaller than Λ. The same principle (cf. [6]) applies to the standard model and it is governed by the following functional

$$\text{Trace}\left(\varpi\left(\frac{D}{\Lambda}\right)\right) + \langle \psi, D\psi \rangle \tag{5.11}$$

whose asymptotic expansion [6] gives our original Lagrangian $\mathcal{L}$ (5.1) together with a Weyl gravity term and a term in $r\varphi^2$, the only term which could be added to $\mathcal{L}$ without changing the standard model. For the physical interpretation of this result see [6].

The finite geometry $(\mathcal{A}_F, \mathcal{H}_F, D_F)$ above was dictated by the experimental results, encapsulated in the details of the standard model. One still has to understand its conceptual significance using as a tool the analogue of Lie groups in noncommutative geometry, i.e. the quantum groups. A simple fact (cf. [26]) is that the spin cover Spin(4) of $SO(4)$ is not the maximal cover in the quantum group category. One has $\mathrm{Spin}(4) = SU(2) \times SU(2)$ and the group $SU(2)$ admits according to Lusztig a finite cover of the form (Frobenius at ∞):

$$1 \to H \to SU(2)_q \to SU(2) \to 1 \tag{5.12}$$

where q is a root of unity $q^m = 1$, m odd. The simplest case is $m = 3$, $q = \exp\left(\frac{2\pi i}{3}\right)$. The finite quantum group H has a finite dimensional Hopf algebra very similar to $\mathcal{A}_F$, and the spinor representation of H defines a bimodule whose structure is very similar to the $\mathcal{A}_F$–bimodule $\mathcal{H}_F$. This suggests to extend spin geometry [25] to quantum covers of the spin group. This requires for the description of principal G bundles to introduce a minimum of noncommutativity (of the type $C^\infty(M) \otimes \mathcal{A}_F$) in the algebras of functions.

At last let us mention that we neglected the important difference between Riemannian and Lorentzian signatures in this exposé.

References

1. M.F. Atiyah: K-theory and reality, *Quart. J. Math. Oxford* (2), **17** (1966), 367-386.
2. R. Beals and P. Greiner: Calculus on Heisenberg manifolds, *Annals of Math. Studies* **119**, Princeton Univ. Press, Princeton, N.J., 1988.
3. T.P. Branson: An anomaly associated with 4-dimensional quantum gravity, to appear.
4. T.P. Branson and B. Ørsted: Explicit functional determinants in four dimensions, *Proc. Amer. Math. Soc.*, **113** (1991), 669-682.
5. A.R. Bernstein and F. Wattenberg: Non standard measure theory. In *Applications of model theory to algebra analysis and probability*, Edited by W.A.J. Luxenburg Halt, Rinehart and Winstin (1969).
6. A. Chamseddine and A. Connes: The spectral action principle, to appear.
7. A. Connes: Noncommutative geometry, Academic Press (1994).
8. A. Connes: Cyclic cohomology and the transverse fundamental class of a foliation, Geometric methods in operator algebras, (Kyoto, 1983), pp. 52-144, *Pitman Res. Notes in Math.* **123** Longman, Harlow (1986).
   reality, *Journal of Math. Physics* **36** n.11 (1995).
9. A. Connes: Gravity coupled with matter and the foundation of noncommutative geometry.
10. A. Connes and H. Moscovici: Cyclic cohomology, the Novikov conjecture and hyperbolic groups, *Topology* **29** (1990), 345-388.
11. A. Connes and H. Moscovici: The local index formula in noncommutative geometry, GAFA, **5** (1995), 174-243.
12. A. Connes and H. Moscovici: Hypoelliptic Dirac operator, diffeomorphisms and the transverse fundamental class.

13. A. Connes, D. Sullivan and N. Teleman: Quasiconformal mappings, operators on Hilbert space, and local formulae for characteristic classes, *Topology*, Vol.33 n.4 (1994), 663-681.
14. T. Damour and J.H. Taylor: Strong field tests of relativistic gravity and binary pulsars, *Physical Review D*, Vol.45 n.6 (1992), 1840-1868.
15. J. Dixmier: Existence de traces non normales, *C.R. Acad. Sci. Paris*, Ser. A-B **262** (1966).
16. S. Doplicher, K. Fredenhagen and J.E. Roberts: Quantum structure of space time at the Planck scale and Quantum fields, to appear in CMP.
17. J. Fröhlich: The noncommutative geometry of two dimensional supersymmetric conformal field theory, *Preprint ETH* (1994).
18. M. Gromov: Carnot–Caratheodory spaces seen from within, Preprint IHES/M/94/6.
19. M. Hilsum, G. Skandalis: Morphismes K-orientés d'espaces de feuilles et fonctorialité en théorie de Kasparov, *Ann. Sci. Ecole Norm. Sup.* (4) **20** (1987), 325-390.
20. W. Kalau and M. Walze: Gravity, noncommutative geometry and the Wodzicki residue, *J. of Geom. and Phys.* **16** (1995), 327-344.
21. G. Kasparov: The operator K-functor and extensions of C^*-algebras, *Izv. Akad. Nauk. SSSR Ser. Mat.*, **44** (1980), 571-636.
22. C. Kassel: Le résidu non commutatif, *Séminaire Bourbaki, exposé* 708, *Astérisque* Vol. 88-89.
23. D. Kastler: The Dirac operator and gravitation, *Commun. Math. Phys.* **166** (1995), 633-643.
24. R.C.Kirbi: Stable homeomorphisms and the annulus conjecture, *Ann. Math.* **89** (1969), 575–582.
25. B. Lawson and M.L. Michelson: *Spin Geometry*, Princeton 1989.
26. Y. Manin: Quantum groups and noncommutative geometry, *Centre Recherche Math. Univ. Montréal* (1988).
27. J. Milnor and D. Stasheff: Characteristic classes, *Ann. of Math. Stud.*, **76** Princeton University Press, Princeton, N.J. (1974).
28. S.P. Novikov: Topological invariance of rational Pontrjagin classes, *Doklady A.N. SSSR*, **163** (1965), 921-923.
29. S. Power: Hankel operators on Hilbert space, *Res. Notes in Math.*, **64** Pitman, Boston, Mass. (1982).
30. M.A. Rieffel: Morita equivalence for C^*-algebras and W^*-algebras, *J. Pure Appl. Algebra* **5** (1974), 51-96.
31. M.A. Rieffel: C^*-algebras associated with irrational rotations, *Pacific J. Math.* **93** (1981), 415-429; MR 83b:46087.
32. B. Riemann: Mathematical Werke, Dover, New York (1953).
33. E. Stein: Singular integrals and differentiability properties of functions, *Princeton Univ. Press, Princeton, N.J.* (1970).
34. J. Stern: Le problème de la mesure, *Séminaire Bourbaki*, Vol. 1983/84, Exp. 632, pp. 325-346, *Astérisque* N. 121/122, *Soc. Math. France, Paris* (1985).
35. D. Sullivan: Hyperbolic geometry and homeomorphisms, in *Geometric Topology, Proceed. Georgia Topology Conf. Athens*, Georgia (1977), 543-555.
36. D. Sullivan: Geometric periodicity and the invariants of manifolds, *Lecture Notes in Math.* **197**, Springer (1971).
37. M. Takesaki: Tomita's theory of modular Hilbert algebras and its applications, *Lecture Notes in Math.* **128**, Springer (1970).
38. Weinberg: Gravitation and Cosmology, John Wiley and Sons, New York London (1972).
39. M. Wodzicki: Noncommutative residue, Part I. Fundamentals K-theory, arithmetic and geometry, *Lecture Notes in Math.*, 1289, Springer-Berlin (1987).

SHIFT OPERATOR FOR THE DISCRETE $SL(2)$ CURRENT ALGEBRA

LUDWIG FADDEEV
St.Petersburg Branch of Steklov Mathematical Institute,
Fontanka 27, St.Petersburg 191011, Russia
and
Research Institute for Theoretical Physics,
University of Helsinki, Siltavuorenpenger 20C,
SF 00014 Helsinki, Finland

AND

ALEXANDER VOLKOV
St.Petersburg Branch of Steklov Mathematical Institute,
Fontanka 27, St.Petersburg 191011, Russia
and
Physique - Mathématique, Université de Montpellier II
Pl. E. Bataillon, Case 50, 34095 Montpellier Cédex 05, France

Abstract. The spatial shift operator is constructed for the discrete $SL(2)$ current algebra. The main result can be considered as a lattice deformation of the Sugawara construction. Explicit formulae use the q-exponent or quantum dilogarithm.

The concrete result, advertised in this lecture, belongs to a rapidly developed subject of quantum integrable models on $1+1$ dimensional discrete space–time. It is instructive to say some general words on this development.

First idiomatic example was proposed by Hirota [1]. The recent development is connected with hamiltonian structure [2, 3], geometric interpretation [4] and, most important, quantization [2]. The arguments in support of this activity are numerous. The most prosaic one is the need for regularization to make the quantization consistent. The most ambitions one refers to natural discreteness of space–time at the level of Planck length. In between lie other arguments not the least of which is the practical question of con-

D. Sternheimer et al. (eds.), Deformation Theory and Symplectic Geometry, 35–42.

struction of quantum cellular automata (see e.g. [5]). The discretization of both space and time (in contradistinction to Hamiltonian lattice approach, when only space is discretized) makes approach more symmetric. Of course, the time dependence is given not by the hamiltonian H but by the one step evolution operator

$$U = e^{-iH\Delta} , \tag{1}$$

where Δ is a length of lattice site. One of attractive features of recent results is the fact, that U looks even easier, than exactly integrable hamiltonian itself. We finish this introduction by citing two less evident reasons for discrete models.

1. Lattice models happen to be more universal than the continuous ones. It becomes clear, that Hirota model properly parametrized has several different continuous limits, both massive and massless. For instance, KdV equation, which is conformally invariant, and Sine–Gordon equation, which is massive, were shown in [6] being the continuous limits of Hirota model with different choice of scaling for space and time coordinates. This shows, in particular, that some version of Virasoro symmetry is realized in discrete space–time and extends to massive case.

2. Even more mysterious duality of integrable models begins to unravel recently. It was observed, that in the framework of Bethe Ansatz approach to quantum integrable models the set of transfer matrices $T_i(\lambda)$, labeled by spin $i = 0, 1/2, 1, \ldots$ of auxiliary space, satisfy the difference equation in variable j and discrete shifts $\lambda \to q\lambda$ of rapidity λ; this equation in form essentially repeats the Heisenberg equation of motion of original model. Thus the transformation of (quantum) dynamical variables $\Phi(n,m)$, given on space–time lattice into set of (classical) transfer matrices $T_{ij} = T_i(q^j\lambda)$, is kind of duality transformation. This observation ([7]–[11]) still waits for more explanation.

We finish this long introduction and come to the main subject of the lecture. We shall present an explicit expression for a shift operator for the quantum dynamical system described in terms of discretized current algebra, corresponding to the group $SL(2)$. In other words, this system is a discretized WZNW model in one chiral sector. Equation of motion in continuous case is trivial, each component of current just moves in light like direction, satisfying the equation

$$\partial_t f + \partial_x f = 0. \tag{2}$$

In discrete case this is realized by taking the time shift being equal to the space shift. So we can stick to the algebra, given on a space chain and construct this shift

$$f_n \to f_{n+1} \tag{3}$$

for each component of quantum discretized current.

The explicit formulas and derivation are rather involved, so we shall bound ourselves with the description of the result referring for details to our preprint [12].

It is instructive to consider first the example of the abelian current algebra, which we discussed earlier [13]. It will allow us to introduce main ideas and technical tools in a simpler situation.

Classical continuous current algebra is generated by dynamical variable $p(x)$ given on a circle $0 \leq x \leq 2\pi$ and satisfying the Poisson relation

$$\{p(x), p(y)\} = \gamma\delta'(x-y) \ . \tag{4}$$

The Virasoro algebra is defined by variable

$$t(x) = p^2(x) + p'(x) \ . \tag{5}$$

The hamiltonian

$$H = \frac{1}{2\gamma}\int p^2 dx = \tag{6}$$

$$= \frac{1}{2\gamma}\int t(x)dx \tag{7}$$

leads to equation of motion

$$\partial_t p(x,t) + \partial_x p(x,t) = 0 \ , \tag{8}$$

already referred to above.

Corresponding discrete quantum system employs algebra of generators w_n with relations

$$\begin{aligned} w_n w_{n+1} &= q^2 w_{n+1} w_n \\ w_n w_{n+1} &= w_{n+1} w_n \quad |n-m| \geq 2 \end{aligned} \tag{9}$$

and periodicity conditions

$$w_{n+N} = w_n \ , \tag{10}$$

where N is a length of a chain.

Here q is expressed via Planck constant $\hbar$ and classical coupling constant γ as follows

$$q = e^{i\hbar\gamma} \ . \tag{11}$$

In classical ($\hbar \to 0$) and continuous limit

$$w_n = 1 + \Delta p(x) \ , \tag{12}$$

where $x = n\Delta$ and Δ is a lattice spacing one gets Poisson relation (4) from the quantum relations (9).

The discrete Virasoro variables are given by

$$t_n = w_n^{-1} + w_{n+1} + q w_n^{-1} w_{n+1} \; . \tag{13}$$

The operator U, realizing the shift

$$U w_n U^{-1} = w_{n+1} \tag{14}$$

can be written in two ways, corresponding to two formulas for the hamiltonian

$$U = \overrightarrow{\prod} r(w_n) = \tag{15}$$

$$= \overrightarrow{\prod} s(t_n) \tag{16}$$

where the functions $r(w)$ and $s(w)$ satisfy difference relations

$$\frac{r(qw)}{r(q^{-1}w)} = \frac{1}{w} \; ; \quad \frac{s(qw)}{s(q^{-1}w)} = \frac{1}{1+w} \tag{17}$$

and are connected by relation

$$r(w) = s(w)s(w^{-1}) \; . \tag{18}$$

The multiplicative locality in (15) and (16) corresponds to the additive locality of classical expressions (6) and (7).

We do not discuss how the periodicity is taken into account in (15)–(16), see [6] for details. We also do not touch a subtle problem of defining the solutions of equations (17) for the case $|q| = 1$, which interests us primarily. Details on this are in [14, 15]. So we leave ourselves with pure algebraic aspects. It is easy to find, that

$$r_n = r(w_n) \tag{19}$$

satisfy the braid relation

$$r_n r_{n+1} r_n = r_{n+1} r_n r_{n+1} \tag{20}$$

and shift property follows from this. Now, as was shown in [16], the variables

$$s_n = s(t_n) \tag{21}$$

also satisfy simple relations. Here they are:

$$\begin{aligned} s_{n+1}s_{n-1}s_n s_{n+1} &= s_{n-1}s_{n+1}s_n \\ s_{n-1}s_n s_{n+1}s_{n-1} &= s_n s_{n-1}s_{n+1} \ . \end{aligned} \tag{22}$$

Just from this relations it follows, that the product in a ascending order

$$U_1 = \overrightarrow{\prod} s_n \tag{23}$$

is a shift

$$s_n U_1 = U_1 s_{n-1} \ , \tag{24}$$

whereas the product in descending order

$$U_2 = \overleftarrow{\prod} s_n \tag{25}$$

is a shift by two lattice cites

$$s_n U_2 = U_2 s_{n+2} \ . \tag{26}$$

This observation will be most important in what follows.

Now we turn to the main subject and begin with the brief discussion of the $sl(2)$ discrete current algebra. For more details see [17]–[20]. This algebra is generated by variables constituting a 2×2 matrix J_n, attached to each lattice cite n

$$J_n = \begin{pmatrix} J_n^{11} & J_n^{12} \\ J_n^{21} & J_n^{22} \end{pmatrix} \ . \tag{27}$$

The commutation relations involve the usual $sl(2)$ $\mathcal{R}$–matrices $\mathcal{R}_+$ and $\mathcal{R}_-$, where $\mathcal{R}_+$ is given by

$$\mathcal{R}_+ = \begin{pmatrix} q^{1/2} & & & \\ & q^{-1/2} & q^{1/2} - q^{-3/2} & \\ & & q^{-1/2} & \\ & & & q^{1/2} \end{pmatrix} \tag{28}$$

and

$$\mathcal{R}_- = \mathrm{P}\mathcal{R}_+^{-1}\mathrm{P} \ , \tag{29}$$

where P is a permutation matrix. In usual notations

$$J_n^1 = J_n \otimes I \ , \quad J_n^2 = I \otimes J_n \tag{30}$$

for 4×4 matrices J the relations take form

$$\mathcal{R}_+ J_n^1 J_n^2 \mathcal{R}_- = J_n^2 J_n^1 \tag{31}$$

$$J_n^1 J_{n+1}^2 = J_{n+1}^2 \mathcal{R}_+ J_n^1 \ . \tag{32}$$

In classical continuous limit these relations reduce to the ordinary Poisson brackets for current $j(x)$

$$J_n = I + i\Delta j(x)\ , \quad x = n\Delta \tag{33}$$

with classical value of central charge

$$\kappa = 1/\gamma\ . \tag{34}$$

We prefer to deal with an extension of this algebra; it is described by two triangular matrices (in close analogy with $L_\pm$ matrices instead of $L = L_+L_-^{-1}$ in the RTF description [21] of finite dimensional quantum Lie algebra). These matrices are attached to the vertices of more refined chain, when each lattice cite is divided in two. The explicit expression is as follows

$$B_{2n} = \begin{pmatrix} \alpha_{2n+1/2}^{-1/2} & 0 \\ 0 & \alpha_{2n+1/2}^{1/2} \end{pmatrix} \begin{pmatrix} 1 & \beta_{2n} \\ 0 & 1 \end{pmatrix} \tag{35}$$

$$C_{2n} = \begin{pmatrix} \alpha_{2n-1/2}^{1/2} & 0 \\ 0 & \alpha_{2n-1/2}^{-1/2} \end{pmatrix} \begin{pmatrix} 1 & 0 \\ \beta_{2n-1} & 1 \end{pmatrix}. \tag{36}$$

Here we attached variables β_n to vertices of chain and variables $\alpha_{n+1/2}$ to the edge between n and $n+1$.

The commutation relations

$$\mathcal{R}_+ \overset{1}{B}_n \overset{2}{B}_n = \overset{2}{B}_n \overset{1}{B}_n \mathcal{R}_+ \tag{37}$$

$$\overset{1}{C}_n \overset{2}{C}_n \mathcal{R}_- = \mathcal{R}_- \overset{2}{C}_n \overset{1}{C}_n \tag{38}$$

$$\overset{1}{B}_n \overset{2}{C}_{n+1} = \overset{2}{C}_{n+1} \mathcal{R}_+ \overset{1}{B}_n \tag{39}$$

$$\overset{1}{C}_n \overset{2}{B}_{n+1} = \overset{2}{B}_{n+1} \mathcal{R}_-^{-1} \overset{1}{C}_n \tag{40}$$

can be written in terms of generators α and β:

$$\alpha_{n+1/2}\alpha_{n-1/2} = q^2 \alpha_{n-1/2}\alpha_{n+1/2} \tag{41}$$

$$\beta_n \alpha_{n-1/2} = q^2 \alpha_{n-1/2}\beta_n \tag{42}$$

$$\alpha_{n+1/2}\beta_n = q^2 \beta_n \alpha_{n+1/2} \tag{43}$$

$$[\beta_{n-1}, \beta_n] = (q - q^{-1})\alpha_{n-1/2}\ . \tag{44}$$

Incidentally the matrix

$$J_n = B_{2n}C_{2n-1} \tag{45}$$

satisfy relations (20)-(22) so the last formula defines the reduction of $B-C$ algebra to J algebra, which contains one generator less at each lattice site.

It is clear from (45), that on our refined lattice we need a shift by two cites to describe the physical shift. So the property of operator U_2 was instrumental for the construction below.

It remains to write down the explicit formulas. The reasoning, checking and discussion is done in [12], to which we refer for details.

The combination

$$t_{n-1/2} = q + q^2 \beta_n \alpha_{n-1/2}^{-1} \beta_{n-1} \tag{46}$$

has the property of discrete Virasoro algebra. The shift operator W is given by local product

$$W = \overleftarrow{\prod} s(t_{n+1/2}) r(-q\alpha_{2n+1/2}\alpha_{2n-1/2}) s(t_{n-1/2}) \, . \tag{47}$$

The classical continuous limit of this formula turns into the usual Sugavara construction.

In conclusion we stress, that the use of multiplicative locality as in (15)-(15) and properties of solutions of differences equations (17) with noncommutative arguments, as in (20)-(22), is the main feature of our approach. In considered examples we dealt only with massless case and so only ingredients of finite dimensional quantum Lie algebra were used. In more general case one is to use the affine algebra and introduce the spectral parameter. However we know, that this can be done also in terms of solutions of equations (17). But this subject goes beyond the goal of this lecture and we return to it elsewhere.

References

1. R.Hirota, Journ. Phys.Soc. Japan, **43** (1977) 2079.
2. L.Faddeev, A.Volkov, Lett. Math. Phys., **32** (1994) 125.
3. F. Nijhoff, H. Capel, Physics Letters, **A163** (1992) 49.
4. A.Bobenko, N.Kunz, U.Pinkal, Physics Letters, **A177** (1993) 399-404.
5. D. Meyer, quant-ph/9605023.
6. A. Volkov, hep-th/9509024.
7. A. Kuniba, T. Nakanishi, J. Suzuki, Int. Journ. Mod. Phys. **A9** (1994) 5215.
8. R. Kashaev, N. Reshetikhin, hep-th/9507065.
9. L. Faddeev in XI-th International Congress of Mathematical Physics, International Press 1995, p.513.
10. A. Zabrodin, hep-th/9610039.
11. P.Wiegmann, cond-mat/9610132.
12. L. Faddeev, A. Volkov, hep-th/9606088.
13. L.D.Faddeev, A.Yu.Volkov, Phys.Lett **B315** (1993) 311.
14. L. Faddeev, Lectures delivered at the International School of Physics "Enrico Fermi", 127th course, held in Villa Monastero, Varenna, Italy, from 28 June to 8 July 1994, hep-th/9408041.

15. L. Faddeev, Lett. Math. Phys. **34** (1995) 249-254.
16. A. Volkov, hep-th/9603003.
17. A. Alekseev, L. Faddeev, M. Semenov-Tjan-Shansky, Con. Math. Phys. **149** (1992) 335.
18. L.Faddeev in "New Symmetry Principles in Quantum Field Theory", ed. J.Frohlich et al, NATO ASI Series, Physics vol.295, (1992) p.159.
19. F. Falceto, K. Gawedsky, J. Geom. Phys. **11** (1993) 251.
20. A. Alekseev, L. Faddeev, J. Froelich, V. Schomerus, q-alg/9604017.
21. N. Reshetikhin, L. Takhtajan, L. Faddeev, Algebra i Analiz, v.I(1) (1989) 178-206.

NAMBU MECHANICS, N-ARY OPERATIONS AND THEIR QUANTIZATION

MOSHÉ FLATO*
Département de Mathématiques, Université de Bourgogne
BP 400, F-21011 Dijon Cedex, France

GIUSEPPE DITO
Research Institute for Mathematical Sciences
Kyoto University, Sakyo-ku, Kyoto 606-01, Japan

AND

DANIEL STERNHEIMER
CNRS & Département de Mathématiques
Université de Bourgogne
BP 400, F-21011 Dijon Cedex, France

Abstract. We start with an overview of the "generalized Hamiltonian dynamics" introduced in 1973 by Y. Nambu, its motivations, mathematical background and subsequent developments – all of it on the classical level. This includes the notion (not present in Nambu's work) of a generalization of the Jacobi identity called Fundamental Identity. We then briefly describe the difficulties encountered in the quantization of such n-ary structures, explain their reason and present the recently obtained solution combining deformation quantization with a "second quantization" type of approach on $\mathbb{R}^n$. The solution is called "Zariski quantization" because it is based on the factorization of (real) polynomials into irreducibles. Since we want to quantize composition laws of the determinant (Jacobian) type and need a Leibniz rule, we need to take care also of derivatives and this requires going one step further (Taylor developments of polynomials over polynomials). We also discuss a (closer to the root, "first quantized") approach in various circumstances, especially in the case of covariant star products (exemplified by the case of $\mathfrak{su}(2)$). Finally we address the question of equivalence and triviality of such deformation quantizations of a new type (the deformations of algebras are more general than those considered by Gerstenhaber).

* The talk was presented by Moshé Flato. The work reported here benefited from scientific contacts supported by the EU Network Contract ERBCHRXCT940701.

D. Sternheimer et al. (eds.), Deformation Theory and Symplectic Geometry, 43–66.

1. Introduction

In 1973, after having kept this very original idea in his files for many years since he did not arrive at what he could have considered a really satisfactory paper, Nambu decided to publish [26] the result "as is". The underlying idea of this new formalism was that in statistical mechanics the basic result is Liouville theorem, which follows from but does not require Hamiltonian dynamics. In geometrical terms, symplectic structures are unimodular but the latter are more general. The lowest dimensional example to start with is $\mathbb{R}^3$, with triplets of dynamical variables and the Poisson bracket of two functions replaced by a "triple bracket", the Jacobian of three functions.

The very mention of the notion of triplets (even more so in those days) is reminiscent of quarks for which satisfactory statistics explaining e.g. confinement is yet to be found – and this was in the back of Nambu's mind. But in order to even think about such a connection one needs to solve the problem of quantization and Nambu was not able to get very far in solving this question. In fact quantization proved to be a very difficult question and a solution was given only last year [12], at least for polynomials on $\mathbb{R}^n$, using an elaborate construction based on arithmetic properties of polynomials and methods of second quantization.

Nambu's paper was followed by a number of papers [1, 5] [25] which essentially "chilled" the idea from the physical point of view by showing that the "classical part" of it could be cast into the framework of classical mechanics with Dirac constraints [9], for which there is a usual quantization scheme. The latter is more or less what Nambu got (and was not satisfied with). Though perfectly correct, these papers had for unwanted result that the whole formalism remained in limbo for another twenty years (and we are in part responsible for that).

On the mathematical and mathematical physics side people had recurrently been interested in Jordan algebras, which Nambu had also considered in his abovementioned attempts to quantization and (justly, it turns out) rejected: nonassociativity is not the solution. More recently (re)appeared a much more general framework, operads [21, 24, 27]. The solution we shall present here is affiliated with that framework, but we shall not develop here this point which would take us too far.

We had never quite forgotten Nambu's interesting formalism and in 1992, independently [14, 31, 33], the underlying structure of n-ary brackets was studied and the analogue of the Jacobi identity for Poisson brackets (called *Fundamental Identity*, in short FI) discovered. That identity, which somehow Nambu (and others) had missed in the 70's, together with an explicit expression of the Leibniz rule, was the basis of the modern developments. It is truly fundamental in many respects. In particular it selects the really interesting generalization of Poisson brackets to n-tuples from other a priori possible generalizations; the selection is based on the fact that a n-ary bracket of conserved quantities is automatically a conserved quantity. (It should be said that other possibilities of n-ary brackets exist in the literature since years; however, in spite of the fact that these possibilities are nontrivial and quite rich mathematically, their utilization in "new mechanics" kind of models is more than problematic and therefore we never considered them). From that identity emerged the notion of Nambu-Poisson manifolds [33], a manifold endowed with a n-bracket satisfying the obvious properties of skew-symmetry and Leibniz rule and the less obvious FI. Locally such manifolds are [18] foliated by manifolds on which the structure is given by a functional determinant.

In this paper we shall first present (Section 2) the earlier developments, until and including the Fundamental Identity. Then (Section 3) we shall describe in great lines the abovementioned solution to quantization of such n-ary operations based on a subtle adaptation of deformation quantization [2, 16, 34]. The basic idea is to replace in the computation of the determinant the ordinary product of (real) polynomials by a symmetrized Moyal star product of their decomposition into irreducibles (what we call their Zariski product, because of the decomposition into irreducible polynomials) – but the full solution is not so simple because of linearity requirements and the fact that the totally symmetrized Moyal product of linear functions is the usual product. In particular (which is consistent with the Barr triviality of Harrison cohomology [20] for algebras of polynomials) we get algebra deformations which are no more of the type studied by Gerstenhaber which we call here "DrG-deformations" (as a tribute to the fundamental contribution of "Dr. G." [19]): for the new type of deformations we have only $\mathbb{R}$-linearity instead of $\mathbb{R}[[\nu]]$-linearity (ν being the deformation parameter), and it is a *generalized deformation* in the sense that the limit for $\nu = 0$ of the deformed algebra is the original algebra. The algebra product which we "deform" is here, in the spirit of second quantization, that of an algebra of "fields", the algebra of the semi-group generated by irreducible polynomials (polynomials over the set of normalized irreducible polynomials). Moreover, since we need to quantize composition laws of the functional determinant (Jacobian) type and need a Leibniz rule, we must also take care of derivatives and this requires going one step further (Taylor developments of polynomials over polynomials).

When we deal with covariant star products for which the totally symmetrized star product of elementary factors is no more trivial, some variants can be developed, of the "first quantized" type; we present them here in Section 4.1, essentially in the case of $\mathfrak{su}(2)$ [11]. A similar procedure can be introduced also on $\mathbb{R}^m$ when the ordering used in the corresponding

"Zariski" product (e.g. of standard type) is not the same as the ordering for the star product. A second part of the last Section is devoted to some remarks on the important question of equivalence of such generalized deformations (which is not straightforward) and to the related (a little simpler) question of triviality, already tackled in [11] where the generalized deformations introduced are not trivial relatively to the more natural notion of "strong triviality" (though they are trivial for the less demanding "weak triviality"). The physical aspect of triviality requires to address also the problem of spectrality of observables, by which we conclude that section.

2. Nambu mechanics and its foundations

2.1. NAMBU'S ORIGINAL PAPER (CLASSICAL PART)

Having in mind the abovementioned motivation, Nambu [26] started with the following "Hamilton equations" on $\mathbb{R}^3$ of the form:

$$\frac{dr}{dt} = \nabla g(r) \wedge \nabla h(r), \quad r = (x, y, z) \in \mathbb{R}^3, \tag{1}$$

where x, y, z are the dynamical variables and g, h are two functions of r. Liouville theorem follows directly from the identity $\nabla \cdot (\nabla g(r) \wedge \nabla h(r)) = 0$, which tells us that the velocity field in Eq. (1) is divergenceless. From (1) follows that the evolution of a function f on $\mathbb{R}^3$ is given by:

$$\frac{df}{dt} = \frac{\partial(f, g, h)}{\partial(x, y, z)}, \tag{2}$$

where the right-hand side is the Jacobian of the mapping $\mathbb{R}^3 \to \mathbb{R}^3$ given by $(x, y, z) \mapsto (f, g, h)$.

For example, in this "baby model for integrable systems", Euler equations for the angular momentum of a rigid body are obtained when the dynamical variables are taken to be the components of the angular momentum vector $L = (L_x, L_y, L_z)$, and g and h are, respectively, the total kinetic energy $\frac{L_x^2}{2I_x} + \frac{L_y^2}{2I_y} + \frac{L_z^2}{2I_z}$ and the square of the angular momentum $L_x^2 + L_y^2 + L_z^2$. Other examples can be given, in particular Nahm's equations for static $\mathfrak{su}(2)$ monopoles, $\dot{x}_i = x_j x_k$ $(i, j, k = 1, 2, 3)$ in $\mathfrak{su}(2)^* \sim \mathbb{R}^3$, with $h = x_1^2 - x_2^2$, $g = x_1^2 - x_3^2$, etc.

In this framework the analogues of canonical transformations are mappings $(x, y, z) \mapsto (x', y', z')$ with $\frac{\partial(x', y', z')}{\partial(x, y, z)} = 1$. The linear ones generate $SL(3, \mathbb{R})$. The Hamilton equations generate infinitesimal canonical transformations, and two sets of "Hamiltonians" h, g and h', g' generate the same transformation if they are related by a "gauge transformation" $\frac{\partial(h', g')}{\partial(h, g)} = 1$.

Here the principle of least action, which states that the classical trajectory C_1 is an extremal of the action functional $A(C_1) = \int_{C_1}(pdq - Hdt)$, is replaced by a similar one [33] with a 2-dimensional cycle C_2 and "action functional" $A(C_2) = \int_{C_2}(xdy \wedge dz - hdg \wedge dt)$ (which bears some flavor of strings and some similitude with the cyclic cocycles of Connes [6]).

Expression (2) was easily generalized to n functions $f_i, i = 1, \ldots, n$. As was already noticed by Nambu (and in apparent contradistinction with the definition of Poisson brackets), the correct way to proceed (see [26, 1]) is not by a direct sum of m triples, where the linear canonical transformations are $SL(3, \mathbb{R})^m$ and the pairs of "Hamiltonians" (h_i, g_i), $i = 1, \ldots, m$, are no longer constants of motion. One should instead introduce a n-tuple of functions on $\mathbb{R}^n$ with composition law given by their Jacobian, with linear canonical transformations $SL(n, \mathbb{R})$ and a corresponding $(n-1)$-form which is the analogue of the Poincaré-Cartan integral invariant. The Jacobian has to be interpreted as a generalized Poisson bracket: it is skew-symmetric with respect to the f_i's and a derivation of the algebra of smooth functions on $\mathbb{R}^n$, i.e., the Leibniz rule is verified in each argument (e.g. $\{f_1 f_2, f_3, \ldots, f_{n+1}\} = f_1\{f_2, \ldots, f_{n+1}\} + \{f_1, f_3, \ldots, f_{n+1}\}f_2$, etc.). Hence there is a complete analogy with the Poisson bracket formulation of Hamilton equations (including the fact that the components of the $(n-1)$-tuple of "Hamiltonians" $(f_2, \ldots f_n)$ are constants of motion) except, at first sight, for the equivalent of the Jacobi identity which seems to be lacking.

2.2. NAMBU'S MECHANICS AS CONSTRAINED MECHANICS

As we said in the Introduction, it turns out [1] that Nambu mechanics on $\mathbb{R}^n$ can be cast into the framework of classical mechanics with Dirac constraints [9] on $\mathbb{R}^{2n}$. Take $n = 3$ to fix ideas. Motivated by (1), which can be obtained using the intuitive Hamiltonian $\mathcal{H} = p \cdot (\nabla g(q) \times \nabla h(q))$, $\quad q = r = (x, y, z) \in \mathbb{R}^3$, we consider the "Nambu Lagrangian" $L_N = h. \sum_{i=1}^{3} \dot{q}_i \frac{\partial g}{\partial q_i}$ for which the Euler-Lagrange equations can be written as $\frac{dg}{dt}\nabla h = \frac{dh}{dt}\nabla g$. Then (in an open set $\Omega \in \mathbb{R}^3$ where $\nabla h \times \nabla g$ is nonzero, by vector multiplication by ∇h and ∇g) we see that h and g are constants of motion, as required.

But now in addition to the equation for $\dot{q}$ there is the "other half" for $\dot{p}$ and thus we need [1] to introduce three Dirac constraints $\phi_i(q, p) = p_i - h(q)\partial_i g(q)$, $i = 1, 2, 3$, where $\partial_i = \frac{\partial}{\partial q_i}$ (these are primary constraints which give a pair of second class constraints and one first class constraint; there are no secondary constraints). We get (with 0 for "Hamiltonian") $\dot{q}_i = u_i$ and $\dot{p}_i = \sum_{j=1}^{3} u_i \partial_i(h(\partial_j g))$ for some functions u_i to be determined. Expressing $\dot{\phi}_i = 0$ we get $\sum_{j=1}^{3} u_i(\partial_i h \partial_j g - \partial_j h \partial_i g) = 0$ hence

$$\dot{q}_i = u_i = v(t, q)[\nabla h \times \nabla g]_i \tag{3}$$

with an arbitrary function v and finally, with a position-dependent time rescaling $\tau = \int_0^t v(s, q(s, q_0))ds$ we obtain $\frac{dq}{d\tau} = \nabla h \times \nabla g$. Thus for a fixed v (supposed to be nonzero in Ω) we have a one-to-one correspondence between solutions of (1) and (3) which becomes an identity by a time rescaling. Since the choice of the time axis is arbitrary, the two formulations contain the same dynamical information. $\mathcal{H}_T = v(t,q)\ p \cdot (\nabla h \times \nabla g)$ is now the total Hamiltonian and it vanishes weakly since it is proportional to the first class constraint $\phi = \mathcal{H} = [\nabla h \times \nabla g] \cdot [p - h\nabla g]$ (the effect of the two second class constraints is to reduce the problem to some $\mathbb{R}^4 \subset \mathbb{R}^6$). We can now try and quantize along the lines of constrained mechanics as proposed by Dirac and recover the usual Heisenberg quantization already proposed by Nambu. But such an approach is not unambiguous and it is not a quantization of the triple bracket. As we shall see in Section 3, there is a much more fruitful and original approach which applies directly to all n-tuple brackets ($n \geq 2$).

2.3. FUNDAMENTAL IDENTITY AND NAMBU-POISSON MANIFOLDS

2.3.1. *Nambu-Poisson brackets and the Fundamental Identity*

In the usual Poisson formulation, the Jacobi identity is the infinitesimal form of the Poisson theorem which states that the bracket of two integrals of motion is also an integral of motion. If we want a similar theorem for Nambu mechanics there must be an infinitesimal form of it which will provide a generalization of the Jacobi identity. Denote by $\{f, g, h\}$ the Jacobian appearing in (2). Let $\phi_t : r \mapsto \phi_t(r)$ be the flow for (1). Then a generalization of the Poisson theorem would imply that ϕ_t is a "canonical transformation" for the generalized bracket:

$$\{f_1 \circ \phi_t, f_2 \circ \phi_t, f_3 \circ \phi_t\} = \{f_1, f_2, f_3\} \circ \phi_t.$$

Differentiation of this equality with respect to t yields the desired generalization of the Jacobi identity:

$$\begin{aligned}\{\{g,h,f_1\},f_2,f_3\} + \{f_1,\{g,h,f_2\},f_3\} + \{f_1,f_2,\{g,h,f_3\}\} \\ = \{g,h,\{f_1,f_2,f_3\}\}, \quad \forall g,h,f_1,f_2,f_3 \in C^\infty(\mathbb{R}^3).\end{aligned}$$

This identity and its generalization to $\mathbb{R}^n$, called Fundamental Identity (FI), was introduced by Flato, Frønsdal [14] and Takhtajan [33] as a consistency condition for Nambu Mechanics (this consistency condition was also formulated in [31]) and allows a generalized Poisson theorem: the generalized bracket of n integrals of motion is an integral of motion. It turns out that the Jacobian on $\mathbb{R}^n$ satisfies the FI. We are thus lead to the following generalization, denoting by S_n the group of permutations of the set $\{1, \ldots, n\}$ and by $\epsilon(\sigma)$ the sign of the permutation $\sigma \in S_n$:

Definition 1 *A Nambu bracket of order n on a m-dimensional manifold M ($2 \leq n \leq m$) is defined by a n-linear map on $A = C^\infty(M)$ taking values in A:*

$$\{\cdot, \ldots, \cdot\} : A^n \to A,$$

such that the following relations are verified $\forall f_0, \ldots, f_{2n-1} \in A$:

a) Skew-symmetry

$$\{f_1, \ldots, f_n\} = \epsilon(\sigma)\{f_{\sigma_1}, \ldots, f_{\sigma_n}\}, \quad \forall \sigma \in S_n; \tag{4}$$

b) Leibniz rule

$$\{f_0 f_1, f_2, \ldots, f_n\} = f_0\{f_1, f_2, \ldots, f_n\} + \{f_0, f_2, \ldots, f_n\} f_1; \tag{5}$$

c) Fundamental Identity

$$\begin{aligned} &\{f_1, \ldots, f_{n-1}, \{f_n, \ldots, f_{2n-1}\}\} \\ &\quad = \{\{f_1, \ldots, f_{n-1}, f_n\}, f_{n+1}, \ldots, f_{2n-1}\} \\ &\qquad + \{f_n, \{f_1, \ldots, f_{n-1}, f_{n+1}\}, f_{n+2}, \ldots, f_{2n-1}\} \\ &\qquad + \cdots + \{f_n, f_{n+1}, \ldots, f_{2n-2}, \{f_1, \ldots, f_{n-1}, f_{2n-1}\}\}. \end{aligned} \tag{6}$$

Properties a) and b) imply that there exists a n-vector field η on M such that:

$$\{f_1, \ldots, f_n\} = \eta(df_1, \ldots, df_n), \quad \forall f_1, \ldots, f_n \in A. \tag{7}$$

2.3.2. *Nambu-Poisson manifolds and associated Nambu mechanics*

Of course the FI imposes constraints on η, analyzed in [33]. A n-vector field on M is called a Nambu tensor if its associated Nambu bracket defined by (7) satisfies the FI. This brings us to:

Definition 2 *A Nambu-Poisson manifold (M, η) is a manifold M on which is defined a Nambu tensor η. Then M is said to be endowed with a Nambu-Poisson structure.*

The dynamics associated with a Nambu bracket on M is specified by $n-1$ Hamiltonians $H_1, \ldots, H_{n-1} \in A$ and the time evolution of $f \in A$ is given by:

$$\frac{df}{dt} = \{H_1, \ldots, H_{n-1}, f\}. \tag{8}$$

Suppose that the flow ϕ_t associated with (8) exists and let U_t be the one-parameter group acting on A by $f \mapsto U_t(f) = f \circ \phi_t$. It follows from the FI that:

Proposition 1 *The one-parameter group U_t is an automorphism of the Nambu bracket structure on A.*

Definition 3 *$f \in A$ is called an integral of motion for the system defined by (8) if it satisfies $\{H_1, \ldots, H_{n-1}, f\} = 0$.*

It follows from the FI that a Poisson-like theorem exists for Nambu-Poisson manifolds:

Proposition 2 *The Nambu bracket of n integrals of motion is also an integral of motion.*

For the case $n = 2$, the FI is the Jacobi identity and one recovers the usual definition of a Poisson manifold. On $\mathbb{R}^2$, the canonical Poisson bracket of two functions $\mathcal{P}(f, g)$ is simply their Jacobian, and Nambu defined his bracket on $\mathbb{R}^n$ as a Jacobian of n functions $f_1, \ldots, f_n \in C^\infty(\mathbb{R}^n)$ of n variables $x_1, \ldots, x_n$:

$$\{f_1, \ldots, f_n\} = \sum_{\sigma \in S_n} \epsilon(\sigma) \frac{\partial f_1}{\partial x_{\sigma_1}} \cdots \frac{\partial f_n}{\partial x_{\sigma_n}}, \tag{9}$$

which gives the *canonical Nambu bracket* of order n on $\mathbb{R}^n$. Other examples of Nambu-Poisson structures have been found [3, 4]. One of them is a generalization of linear Poisson structures and is given by the following Nambu bracket of order n on $\mathbb{R}^{n+1}$:

$$\{f_1, \ldots, f_n\} = \sum_{\sigma \in S_{n+1}} \epsilon(\sigma) \frac{\partial f_1}{\partial x_{\sigma_1}} \cdots \frac{\partial f_n}{\partial x_{\sigma_n}} x_{\sigma_{n+1}}. \tag{10}$$

In general any manifold endowed with a Nambu-Poisson structure of order n is *locally foliated* [18] by Nambu-Poisson manifolds of dimension n endowed with the canonical Nambu-Poisson structure (9). In particular it is shown in [18] that any Nambu tensor is decomposable (this fact was conjectured by Takhtajan, eventually proved in a very elegant way by Gautheron and then discovered by chance by Takhtajan to follow from a result going back to 1923, reproduced in a textbook by Schouten!).

We now have a complete parallel between Poisson classical mechanics and Nambu classical mechanics: we have a bracket (the Nambu n-bracket, with $n = 2$ in the former case) satisfying the required properties (skew symmetry, Leibniz rule and the FI), equations of evolution of a similar form (8) with solutions which can be written as $f_t = \exp(t\{H_1, \ldots, H_{n-1}, .\})f_0$, and an *action functional* for which the classical trajectory is an extremal [33]. That action is the integral $A(C_{n-1}) = \int_{C_{n-1}} \omega^{(n-1)}$ over $(n-1)$-chains $C_{(n-1)}$ in the extended phase space $M \times \mathbb{R}$ (the latter factor stands for the time axis) of the generalized Poincaré-Cartan integral invariant action form $\omega^{(n-1)} = x_1 dx_2 \wedge \ldots \wedge dx_n - H_1 dH_2 \wedge \ldots \wedge dH_{n-1} \wedge dt$. We refer to [33] for more details (especially for the case $n = 3$).

3. Quantization of Nambu mechanics

3.1. FIRST ATTEMPTS AND PROBLEMS RAISED

As realized already by Nambu ([26]), the first problem in the intriguing question of quantization (of such a system of equations) is how to define a quantization. A simplistic answer is to say that, since the classical system can be treated as a (Dirac) constrained Hamiltonian system, it is enough to quantize the latter. Even within this framework a rigourous solution (in particular due to the presence of a first class constraint which turns the physical manifold into a Poisson, not symplectic, manifold) is not so easy to obtain. But there is more: how can we be sure that the classical equivalence can be carried over to the quantum case? Quantum anomalies pop up like mushrooms nowadays, so one must attack directly the problem of quantization of the original classical system.

Nambu [26], assuming only (4) and (5), could not find a satisfactory solution. Assuming the "usual framework" (operator algebras) to quantize (8) he arrives at the "disappointing" (albeit not so surprising in view of the classical equivalence with a constrained system) result that the quantum analogue of (8) is equivalent to a Heisenberg equation. Indeed, skew-symmetry (4) gives what is called in [33] ($n = 3$ case) the Nambu–Heisenberg commutation relation:

$$[A_1, A_2, A_3] \equiv \sum_{\sigma \in S_3} \epsilon(\sigma) A_{\sigma_1} A_{\sigma_2} A_{\sigma_3} = cI, \tag{11}$$

where c is a constant and I is the unit operator. Representations could be found (a more complete treatment is given in [33]), but already the Leibniz rule (5) brings back to the Heisenberg case. Using associative algebras to find realizations of the Poisson bracket for the canonical triplet leads again nowhere. He thus naturally turns to the use of nonassociative algebra. The Cayley–Dickson algebra (an algebra with alternating associator and 8 generators) is similarly rejected. So Nambu turns to Jordan algebras with the same conclusion, which is no surprise since all but one of them (the exceptional algebra, 3×3 matrices with Cayley coefficients) arise from associative algebras and thus bring back to Heisenberg formalism. Takhtajan [33], with the additional constraint of the FI, could certainly do no better. His main positive contribution there (for the quantization problem) was to find an explicit realization of the Nambu–Heisenberg relation (the case of general n was also discussed).

3.2. DEFORMATION QUANTIZATION

Deformation quantization was introduced [16] a few years after Nambu's paper. It is well adapted to constrained systems but (for the same reasons as in the previous subsection) using it in this way for Nambu mechanics will not give anything new. A more radical approach is needed if one wants to get a solution expressing the specificity of Nambu mechanics.

As we have mentioned in Section 2.3.2, the Jacobian is the typical case of Nambu bracket. It is thus natural to try and replace in (9) usual products of observables by star products. For the sake of self-completeness we shall present here a very brief overview of deformation quantization (see also [17, 34]).

Let M be a Poisson manifold, a differentiable manifold endowed with a (possibly degenerate) 2-tensor τ satisfying $[\tau, \tau] = 0$ in the sense of Schouten-Nijenhuis brackets. The latter ensures that the corresponding Poisson bracket $\mathcal{P}(f, g) = \tau(df \wedge dg)$ of two functions $f, g \in A = C^\infty(M)$ is a Lie algebra law. We denote by $A[[\nu]]$ the space of formal power series in the parameter ν with coefficients in A. A *star product* $*_\nu$ on M is an associative (non-Abelian) deformation of the usual product of the algebra A defined on A by:

$$f *_\nu g = \sum_{r \geq 0} \nu^r C_r(f, g), \quad \forall f, g \in A, \tag{12}$$

where $C_0(f, g) = fg$, $f, g \in A$, and $C_r\colon A \times A \to A$ $(r \geq 1)$ are bidifferential operators (bipseudodifferential operators can sometimes be considered) on A satisfying:

a) $*_\nu$ extends by linearity in ν to $A[[\nu]]$ and the associativity condition $(f *_\nu g) *_\nu h = f *_\nu (g *_\nu h)$ is satisfied for all $f, g, h \in A[[\nu]]$.

b) $C_1(f, g) - C_1(g, f) = 2\mathcal{P}(f, g)$, $f, g \in A$.

(12) and condition a) express that we have an associative algebra deformation in the sense of Gerstenhaber [19] (in short, a *DrG-deformation*) while condition b) ensures that the corresponding commutator $[f, g]_{*_\nu} \equiv (f *_\nu g - g *_\nu f)/2\nu$ is a DrG-deformation of the Lie algebra $(A, \mathcal{P})$.

Equivalence of two DrG-deformations is defined in the associative case (two star products $*_\nu$ and $*'_\nu$) by the existence of a formal series of (differential) operators $T = \sum_{r \geq 0} \nu^r T_r$ with $T_0 = Id$ such that $Tf * Tg = T(f *' g)$, $f, g \in A[[\nu]]$. By equivalence one may consider only star products "vanishing on constants" (i.e. $C_r(f, c) = C_r(c, f) = 0$, $r \geq 1$, $c \in \mathbb{R}$, $f \in A$, since by [20] a deformation of a unital algebra is unital and equivalent to a deformation with the same unit) and assume $C_1 = \mathcal{P}$. A deformation is said trivial if it is equivalent to the original product (resp. bracket, for Lie algebra deformations); equivalence classes of DrG-deformations are classified, at each order in ν, by the second Hochschild (resp. Chevalley) cohomology spaces $H^2(A)$ (resp. $\tilde{H}^2(A)$) of A valued in itself with the natural action (product, resp. adjoint).

When τ is everywhere nondegenerate, M is *symplectic* (we denote by ω the symplectic 2-form on M, inverse of τ). In this case (and similarly for regular Poisson manifolds; cf. e.g. [13, 7, 34]) there always exist a star product, $\mathcal{P}$ is a nontrivial Hochschild 2-cocycle and $\dim\tilde{H}^2(A) = 1 + b_2$ (where $b_2 = \dim H^2(M)$, the second Betti number of M), which classifies at each order in ν inequivalent 1-differentiable deformations [15] of $(A, \mathcal{P})$ vanishing on constants (they are given by deformations of the symplectic structure). Then [13, 8] equivalence classes of star products are in one-to-one correspondence with series $[\omega] + \nu[H^2(M)][[\nu]]$. In the "flat case" $b_2 = 0$, star products are (up to equivalence) unique and given by the Moyal product $*_M$ [2] which can be expressed on $\mathbb{R}^{2n}$ by $f *_M g = \exp(\nu\mathcal{P})(f, g)$. It corresponds to the Weyl (totally symmetric) ordering of operators in quantum mechanics, the deformation parameter being here $\nu = i\hbar/2$.

A given Hamiltonian $H \in A$ determines the time evolution of an observable $f \in A$ by the Heisenberg equation $\frac{df_t}{dt} = [H, f_t]_{*_\nu}$, the corresponding one-parameter group of time evolution being given by the star exponential defined by:

$$\exp_*\left(\frac{tH}{2\nu}\right) \equiv \sum_{r\geq 0} \frac{1}{r!}\left(\frac{t}{2\nu}\right)^r (*H)^r, \tag{13}$$

where $(*H)^r = H * \cdots * H$ (r factors). Then the solution to the Heisenberg equation above can be expressed as $f_t = \exp_*(tH/2\nu) * f * \exp_*(-tH/2\nu)$. In many examples (cf. e.g. [2]), the star exponential is convergent as a series in the variable t in some interval ($|t| < \pi$ for the harmonic oscillator in the Moyal case) and converges as a distribution on M for fixed t. Then it makes sense to consider a Fourier-Dirichlet expansion of the star exponential:

$$\exp_*\left(\frac{tH}{2\nu}\right)(x) = \int \exp(\lambda t/2\nu)d\mu(x;\lambda), \quad x \in M, \tag{14}$$

the "measure" μ being interpreted as the Fourier transform (in the distribution sense) of the star exponential in the variable t. Equation (14) permits to define [2] the spectrum of the Hamiltonian H as the support Λ of the measure μ. In the discrete case, $\exp_*(tH/2\nu)(x) = \sum_{\lambda\in\Lambda}\exp(\lambda t/2\nu)\pi_\lambda(x)$, $x \in M$. The functions π_λ on M are interpreted as eigenstates of H associated with the eigenvalues λ, and satisfy $H * \pi_\lambda = \pi_\lambda * H = \lambda\pi_\lambda$, with $\pi_\lambda * \pi_{\lambda'} = \delta_{\lambda\lambda'}$ and $\sum_{\lambda\in\Lambda}\pi_\lambda = 1$.

In the Moyal case, the Feynman path integral can be expressed [28] as the Fourier transform over momentum space of the star exponential. In field theory, where the normal star product (which is the exponential of "half of the Poisson bracket" in the variables $p \pm iq$) is relevant, the Feynman path integral is given (up to a multiplicative factor) [10] by the star exponential.

We have limited ourselves here to some highlights (most of which will be needed later) of deformation quantization to give the flavour of how it provides a completely autonomous quantization scheme of a classical Hamiltonian system. See [32] for a more extensive list of relevant references on the subject. We shall now use it as a tool for the quantization of Nambu-Poisson structures.

3.3. ZARISKI PRODUCT: A GENERALIZED ABELIAN DEFORMATION

Our starting point here is a simple remark: the Jacobian of n functions on $\mathbb{R}^n$ is a Nambu bracket because the usual product of functions is Abelian, associative, distributive and respects the Leibniz rule. This is what permits us to work out the functional determinant and the required properties of a Nambu bracket (including the FI) will be satisfied. Therefore, if we replace the usual product in the Jacobian by a new product having the preceding properties, we get a "modified Jacobian" which is still a Nambu bracket in the sense of Definition 1 – but which has "built in" quantization if the new product has it. In [12] the interested reader will find more in detail how we arrived at the solution which we shall now describe.

3.3.1. *Deformation of the usual product law*

To fix ideas, take the space N of polynomials on $\mathbb{R}^3$ and denote by $*$ the Moyal star product in (e.g.) the first two variables. To get an Abelian product the natural idea is to symmetrize the Moyal product. But if one does it brutally (i.e. $\frac{1}{2}(a * b + b * a),\ a, b \in N$) one loses associativity (this is the star analogue of the Jordan algebra formalism already studied by Nambu).

One can then try, using linearity, to bring down the problem to monomials and then to perform total symmetrization in star monomials (star products of the coordinates functions): symmetrization will give back the usual product. (This however would not necessarily be true with different orderings or nonlinear elementary factors; we shall come back to this point in Section 4).

But algebra tells us there is another way to decompose polynomials: into irreducible factors, which can be of any degree. The latter is true for polynomials in several complex variables and a fortiori in the real domain, which is of interest for us since we want e.g. the harmonic oscillator Hamiltonian $p^2 + q^2$ to be "elementary". Symmetrization adapted to the number of irreducible factors will restore associativity of the product, but at the same time there will be "built in" quantization because the factors can be of any degree. That decomposition is the core of Zariski topology on the manifold ($\mathbb{R}^3$ here), hence the name of Zariski product.

More precisely, we start with $N = \mathbb{R}[x_1, x_2, x_3]$ and consider $\mathcal{S}(N) = \bigoplus_{n=1}^{\infty} N^{\overset{n}{\otimes}}$, its *symmetric* tensor algebra without scalars (a kind of Fock space over N). We denote by N_1 the semi-group of normalized polynomials, those for which the "maximal" monomial (defined in a natural manner: highest total degree, then maximal with respect to the lexicographic ordering in the variables (x_1, x_2, x_3)) has coefficient 1, with the convention that $0 \in N_1$. In $N[\nu]$ we consider similarly the semi-group (under usual product) N_1^ν of normalized polynomials, those for which the non-vanishing coefficient of lowest degree in ν belongs to N_1. Now we define a map $\alpha : N_1^\nu \to \mathcal{S}(N)$ by $\alpha(P) = P_1 \otimes \cdots \otimes P_n$ where $P_1 \cdots P_n$ is the decomposition into irreducible factors of the coefficient of degree 0 in ν of $P \in N_1^\nu$ and $\otimes$ denotes the *symmetric* tensor product. In other words, α takes into account only the classical part (which can be 0) of the polynomial in N_1^ν, and then replaces the usual product by the symmetric tensor product.

In order to get our "built in" quantization we use now an *"evaluation map"* $T : \mathcal{S}(N) \to N[\nu]$ defined by replacing the symmetric tensor product by a symmetrized Moyal product. In the above notations the result of all this is

$$T(\alpha(P)) = \frac{1}{n!} \sum_{\sigma \in S_n} P_{\sigma_1} * \cdots * P_{\sigma_n} \tag{15}$$

and we can define an Abelian product by

$$P \times_\alpha Q = T(\alpha(P) \otimes \alpha(Q)), \quad \forall P, Q \in N_1^\nu. \tag{16}$$

Obviously the product (16) maps $N_1^\nu \times N_1^\nu$ into N_1^ν and if $P, Q \in N_1$, then $P \times_\alpha Q|_{\nu=0} = PQ$. In this sense, $\times_\alpha$ is a generalized deformation of the usual product on the semi-group N_1, but this is far from being a DrG-deformation of the algebra N. In [12] simple examples are given to make this more explicit. We only have a deformed Abelian semi-group law, and if we plug this law into the Jacobian then the Leibniz rule and the FI will fail. What is lacking (and obviously so since addition and decomposition into irreducible factors of polynomials are highly noncommutative operations) is distributivity of the product $\times_\alpha$ with respect to addition.

In fact it follows from general results of Harrison cohomology (cf. e.g. [20] and references therein; compare also with Gelfand's theorem on the realization of Abelian involutive Banach algebras as algebras of functions over the spectrum) that Abelian DrG-deformations of algebras of polynomials (or even functions [30]) are trivial. To overcome this difficulty and restore distributivity we use a trick inspired by the second quantization procedure. Formally, let us look at "functions" on N_1 (e.g. formal series). Intuitively we get a deformed coproduct and the dual of this space of "functions" (polynomials on polynomials) will then have a product and a deformed product, both of which will be distributive with respect to the vector space addition. Now the product of polynomials is again a polynomial. So in the end we are getting some deformed product on an algebra generated by the polynomials.

3.3.2. *The Zariski algebra and its deformations*
The "Zariski" algebra generated by the polynomials is nothing but the algebra $\mathcal{Z}_0$ of the semi-group N_1, i.e. the free Abelian algebra generated by the set of normalized real irreducible polynomials $N_1^{irr} \subset N_1$ as building blocks. If we denote by Z_u the element of $\mathcal{Z}_0$ defined by $u \in N_1$, the *classical Zariski product* $\bullet^z$ in $\mathcal{Z}_0$ is given by $Z_u \bullet^z Z_v = Z_{uv}$, $u, v \in N_1$. Defining $Z_{cu} = cZ_u$ for $c \in \mathbb{R}$, we can extend the above product to all $u \in N$ and get a multiplicative injection of N into $\mathcal{Z}_0$ which is non-additive, i.e. $Z_{u+v} \neq Z_u + Z_v$ (the addition in $\mathcal{Z}_0$ is *not* related to the addition in N). It is the algebra $\mathcal{Z}_0$ that we shall now quantize.

We consider the space $\mathcal{Z}_\nu = \mathcal{Z}_0[\nu]$ of polynomials in ν with coefficients in $\mathcal{Z}_0$ and inject N_1^ν into $\mathcal{Z}_\nu$ by $\zeta : (\sum_{r\geq 0} \nu^r u_r) \mapsto \sum_{r\geq 0} \nu^r Z_{u_r}$ with $u_0 \in N_1$ and $u_i \in N, i \geq 1$. Using this injection we can define a *deformed Zariski product* $\bullet^z_\nu$ based on $\times_\alpha$, first on basis elements Z_u, $u = u_1 \cdots u_m$, $u_j \in N_1^{irr}$, $j = 1, \ldots, m$, and Z_v, by $Z_u \bullet^z_\nu Z_v = \zeta(u \times_\alpha v)$; then we extend it by linearity to all of $\mathcal{Z}_\nu$, taking into account the requirement that the product $\bullet^z_\nu$ annihilates the non-zero powers of ν:

$$(\sum_{r\geq 0} \nu^r A_r) \bullet^z_\nu (\sum_{s\geq 0} \nu^s B_s) = A_0 \bullet^z_\nu B_0, \quad \forall A_r, B_s \in \mathcal{Z}_0, r, s \geq 0. \tag{17}$$

The $\bullet^z_\nu$ product is obviously associative (because $\times_\alpha$ is), distributive with respect to addition in $\mathcal{Z}_\nu$, and Abelian. Its limit for $\nu = 0$ is $\bullet^z$. Thus we have achieved a first step:

Theorem 1 *The vector space $\mathcal{Z}_\nu$ endowed with the product $\bullet^z_\nu$ is an Abelian algebra which is a (generalized) deformation of the Abelian algebra $(\mathcal{Z}_0, \bullet^z)$.*

The algebra $\mathcal{Z}_\nu$ with product $\bullet^z_\nu$ provides thus an Abelian deformation of the algebra $\mathcal{Z}_0$, a fact interesting *per se* because it gives a first example of a non-trivial Abelian deformation, however generalized and therefore not necessarily classified by the Harrison cohomology (defined on the subcomplex of the Hochschild complex consisting of symmetric cochains [20]).

3.4. ARITHMETIC "SECOND" QUANTIZATION OF JACOBIAN IN $\mathbb{R}^3$

3.4.1. *Problems with the definition of derivatives*
The next step towards our goal of deforming the Nambu bracket on $\mathbb{R}^3$ is to define the *derivatives* δ_i, $1 \leq i \leq 3$, on $\mathcal{Z}_0$ and extend them to $\mathcal{Z}_\nu$. This would allow to define first the classical Nambu bracket on $\mathcal{Z}_0$, and the quantum one on $\mathcal{Z}_\nu$. But this is not so simple to achieve. The straightforward

definition $\delta_i Z_u = Z_{\partial_i u}$, $\forall u \in N$, where ∂_i is the usual derivative with respect to x^i, does not satisfy the Leibniz rule because of the different nature of the addition in N and in $\mathcal{Z}_0$ (except on the diagonal, a remark relevant for the analogue of the star exponential (13), which we shall consider in Section 4.3.1).

The next natural choice is then to "build in" the Leibniz rule by using the above definition of the derivatives δ_i only for irreducible polynomials $u \in N_1^{irr}$ and postulating the Leibniz rule on the product of irreducible polynomials $v = v_1 v_2 \cdots v_m$:

$$\delta_i \, Z_{v_1 v_2 \cdots v_m} = Z_{(\partial_i v_1) v_2 \cdots v_m} + Z_{v_1 (\partial_i v_2) \cdots v_m} + \cdots + Z_{v_1 v_2 \cdots (\partial_i v_m)}.$$

Obviously, the maps δ_i are derivations on the algebra $\mathcal{Z}_0$, but they are not commuting maps, i.e. the Frobenius property (trivially verified with the straightforward definition above) $\delta_i \delta_j = \delta_j \delta_i$, $i \neq j$, is not verified. This comes from the fact that when one takes the derivatives of an irreducible polynomial u, the polynomials $\partial_i u$, $1 \le i \le 3$, do not necessarily factorize out into the same number of factors (a simple example is given in [12]).

A consequence of this fact is the following: If one defines the classical Nambu bracket on $\mathcal{Z}_0$ by replacing, in the Jacobian, the usual product by $\bullet^z$ and the usual partial derivatives by the maps δ_i, this new bracket will not satisfy the FI. There will be anomalies in the FI (even at this classical, or "prequantized" level) due to terms which cannot cancel out each other because the Frobenius property is not satisfied on $\mathcal{Z}_0$.

3.4.2. *Taylor series algebra: a solution (case $n = 3$)*
Denote by $\mathcal{E} = \mathcal{Z}_0[y^1, y^2, y^3]$, the algebra of polynomials in some real variables (y^1, y^2, y^3) with coefficients in $\mathcal{Z}_0$. The Taylor series development of translated polynomials $x \mapsto u(x+y)$ can now be written $u(x+y) = u(x) + \sum_i y^i \partial_i u(x) + \frac{1}{2} \sum_{i,j} y^i y^j \partial_{ij} u(x) + \cdots$. These translated polynomials are multiplied by $(uv)(x+y) = u(x+y)v(x+y)$. In accordance with our setup we shall look instead at "Taylor series" in $\mathcal{E}$, for $u \in N_1$:

$$J(Z_u) = Z_u + \sum_i y^i Z_{\partial_i u} + \frac{1}{2} \sum_{i,j} y^i y^j Z_{\partial_{ij} u} + \cdots = \sum_n \frac{1}{n!} (\sum_i y^i \partial_i)^n (Z_u), \tag{18}$$

where $\partial_i u$, $\partial_{ij} u$, etc. are the usual derivatives of $u \in N_1 \subset N$ with respect to the variables x^i, x^i and x^j, etc., $\partial_i Z_u \equiv Z_{\partial_i u}$ and, since in general the derivatives of $u \in N_1$ are in N, one has to factor out the appropriate constants in $Z_{\partial_i u}$, $Z_{\partial_{ij} u}$, etc. (i.e. $Z_{\lambda u} \equiv \lambda Z_u$, $u \in N_1$, $\lambda \in \mathbb{R}$). J defines an additive map from $\mathcal{Z}_0$ to $\mathcal{E}$ (to say that J is multiplicative is tantamount to the Leibniz property).

That algebra is however too large and for our purpose we need to restrict to the smaller algebra $\mathcal{A}_0$, the subalgebra of $\mathcal{E}$ generated by elements of the form (18). We shall denote by $\bullet$ the product in $\mathcal{A}_0$ which is naturally induced by the product in $\mathcal{E}$. In order to define the (classical) Nambu-Poisson structure on $\mathcal{A}_0$, we need a correct definition of the derivative of an element of $\mathcal{A}_0$. Notice that the derivative $\partial_i u(x+y)$ is again a Taylor series of the form $\partial_i u(x) + \sum_j y^j \partial_{ij} u(x) + \cdots$. We shall thus define the derivative Δ_a, $1 \leq a \leq 3$, of an element of the form (18) by the natural extension to $\mathcal{A}_0$ of the previous straightforward definition, i.e.,

$$\Delta_a(J(Z_u)) = J(Z_{\partial_a u}) = Z_{\partial_a u} + \sum_i y^i Z_{\partial_{ai} u} + \frac{1}{2} \sum_{i,j} y^i y^j Z_{\partial_{aij} u} + \cdots, \quad (19)$$

for $u \in N_1$, $1 \leq a \leq 3$. One can look at definition (19) of Δ_a as the restriction, to the subset of elements of the form $J(Z_u)$, of the formal derivative with respect to y^a in the ring $\mathcal{E} = \mathcal{Z}_0[y^1, y^2, y^3]$. Since $\Delta_a(J(Z_u)) = J(Z_{\partial_a u})$, we have $\Delta_a(\mathcal{A}_0) = \mathcal{A}_0$ and we get a family of maps $\Delta_a : \mathcal{A}_0 \to \mathcal{A}_0$, $1 \leq a \leq 3$, restriction to $\mathcal{A}_0$ of the derivations with respect to y^a, $1 \leq a \leq 3$, in $\mathcal{E}$. We summarize the properties of Δ_a in the following:

Lemma 1 *The maps $\Delta_a : \mathcal{A}_0 \to \mathcal{A}_0$, $1 \leq a \leq 3$, defined by Eq. (19) constitute a family of commuting derivations (satisfying the Leibniz rule) of the algebra $\mathcal{A}_0$.*

The definition of derivatives on $\mathcal{A}_0$ leads to the following natural definition of the classical Nambu bracket on the Abelian algebra $\mathcal{A}_0$ (it is obvious that this defines a Nambu-Poisson structure on $\mathcal{A}_0$).

Definition 4 *The classical Nambu bracket on $\mathcal{A}_0$ is the trilinear map taking values in $\mathcal{A}_0$ given, $\forall A, B, C \in \mathcal{A}_0$, by:*

$$(A, B, C) \mapsto [A, B, C]_\bullet \equiv \sum_{\sigma \in S_3} \epsilon(\sigma) \Delta_{\sigma_1} A \bullet \Delta_{\sigma_2} B \bullet \Delta_{\sigma_3} C. \quad (20)$$

Now that we have a classical Nambu-Poisson structure on $\mathcal{A}_0$, we shall construct a quantum Nambu-Poisson structure by defining a generalized Abelian deformation $(\mathcal{A}_\nu, \bullet_\nu)$ of $(\mathcal{A}_0, \bullet)$. The construction is based on the map α introduced in Section 3.3.1 and we shall extend the definition of the product $\bullet_\nu^z$ defined in Section 3.3.2 to the present setting for the Nambu-Poisson structure on $\mathcal{A}_0$.

Let $\mathcal{E}[\nu]$ be the algebra of polynomials in ν with coefficients in $\mathcal{E}$. We consider the subspace $\mathcal{A}_\nu$ of $\mathcal{E}[\nu]$ consisting of series $\sum_{r \geq 0} \nu^r A_r$ for which the coefficient A_0 is in $\mathcal{A}_0$. Then we define a map $\bullet_\nu : \mathcal{A}_\nu \times \mathcal{A}_\nu \to \mathcal{E}[\nu]$ by extending the product $\bullet_\nu^z$ defined previously, with $u, v \in N_1$ (it is sufficient to define it on $\mathcal{A}_0$ since $\bullet_\nu^z$ annihilates the non-zero powers of ν):

$$J(Z_u) \bullet_\nu J(Z_v) = Z_u \bullet_\nu^z Z_v + \sum_i y^i (Z_{\partial_i u} \bullet_\nu^z Z_v + Z_u \bullet_\nu^z Z_{\partial_i v}) + \cdots. \quad (21)$$

Actually $\bullet_\nu$ defines a product on $\mathcal{A}_\nu$. Moreover, for $A = \sum_{r \geq 0} \nu^r A_r$ and $B = \sum_{s \geq 0} \nu^s B_s$ in $\mathcal{A}_\nu$, we have $A \bullet_\nu B = A_0 \bullet_\nu B_0$ and the coefficient of ν^0 of the latter is $A_0 \bullet B_0$ which is in $\mathcal{A}_0$ since $A_0, B_0 \in \mathcal{A}_0$. This shows that $\bullet_\nu$ is actually a product on $\mathcal{A}_\nu$ which is Abelian by definition. Finally, for $\nu = 0$, we have $A \bullet_\nu B|_{\nu=0} = A_0 \bullet B_0$. Thus:

Theorem 2 *The vector space $\mathcal{A}_\nu$ endowed with the product $\bullet_\nu$ is a (generalized) Abelian algebra deformation of the Abelian algebra $(\mathcal{A}_0, \bullet)$.*

The derivatives Δ_a, $0 \leq a \leq 3$, are naturally extended to $\mathcal{A}_\nu$. Every element $A \in \mathcal{A}_\nu$ can be written as $A = \sum_I y^I A_I$, where $I = (i_1, \ldots, i_n)$ is a multi-index and $A_I \in \mathcal{Z}_\nu$. Then we have, for $A, B \in \mathcal{A}_\nu$, $A \bullet_\nu B = \sum_{I,J} y^I y^J A_I \bullet_\nu^z B_J$. Since $(\mathcal{Z}_\nu, \bullet_\nu^z)$ is an Abelian algebra and the derivative Δ_a acts as a formal derivative with respect to y^a on the product $A \bullet_\nu B$, the usual properties (linearity, Leibniz, Frobenius) of a derivative are still satisfied on $\mathcal{A}_\nu$ and we can define the quantum Nambu bracket on $\mathcal{A}_\nu$.

Definition 5 *The quantum Nambu bracket on $\mathcal{A}_\nu$ is the trilinear map taking values in $\mathcal{A}_\nu$ defined, $\forall A, B, C \in \mathcal{A}_\nu$, by:*

$$(A, B, C) \mapsto [A, B, C]_{\bullet_\nu} \equiv \sum_{\sigma \in S_3} \epsilon(\sigma) \Delta_{\sigma_1} A \bullet_\nu \Delta_{\sigma_2} B \bullet_\nu \Delta_{\sigma_3} C. \qquad (22)$$

It is now straightforward to show:

Theorem 3 *The quantum Nambu bracket (22) endows $\mathcal{A}_\nu$ with a Nambu-Poisson structure, a (generalized) deformation of the classical Nambu structure on $\mathcal{A}_0$.*

3.4.3. *General case*

The above procedure can be easily generalized to $\mathbb{R}^n$, $n \geq 2$. The only non-straightforward modification to be done appears in the evaluation map. One has to distinguish the cases n even and n odd. If $n = 2p$, $p \geq 1$, then one replaces the partial Moyal product in (15) by the usual Moyal product on $\mathbb{R}^{2p}$. If $n = 2p+1$, $p \geq 1$, one uses the partial Moyal product $*_{1\cdots 2p}$ on (e.g.) the hyperplane defined by $x_{2p+1} = 0$. The other definitions and properties are directly generalized to $\mathbb{R}^n$. More generally, on a manifold M endowed with a star product (e.g. a regular Poisson manifold), the same procedure can be developed, using the a priori given star product in Formula (15).

The question of spectrum of observables can be treated for Zariski products [12] as we have done for deformation quantization. We shall not develop this point here but shall briefly discuss the question in Section 4.3.1 together with the "first quantized" approach to which we devote Section 4.

Note that the canonical Nambu-Poisson structure of order 2 on $\mathbb{R}^2$ is the usual Poisson structure; there our procedure gives a quantization of the Poisson bracket $\mathcal{P}$ different from Moyal, however not on $N[\nu]$ but on $\mathcal{A}_\nu$;

this quantization will be somewhat like in field theory. The same applies to $\mathbb{R}^{2n}$ by starting with a sum of Poisson brackets on the various $\mathbb{R}^2$.

4. Generalized deformations of the "first quantized" type; equivalence, triviality and spectrality for generalized deformations

At the beginning of Section 3.3.1, we mentioned another possibility for getting a generalized deformation, already at the level of the algebra N of real polynomials over $\mathbb{R}^n$. In [11] this was developed in the covariant case, which we shall present below. Let us first present in an even simpler context how this can be done.

Take $\mathbb{R}^2$, with coordinates p and q. If p (resp. q) corresponds to an operator P (resp. Q), then antistandard ordering associates to a monomial $q^j p^k$ the operator $Q^j P^k$ and we define in this way a star product $*_S$ which [17] is nothing but the product of symbols of pseudodifferential operators. It is equivalent to $*_M$ (Moyal) as a star product. The same can be done with standard (first P) and normal (with $P \pm iQ$) orderings, etc. Now change the product $\times_\alpha$ in (15) into a "sun" product $\odot_{MS}$ defined on monomials by $(q^{j_1}p^{k_1}) \odot_{MS} (q^{j_2}p^{k_2}) = q^{j_1+j_2} *_M p^{k_1+k_2}$. This is obviously an Abelian product but now it has some "built in" quantization. We extend it by linearity to all of $N[[\nu]]$ with the convention that it annihilates nonzero powers of ν. We get in this way a generalized deformation, an associative, distributive and Abelian product with which we can proceed as before, but this time at the level of the algebra of polynomials on $\mathbb{R}^2$. If now we consider some $\mathbb{R}^2 \subset \mathbb{R}^3$, we can quantize Nambu brackets at this "quantum mechanical" level.

More generally (and in rough terms), if $*$ is some star product on some manifold M and $N \subset C^\infty(M)$ is some algebra, $\mathcal{S}$ the symmetric algebra over it, given some map $\alpha : N[[\nu]] \to \mathcal{S}$ defined in a way similar to the previous case (e.g. by performing some ordering operation on basis elements of N and extending to formal series with the convention that it annihilates nonzero powers of ν) and an evaluation map T defined by replacing $\otimes$ by $*$, we can introduce a "sun" product $\odot_\nu = T\alpha$. The question which has to be addressed is how trivial is this type of quantization, but before we shall present the more sophisticated form studied in [11].

4.1. COVARIANT STAR PRODUCTS AND ZARISKI QUANTIZATION

Let N be the algebra of real polynomials over $\mathbb{R}^n$ ($n = 3$ in the following) and define a sun product $\odot_\nu = T\alpha$ where α: $N[\nu] \to \mathcal{S}$ maps the monomial $x_1^{k_1} \cdots x_n^{k_n} \in N$ into a similar expression belonging to the symmetric tensor algebra $\mathcal{S}$ over N obtained by replacing the product of the coordinates x_i ($i = 1, \ldots, n$) by the symmetrized tensor product $\otimes$; we extend by linearity to N and to $N[\nu]$ by the convention that α annihilates nonzero powers of ν (in other words, we extend by linearity to $N[\nu]$ and compose with the canonical projection of $N[\nu]$ on N). The evaluation map $T\colon \mathcal{S} \to N[\nu]$ is defined, for $F_i \in N$, $1 \le i \le k$, $\forall k \ge 1$, by

$$T(F_1 \otimes \cdots \otimes F_k) = \frac{1}{k!} \sum_{\sigma \in S_k} F_{\sigma_1} * \cdots * F_{\sigma_k} \tag{23}$$

where $*$ is some star product on N. We call $\odot_\nu$ the *sun product associated with the star product* $*$. It is obviously Abelian and associative.

In the following we consider $\mathbb{R}^3$ as the dual of the $\mathfrak{su}(2)$-Lie algebra and take $*_M$ to be the Moyal product on $\mathbb{R}^6$. The Lie algebra $\mathfrak{su}(2)$ can be realized with the functions $L_i(p,q) = \sum_{1 \le j,k \le 3} \varepsilon_{ijk} p_j q_k$, $1 \le i,j,k \le 3$, on $\mathbb{R}^6$, where ε_{ijk} is the totally skew-symmetric tensor with $\varepsilon_{123} = 1$:

$$[L_i, L_j]_M \equiv \frac{1}{2\nu}(L_i *_M L_j - L_j *_M L_i) = \sum_{1 \le k \le 3} \varepsilon_{ijk} L_k. \tag{24}$$

One can easily show that for any real polynomial $F \in N(\mathbb{R}^3)$, the polynomial on $\mathbb{R}^6$ defined by $F(L_1, L_2, L_3)$ satisfies $(1 \le i,j,k \le 3)$

$$L_i *_M F = L_i F + \nu \sum_{1 \le j,k \le 3} \varepsilon_{ijk} L_k \frac{\partial F}{\partial L_j} + \nu^2 \Big(2 \frac{\partial F}{\partial L_i} + \sum_{1 \le j \le 3} L_j \frac{\partial^2 F}{\partial L_i \partial L_j} \Big). \tag{25}$$

Therefore $L_{i_1} *_M \cdots *_M L_{i_k}$ is a polynomial in (L_1, L_2, L_3) and any polynomial in (L_1, L_2, L_3) can be expressed as a $*_M$-polynomial of the L_i's, so that the product $F *_M G$ of two polynomials F, G in (L_1, L_2, L_3) is again a polynomial in (L_1, L_2, L_3). Hence from the Moyal product $*_M$ on $\mathbb{R}^6$, we get a star product on $\mathbb{R}^3$ satisfying Eq. (25) for any polynomial F. It is actually an invariant (and covariant) star product on $\mathfrak{su}(2)^* \sim \mathbb{R}^3$. Herebelow we shall denote by $*$ this star product and by $\odot_\nu$ the associated $\odot_\nu$-product. Since $L_i * L_j = L_i L_j + \nu \sum_{1 \le k \le 3} \varepsilon_{ijk} L_k + 2\nu^2 \delta_{ij}$, $1 \le i,j \le 3$, and $L_i \odot_\nu L_j = L_i L_j + 2\nu^2 \delta_{ij}$, this $*$-product does provide quantum terms.

Let us write $F * G = \sum_{r \ge 0} \nu^r C_r(F,G)$, where $C_1(F,G) = \mathcal{P}(F,G) \equiv \sum_{1 \le i,j,k \le 3} \varepsilon_{ijk} L_k \frac{\partial F}{\partial L_i} \frac{\partial G}{\partial L_j}$ is the standard Poisson bracket on $\mathfrak{su}(2)^* \sim \mathbb{R}^3$. We denote by Δ the Laplacian operator: $\Delta = \sum_{1 \le k \le 3} \frac{\partial^2}{\partial L_k^2}$. From (25) we get $C_2(L_i, F) = (2 + \mathcal{D})(\frac{\partial F}{\partial L_i})$, in terms of the differential operator (constant on the subspace $H_n \subset N$ of homogeneous polynomials of degree n) $\mathcal{D} = \sum_{1 \le k \le 3} L_k \frac{\partial}{\partial L_k}$. Complicated calculations give $\forall F, G \in H_n$, $F \odot_\nu G = \sum_{r \ge 0} \nu^{2r} a(2n, r) \Delta^r (FG)$, with $a(n, 0) = 1 = a(0, r)$ and $\forall n \ge 1$

$a(n,r) = \frac{1}{n}((n-2r)a(n-1,r) + (n-2r+2)a(n-1,r-1))$. After some more calculations [11] we obtain

$$a(n,r) = \sum_{\substack{j_1+\cdots+j_{n-2r+2}=r \\ j_1,\ldots,j_{n-2r+2}\geq 0}} \gamma_{j_1}\gamma_{j_2}\tau_{j_3}\cdots\tau_{j_{n-2r+2}}, \quad n \geq 2r, \tag{26}$$

where $\gamma_n = E_{2n}/(2n)!$ and $\tau_n = 2^{2n+2}(2^{2n+2}-1)B_{2n+2}/(2n+2)!$, the constants E_k (resp. B_k) being the Euler (resp. Bernoulli) numbers. Let $B(k,p)$ be the set of partitions of k of length p, i.e. the number of ways to write k as a sum of p strictly positive integers: $k = n_1 + \cdots + n_p$, with $n_1 \geq \cdots \geq n_p$, and $A_k = \sum_{\substack{i+j=k \\ i,j\geq 0}} \gamma_i\gamma_j$, $k \geq 0$. In the above notations we then get:

Theorem 4 *The sun product $\odot_\nu$ associated with the invariant star product on $\mathfrak{su}(2)^*$ defined by Eq. (25) admits the following form:*

$$F\odot_\nu G = FG + \sum_{r\geq 1} \nu^{2r}\eta_r(FG), \quad F,G \in N \quad (\textit{polynomials on } \mathbb{R}^3), \tag{27}$$

where $\eta_r\colon N \to N$, $r \geq 1$, are differential operators given by

$$\eta_r(F) = \left(A_r + \sum_{p=1}^{r} z_{p,r}\mathcal{D}(\mathcal{D}-1)\cdots(\mathcal{D}-p+1)\right)\Delta^r(F), \quad F \in N, \tag{28}$$

with $z_{p,r} = \sum_{k=p}^{r} A_{r-k}\sum_{(n_1,\ldots,n_p)\in B(k,p)}(m_1!\cdots m_r!)^{-1}\tau_{n_1}\cdots\tau_{n_p}$. This $\odot_\nu$-product has a unique extension to $F,G \in C^\infty(\mathbb{R}^3)$.

4.2. TWO NOTIONS OF EQUIVALENCE AND TRIVIALITY

4.2.1. *Sun products*

Definition 6 *Two products $\odot_\nu$ and $\odot'_\nu$ are said to be A-equivalent, if there exists a $\mathbb{R}[[\nu]]$-linear (formally invertible) map $S_\nu\colon N[[\nu]] \mapsto N[[\nu]]$ of the form $S_\nu = \sum_{r\geq 0}\nu^r S_r$, where the $S_r\colon N \mapsto N$, $r \geq 1$, are differential operators and $S_0 = Id$, such that*

$$S_\nu(F\odot_\nu G) = S_\mu(F)\odot'_\nu S_\mu(G)\big|_{\mu=\nu}, \quad \forall F,G \in N. \tag{29}$$

We say that they are B-equivalent if the map S_ν satisfies

$$S_\nu(F\odot_\nu G) = S_\nu(F)\odot'_\nu S_\nu(G), \quad \forall F,G \in N. \tag{30}$$

It is straightforward to check that both notions are indeed equivalence relations (between $\odot_\nu$-products). Note that there is no order relation between both; in particular two $\odot_\nu$-products which are B-equivalent are not necessarily A-equivalent. Relation (30) can be written as $S_\nu(F\odot_\nu G) = F\odot'_\nu G$ and relation (29) amounts to

$$\sum_{r,s\geq 0} \nu^{r+s} S_s(\rho_r(FG)) = \sum_{r,s,s'\geq 0} \nu^{r+s+s'} \rho'_r(S_s(F)S_{s'}(G)), \quad F,G \in N, \tag{31}$$

where ρ_r (resp. ρ'_r) are the cochains of the product $\odot_\nu$ (resp. $\odot'_\nu$).

Triviality is equivalence with the original product of the algebra, the usual product. That is:

Definition 7 *A sun product $\odot_\nu$ is said strongly (resp., weakly) trivial if it is A-equivalent (resp., B-equivalent) with the usual product.*

In other words, if we take for $\odot'_\nu$ the usual product, $\odot_\nu$ is strongly trivial if there exists a (formally invertible) $\mathbb{R}[[\nu]]$-linear map $S_\nu\colon N[[\nu]] \mapsto N[[\nu]]$ of the form $S_\nu = \sum_{r\geq 0} \nu^r S_r$, where the $S_r\colon N \mapsto N$, $r \geq 1$, are differential operators and $S_0 = Id$, such that (denoting by $\cdot$ the usual product):

$$S_\nu(F\odot_\nu G) = S_\nu(F)\cdot S_\nu(G), \quad \forall F,G \in N. \tag{32}$$

$\odot_\nu$ is weakly trivial if (with the same notations)

$$S_\nu(F\cdot G) = F\odot_\nu G, \quad \forall F,G \in N, \tag{33}$$

or equivalently $F\cdot G = S_\nu^{-1}(F\odot_\nu G)$, which corresponds to taking for $\odot'_\nu$ (instead of $\odot_\nu$) in (30) the usual product. We have here manifest symmetry in the substitution of the usual product. If in (29) we take for $\odot_\nu$ the usual product we get $S_\nu(F\cdot G) = S_\mu(F)\odot_\nu S_\mu(G)|_{\mu=\nu}$, which (using (31)) is equivalent to (32) with S^{-1} instead of S. From Definition 7 we see that a strongly trivial sun product is weakly trivial (hence the terminology). As a matter of fact one proves easily:

Proposition 3 *A $\odot_\nu$-product is strongly trivial if and only if it coincides with the usual product. A $\odot_\nu$-product is weakly trivial whenever its cochains are given by differential operators.*

The proof of the first part is by elementary calculation. For the second part one writes (for $F,G \in N$) $F\odot_\nu G = \rho(F\cdot G)$ where $\rho = Id + \sum_{r\geq 1}\nu^r\rho_r$ is formally invertible and the ρ_r's are differential operators acting on N. From Theorem 4 we get:

Corollary 1 *The sun product constructed in Section 4.1 is weakly trivial but not strongly trivial.*

4.2.2. *Zariski products*

The previous treatment (with the obvious modifications) applies to the case of Zariski quantization. But in this case we see easily that *the Zariski* *product $\bullet_\nu$ is never trivial in either sense* (except when the defining product $*_\nu$ is not a star product but the usual product).

It is true that the map $T\alpha$ intertwines the Zariski product $\bullet_\nu$ with the usual product $\bullet$ on the classical algebra $\mathcal{A}_0$. But it is neither invertible as a formal series nor acting by differential operators. Even if one defines S_ν by taking the $\mathbb{R}[[\nu]]$-linear extension of the restriction of $T \circ \alpha$ to $\mathcal{A}_0$, one gets an invertible map but not an intertwining map (in the sense of B-equivalence) as S_ν is not given by differential operators on $\mathcal{A}_0$. We restrict the intertwining operators to be defined by *differentiable* cochains (in spite of the fact that in general the deformations considered are not defined by differentiable cochains) because eventually the non-triviality for the Abelian generalized deformations should be linked with (differentiable) Hochschild or Harrison cohomologies. A step in this direction was very recently made by Pinczon [29] who introduced "superdeformations" where the deformation parameter no longer commutes with the original algebra and which are nevertheless classified by Hochschild-type cohomologies.

4.3. SPECTRALITY AND CONCLUDING REMARKS

4.3.1. *Exponentials and spectrum*

As we recalled in Section 3.2, the spectrum of an observable f is obtained, in deformation quantization, via the $*$-exponential (13). If we take a DrG-deformation of the form (12) which is trivial with intertwining operator T (note that a star product is never trivial because the Poisson bracket is a nontrivial Hochschild 2-cocycle), the corresponding exponential is the inverse image of the usual exponential of Tf, hence the spectrum is continuous.

For a $\odot_\nu$ product (e.g. the case of $SU(2)$) we can define $\exp_\odot(f)$ like in the star case (13) with $\odot_\nu$ instead of $*$. This exponential coincides with the star exponential for linear elements like L_3 in the case of $SU(2)$ but not for expressions like $H_1 = \frac{1}{2}(L_1^2 + L_2^2 + L_3^2)$. In particular the Fourier decomposition of $\exp_\odot(L_3)$ is discrete but for H_1 the spectra do not coincide. Now if we take a weakly trivial $\odot_\nu$ product, the intertwiner S is not enough to trivialize the spectrum, in contradistinction with the previous case.

For the Zariski product $\bullet_\nu^z$ we define similarly the corresponding exponential. Here, in addition to linear elements, the star and $\bullet_\nu^z$ exponentials coincide for irreducible polynomials, giving e.g. the usual discrete spectrum for the harmonic oscillator Hamiltonian.

4.3.2. *Quantized Nambu Bracket*

A quantization of the classical Nambu bracket is achieved by replacing the usual product by the sun product. Due to the properties of the $\odot_\nu$-product, it is easy to see that actually the quantized Nambu bracket is given by:

$$[F, G, H]_{\odot_\nu} = T(\alpha(\{F, G, H\})), \quad F, G, H \in C^\infty(\mathbb{R}^3),$$

where $\{F, G, H\}$ denotes the classical Nambu bracket on $\mathbb{R}^3$, i.e., the Jacobian. Though the Leibniz rule is not satisfied for the $\odot_\nu$-product, this quantized Nambu bracket does satisfy the Fundamental Identity. Indeed only the weaker form of the Leibniz rule

$$\begin{aligned} &F \odot_\nu \left(\frac{\partial}{\partial L_i}(G \odot_\nu H) - G \odot_\nu \frac{\partial H}{\partial L_i} - \frac{\partial G}{\partial L_i} \odot_\nu H\right) \\ &= T\left(\alpha\left(F\left(\frac{\partial}{\partial L_i}(GH) - G\frac{\partial H}{\partial L_i} - \frac{\partial G}{\partial L_i} H\right)\right)\right) = 0, \end{aligned}$$

is required to ensure that the Fundamental Identity is verified by the quantized Nambu bracket.

Finally let us mention that, as for the $\odot_\nu$-product case, the quantized Nambu bracket is weakly trivial, but not strongly trivial.

References

1. Bayen, F., Flato, M.: Remarks concerning Nambu's generalized mechanics, *Phys. Rev. D* **11** (1975), 3049–3053.
2. Bayen, F., Flato, M., Frønsdal, C., Lichnerowicz, A., Sternheimer, D.: Deformation Theory and Quantization: I. Deformations of Symplectic Structures; II. Physical Applications; *Ann. Phys.* **111** (1978), 61–110 and 111–151.
3. Chatterjee, R.: Dynamical Symmetries and Nambu Mechanics, *Lett. Math. Phys.* **36** (1996), 117–126.
4. Chatterjee, R., Takhtajan, L.: Aspects of Classical and Quantum Nambu Mechanics, *Lett. Math. Phys.* **37** (1996), 475–482.
5. Cohen, I., Kálnay, A.J.: On Nambu's generalized Hamiltonian mechanics, *Int. J. Theor. Phys.* **12** (1975), 61–67; Cohen, I.: Generalization of Nambu's mechanics, *ibid.*, 69–78; Garcia Sucre, M., Kálnay, A.J.: On the statistics consistent with Nambu's new quantum rules. Quarks? *ibid.*, 149–151; Kálnay, A.J.: On the new Nambu mechanics and its quantization, *Géométrie symplectique et physique mathématique*, Editions du Centre National de la Recherche Scientifique, Paris (1975), 17–21.
6. Connes, A.: *Noncommutative geometry*, Academic Press, San Diego (1994).
7. Dazord P.: Construction explicite de produits étoile sur les variétés de Poisson régulières, Preprint IGD Lyon (1997).
8. Deligne, P.: Déformations de l'Algèbre des Fonctions d'une Variété Symplectique: Comparaison entre Fedosov et De Wilde, Lecomte, *Selecta Mathematica*, New Series, **1** (1995) 667–697.
9. Dirac, P. A. M.: Lectures on Quantum Mechanics, Belfer Graduate School of Sciences Monograph Series No. 2, Yeshiva University, New-York (1964).
10. Dito, J.: Star-Product Approach to Quantum Field Theory: The Free Scalar Field, *Lett. Math. Phys.* **20** (1990) 125–134. [Dito, Joseph ≡ Dito, Giuseppe].
11. Dito, G., Flato, M.: Generalized Abelian Deformations: Application to Nambu Mechanics, *Lett. Math. Phys.* **39** (1997), 107–125.

12. Dito, G., Flato, M., Sternheimer, D., Takhtajan, L.: Deformation Quantization and Nambu Mechanics, *Commun. Math. Phys.* **153** (1997), 1–22.
13. Fedosov, B.V.: *Deformation Quantization and Index Theory*, Akademie Verlag, Berlin (1996).
14. Flato, M., Frønsdal, C.: Unpublished (1992).
15. Flato, M., Lichnerowicz, A., Sternheimer, D.: Déformations 1-différentiables d'algèbres de Lie attachées à une variété symplectique ou de contact, *C.R. Acad. Sci. Paris*, **279** (1974), A 877–881 and *Compositio Mathematica*, **31** (1975), 47–82.
16. Flato, M., Lichnerowicz, A., Sternheimer, D.: Crochets de Moyal-Vey et quantification, *C.R. Acad. Sci. Paris*, **283** (1976), A 19–24.
17. Flato, M., Sternheimer, D.: Star-products, quantum groups, cyclic cohomology and pseudo-differential calculus, *American Mathematical Society Contemporary Mathematics* (P. Sally et al. eds.) **175**, (1994), 53–72.
18. Gautheron, Ph.: Some remarks concerning Nambu mechanics, *Lett. Math. Phys.* **37** (1996), 103–116.
19. Gerstenhaber, M.: On the deformation of rings and algebras, *Ann. Math.* **79** (1964), 59–103.
20. Gerstenhaber, M., Schack, S.: Algebraic Cohomology and Deformation Theory, in: Hazewinkel, M. and Gerstenhaber, M. (eds.), *Deformation Theory of Algebras and Structures and Applications*, Kluwer, Dordrecht (1988), pp. 11–264.
21. Ginzburg, G., Kapranov, M.: Koszul duality for operads, *Duke Math. J.* **76** (1994), 203–272.
22. Gnedbaye, A.V.: Les algèbres k-aires et leurs opérades, *C. R. Acad. Sci. Paris Sér.I Math.* **321** (1995), 147–152.
23. Gutt, S.: Unpublished (1992).
24. Loday, J..-L.: La renaissance des opérades, *Séminaire Bourbaki*, Exposé 792 (Novembre 1994).
25. Mukunda, N., Sudarshan, E.C.G.: Relation between Nambu and Hamiltonian mechanics, *Phys. Rev. D* **13** (1976), 2846–2850.
26. Nambu, Y.: Generalized Hamiltonian Dynamics, *Phys. Rev. D* **7** (1973), 2405–2412.
27. Mathieu, O.: The Symplectic Operad, in: "Functional Analysis on the Eve of the 21st Century, In Honor of I.M. Gelfand", *Progress in Mathematics* **131** (Eds. S. Gindinkin, J. Lepowsky and R.L. Wilson), Birkhäuser, Boston (1995), 223–243.
28. Pankaj Sharan: $*$-Product Representation of Path Integrals, *Phys. Rev. D* **20** (1979), 414–426.
29. Pinczon, G. : Non commutative deformation theory, Dijon preprint to be published in *Lett. Math. Phys.* (1997).
30. Pinczon, G.: On the equivalence between continuous and differential deformation theories, *Lett. Math. Phys.* **39** (1997), 143–156; see also Gutt, S.: On some second Hochschild cohomology spaces for algebras of functions on a manifold, *ibid.* **39** (1997), 157–162.
31. Sahoo, D., Valsakumar, M. C.: Nambu mechanics and its quantization. *Phys. Rev. A* **46** (1992), 4410–4412.
32. Sternheimer, D.: Star Products: their ubiquity and unicity, in: *Modern Group Theoretical Methods in Physics: Proceedings of the Conference in Honour of Guy Rideau*, J. Bertrand et al. eds, Math. Physics Studies vol. 18, Kluwer Acad. Publ. (1995), pp. 255–265.
33. Takhtajan, L.: On Foundation of the Generalized Nambu Mechanics, *Commun. Math. Phys.* **160** (1994), 295–315.
34. Weinstein, A.: Deformation quantization, *Séminaire Bourbaki*, Exposé 769 (Juin 1994).

THE GEOMETRIZATION PRINCIPLE

JÜRG FRÖHLICH
Theoretical Physics, ETH-Hönggerberg
CH-8093 Zürich, Switzerland

In attempts to reconcile quantum theories of matter with the general theory of relativity, which describes the gravitational field created by matter distributed in space-time, in the form of a unified "quantum theory of space-time-matter", one finds that, plausibly, there is a fundamental uncertainty in the location of events in space-time, the length scale of this uncertainty being bounded below by the Planck length l_P; see e.g. [2].

The number of events in a space-time region of volume v is expected not to exceed const. $\left(v\, l_P^{-d-1}\right)$, where $d \geq 3$ is the number of space dimensions, and one can argue that the number of distinct states of a quantum-mechanical subsystem localized in a spatial region bounded by a surface of area A ought to be bounded by $\exp\left(\text{const.}\, A\, l_P^{-d+1}\right)$.

Assuming that these bounds are fundamentally correct, one concludes that, in a unified "quantum theory of space-time-matter", space, time and matter loose their individuality, and that classical space-time is merely an approximate notion, emerging only in some limiting regimes. One would expect that a fundamental quantum theory of space-time-matter must be based on a *deformation* of quantum theory and classical geometry.

A classical, smooth manifold, M, can be described in terms of the *algebra, $C^\infty(M)$, of smooth functions on M. According to A. Connes [1], a non-commutative space should be described by a *non-abelian* *algebra, $\mathcal{A}$, which has a faithful *representation by bounded operators on a separable Hilbert space $\mathcal{H}$. The differential topology and geometry of the non-commutative space described by $(\mathcal{A}, \mathcal{H})$ can be encoded in spectral data

$$(\mathcal{A}, \mathcal{H}, D_1, \dots, D_n, \gamma) \qquad (*)$$

where $D_1, \dots, D_n$ are selfadjoint operators on $\mathcal{H}$, and γ is a selfadjoint involution on $\mathcal{H}$, with the properties:

(1) $[D_i, a]$ is a bounded operator on $\mathcal{H}$, for all $a \in \mathcal{A}$ and all $i = 1, \dots, n$;
(2) the operator $\exp\left(-\beta \sum_{i=1}^n D_i^2\right)$ is trace-class, for arbitrary $\beta > 0$;

D. Sternheimer et al. (eds.), Deformation Theory and Symplectic Geometry, 67–68.

(3) $D_i D_j + D_j D_i = 0$, for all i, j, with $i \neq j$;
(4) $\gamma D_i + D_i \gamma = 0$, for $i = 1, \dots, n$, and $[\gamma, a] = 0$, for all $a \in \mathcal{A}$.

Spectral data of the form (∗), with $n = 1$, have been introduced and used to explore the topology and geometry of non-commutative spaces by Connes; see [1] and references given there. Generalizations to larger values of n have been studied in some detail in [4, 5].

Spectral data of the form (∗), with properties (1) – (4), define models of *supersymmetric quantum theory*, in the sense of Witten [7]. They enable one to construct natural generalizations of classical geometry to a geometry of non-commutative spaces. The data (∗) with $n = 1$ yield a generalization of spin^c geometry [1]; when $n = 2$ one obtains a generalization of Kähler geometry; while $n = 4$ corresponds to a generalization of hyper-Kähler geometry. Symplectic and complex Hermitian geometry arise from data (∗) with two supersymmetry generators, $D_1, D_2 (n = 2)$, by an *explicit breaking* of the supersymmetry algebra; see [4, 5].

In [5], several attempts to make use of Connes' non-commutative geometry in exploring various caricatures of a "quantum theory of space-time-matter", such as superstring theory [6], are described in some detail. One upshot (for some simple examples, see [5, 3]) of the analysis in [5] is the following:

"Geometrization Principle": *A fundamental quantum theory of space-time-matter has solutions yielding spectral data analogous to* (∗) *from which models of non-commutative space(-time) can be reconstructed.*

Acknowledgements. I thank A. Connes for inspiration and guidance and my collaborators, A.H. Chamseddine, G. Felder, K. Gawędzki, O. Grandjean and A. Recknagel, for their help and for all they have taught me. I thank my friends Daniel and Moshé for their patience.

References

1. A. Connes, *Noncommutative Geometry*, Academic Press: New York, Boston, 1994.
2. S. Doplicher, K. Fredenhagen, and J.E. Roberts, Commun. Math. Phys. **172**, 187 (1995).
3. J. Fröhlich and K. Gawędzki, in: CRM Proceedings and Lecture Notes, vol. 7, pg. 57 (1994).
4. J. Fröhlich, O. Grandjean, and A. Recknagel, *Supersymmetry and (Non-Commutative) Differential Geometry*, submitted to Commun. Math. Phys.
5. J. Fröhlich, O. Grandjean, and A. Recknagel: Notes to lectures by J. Fröhlich at Les Houches, summer 1995; to appear in the proceedings, A. Connes, K. Gawędzki, and J. Zinn-Justin (eds.).
6. M. Green, J. Schwarz, and E. Witten, *Superstring Theory*, vols. I, II, Cambridge University Press: Cambridge, London, New York 1987.
7. E. Witten, Nucl. Phys. B **202**, 253 (1982); J. Diff. Geometry **17**, 661 (1982).

Editors' Note: These two pages are a short presentation of recent extensive work by the author, published and to be published.

GENERALIZATION AND DEFORMATIONS OF QUANTUM GROUPS

CHRISTIAN FRØNSDAL
Physics Department, University of California,
Los Angeles CA 90095-1547, USA

Abstract. A large family of "standard" coboundary Hopf algebras is investigated. The existence of a universal R-matrix is demonstrated for the case when the parameters are in general position. Algebraic surfaces in parameter space are characterized by the appearance of certain ideals; in this case the universal R-matrix exists on the associated algebraic quotient. In special cases the quotient is a "standard" quantum group; all familiar quantum groups including twisted ones are obtained in this way. In other special cases one finds new types of coboundary bi-algebras.

The "standard" universal R-matrix is shown to be the unique solution of a very simple, linear recursion relation. The classical limit is obtained in the case of quantized Kac-Moody algebras of finite and affine type.

Returning to the general case, we study deformations of the standard R-matrix and the associated Hopf algebras. A preliminary investigation of the first order deformations uncovers a class of deformations that encompasses the quantization of all Kac-Moody algebras of finite and affine type. The corresponding exact deformations are described as generalized twists, $R_\epsilon = (F^t)^{-1}RF$, where R is the standard R-matrix and the cocycle F (a power series in the deformation parameter ϵ) is the solution of a linear recursion relation of the same type as that which determines R. Included here is the universal R-matrix for the elliptic quantum groups associated with $\widehat{sl(N)}$, a big surprise!

1. Introduction

The quantum groups that have so far found employment in physics are very special (characterized by a single "deformation" parameter q). Deformation theory is a means of attacking the classification problem; at the same time

D. Sternheimer et al. (eds.), Deformation Theory and Symplectic Geometry, 69–84.

it offers a wider horizon against which to view the whole subject. The new quantum groups discovered this way (the deformations of the twisted ones) are dramatically different; the physical applications should be interesting.

Let $\mathfrak{g}$ be a simple Lie algebra over $\mathbb{C}$. A structure of coboundary Lie bialgebra on $\mathfrak{g}$ is determined by a "classical" r-matrix; an element $r \in \mathfrak{g} \otimes \mathfrak{g}$ that satisfies the classical Yang-Baxter relation.

It is widely believed that there corresponds, to each such r-matrix, via a process of "quantization", a unique quantum group [7]. Somewhat more precisely, one expects that there exists a Hopf algebra deformation $\tilde{U}(\mathfrak{g})$ of $U(\mathfrak{g})$, and an element $R \in \tilde{U}(\mathfrak{g}) \otimes \tilde{U}(\mathfrak{g})$, such that $\Delta R = R\Delta'$, where Δ is the coproduct of $\tilde{U}(\mathfrak{g})$ and Δ' is the opposite coproduct, satisfying the (quantum) Yang-Baxter relation

$$R_{12}R_{13}R_{23} = R_{23}R_{13}R_{12}\,; \tag{1}$$

such that r can be recovered by an expansion of expanding R with respect to a parameter $\hbar$:

$$R = 1 + \hbar r + o(\hbar^2)\,. \tag{2}$$

Till now, this program had been realized for r-matrices of a class that we call "standard".

The R-matrices associated with the multiparameter quantum groups discovered by Reshetikhin [17] and others [19, 20] are all included in the rubrique "standard". The principal characteristic of a standard R-matrix is that it "commutes with Cartan": $[H_a \otimes 1 + 1 \otimes H_a, R] = 0$. Until now, non-standard R-matrices were known only in the fundamental representation [5, 12].

The aim of this work is to use deformation theory to discover the so far unknown quantum groups that are alleged to be associated with non-standard r-matrices. Progress achieved in the present paper is due to the discovery of a differential complex associated with the Yang-Baxter relation; it is barely mentioned here because of space limitations; see [10].

The existence of a universal R-matrix for quantized Lie algebras and for Kac-Moody algebras, was known [6, 9, 16, 18, 21]. But it turns out that the representation (7) for a standard, universal R-matrix that satisfies the Yang-Baxter equation makes sense in a context that is much wider than quantized Kac-Moody algebras. See Definition 2.1, where we introduce an algebra $\mathcal{A}$ with generators $e_\alpha, e_{-\alpha}, H_a$. A key point is to refrain from introducing, *a priori*, any relations of the Serre type among the e_α (generalized Chevalley-Drinfeld generators). The algebras of ultimate interest are obtained by identifying an appropriate ideal $I \subset \mathcal{A}$ that intersects $\mathcal{A}^0$ trivially, and passing to the quotient $\mathcal{A}' = \mathcal{A}/I$.

Theorem 2 asserts that the Yang-Baxter relation for the standard R-matrix on $\mathcal{A}$ is equivalent to a simple, linear recursion relation (8) for the coefficients $t_{(\alpha)}^{(\alpha')}$ in the formula (7) for the Universal R-matrix. The integrability of this recursion relation is related to the first cohomology group on quantum planes. Generically, all one-forms are exact; when the parameters of $\mathcal{A}$ are in general position there exists a unique set of coefficients $t_{(\alpha)}^{(\alpha')}$ such that the standard R-matrix satisfies the Yang-Baxter relation.

Obstructions to the solution of the recursion relation (8), and thus to the Yang-Baxter relation on $\mathcal{A}$, exist on certain hyper-surfaces in the space of parameters of $\mathcal{A}$; they are detected by the presence of "constants". A Serre relation is a special type of constant. The obstructions (that is, the constants) generate an ideal $I \subset \mathcal{A}$, and a unique standard R-matrix, satisfying Yang-Baxter, exists on $\mathcal{A}/I$.

We specialize to quantized Kac-Moody algebras and calculate the classical limit; that is, the r-matrix associated with the standard R-matrix.

Deformations. The rest of the paper (from Section 6) is a study of the deformations of the standard R-matrix, satisfying the Yang-Baxter relation, but in the wider context of the bialgebras $\mathcal{A}$ and $\mathcal{A}' = \mathcal{A}/I$ described above. We begin by calculating a class of first order deformations of the standard R-matrix on $\mathcal{A}/I$ for any ideal of obstructions $I \subset \mathcal{A}$. We specialize to the case of simple quantum groups and obtain quantizations of all the simple Lie bialgebras of [2].

Let R be the R-matrix of a coboundary Hopf algebra $\mathcal{A}'$, and $F \in \mathcal{A}' \otimes \mathcal{A}'$, invertible. Then

$$\tilde{R} := (F^t)^{-1} R F \tag{3}$$

satisfies the Yang-Baxter relation if F satisfies the cocycle condition [13]

$$((1 \otimes \Delta_{21})F)F_{12} = ((\Delta_{13} \otimes 1)F)F_{31}. \tag{4}$$

Applying this to our context, we find that the relation (4) is equivalent to a simple, linear recursion relation that can be reduced to the same form as the recursion relation for the coefficients in the expansion of R. It has a unique solution that can be expressed directly in terms of these coefficients. Just as in the standard case, this leads to a simple equation for the classical r-matrix, from which the latter is determined to all orders.

In Section 8 we specialize to the case of deformed quantized, affine Kac-Moody algebras and take the classical limit, to recover the esoteric, affine r-matrices of the full Kac-Moody algebras, with their central extensions.

Finally, in Section 9, we deal with a very special case, to discover that the elliptic R-matrices of $sl(N)$ also arise as the classical limit of certain deformed quantum groups.

2. Standard Universal R-matrices

Definition 2.1. Let M, N be two countable sets, φ, ψ two maps,

$$\varphi: \ M \times M \to \mathbb{C}, \psi: \ M \times N \to \mathbb{C}, \quad a, b \mapsto \varphi^{ab}, a, \beta \mapsto H_a(\beta). \tag{5}$$

Let $\mathcal{A}$ or $\mathcal{A}(\varphi, \psi)$ be the universal, associative, unital algebra over $\mathbb{C}$ with generators $\{H_a\}\, a \in M$, $\{e_{\pm\alpha}\}\, \alpha \in N$, and relations

$$[H_a, H_b] = 0\,, [H_a, e_{\pm\beta}] = \pm H_a(\beta) e_{\pm\beta}\,,$$

$$[e_\alpha, e_{-\beta}] = \delta_\alpha^\beta (e^{\varphi(\alpha,\cdot)} - e^{-\varphi(\cdot,\alpha)})\,, \tag{6}$$

$\varphi(\alpha, \cdot) = \varphi^{ab} H_a(\alpha) H_b$, $\varphi(\cdot, \alpha) = \varphi^{ab} H_a H_b(\alpha)$, $e^{\varphi(\alpha,\cdot)+\varphi(\cdot,\alpha)} \neq 1, \quad \alpha \in N$.

The last condition on the parameters is included in order to avoid having to make some rather trivial exceptions. The absence of Serre relations among the e_α or among the $e_{-\alpha}$ is fundamental.

The free subalgebra generated by $\{e_\alpha\}\, \alpha \in N$ (resp. $\{e_{-\alpha}\}\, \alpha \in N$) will be denoted $\mathcal{A}^+$ (resp. $\mathcal{A}^-$). The subalgebra generated by $\{H_a\}\, a \in M$ is denoted $\mathcal{A}^0$. If necessary we may assume that M is finite.

Definition 2.2. A standard R-matrix is a formal series of the form

$$R = e^{\varphi}(1 + e_{-\alpha} \otimes e_\alpha + \sum_{k=2}^{\infty} t_{\alpha_1 \ldots \alpha_k}^{\alpha'_1 \ldots \alpha'_k} e_{-\alpha_1} \ldots e_{-\alpha_k} \otimes e_{\alpha'_1} \ldots e_{\alpha'_k}). \tag{7}$$

In this formula, and in others to follow, summation over repeated indices is implied. For fixed $(\alpha) = \alpha_1, \ldots, \alpha_k$ the sum over (α') runs over the permutations of (α). The coefficients $t_{(\alpha)}^{(\alpha')}$ are in $\mathbb{C}$.

Theorem 2. *The standard R-matrix (7), on $\mathcal{A}$, satisfies the Yang-Baxter relation (1) if and only if the coefficients $t_{(\alpha)} = t_{(\alpha)}^{(\alpha')} e_{\alpha'_1} \ldots e_{\alpha'_l}$ satisfy the following recursion relation*

$$[t_{\alpha_1 \ldots \alpha_\ell}, e_{-\gamma}] = e^{\varphi(\gamma,\cdot)} \delta_{\alpha_1}^\gamma t_{\alpha_2 \ldots \alpha_\ell} - t_{\alpha_1 \ldots \alpha_{\ell-1}} \delta_{\alpha_\ell}^\gamma e^{-\varphi(\cdot,\gamma)}\,. \tag{8}$$

That (8) is a necessary condition for R to satisfy the Yang-Baxter relation is the easy part, to show that it is sufficient takes a long calculation. A most interesting aspect is the integrability of Eq.(8).

We define two differential operators, $\vec{\partial}_{-\gamma}$ and $\overleftarrow{\partial}_{-\gamma}$, on $\mathcal{A}^+$, by

$$[X, e_{-\gamma}] = e^{\varphi(\gamma,\cdot)} \vec{\partial}_{-\gamma} X - X \overleftarrow{\partial}_{-\gamma} e^{-\varphi(\cdot,\gamma)}\,,$$

$X \in \mathcal{A}^+$; note that $\overleftarrow{\partial}_{-\gamma}$ operates from the right. These definitions are equivalent to the rules

$$\vec{\partial}_{-\gamma} e_\alpha = \delta^\gamma_\alpha + e^{-\varphi(\gamma,\alpha)} e_\alpha \vec{\partial}_{-\gamma}\,, \quad e_\alpha \overleftarrow{\partial}_{-\gamma} = \delta^\gamma_\alpha + e^{-\varphi(\alpha,\gamma)} \overleftarrow{\partial}_{-\gamma} e_\alpha,$$

then it can be shown that (8) is equivalent to

$$\vec{\partial}_{-\gamma} t_{\alpha_1 \ldots \alpha_\ell} = \delta^\gamma_{\alpha_1} t_{\alpha_2 \ldots \alpha_\ell}\,.$$

This equation is of the form

$$\partial_i X = Y_i,$$

with X and Y_i in the algebra $\mathcal{A}_+$, freely generated by $\{e_\alpha\}$. We show that, when the parameters are in general position, then this equation is always integrable. for any set of Y_i's. The existence of Y_i for which the equation is not integrable is tantamount to the existence of "constants", non-zero elements annihilated by the derivations. On special subvarieties of parameter space, where constants appear, they generate an ideal I_+ of $\mathcal{A}$. At the same time an ideal I_- appears in $\mathcal{A}_-$, and the Yang-Baxter relation can be satisfied, not on $\mathcal{A}$, but on the quotient by $I = I_+ \cup I_-$. Serre ideals are a special case and quantized Kac-Moody algebras are included in a much larger family of coboundary Hopf algebras.

3. Obstructions and Generalized Serre Relations

We have been concerned with the construction of a standard R-matrix, Definition (2.2), that satisfies the Yang-Baxter relation (1), on an algebra $\mathcal{A}$, Definition (2.1). The relations of $\mathcal{A}$ involve parameters; when these parameters are in general position then the recursion relation (8) has a unique solution that provides the unique standard R-matrix on $\mathcal{A}$ that satisfies the Yang-Baxter relation. At certain hypersurfaces in parameter space we encounter obstructions, but all these obstructions can be overcome by the introduction of additional relations in the definition of $\mathcal{A}$ or, what is the same, by replacing $\mathcal{A}$ by a quotient $\mathcal{A}/I$, where I is the ideal generated by the constants.

Theorem 3. *Let $I \subset \mathcal{A}$ be the ideal generated by the constants in $\mathcal{A}^+$ and the constants in $\mathcal{A}^-$, and let $\mathcal{A}'$ be the quotient $\mathcal{A}/I$. Interpret the standard, universal R-matrix (7) as an element of $\mathcal{A}' \otimes \mathcal{A}'$. The Yang-Baxter relation for the standard R-matrix on $\mathcal{A}'$ is equivalent to the recursion relation*

$$[t_l, 1 \otimes e_{-\gamma}] \;=\; (e_{-\gamma} \otimes e^{\varphi(\gamma,\cdot)}) t_{l-1} - t_{l-1}(e_{-\gamma} \otimes e^{-\varphi(\cdot,\gamma)}), \tag{9}$$

$$t_l \;:=\; t^{(\alpha')}_{(\alpha)} e_{-\alpha_1} \ldots e_{-\alpha_l} \otimes e_{\alpha'_1} \ldots e_{\alpha'_l}. \tag{10}$$

This relation, with the initial condition $t_1 = e_{-\alpha} \otimes e_\alpha$, *is integrable and determines all the* $t_l, l > 1$.

4. The Standard Classical r-matrix for Simple Lie Algebras

We shall now specialize, by stages, until we arrive at simple quantum groups, where a limiting process relates the R-matrix to a classical r-matrix.

Suppose that

$$\mathrm{Card}(N) \ := \ \ell < \infty.$$

Suppose next that the ideal I (generated by the constants of $\mathcal{A}$) is generated by a complete set of Serre relations; that is, for each pair $\alpha, \beta \in N$, $\alpha \neq \beta$, there is a smallest positive integer $k_{\alpha\beta}$ such that in $\mathcal{A}/I$,

$$\sum_{m=0}^{k_{\alpha\beta}} Q_m^{(\alpha,\beta)} (e_\alpha)^m e_\beta (e_\alpha)^{k_{\alpha\beta}-m} = 0\,, \tag{11}$$

with coefficients $Q_m^{(\alpha\beta)}$ in the field. The left side, as an element of $\mathcal{A}^+$, is a constant, and thus belongs to I, provided

$$e^{\varphi(\alpha,\beta)+\varphi(\beta,\alpha)+(k_{\alpha\beta}-1)\varphi(\alpha,\alpha)} = 1\,, \tag{12}$$

and the coefficients are

$$Q_m^k = (-)^m e^{m\varphi(\alpha,\beta)} \ q^{m(m-1)/2} \binom{k}{m}_q\,, \quad q := e^{\varphi(\alpha,\alpha)}.$$

We specialize further by supposing that the exponent in (12) vanish,

$$\varphi(\alpha,\beta) + \varphi(\beta,\alpha) = (1 - k_{\alpha\beta})\varphi(\alpha,\alpha)\,, \quad \alpha \neq \beta. \tag{13}$$

The form $(\cdot,\cdot)$ defined by

$$(\alpha,\beta) = \varphi(\alpha,\beta) + \varphi(\beta,\alpha) \tag{14}$$

will be called the restricted Killing form, and the matrix with components

$$A_{\alpha\beta} \ := \frac{(\alpha,\beta) + (\beta,\alpha)}{(\alpha,\alpha)} \tag{15}$$

will be called the generalized Cartan matrix; that it is symmetrizable. A restriction on $\mathrm{Card}(M)$ brings us to quantized Kac-Moody algebras.

Definition 4. Let $\mathcal{A}'$ be as above; that is, the quotient of an algebra $\mathcal{A}$ as per Definition 2.1, with parameters satisfying (12), by the ideal generated by the Serre relations (11). We shall say that $\mathcal{A}'$ is a quantized Kac-Moody algebra of finite type if (i) Card M = Card $N =< \infty$, and (ii) the (symmetrizable) generalized Cartan matrix

$$A_{\alpha\beta} = \frac{\varphi(\alpha, \beta) + \varphi(\beta, \alpha)}{\varphi(\alpha, \alpha)} \tag{16}$$

is positive definite with $A_{\alpha\beta} \in \{0, -1, \ldots\}$, $\alpha \neq \beta$. We shall say that $\mathcal{A}'$ is a quantized Kac-Moody algebra of affine type if (i) Card M = 2 + Card $N < \infty$, and (ii) the generalized Cartan matrix is positive semi-definite with $A_{\alpha\beta} \in \{0, -1, \ldots\}$, $\alpha \neq \beta$, all its principal minors positive definite.

The rest of this section deals with Kac-Moody algebras of finite type. The semi-classical limit of R is defined by replacing

$$\varphi(\cdot,\cdot) \to \hbar\varphi(\cdot,\cdot)\,, e_\alpha \to \kappa\, e_\alpha\,, \quad e_{-\alpha} \to \kappa' e_{-\alpha}\,, \quad \kappa\kappa' = \hbar\,, \quad \alpha \in N\,, \tag{17}$$

and developing the exponentials to first order in $\hbar$. Then Eq. (6) becomes

$$[e_\alpha, e_{-\beta}] = \delta_{\alpha\beta}(\varphi(\alpha,\cdot) + \varphi(\cdot,\alpha)) =: \delta_{\alpha\beta} H_{(\alpha)}\varphi(\alpha,\alpha)\,. \tag{18}$$

It follows from (18) and (6) that

$$[H_{(\alpha)}, e_\beta] = A_{\alpha\beta} e_\beta\,, \quad \alpha, \beta \in N\,. \tag{19}$$

The (classical) r-matrix r associated with the R-matrix (7) is defined by

$$R = 1 + \hbar r + o(\hbar^2)\,. \tag{20}$$

Two terms in r are obvious: $r = \varphi + \sum e_{-\alpha} \otimes e_\alpha + ?$, with the sum extending over simple roots. Evaluating the remaining terms is more difficult, because a) we do not have a sufficiently explicit expression for the coefficients $t_{(\alpha)}^{(\alpha')}$ and b) because all these coefficients are singular in the classical limit. Both these difficulties are avoided by the recursion relation (9), as we shall see later. The result, which was known by indirect means, with a particular normalization of the non-simple roots, is that

$$r = \varphi + \sum_{\alpha \in \Delta^+} E_{-\alpha} \otimes E_\alpha\,, \tag{21}$$

where Δ^+ is the set of positive roots. This is what we call the standard r-matrix for a simple Lie algebra. It satisfies the classical Yang-Baxter relation $[r_{12}, r_{13} + r_{23}] + [r_{13}, r_{23}] = 0$ and $r + r^t = \hat{K}$, the Killing form of g. In the list of (constant) r-matrices obtained by Belavin and Drinfeld [2], (21) is the simplest. The quantum groups to which these r-matrices are associated are the twisted quantum groups of Reshetikhin and others [17, 19, 20].

5. The Standard Classical r-matrix for affine Kac-Moody Algebras

A quantized affine Kac-Moody algebra can be described as follows. Let $\hat{\mathcal{A}}'$ be as above, with parameters satisfying (12) and Serre relations (11), with root generators $\{e_{\pm\alpha}\}\,\alpha = 0,\ldots,l$ and Cartan generators $H_1,\ldots,H_l$ and c, d, such that the subset that consists of $\{e_{\pm\alpha}\}\,\alpha \neq 0$ and $H_1,\ldots,H_l$ generates a subalgebra $\mathcal{A}'$ that is a quantized Kac-Moody algebra of finite type. Let $\hat{\varphi}$ refer to $\hat{\mathcal{A}}'$ and φ to $\mathcal{A}'$, and suppose that

$$\hat{\varphi} = \varphi + u\, c \otimes d + (1-u)\, d \otimes c, \quad [d, e_{\pm\alpha}] = \pm\delta_\alpha^0\, e_{\pm 0}. \tag{22}$$

with some $u \in \mathbb{C}$. Suppose that c is central and that the extra root defined by $[H_a, e_0] = H_a(0)e_0$ is such as to make the generalized Cartan matrix of $\hat{\mathcal{A}}'$ positive semi-definite with all its principal minors positive. Then $\hat{\mathcal{A}}'$ is a quantized affine Kac-Moody algebra.

Consider a quantized affine Kac-Moody algebra $\hat{\mathcal{A}}'$, with generators $e_{\pm 0},\ldots,e_{\pm l}$ and $H_1,\ldots,H_l,c,d$. Renormalize as in (17) and pass to the classical limit.

After the renormalization (17) t_n is of order $\hbar$ and the classical r-matrix is defined by (20). The Yang-Baxter relation for R is equivalent to the recursion relation (1). To lowest order in $\hbar$ it is

$$[1 \otimes e_{-\gamma} + e_{-\gamma} \otimes 1,\ r] = \varphi(\cdot,\gamma) \wedge e_{-\gamma}.$$

This is easy to solve and the result is that

$$r = \varphi + \sum E_{-i} \otimes E_i - C_0 + \frac{1}{1-x^k}\sum_0^{k-1} x^j C_j. \tag{23}$$

Here $C = \sum_i C_i$ is the decomposition of the Casimir element according to the action of a Coxeter automorphism. This result applies to the full Kac-Moody algebra; projected on the loop algebra it is in agreement with [2], in the notation of [14].

6. First Order Deformations

Quantum groups can be understood as deformations of the Hopf structure associated with Lie algebras or Kac-Moody algebras. The point of view that emphasizes the direct connection between quantum groups and Lie groups, as well as the deep roots of quantum groups in deformation theory and in the theory of $*$-products, has been shown to lead to profound insight into their general structure [4, 3, 8]. Here we use deformation theory with a different purpose. The initial structure is the bialgebra associated with a standard R-matrix, with a fixed set of parameters. The deformed structure is a bialgebra equipped with an R-matrix that is non-standard and that does not commute with the Cartan sub-algebra. We emphasize that the context is more general than quantized Kac-Moody algebras.

A deformation of the standard R-matrix is a formal series

$$R_\epsilon = R + \epsilon R_1 + \epsilon^2 R_2 + \dots . \tag{24}$$

Here R is a standard R-matrix on $\mathcal{A}' = \mathcal{A}/I$ with any choice of parameters and the ideal I determined by them. The coefficients $t_{(\alpha)}^{(\alpha')}$ of R are determined by the Yang-Baxter relation, and we attempt to find $R_1, R_2, \dots$ so that R_ϵ will satisfy the same relation to each order in ϵ. To make this program precise, we must specify the nature of the leading term; the remainder should then be more or less unique. Recall further that R is driven by the linear term; by virtue of the Yang-Baxter relation, R is completely determined by the term $e_{-\alpha} \otimes e_\alpha$. It is natural to study deformations that are driven by a similar term,

$$S(e_\sigma \otimes e_{-\rho}), \quad S \in \mathcal{A}^0 \otimes \mathcal{A}^0 . \tag{25}$$

(Other driving terms, for example $S(e_\sigma \otimes e_\rho$, have also been investigated, but with meager results.

Theorem 6.1. *Let R be the standard R-matrix described in Theorem 3. Suppose that $R + \epsilon R_1$ is a first order deformation, satisfying the Yang-Baxter relation to first order in ϵ. Suppose also that the term of lowest grade in R_1 has the form (25); then the parameters satisfy*

$$e^{\varphi(\cdot,\rho)+\varphi(\sigma,\cdot)} = 1 . \tag{26}$$

Conversely, when the parameters are in general position on this surface, then there exists a unique first order deformation such that the term of lowest grade has the form (25), namely

$$R_1 = R(Ke_\sigma \otimes Ke_{-\rho}) - (Ke_{-\rho} \otimes Ke_\sigma)R , \quad K := e^{\varphi(\cdot,\rho)}. \tag{27}$$

The essential deformations of the standard classical Lie bi-algebras have been calculated [11] and found to coincide with the classification of Lie bialgebras in [2].

Theorem 6.2. *The first order deformations of the standard R-matrix described in Theorem 6.1, upon specialization to a simple quantum group, are in one-to-one correspondence, with the first order essential deformations of the associated standard r-matrix [11], modulo variations of the parameters.*

One concludes that the class of deformations investigated here is wide enough to encompass the quantization of all simple Lie bialgebras. We shall see that the affine Kac-Moody algebras are provided for also.

7. Exact Deformations

We return to the first order deformations described in Theorem 6.1. A deformation of this type, involving a single pair (ρ, σ) for which (26) holds, is called an elementary deformation. To each elementary, first order deformation, there is an exact deformation (to all orders in ϵ), that can be expressed in closed form. To first order in ϵ, the problem being then linear, one obtains a more general space of deformations by adding the contributions of several such pairs,

$$R_1 = \sum_{(\sigma,\rho)\in[\tau]} (Rf_\sigma \otimes f_{-\rho} - f_{-\rho} \otimes f_\sigma R). \tag{28}$$

Here the sum is over a subset $[\tau]$ of the pairs (σ, ρ); $\sigma \in \hat{\Gamma}_1$, $\rho \in \hat{\Gamma}_2$, where $\hat{\Gamma}_{1,2}$ are subsets of the set of positive generators, and

$$f_\sigma := e^{-\varphi(\sigma,\cdot)} e_\sigma, \quad f_{-\rho} := e_{-\rho} e^{\varphi(\cdot,\rho)}, \quad e^{\varphi(\sigma,\cdot)+\varphi(\cdot,\rho)} = 1, \quad (\sigma, \rho) \in [\tau].$$

Not all compounded, first order deformations lift to exact deformations.

The deformed co-product was also calculated to first order in ϵ, and the results suggest an approach to the exact deformations. The formula (27) for R_1, as well as the expression for the first order deformation of the coproduct, both suggest that the deformation be formulated as a twist [7]. Additional support for this is found in the fact that the exact, elementary deformations mentioned above are also of this type. For the following result $\mathcal{A}'$ is any coboundary Hopf algebra.

Theorem 7.1. *Let R be the R-matrix, Δ the coproduct, of a coboundary Hopf algebra $\mathcal{A}'$, and $F \in \mathcal{A}' \otimes \mathcal{A}'$, invertible, such that*

$$((1 \otimes \Delta_{21})F)F_{12} = ((\Delta_{13} \otimes 1)F)F_{31}. \tag{29}$$

Then

$$\tilde{R} := (F^t)^{-1} RF \tag{30}$$

(a) satisfies the Yang-Baxter relation and (b) defines a Hopf algebra $\tilde{\mathcal{A}}$ with the same product and with co-product

$$\tilde{\Delta} = (F^t)^{-1} \Delta F^t. \tag{31}$$

The general case of compound deformations is much more complicated. The calculations are manageable only so long as F can be constructed from elements of the type $f_\sigma \otimes f_{-\rho}$ only, with the factors in this order.

Theorem 7.2. *Let Γ_1, Γ_2 be subalgebras of $\mathcal{A}'^+$, generated by subsets $\hat{\Gamma}_1, \hat{\Gamma}_2$ of the generators, and $\tau : \Gamma_1 \to \Gamma_2$ an algebra isomorphism. Let $F \in \mathcal{A}' \otimes \mathcal{A}'$ be a formal series of the form*

$$F = 1 + \sum_{n=1}^{\infty} \epsilon^n F_n, \quad F_n = \sum F^{(\rho)}_{(\sigma)} f_{\sigma_1} \cdots f_{\sigma_n} \otimes f_{-\rho_1} \cdots f_{-\rho_n}. \tag{32}$$

The second sum is here over all $\sigma_i \in \hat{\Gamma}_1, \rho_i \in \hat{\Gamma}_2$. (!) Note that F_n is a power series in ϵ. Suppose that F satisfies (29), and that

$$F_0 = 1, \quad F_1 = -\sum_{\tau^m \sigma = \rho} \epsilon^{m-1} (f_\sigma \otimes f_{-\rho}), \tag{33}$$

then F_n satisfies

$$(1 \otimes K_\rho \partial_\rho) F_n + \sum_{\tau^m \sigma = \rho} \epsilon^m \, [1 \otimes f_\sigma, F_n] + \sum_{\tau^m \sigma = \rho} \epsilon^{m-1} (f_\sigma \otimes K^\sigma) F_{n-1} = 0. \tag{34}$$

With F_0 and F_1 thus fixed, F_n are determined recursively and uniquely. (The operator $K_\rho \partial_\rho$ is the derivation of $\mathcal{A}'_-$ that replaces $f_{-\rho}$ by K_ρ.)

The converse, that the solution of (34) with $F_0 = 1$ and F_1 given by (33) satisfies (29) (and therefore gives a solution of the Yang-Baxter relation) was proved only in the special case that $\hat{\Gamma}_1 \cap \hat{\Gamma}_2$ is empty. Further direct computation supports the idea that R_ϵ always has the form $(F^t)^{-1} R F^t$, with F of the form assumed in (32). This is strong support for the belief that the solution of the recursion relation (34), which was proved to exist always, actually furnishes the solution to the problem of exact deformations in the general case. The results stated in Theorems 6.1 and 6.2 may also be considered as strong evidence. As we shall see, additional favorable evidence comes from an examination of the classical limit. The solution of the recursion relation is simplified by the following observation.

Theorem 7.3. *Suppose that F^m, for $m = 1, 2, ...$, is an element of $\mathcal{A} \otimes \mathcal{A}$ of the form*

$$F^m = \sum_{n, \rho = \tau^m \sigma} \bar{t}^{(\sigma')}_{(\sigma)} f_{\sigma_1} \cdots f_{\sigma_n} \otimes f_{-\tau^m \sigma'_1} \cdots f_{-\tau^m \sigma'_n}, \quad F^0 = 1,$$

satisfying

$$[1 \otimes f_\sigma, F^m] = -\epsilon^m \Big((f_{\tau^{-m}\sigma} \otimes K_\sigma) F^m - F^m (f_{\tau^{-m}\sigma} \otimes K^\sigma) \Big), \tag{35}$$

then $F = F^1 F^2 \ldots$ *satisfies (34).*

8. Esoteric r-matrices

a) Quantized Kac-Moody algebra of finite type.

Proposition 8.1. *If $\mathcal{A}'$ is a quantized Kac-Moody algebra of finite type, then $\hat{\Gamma}_1$ is a proper subset of the set of positive generators and $\tau^{m+1}\hat{\Gamma}_1 \cap \hat{\Gamma}_1$ is a proper subset of $\tau^m \hat{\Gamma}_1 \cap \hat{\Gamma}_1$.*

In the classical limit

$$R_\epsilon = 1 + \hbar r_\epsilon + o(\hbar^2) \quad r_\epsilon = r + \epsilon + o(\epsilon^2). \tag{36}$$

In the case of an exact elementary deformation R_ϵ, the associated exact deformation r_ϵ of r coincides with the first order,

$$r_\epsilon = r + \epsilon r_1. \tag{37}$$

Consider the general case of an exact deformation of R of the form postulated in Theorem 7.2. Define X_ϵ by

$$F = 1 + \hbar X_\epsilon + o(\hbar^2), \tag{38}$$

so that

$$r_\epsilon = r + X_\epsilon - X_\epsilon^t. \tag{39}$$

From the fact that the coefficients in the expansion of F^m are the same as the coefficients in the expansion (7) of the standard R-matrix, and the known classical limit of the standard R-matrix for a Kac-Moody algebra of finite type, we get without calculation that

$$X_\epsilon = -\sum_m \sum_{E_i \in \Gamma_1} \epsilon^{nm} E_i \otimes E_{-\tau^m i}, \tag{40}$$

in which n is the height of E_i. Consequently,

$$r_\epsilon = r - \sum_m \sum_{\substack{E_i \in \Gamma_1 \\ E_j = \tau^m E_i}} \epsilon^{nm} E_i \wedge E_{-j}. \tag{41}$$

The sums are finite, by Proposition 8.1. A renormalization exists that reduces the numerical coefficients to unity ($\epsilon \in \mathbb{C}$); the result is in complete agreement with [2].

b) Quantized Kac-Moody algebra of affine type.

Let $\mathcal{A}'$ be a quantized Kac-Moody algebra of affine type. Two cases should be distinguished. If the subsets $\hat{\Gamma}_{1,2}$ of positive roots do no include the imaginary root e_0, then the formula (41) applies without change, except that now r is one of the standard affine r-matrices determined earlier. There is nothing more to be said about this and we turn our attention to the other case.

What merits special attention is the possibility that the first order deformation (28) may include one of the following

$$e_0 \wedge e_{-\rho} = \mu(E_- \otimes e_{-\rho}) - \lambda(e_{-\rho} \otimes E_-), \tag{42}$$

or

$$e_\sigma \wedge e_{-0} = \lambda^{-1}(e_\sigma \otimes E_+) - \mu^{-1}(E_+ \otimes e_\sigma), \tag{43}$$

with

$$\varphi(\cdot, \rho) + \varphi(0, \cdot) = 0, \quad \text{resp.} \quad \varphi(\cdot, 0) + \varphi(\sigma, \cdot) = 0, \tag{44}$$

Proposition 8.2 *If $\mathcal{A}'$ is a quantized Kac-Moody of affine type, then either the statement about $\hat{\Gamma}_1$ in Proposition 8.1 continues to hold, or $\mathcal{A}'$ is of type $A_{N-1}^{(1)}$, $\hat{\Gamma}_1$ consists of all the positive generators, and τ generates the cyclic group of order N.*

In this section we exclude the exceptional case. This means that the classical limit of Γ_1 is a finite dimensional Lie algebra, so that (41) can be applied directly, since the sum is finite.

Alternatively, the classical limit can be found with the help of the recursion relation (35). This implies that $X^m = \sum_{n=0,1,\ldots} \epsilon^{mn} X_n^m$ (a finite sum) is the unique solution (of the form that appears in (40)) of

$$[1 \otimes f_\rho + \epsilon^m f_\sigma \otimes 1, X^m] = \epsilon^m f_\sigma \otimes (\varphi + \varphi^t)(\rho), \quad \tau^m \sigma = \rho \in \hat{\Gamma}_2. \tag{45}$$

9. Universal Elliptic R and r-matrices

Here we consider the exceptional case (Proposition 8.2) in which $\hat{\Gamma}_1$ contains all the generators of $\mathcal{A}'^+$, $\mathcal{A}'$ is of type $A_{N-1}^{(1)}$ and $\tau^N = 1$.

The infinite product

$$F = F^1F^2\dots$$

cannot be given anything more than a formal significance in the structural context but, as will be shown below, in a finite dimensional representation the question of convergence (with $\epsilon \in \mathbb{C}$) is not difficult. We define F^m by the (always uniquely integrable) relation (35), In the classical limit $F^m = 1 + \hbar X^m + o(\hbar^2)$ and X^m satisfies (45). We shall solve these relations in the case of $\widehat{sl(2)}$. Set

$$[f_1, f_{-1}] = (\varphi + \varphi^t)(1) =: H, \quad \tau f_1 = f_0, \quad \tau f_0 = -f_1,$$

and

$$X^m = \Psi^m + A^m(f_1 \otimes f_{-1} + f_0 \otimes f_{-0}) + C^m(f_1 \otimes f_{-0} + f_0 \otimes f_{-1})$$

and impose (45). The result is, with $x = \lambda/\mu = e^{2\pi i u}$,

$$\begin{aligned}(i/2)(X - X^t) &= \sum_{n=1}^{\infty}\Big\{\frac{-\epsilon^{2n}}{1+\epsilon^{2n}}(\varphi + \varphi^t)\sin 2n\pi u \\ &\quad + \frac{\epsilon^{4n-2}}{1-\epsilon^{4n-2}}(x f_1 \otimes f_{-1} + \frac{1}{x} f_{-1} \otimes f_1)\sin(2n-1)\pi u \\ &+ \frac{\epsilon^{2n-1}}{1-\epsilon^{4n-2}}(\sqrt{1/\mu\lambda}\, f_1 \otimes f_1 + \sqrt{\mu\lambda}\, f_{-1} \otimes f_{-1})\sin(2n-1)\pi u\Big\}. \qquad (46)\end{aligned}$$

The trigonometric r-matrix is

$$\frac{i}{2}\Big(\frac{1}{\tan\pi u}(\varphi + \varphi^t) + \frac{1}{\sin\pi u}(\sqrt{x}\, f_1 \otimes f_{-1} + \sqrt{1/x}\, f_{-1} \otimes f_1)\Big).$$

Adding, one finds the series expansion of elliptic functions, and complete agreement with the elliptic r-matrices of [2]. To transform to their notation replace

$$f_1 \to \sqrt{\lambda}\, e_{12}, \; f_{-1} \to \sqrt{1/\lambda}\, e_{21}.$$

Finally, we shall show that the expression for the Universal Elliptic R-matrix $R_\epsilon = (F^t)^{-1}RF$, $F = F^1F^2\dots$ in terms of an infinite product is both meaningful and usable, by projecting on a finite dimensional representation. We limit ourselves to the fundamental representation of $sl(2)$. After rescaling of the generators F^m and R_ϵ take the form

$$F^m = \begin{pmatrix} a^m & & & d^m \\ & b^m & c^m & \\ & c^m & b^m & \\ d^m & & & a^m \end{pmatrix}, \quad R_\epsilon = \begin{pmatrix} a & & & d \\ & b & c & \\ & c & b & \\ d & & & a \end{pmatrix}.$$

The matrix elements are completely determined by the recursion relation (35) and the result is that

$$a+d \; : \; a-d \; : \; b+c \; : \; b-c \qquad (47)$$
$$= \frac{dn(u+\rho)}{dn(u-\rho)} \; : \; 1 : \frac{cn(u+\rho)}{cn(u-\rho)} : \frac{sn(u+\rho)}{sn(u-\rho)}.$$

A modular transformation brings this into agreement with Baxter [1].

Acknowledgements

I am indebted to Moshe Flato for a detailed and constructive criticism of the original manuscript, and Masaki Kashiwara and Tetsuji Miwa for fruitful discussions. This work was completed at the Research Institute for Mathematical Sciences, Kyoto University. I thank the director, Professor Huzihiro Araki, for making the pleasant and productive stay in Kyoto possible, and for friendly hospitality during my visit. I thank the staff of the International Office of the Institute for many services, and the Japanese Ministry of Education, Science, Sports and Culture for financial support.

References

1. R.J. Baxter, Partition Function of the eight-Vertex Model, *Annals of Physics* **70** (1972) 193-228.
2. A.A. Belavin and V.G. Drinfeld, Triangle Equation and Simple Lie Algebras, *Sov. Sci. Rev. Math.* **4** (1984) 93-165.
3. F. Bidegain and G. Pinczon, Quantization of Poisson Lie groups and applications, Dijon preprint 1995.
4. P. Bonneau, M. Flato, M. Gerstenhaber and G. Pinczon, The hidden group structure of quantum groups: strong duality, rigidity, and preferred deformations, *Commun. Math. Phys.* (1994) 125-156.
5. E. Cremmer and J.-L. Gervais, *Commun. Math. Phys.* **134** (1990).
6. V.G. Drinfeld, Quantum Groups, in *Proceedings of the International Congress of Mathematicians, Berkeley*, A.M. Gleason, ed. (American Mathematical Society, Providence 1987.)
7. V.G. Drinfeld, in "Quantum Groups", P.P. Kulish, ed. Proceedings of the Workshop held in Euler International Mathematical Institute, Leningrad, Fall 1990.
8. P. Etingof and D. Kazhdan, Quantization of Poisson algebraic groups and Poisson homogeneous spaces, Harvard preprint, Oct 20, 1995 (q-alg/9510020).
9. I.B. Frenkel, N.Yu. Reshetikhin, Quantum Affine Algebras and Holonomic Difference Equations, *Commun. Math. Phys.* (1992) **146** 1-60.
10. C.Frønsdal, to appear in RIMS publications.
11. C. Frønsdal, Cohomology and Quantum Groups, in Proceedings of the Karpacz Winter School, February 1994.
12. C. Frønsdal and A. Galindo, *Contemporary Mathematics*, **175** (1994) 73-88.
13. M. Gerstenhaber, *Ann. Math.* **78**(1963) 267-288.
14. M. Jimbo, *Commun. Math. Phys.* **102** (1986) 537-547.
15. A.N. Kirillov, N.Yu. Reshetikhin, q-Weyl group and a multiplicative formula for the universal R-matrix, *Commun. Math. Phys.* **134** (1991) 421-431.

16. S.Z. Levendorski and Y.S. Soibelman, Some applications of quantum Weyl group I. The multiplicative formula for universal R-matrix for simple Lie algebras. *J. Geom. Phys.* **7** (1991) 1-14.
17. N.Yu. Reshetikhin, *Lett. Math. Phys.* **20** (1990) 331-336.
18. M. Rosso, An analogue of P.B.W. theorem and the universal R-matrix for $U_h sl(N)$, *Commun. Math. Phys.* **124** (1989) 307-319.
19. A. Schirrmacher, *Z. Phys.* C **50** (1991) 321.
20. A. Sudbery, *J. Phys. A (Math. Gen.)* **23** (1990) L697.
21. T. Tanisaki, *Int. J. Mod. Phys. A*, Suppl. 1B (1992) 941-961.

QUANTIZED INTERSECTION CARDINALITIES

MURRAY GERSTENHABER
Dept. of Mathematics, University of Pennsylvania
Philadelphia, PA 19104-6395, USA
mgersten@mail.sas.upenn.edu *or* murray@math.upenn.edu

Abstract. Let U be a finite ordered set. Functions $f(S,T)$ of pairs of subsets of U invariant under the operation of the symmetric group on U, e.g., gravitational force between equal masses placed at the vertices of a cube in n-space, can be quantized, becoming invariant under the Hecke algebra.

Let U be a finite set of n elements with a fixed linear order. Every subset S of U can then be represented by a vertex of the unit n-cube by assigning to S the point whose i-th coordinate is 1 if the i-th element of U is in S and 0 otherwise. Any "interaction" between vertices which depends only on the Euclidean distance between them, e.g., gravitational or electrical forces between equal masses or charges placed at the vertices of the cube, is then just a function of the cardinality of the intersection of the corresponding sets. For if the sets are S and T, then the square of the distance between the points is $n - \#(S \cap T)$. These functions of intersection cardinalities form a linear space of dimension $n+1$. We show that they can be quantized in a meaningful way.

At first it may seem unreasonable to quantize something so basic as the cardinality of the intersection of two sets. On the other hand, any permutation of U induces a permutation on the family of all subsets of U. If N is a family of subsets of U stable under the operation of S_n then $\mathbb{C}N$ becomes a module over the group algebra $\mathbb{C}S_n$. A function f defined on pairs of sets in N may be written as a matrix whose rows and columns are indexed by the sets S, T in N and whose entry in the S,T place is just $f(S,T)$ and so f may be viewed as a linear transformation from $\mathbb{C}N$ to itself. If $f(S,T)$ depends only on $\#(S \cap T)$ then this linear transformation commutes with the operation of S_n on $\mathbb{C}N$ and hence is a $\mathbb{C}S_n$ module endomorphism of $\mathbb{C}N$.

When N is the family of all subsets of some fixed cardinality k, which

D. Sternheimer et al. (eds.), Deformation Theory and Symplectic Geometry, 85–102.

we will denote by $N(k)$, then the converse is also true. For if S, T, S', T' are subsets of U all of the same cardinality k, then the necessary and sufficient condition that there exist a permutation $\sigma \in \mathcal{S}_n$ with $\sigma S = S'$, $\sigma T = T'$ is that $\#(S \cap T) = \#(S' \cap T')$. Now the group algebra of $\mathcal{S}_n$ deforms in a natural way to the Hecke algebra $\mathcal{H}_n = \mathcal{H}_n(q)$ and it is possible simultaneously to deform $\mathbb{C}N$ to a module over $\mathcal{H}_n$. (In principal this is trivial, since as long as q is not a root of unity $\mathcal{H}_n$ is isomorphic to $\mathbb{C}\mathcal{S}_n$, although finding explicit formulas may be difficult.) We can therefore define quantization of an $\mathcal{S}_n$ endomorphism f of $\mathbb{C}N$ to mean deformation of f to an endomorphism of the deformed module.

In this note we give an inductive construction of formulas for quantization of $\mathcal{S}_n$ invariant functions $f(S,T)$ of pairs of subsets of U and a sketch of the proof of their correctness. Ordinarily quantizations meet obstructions and involve arbitrary choices but the procedure here avoids both because of the existence of certain canonical bases (cf. [5]) for the modules involved. However, the quantization does depend both on U and on its ordering, for if $S, T \subset U$ and if U is embedded as the initial segment of some larger ordered set U' then the quantized intersection cardinality of S and T relative to U will generally be different from that relative to U'.

One can not expect the quantized intersection cardinality, which we will denote by $\#_q(S,T)$, to behave too much like the ordinary one. It is symmetric, but in general $\#_q(S,S) \neq \#S$, and when q is real the quantized intersection cardinality of disjoint sets may even become negative no matter how close q is to 1. (On the other hand, in the few cases computed it always has a well-defined limit when $q \to \infty$, but this is often 0.) If q is allowed to be complex, then (as expected) $\#_q(S,T)$ will generally be undefined when q is a $2m$-th root of unity with $m \leq n$ (a "bad" root) where $n = \#U$. Most important, of course, is that the $\mathcal{S}_n$ symmetry is lost. If equal particles at the vertices of the n-cube actually moved under the influence of a gravity as quantized here, their positions would cease to be symmetric as soon as some linear order were placed on them to make the quantization unambiguous. We do not know if the quantized forces have potentials.

It should be possible to extend the idea of quantization to functions of more than two subsets of U. We can view the invariant functions of two variables as functions on $\mathbb{C}N \otimes \mathbb{C}N$ which vanish on all components of that $\mathcal{S}_n$ module except the invariants. (Note that a permutation module is always isomorphic to its own dual because the matrix of the dual is the transpose inverse of the original, but for a permutation matrix the transpose is the inverse.) This extends to higher tensor powers. The problem now is to let the Hecke algebra operate on the same underlying vector space and to reinterpret the meaning of invariance. It should also be possible to extend the present ideas to certan lattices obtained by repeated reflections.

A technical note: We have been careful at the start to use the smallest ring of coefficients over which our assertions make sense. (With these one can reduce modulo primes; there is the hope that one can ultimately get information about characteristic zero by considering all positive characteristics.) However, nothing significant here will be lost if the reader assumes throughout that the coefficients are in $\mathbb{C}$. We wish to thank H. Wilf for bringing to our attention the beautiful properties of intersection cardinality matrices, cf. [3].

Acknowledgements

I would like to extend my sincere thanks to my co-organizers of the conference, to the ETH for the hospitality at Monte Verità, and particularly to Moshé who, in addition to mathematical guidance was a superb host.

1. Canonical modules and canonical bases

In this section we recapitulate some results of [5]. Let V be a free $\mathbb{Z}$ module of rank d with a fixed basis $x_1, \dots, x_d$. Its r-th tensor power, here denoted simply V^r, has a basis consisting of the non-commutative monomials $x_{i_1}x_{i_2}\dots x_{i_r}$. The direct sum of all the V^r with concatenation as multiplication (i.e., the tensor algebra of V) is just the non-commutative polynomial algebra in $x_1, \dots, x_d$. The monomial generators $\mu = x_{i_1}x_{i_2}\dots x_{i_n}$ can be put into a natural one-to-one correspondence with the ordered partitions of U into d subsets $(U_1, \dots, U_d)$, some of which may be empty, as follows: When μ is given put the i-th element of U into U_j if the i-th factor of μ is x_j.

When $d = 2$ we will denote the generators simply by x, y. In this case the partition consists of two complementary subsets so we only have to specify one of them and will adopt the following convention: A monomial μ in x and y corresponds to the subset S of U marked by those places in which the factor is y. Thus, for example, if $U = \{1, 2, 3, 4, 5\}$ then $yxxyy$ corresponds to the subset $\{1, 4, 5\}$.

The operation of S_n on U (which is well-defined, since U is ordered) induces an operation on various families N of sets built from U. Amongst the examples are:

(i) The family of all subsets of U; it has 2^n elements and will generally be denoted by 2^U.

(ii) The family of all k-element subsets of U, where k is a fixed integer with $0 \le k \le n$. This is just the orbit of any one set in the family and will be denoted by $N(k)$ when U is understood. Geometrically, the $N(k)$ lie on parallel hyperplanes perpendicular to the line on which all coordinates are equal.

(iii) Choosing a partition $(n_1, n_2, \ldots, n_d)$ of n, let $N(n_1, \ldots, n_d)$ be the family of all ordered partitions of U into d subsets $(U_1, \ldots, U_d)$ with $\#U_i = n_i$. We do not consider this here but it is potentially important if the quantized intersection cardinality is to be extended to more than two sets.

In each case the free $\mathbb{Z}$ module $\mathbb{Z}N$ spanned by the elements of N becomes a module over the group ring $\mathbb{Z}\mathcal{S}_n$. Now $\mathcal{S}_n$ also operates on V^n by permutation of the tensor factors, thereby making V^n into a $\mathbb{Z}\mathcal{S}_n$ module. When, e.g., $d = 2$ it is clear that this representation is the same as that induced on $\mathbb{Z}2^U$. Taking the submodule $V(n-k, k)$ of V^n spanned by those monomials of degree $n-k$ in x and k in y gives a representation equivalent to that on $\mathbb{Z}N(k)$. Analogous assertions hold when V has arbitrary rank d.

One can introduce an inner product $< \cdot, \cdot >$ on V^n by taking the monomials $\mu = x_{i_1}x_{i_2}\cdots x_{i_n}$ as an orthonormal basis. The elements of $\mathcal{S}_n$ then become orthogonal operators. The Lie algebra sl_d, with basis $\{X_i, Y_i, H_i, i = 1, \ldots, d-1\}$, operates on the tensor algebra of V as follows: Write ∂_i for $\partial/\partial x_i$. Then X_i, Y_i, H_i operate respectively as $x_i\partial_{i+1}$, $x_{i+1}\partial_i$, $x_i\partial_i - x_{i+1}\partial_{i+1}$. These operations commute with those of $\mathcal{S}_n$.

For every non-increasing partition $\mathfrak{p} = (n_1, \ldots, n_d)$ of n into d parts (some of which may be zero) denote by $V(\mathfrak{p})$ the $\mathcal{S}_n$ submodule of V^n spanned by all monomials whose degree in x_i is n_i. Let $V(\mathfrak{p}; 0) = \bigcap \ker(X_i \mid V(\mathfrak{p})$. Suppose that $d \leq n$. After extending coefficients to $\mathbb{Z}[1/n!]$ the following hold:

(i) The $V(\mathfrak{p}; 0)$ are simple $\mathcal{S}_n$ modules, no two are isomorphic, and if $d = n$ then the representation is faithful and every simple $\mathcal{S}_n$ module is isomorphic to one of these; we call them the *canonical* $\mathcal{S}_n$ modules. The group algebra $\mathbb{Z}[1/n!]\mathcal{S}_n$ operates on each as its full ring of linear (over $\mathbb{Z}[1/n!]$) endomorphisms.

(ii) V^n is a direct sum of simple modules each of which is isomorphic to one of the $V(\mathfrak{p}; 0)$.

(iii) Each canonical module $V(\mathfrak{p}; 0)$ has a canonical orthonormal basis whose elements are indexed by the standard Young tableaux corresponding to the partition $\mathfrak{p}$.

We illustrate this construction in the case $d = 2$, since that is the only one with which we will actually do any calculations in this note. (For the general case, see [5].)

Suppose that V has rank 2, is spanned by x and y, and so is operated on by sl_2 with basis X, Y, H. We write $V(n-k, k)$ for the submodule spanned by all monomials of total degree $n-k$ in x and k in y. Any non-increasing partition $\mathfrak{p}$ of n into two parts has the form $(n-k, k)$ with $k \leq n/2$; write $V(n-k, k; 0) = \ker(X \mid V(n-k, k))$. This is a simple module of rank $\binom{n}{k} - \binom{n}{k-1}$. (It is meaningful but vanishes for $k > n/2$.) If $\alpha \in V(n-k, k)$ then setting $|\alpha| = n - 2k$ we have $H\alpha = |\alpha|\alpha$.

To construct the canonical orthonormal basis we only have to construct a canonical orthogonal basis; it can subsequently be normalized. Now $V(n-k,k;0), k \leq n/2$, is simple as a $\mathbb{Z}[1/n!]\mathcal{S}_n$ module but may decompose over a smaller ring. View $\mathcal{S}_{n-1}$ for the moment as the subgroup of $\mathcal{S}_n$ operating on $\{2,\ldots,n\}$ and leaving 1 fixed; in this way we can consider $V(n-k,k;0)$ as an $\mathcal{S}_{n-1}$ module. There is then an $\mathcal{S}_{n-1}$ module morphism $\mathcal{P} : V(n-k,k-1;0) \to V(n-k,k;0)$ sending α to $\mathcal{P}\alpha := y\alpha - |\alpha|^{-1}xY\alpha$, and if $2k+1<n$ then there is also an $\mathcal{S}_{n-1}$ module morphism from $V(n-k-1,k;0)$ to $V(n-k,k;0)$ sending α to $x\alpha$. As an $\mathcal{S}_{n-1}$ module, $V(n-k,k;0)$ is the direct sum of these submodules, which, moreover, are orthogonal to each other under the inner product that has been introduced. The mappings $\mathcal{P}$ and multiplication by x both preserve orthogonality. (In the case where V has rank d there are d maps $\mathcal{P}_1,\ldots,\mathcal{P}_d$ generalizing the $\mathcal{P}$ just defined, which actually is $\mathcal{P}_2$; $\mathcal{P}_1$ is multiplication by x.) It follows that we can inductively construct an orthogonal basis for $V(n-k,k;0)$, for if we have those for $V(n-k,k-1;0)$ and $V(n-k-1,k;0)$ then the union of their images under $\mathcal{P}$ and multiplication by x, respectively, is the canonical orthogonal basis for $V(n-k,k;0)$.

Here is the construction for some small values of n and k. The case $k=0$ is trivial since $V(n,0;0)=V(n,0)$; this has a single generator, x^n.

For $n=2$, the only non-trivial possibility with $k \leq n/2$ is $k=1$. Here $V(1,1;0)$ is spanned by the single element $\mathcal{P}x = yx - xy$. The norm of this element is 2, so the orthonormal basis consists of $(1/\sqrt{2})(yx-xy)$.

For $n=3$ the only non-trivial case is still $k=1$. We have $V(2,1;0) = xV(1,1;0) + \mathcal{P}V(2,0)$, so as basis elements we have $x(yx-xy)$ and $yxx - \frac{1}{2}xYx^2 = yxx - \frac{1}{2}x(yx+xy)$. These are orthogonal but must still be normalized.

For $n=4$ we must consider $V(3,1;0) = xV(2,1;0) + \mathcal{P}V(3,0)$ whose canonical orthogonal basis is $xx(yx-xy)$, $x(yxx-\frac{1}{2}x(yx+xy))$, $yxxx - \frac{1}{3}x(yxx+xyx+xxy)$, and $V(2,2;0) = \mathcal{P}V(2,1;0)$ whose basis is $yx(yx-xy) - xYx(yx-xy) = yx(yx-xy) - xy(yx-xy) = (yx-xy)^2$ and (after simplification) $yyxx + xxyy - \frac{1}{2}(yx+xy)^2$.

By induction one finds that up to normalization, the canonical basis of $V(n-1,1;0)$ consists of all $x^i(yx^{n-i-1} - 1/(n-i-1)xYx^{n-i-1})$, $i = 0,1,\ldots,n-2$. Using this one can compute inductively the bases for all $V(n-2,2;0)$ since $V(n-2,2;0) = \mathcal{P}V(n-2,1;0) + x\mathcal{P}V(n-3,1;0) + \ldots + x^{n-4}\mathcal{P}V(2,1;0)$, but the calculations become increasingly difficult. (The same canonical bases can also be obtained by a Gram–Schmidt process which we omit.)

The canonical basis elements of $V(n-k,k;0)$ are naturally indexed by the standard Young tableaux corresponding to the (non-increasing) partition $(n-k,k)$ of n as follows: Each basis element may be viewed as the result of applying to 1 a sequence of n operators each of which is either $\mathcal{P}$ or multiplication by x, with the conditions that $\mathcal{P}$ appears exactly k times and that at any point in the sequence (reading from the right, since operators are written on the left) the number of times $\mathcal{P}$ has been applied can not exceed the number of times that we have multiplied by x. These will be called *generating sequences.* Clearly the right-most operator must be multiplication by x. Construct the corresponding Young tableaux as follows: Place the integer i in the first row of the (two-rowed) diagram, to the immediate right of all other integers already entered, if the i-th operator is multiplication by x and place it in the second row if it is $\mathcal{P}$. The monomials spanning $V(n-k,k)$ have a natural lexicographic order but we will reverse this, putting y before x. The generating sequences in $\mathcal{P}$ and x will similarly be ordered by putting $\mathcal{P}$ before x.

Our construction of the canonical basis is extended to all of V^n as follows: Set $V(n-k,k;r) = Y^r V(n-k+r, k-r; 0)$, $r \leq k$, where we now allow $k > n/2$ as long as $k - r \leq n/2$. Then V^n is the orthogonal direct sum of all the non-zero $V(n-k,k;r)$. Moreover, $Y^r \mid V(n-k+r,k-r;0)$ multiplies the norm of every element by a fixed non-zero constant as long as $r \leq 2k - n$. It therefore is an isomorphism and preserves orthogonality (although Y^r generally does not). We therefore obtain an orthogonal basis for $V(n-k,k;r)$ by transporting that of $V(n-k+r,k-r;0)$.

The summands of $V(n-k,k)$ are ordered by the value of r, with $V(n-k,k;0)$ the first summand and $V(n-k,k;k)$ the last. Ordering the canonical basis elements of $V(n-k,k;r) = Y^r V(n-k+r,k-r;0)$ according to their preimages under Y^r linearly orders the entire canonical basis of $V(n-k.k)$. (This outline actually inverts the proof of the basic facts about V^n, since it is the inductive construction of the canonical basis that is used to prove the other assertions.)

It is important to note that our "canonical" basis is actually not completely canonical. We have embedded $\mathcal{S}_{n-1}$ in $\mathcal{S}_n$ as the subgroup leaving 1 fixed, but one could equally well have chosen to leave n fixed. The canonical basis obtained this way can be obtained from that just constructed by reversing the order of multiplication in the tensor algebra of V. We will call this the "reverse" canonical basis.

2. The quantized canonical basis

To quantize the foregoing construction of canonical modules and bases, instead of coefficients in $\mathbb{Z}[1/n!]\mathcal{S}_n$ we take an extension of $\mathbb{Z}[q,q^{-1}]$, and replace the operation of $\mathcal{S}_n$ with that of the *Hecke algebra* $\mathcal{H}_n = \mathcal{H}_n(q)$. This algebra is a deformation of the group algebra of the symmetric group $\mathcal{S}_n$: Let q be a formal variable and set $\mathbb{Z}_q = \mathbb{Z}[q,q^{-1}]$. Then $\mathcal{H}_n = \mathcal{H}_n(q)$ is a free module of rank $n!$ over $\mathbb{Z}_q$ with basis elements T_w indexed by the elements of $\mathcal{S}_n$ and with multiplication defined as follows. Let s_i, $i = 1, \dots, n-1$ be the transposition $(i, i+1) \in \mathcal{S}_n$ interchanging successive integers; these generate $\mathcal{S}_n$. The *length* $\ell(w)$ of an arbitrary $w \in \mathcal{S}_n$ is the number of factors in a shortest expression of w as a product of the generators s_i. If s is any of these "adjacent interchanges" then in $\mathcal{H}_n$ set

(1) $T_s T_w = T_{sw}$ if $\ell(sw) > \ell(w)$, and we set
(2) $T_s^2 = (q - q^{-1})T_s + 1$.

This is sufficient to define the multiplication completely since it implies that $T_s T_w = (q - q^{-1})T_w + T_{sw}$ when $\ell(sw) < \ell(w)$. Setting $q = 1 + \hbar$ one sees that $\mathcal{H}_n(1 + \hbar)$ is a deformation of $\mathbb{Z}\mathcal{S}_n$.

For any positive or negative integer i, set $i_q := (1 - q^i)/(1 - q)$, and similarly $i_{q^2} := (1-q^{2i})/(1-q^2)$. Set $n_{q^2}! := n_{q^2}(n-1)_{q^2} \cdots 2_{q^2}$, and $\mathbb{Z}_{q,n} := \mathbb{Z}_q[1/n_{q^2}!] = \mathbb{Z}[q, q^{-1}, 1/n_{q^2}!]$. If the coefficients of $\mathcal{H}_n$ are extended to $\mathbb{Z}_{q,n}$ then $\mathcal{H}_n$ becomes a direct sum of matrix algebras. There is, moreover, a canonical correspondence between the matrix blocks of $\mathbb{Z}_n\mathcal{S}_n$ and those of $\mathbb{Z}_{q,n}\mathcal{H}_n$, and corresponding blocks have the same dimension. The existence of canonical bases will give a canonical isomorphism between them. (The assertions about $\mathbb{Z}_n\mathcal{S}_n$ are recaptured in the 'quasi-classical' limit by letting $q \to 1$.) It follows in particular that if we view q as a complex number then there is an isomorphism $\mathcal{H}_n(q) \to \mathbb{C}\mathcal{S}_n$ as long as q is not a bad root of unity. For this condition insures that none of the 'q-numbers' i_{q^2} which must be inverted can vanish. The "Artin presentation" of $\mathcal{S}_n$ (probably due to Serre, but so named by him) is given by generators s_i, $i = 1, \dots, n-1$ representing the transposition $(i, i+1)$ with the following relations:

$$\begin{aligned} s_i s_{i+1} s_i &= s_{i+1} s_i s_{i+1} \\ s_i s_j &= s_j s_i \quad \text{if} \quad |i-j| > 1 \\ s_i^2 &= 1 \end{aligned}$$

Setting $T_i = T_{s_i}$, it follows that we get an equivalent definition of the Hecke algebra $\mathcal{H}_n$ by requiring that the generators T_i satisfy the *braid* and *Hecke* relations

$$\begin{aligned} T_i T_{i+1} T_i &= T_{i+1} T_i T_{i+1} \\ T_i T_j &= T_j T_i \quad \text{if} \quad |i-j| > 1 \\ T_i^2 &= (q - q^{-1})T_i + 1 \end{aligned}$$

(For a short proof of the equivalence cf. [5]. Many authors, e.g., [7] simply adopt the second definition.)

Now let V be a free module of rank d over $\mathbb{Z}_q = \mathbb{Z}[q, q^{-1}]$. Any $d^2 \times d^2$ quantum Yang–Baxter (qYB) matrix quantizing the algebra of polynomial functions on the affine Lie group Sl_d can now be used to define a representation of $\mathcal{H}_n$ on V^n. We will use the simplest one,

$$R = \sum_{\substack{i \neq j \\ i,j=1}}^{d} e_{ii} \otimes e_{jj} + q \sum_{i=1}^{d} e_{ii} \otimes e_{ii} + (q - q^{-1}) \sum_{1 \leq j < i \leq n} e_{ij} \otimes e_{ji}.$$

This has coefficients in $\mathbb{Z}_q$. Let the basis of V be $x_1, \ldots, x_d$, so that of V^2 consists of the $x_i x_j (= x_i \otimes x_j)$ in lexicographic order. Then $(e_{ij} \otimes e_{kl}) x_r x_s = x_i x_k$ if $j = r$, $l = s$ and 0 otherwise. Let (12) operate as the interchange of tensor factors in V^2; as a matrix, we have $(12) = \sum_{i,j} e_{ij} \otimes e_{ji}$. Set

$$\begin{aligned} \overline{R} &= (12)R \\ &= \sum_{\substack{i \neq j \\ i,j=1}}^{d} e_{ji} \otimes e_{ij} + q \sum_{i=1}^{d} e_{ii} \otimes e_{ii} + (q - q^{-1}) \sum_{1 \leq j < i \leq n} e_{jj} \otimes e_{ii}. \end{aligned}$$

This is symmetric. (The transpose of $e_{ij} \otimes e_{kl}$ is $e_{ji} \otimes e_{lk}$, the pair (i, k) being the row index and (j, l) the column index.) It is also *balanced*, i.e., in every term the sum of the column indices equals the sum of the row indices and it is therefore a direct sum of matrices in each of which these "*weights*" are constant. (This concept is meaningful for an arbitrary tensor power of a matric algebra; for the first power, a balanced matrix is simply diagonal.) Let $\overline{R}_i$, $i = 1, \ldots, n-1$ denote the operation of $\overline{R}$ in tensor factors $i, i+1$ of V^n. These matrices satisfy the braid relations, cf. [2]:

$$\begin{aligned} \overline{R}_i \overline{R}_{i+1} \overline{R}_i &= \overline{R}_{i+1} \overline{R}_i \overline{R}_{i+1} \\ \overline{R}_i \overline{R}_j &= \overline{R}_j \overline{R}_i \quad \text{if} \quad |i - j| > 1. \end{aligned}$$

Sending T_i to $\overline{R}_i$ therefore induces a representation of $\mathcal{H}_n$ on V^n. Note that $\overline{R}$ is self-adjoint – its matrix is symmetric relative to the standard basis of V^2 consisting of the $x_i x_j$ in lexicographic order – and therefore so are all the $\overline{R}_i$. Here, for example, is the operation of $\overline{R}$ on V^2 in the case where $d = 2$:

$$\begin{aligned} \overline{R}x^2 &= qx^2 \\ \overline{R}xy &= (q - q^{-1})xy + yx \\ \overline{R}yx &= xy \\ \overline{R}y^2 &= qy^2. \end{aligned}$$

There is a corresponding quantization of the universal enveloping algebra $U_q sl_2$ which now operates on V^n and commutes with the operation of the Hecke algebra on V^n: The unquantized Usl_2 is a Hopf algebra with primitive generators H, X, Y. The Drinfel'd–Jimbo quantization (cf. [1, 6]) replaces H by a pair of invertible group-like generators K, K^{-1} (i.e., $\Delta K = K \otimes K$, $\Delta K^{-1} = K^{-1} \otimes K^{-1}$). We want K to act on the tensor algebra of V formally like q^H, i.e., as the automorphism sending x to qx and y to $q^{-1}y$, which forces the commutation relations $KX = q^2XK$, $KY = q^{-2}YK$. The remaining multiplication and comultiplication rules are given by

$$qXY - q^{-1}YX = (q^{-1} - q)^{-1}(1 - K^2)$$
$$\Delta X = X \otimes 1 + K \otimes X, \qquad \Delta Y = Y \otimes 1 + K \otimes Y$$

(For a full definition one must add the q-analogues of the Serre relations giving, in particular, the nilpotence of $\operatorname{ad} X$ and $\operatorname{ad} Y$, but we do not need these here, nor shall we need the quantized antipode.) The right side of the first relation, which has the correct quasi-classical limit as $q \to 1$, namely H, may seem to pose a problem since we do not assume that $q - q^{-1}$ is invertible. We therefore take the left side, which is well-defined, as an additional generator denoted simply by H. This quantized H should be distinguished from the original unquantized one. Both K and the quantized H are well-defined operators on V^n.

The operations of X, Y on V itself are the same as before quantization, that is, $Xx = 0$, $Xy = x$, $Yx = y$, $Yy = 0$. However, the new comultiplication (which defines the tensor product of modules, in particular extending the operation of U_q to V^n for every n) now specifies that for homogeneous elements α, $\beta \in V^n$ we have

$$X(\alpha\beta) = (X\alpha)\beta + q^{|\alpha|}\alpha X\beta \quad Y(\alpha\beta) = (Y\alpha)\beta + q^{|\alpha|}\alpha Y\beta.$$

From the definitions one can also readily deduce that

$$H\alpha = q|\alpha|_{q^2}\alpha.$$

It follows, in particular, that if $X\alpha = 0$ then $XY\alpha = |\alpha|_{q^2}\alpha$. Most importantly, as the reader should check, the operations of U_q so defined on V^n commute with those of the $\overline{R}_i$ and therefore are $\mathcal{H}$-module morphisms.

The submodule $\ker X$ of V^n is homogeneous, i.e., the direct sum of its components in each of the $V(n-i,i)$. As before, write $\ker X|V(n-i,i) = V(n-i,i;0)$. It is clear that if $\alpha \in V(n-i-1,i;0)$ then $x\alpha \in V(n-i,i;0)$. Now define the quantized $\mathcal{P} : V(n-i,i-1;0) \to V(n-i,i:0)$ by $\mathcal{P}\beta = y\beta - (q|\beta|_{q^2})^{-1}xY\beta$ for $|\beta| \neq 0$ and $\mathcal{P}\beta = 0$ if $|\beta| = 0$. As in the unquantized case, if $n-2i \geq 0$ then $V(n-i,i;0)$ is the orthogonal direct sum of $xV(n-i-1,i;0)$ and $\mathcal{P}V(n-i,i-1;0)$, where the first summand appears only if $n-2i > 0$. With this we can construct the canonical orthogonal basis for $V(n-i,i;0)$ precisely as in the unquantized case.

Here are examples analogous to those in the preceding section. As before, $V(n,0;0) = V(n,0)$ has the single generator x^n. Note that $Yx^n = yx^{n-1} + qxyx^{n-2} + \ldots + q^{n-1}x^{n-1}y$.

The only non-trivial case with $n = 2$ is $V(1,1;0)$, which is spanned by the single element $\mathcal{P}x = yx - q^{-1}xy$. The (square) norm of this element is $1 + q^{-2} = q^{-2}2_{q^2}$.

For $n = 3$ we have $V(2,1;0) = xV(1,1;0) + \mathcal{P}V(2,0)$, so the basis elements are $x(yx - q^{-1}xy)$ and $yxx - (q2_{q^2})^{-1}xYx^2 = yxx - (q2_{q^2})^{-1}(xyx + qxxy)$ whose norm can be written as $q^{-2}3_{q^2}/2_{q^2}$. (It is always the case that this is a unit in the coefficient ring.)

For $n = 4$ we must consider $V(3,1;0) = xV(2,1;0) + \mathcal{P}V(3,0)$ whose canonical orthogonal basis is $xx(yx - q^{-1}xy)$, $x(yxx - (q2_{q^2})^{-1}x(yx + qxy))$, $yxxx - (q3_{q^2})^{-1}x(yxx + qxyx + q^2xxy)$, and $V(2,2;0) = \mathcal{P}V(2,1;0)$ whose basis (after simplification) is $(yx - q^{-2}xy)^2$, $yyxx + q^{-2}xxyy - (q2_{q^2})^{-1}(yx + qxy)^2$.

By induction one finds that up to normalization, the canonical basis of $V(n-1,1;0)$ consists of all $x^i(yx^{n-i-1} - (q(n-i-1)_{q^2})^{-1}xYx^{n-i-1})$, $i = 0,1,\ldots,n-2$. Set $V(n-i,i;r) = Y^rV(n-i+r,i-r;0)$ for $r = 0,\ldots,i$. (They vanish for larger r). These are all $\mathcal{H}_n$ submodules and have the following properties.

If $0 \le r \le n-2i$ then $Y : V(n-i-r,i+r;r) \to V(n-i-r-1,i+r+1;r+1)$ and $X : V(n-i-r-1,i+r+1;r+1) \to V(n-i-r,i+r;r)$ are isomorphisms preserving orthogonality and the compositions XY and YX are multiplication by invertible constants. The canonical orthogonal basis for $V(n-i,i;r)$ is obtained by transporting that of $V(n-i+r,i-r;0)$ by Y^r. (Again, $Y^rV(n-i,i;0) = 0$ for $r > n-2i$, so $V(n-i-r,i+r;r) = 0$ for $r > n-2i$, and $V(n-i,i;0) = 0$ for $n-i < i$.) Every simple $\mathcal{H}_n$-submodule of V^n is isomorphic to exactly one of the $V(n-i,i;0)$. Each of the latter has a canonical orthogonal basis indexed by the two-rowed Young diagrams with rows of length $n-i$ and i, and $\mathcal{H}_n$ acts on each as its full ring of linear endomorphisms. units. Also, $V(n-i,i)$ is the orthogonal direct sum of its submodules $V(n-i,i;r)$, each of which is simple.

One obtains the quantized reverse canonical basis by passing to the opposite of the tensor algebra of V and also replacing q by q^{-1}. Note that this also replaces the qYB matrix by its inverse.

All eigenvalues of the $d^2 \times d^2$ qYB matrix $\overline{R}$ are equal to q or $-q^{-1}$, the former occurring with multiplicity $d(d+1)/2$ and the latter with multiplicity $d(d-1)/2$. It follows that

$$\overline{T} = \frac{2}{q+q^{-1}}\overline{R} - \frac{q-q^{-1}}{q+q^{-1}}I$$

has eigenvalues $+1$ and -1, likewise with multiplicities $d(d+1)/2$ and $d(d-1)/2$, respectively, and therefore is conjugate to the transposition $(1,2)$. Setting $q = \sec\hbar - \tan\hbar$ one has $\overline{T} = \cos\hbar \cdot \overline{R} + \sin\hbar \cdot I$. Letting $\overline{T}_i$, $i = 1, \ldots, n-1$ denote $\overline{T}$ operating on factors $(i, i+1)$ of V^n, these satisfy $\overline{T}_i^2 = 1$ for all i and the modified braid relations

$$\begin{aligned} \overline{T}_i\overline{T}_{i+1}\overline{T}_i - \overline{T}_{i+1}\overline{T}_i\overline{T}_{i+1} &= \sin^2\hbar(\overline{T}_i - \overline{T}_{i+1}) \\ \overline{T}_i\overline{T}_j &= \overline{T}_j\overline{T}_i, \qquad |i-j| > 1. \end{aligned}$$

When q is not a $2m$-th root of unity with $m \leq n$ (a bad root) the Hecke algebra $\mathcal{H}_n(q)$ is isomorphic to the group algebra of the symmetric group $\mathcal{S}_n$. The group generated by the $\overline{T}_i$ has been called the quantum symmetric group in [4] and is generally infinite, although for special choices of the qYB matrix it may be isomorphic to the symmetric group. Its abstract group algebra can therefore be very large, but it is clear that the actual algebra of endomorphisms of V^n induced by its operation is the same same as that of the the Hecke algebra.

3. Intersection cardinality matrices and quantized intersection cardinalities

Since we must now extend coefficients to $\mathbb{Z}_{q,n}[1/n!] = \mathbb{Z}[q, q^{-1}, 1/n_{q^2}!, 1/n!]$, for simplicity let them be extended to $\mathbb{C}$ with q not a bad root of unity. Then there is a canonical linear automorphism of V^n defined by sending each quantized orthonormal basis element to its corresponding unquantized one. This linear automorphism, henceforth denoted by $\gamma(q)$ or simply γ, is an orthogonal transformation carrying $\mathcal{H}_n(q)$ submodules to $\mathcal{S}_n$ submodules. In particular, it carries the module of invariants of the quantum symmetric group to that of $\mathcal{S}_n$; the former is precisely the set of common eigenvectors for the eigenvalue q of all the generators T_i of $\mathcal{H}_n$. Notice that all $V(n-k,k)$ are simultaneously $\mathcal{S}_n$ and $\mathcal{H}_n$ submodules of V^n and consequently stable under γ.

The orthogonal transformation γ of V^n induces, for every r, an orthogonal transformation $\gamma^{\otimes r}$ of $(V^n)^{\otimes r} = V^n \otimes \cdots \otimes V^n$, r times. By the **quantization** of a function $f : (V^n)^{\otimes r} \to \mathbb{C}$ we will mean just the composite $f\gamma^{\otimes r}$. (This formally defines a quantization for all functions on $(V^n)^{\otimes r}$, but we don't know its properties.) Now V^n is a permutation module for $\mathcal{S}_n$ and as such is canonically isomorphic to its own dual. Our interest here is in the case $r = 2$. In that case, $\mathcal{S}_n$ invariant functions on $V^n \otimes V^n$ are the same thing as $\mathcal{S}_n$ invariant elements of $V^n \otimes V^{*n}$, i.e., endomorphisms of V^n which commute with the operation of $\mathcal{S}_n$. Quantization is now simply conjugation by γ. Let U again be the linearly ordered set of n elements and N be a family of subsets of U stable under the permutations of U. An *intersection cardinality matrix* (ICM) will be one whose rows and columns are indexed by sets $S, T \in N$ and whose entries depend only on the intersection cardinality $\#(S \cap T)$. Consider now the case $N = N(k)$, the family of k-element subsets of U. We may suppose that $k \leq n/2$.

The $\binom{n}{k}$-dimensional vector space $\mathbb{C}N(k)$ is an $\mathcal{S}_n$ module isomorphic to $V(k, n-k)$ the decomposition of which into simple submodules we have already described: $V(n-k,k) = \oplus_{i=0}^{k} V(n-k,k;i)$, where $V(n-k,k;i) \cong V(n-k+i, k-i; 0)$. No two of these simple summands are isomorphic so $V(n-k,k)$ has *"simple spectrum"*, i.e., its isotypical components are all simple. It follows from Schur's Lemma that its endomorphisms commute. These have been studied, cf [3]: They form a $k+1$-dimensional linear space, the general element of which has exactly $k+1$ distinct eigenvalues, and the eigenspaces are just the $V(n-k,k;i)$. The dimensions of these are $\binom{n}{r} - \binom{n}{r-1}$, $r = 0, \ldots, k$ where for $i = 0$ the dimension is 1.

In matrix terms, if we consider the $\binom{n}{k} \times \binom{n}{k}$ matrices with rows and columns indexed by the k-element subsets of U, then the necessary and sufficient condition that such a matrix be an $\mathcal{S}_n$ endomorphism of $\mathbb{C}N(k)$ is that its entry in the S,T place depend only on $\#(S \cap T)$. The space of these ICM's has a basis consisting of the matrices J_r, $r = 0, \ldots, k$ whose S,T entry is zero unless S and T have exactly r elements in common, in which case it is 1. Since the J_r are real, symmetric, and commute, they can be simultaneously diagonalized, and hence so can all the ICM's. Moreover, if an ICM has integer entries then its eigenvalues are integers. Denoting the eigenvalue of J_r on $V(n-k,k;i)$ by $\lambda_{r,i}(n,k)$ we have

$$\begin{aligned}\lambda_{r,i}(n,k) &= \textstyle\sum_{r_1+r_2=r} \lambda_{r_1,i}(n-2k+2i, i) \times \lambda_{r_2,k-i}(2k-2i, k-i) \\ &= \textstyle\sum_{r_1+r_2=r} \binom{i}{r_1}\binom{n-2k+i}{i-r_1} \times (-1)^{k-i-r_2}\binom{k-i}{r_2}, \quad 0 \leq r_1 \leq i, 0 \leq r_2 \leq k-i.\end{aligned}$$

The $k+1$ eigenvalues are ordered according to the order of their eigenspaces $V(n-k,k;0), V(n-k,k,1), \ldots, V(n-k,k;k)$.

Denote by $\Gamma(q)$ the matrix of γ relative to the basis $\{N(k)\}$ of $\mathbb{C}N(k)$, and for any intersection cardinality matrix A let $A(q) := \Gamma(q)^{-1} A \Gamma(q)$ denote its quantization. Then A and $A(q)$ clearly have the same eigenvalues, so the foregoing λ clearly are also the eigenvalues of the $J_r(q) := \Gamma(q)^{-1} J_r \Gamma(q)$.

The computation of $A(q)$ can be simplified by writing $\gamma = \phi\psi$ where ϕ is the orthogonal transformation carrying the monomial basis of $N(k)$ (in the reverse lexicographic order which we have adopted) to the unquantized canonical orthonormal basis, and ψ is the orthogonal transformation carrying the quantized orthonormal basis to the monomial basis. The matrices of ϕ, ψ will be denoted Φ, Ψ, respectively. But the (unquantized) canonical orthonormal basis vectors, which are the columns of Φ, are here just the eigenvectors of any ICM. Therefore, denoting by Λ_A the diagonal matrix whose diagonal elements are the eigenvalues of A (in order), we have $\Phi^{-1}A\Phi = \Lambda_A$. Thus, if we know the eigenvalues of A then we have $A(q) = \Psi^{-1}\Lambda_A\Psi$. Of course, Φ and Ψ are both orthogonal, so in particular, $\Psi^{-1} = \Psi^t$. The rows of Ψ are the quantized orthonormal basis vectors. If we have computed a canonical quantized orthogonal but unnormalized basis, then taking these vectors as the rows of a matrix Ψ_0 we can write $\Psi = D\Psi_0$ where D is a diagonal matrix whose diagonal entries are the inverses of the lengths of the rows of Ψ_0. It follows that $A(q) = \Psi_0^t D^2 \Lambda_A \Psi_0$ where the diagonal entries of D^2 are the inverses of the (square) norms of the quantized but unnormalized canonical basis vectors.

By *the* intersection cardinality matrix for a family N of subsets of U we will mean that matrix whose entry in the S,T place is precisely $\#(S \cap T)$. It will be denoted by $J_\#$ and its quantization by $J_\#(q)$.

We give three examples of quantization, where the geometric configurations are, respectively, an equilateral triangle, a square, and the regular octahedron (dual of a cube, having vertices at the centers of the faces of the cube).

1. The simplest example is the quantization of the intersection cardinalities for the family $N(1)$ of one-element subsets of U. Here is the result for $U = \{a, b, c\}$ (in that order), represented respectively by yxx, xyx, xxy. (In this example $n = 3$; the result for arbitrary n will be obvious.) The geometric configuration is the equilateral triangle with vertices at $(1,0,0)$, $(0,1,0)$, $(0,0,1)$ (viewed as column vectors). Here $J_\#$ (which is identical with J_1) is just the identity, and therefore so is $J_\#(q)$. On the other hand, the "disjointness matrix"

$$J_0 = \begin{pmatrix} 0 & 1 & 1 \\ 1 & 0 & 1 \\ 1 & 1 & 0 \end{pmatrix}$$

whose S,T entry is 1 if S and T are disjoint and 0 otherwise, has two distinct eigenvalues. (It is actually somewhat easier to quantize $J_0 + J_1 = J_0 + I$ all of whose entries are equal to 1; to obtain the quantization of J_0 just subtract the identity.)

$$J_0(q) = \frac{3}{3_{q^2}} \begin{pmatrix} 1 & q & q^2 \\ q & q^2 & q^3 \\ q^2 & q^3 & q^4 \end{pmatrix} - I$$

In this case the Hecke algebra has two generators, T_1, T_2 which operate as $\overline{R}_{12}$, $\overline{R}_{23}$, respectively, but one must restrict them to the module spanned by yxx, xyx, xxy. In matrix form relative to this basis we have

$$T_1 = \begin{pmatrix} 0 & 1 & 0 \\ 1 & q-q^{-1} & 0 \\ 0 & 0 & q \end{pmatrix}, \quad T_2 = \begin{pmatrix} q & 0 & 0 \\ 0 & 0 & 1 \\ 0 & 1 & q-q^{-1} \end{pmatrix}.$$

It is easy to check that these commute with $J_0(q)$.

In this example one can give a complete description of the quantum symmetric group, which is generated by $\overline{T}_1$ and $\overline{T}_2$, and of its operation. For setting $\rho := \overline{T}_1$, $\sigma := \overline{T}_1\overline{T}_2$, these also are generators, $\rho^2 = 1$, $\rho\sigma\rho = \sigma^{-1}$, and σ is of infinite order for generic q. Thus the quantum symmetric group, which we will denote by $\mathcal{S}_n(q)$, is here a dihedral group, and is of infinite order except for special values of q. As before, set $q = \sec\hbar - \tan\hbar$. Abbreviating $\sin\hbar, \cos\hbar$ for the moment to S, C, respectively, the representations of ρ and σ relative to the basis yxx, xyx, xxy (representing the transposes of $(1,0,0)$, $(0,1,0)$, $(0,0,1)$) are

$$\rho = \begin{pmatrix} S & C & 0 \\ C & -S & 0 \\ 0 & 0 & 1 \end{pmatrix},$$

$$\sigma = \rho \cdot \begin{pmatrix} 1 & 0 & 0 \\ 0 & S & C \\ 0 & C & -S \end{pmatrix} = \begin{pmatrix} S & SC & C^2 \\ C & -S^2 & -SC \\ 0 & C & -S \end{pmatrix}.$$

The axis of the proper orthogonal transformation σ runs through (the transpose of) $(1+S, C, 1-S)$ which is proportional to $(1, q, q^2)$. The cosine of the angle of rotation is $-(1+S^2)/2$ so if $\hbar$ is real and small the angle is slightly greater than $2\pi/3$ regardless of the sign of $\hbar$, and generally is incommensurable with π. Thus, under the operation of σ, which cyclically renumbers the vertices, the triangle spanned by $(1,0,0)$, $(0,1,0)$, $(0,0,1)$ both "wobbles" and precesses.

For generic q, the closure of the set of images in the special orthogonal group SO_3 of the powers of σ is the circle group $\mathbb{T}$ of rotations around the axis through $(1, q, q^2)$, so the closure of the image of all $\mathcal{S}_n(q)$ is the semidirect product of the two-element group generated by ρ and $\mathbb{T}$. Since the generators commute with the quantized ICM's, the same is true of the entire semidirect product. This commutativity, however, is a formal property and therefore must hold for all q.

If the interpretation of quantized ICM's as quantized forces makes any sense then perhaps one should also allow a change in their directions. In the classical case, the tendency of the three equal masses at the vertices of our equilateral triangle to collapse towards its center can be precisely counterbalanced if the whole configuration rotates with a suitable angular velocity around the axis through $(1,1,1)$. The configuration then remains stable. In the quantized case, we don't know if one should expect a similar assertion for rotation around the axis through $(1, q, q^2)$. (The directions of the forces may influence the answer.) In this example, note finally, that we can generate a triangular lattice by repeated reflection of the original equilateral triangle across its sides. It seems like a daunting problem to see how the entire lattice would behave.

2. The case where U is a two-element set $\{a,b\}$ and $N = 2^U$ is not completely trivial. Writing simply ab for $\{a,b\}$, and similarly omitting braces for other subsets, the subsets are ab, a, b, $\emptyset$, represented by yy, yx, xy, xx. The geometric configuration is the unit square with vertices (in order) at $(1,1)$, $(1,0)$, $(0,1)$, $(0,0)$. In this example the ICM's no longer commute. By contrast, when the family N of subsets of U is of the form $N(k)$ the fact that the ICM's commute implies that all are polynomials in any single one, call it A, which has the maximum number, $k+1$, of distinct eigenvalues. It is then only necessary, in a sense, to quantize this A, which can be done once one knows its eigenvalues and the factor ψ of γ. Since that fails here, we give the quantizations of both $J_\#$ and of the matrix G ("gravity") whose off-diagonal entries are just the inverse squares of the distances between the points of the square configuration, and whose diagonal entries are 0. To shorten the notation, set $t := \sqrt{2/2_{q^2}}$, so $t = 1$ when $q = 1$. We then have

$$J_\# = \begin{pmatrix} 2 & 1 & 1 & 0 \\ 1 & 1 & 0 & 0 \\ 1 & 0 & 1 & 0 \\ 0 & 0 & 0 & 0 \end{pmatrix},$$

$$G = \begin{pmatrix} 0 & 1 & 1 & 1/2 \\ 1 & 0 & 1/2 & 1 \\ 1 & 1/2 & 0 & 1 \\ 1/2 & 1 & 1 & 0 \end{pmatrix},$$

$$\gamma = \begin{pmatrix} 1 & 0 & 0 & 0 \\ 0 & (q+1)/\sqrt{2\cdot 2_{q^2}} & (q-1)/\sqrt{2\cdot 2_{q^2}} & 0 \\ 0 & (-q+1)/\sqrt{2\cdot 2_{q^2}} & (q+1)/\sqrt{2\cdot 2_{q^2}} & 0 \\ 0 & 0 & 0 & 1 \end{pmatrix},$$

whence

$$J_\#(q) = \begin{pmatrix} 2 & t & qt & 0 \\ t & 1 & 0 & 0 \\ qt & 0 & 1 & 0 \\ 0 & 0 & 0 & 0 \end{pmatrix},$$

$$G(q) = \begin{pmatrix} 0 & t & qt & 1/2 \\ t & (1-q^2)/2\cdot 2_{q^2} & q/2\cdot 2_{q^2} & t \\ qt & q/2\cdot 2_{q^2} & (q^2-1)/2\cdot 2_{q^2} & qt \\ 1/2 & t & qt & 0 \end{pmatrix}.$$

In $G(q)$ the diagonal elements are no longer zero, indicating that after quantization the masses may exert some force on themselves, but in an unknown direction. It might be better, therefore, to let the original diagonal elements of G be ∞ so that those of $G(q)$ are the same. On the other hand, the unique non-identity element $\overline{T}$ of the quantum symmetric group operates on the unit square leaving $(0,0)$ and $(1,1)$ fixed but reflecting $(1,0)$ and $(0,1)$ across the line through $(1,q)$. It thus distorts the square, so there may be some reasonable interpretation for the new forces.

3. As final example we take $\#U = 4$ and $N = N(2)$. Let $U = \{a,b,c,d\}$. The six elements of $N(2)$ are ab, ac, ad, bc, bd, cd corresponding to the monomials $yyxx$, $yxyx$, $yxxy$, $xyyx$, $xyxy$, $xxyy$, given in the order which we have adopted, and the geometric configuration of the corresponding six points in 3-space consists of the vertices of a regular octahedron. The unnormalized quantized canonical basis is $\mathcal{PP}xx$, $\mathcal{P}x\mathcal{P}x$, $Y(\mathcal{P}xxx)$, $Y(x\mathcal{P}xx)$, $Y(xx\mathcal{P}x)$, $YY(xxxx)$. It is convenient to remove a common factor of $q+q^{-1}$ from the terms of $YY(xxxx)$ which will only change the normalizing factor in D. With this we have

$$\Psi_0 = \begin{pmatrix} 1 & -q^{-1}2_{q^2}^{-1} & -2_{q^2}^{-1} & -2_{q^2}^{-1} & -q2_{q^2}^{-1} & q^{-2} \\ 0 & 1 & -q^{-1} & -q^{-1} & q^{-2} & 0 \\ q^2 & q^3 & q^4 & -1 & -q & -q^2 \\ 1 & -q^{-1}2_{q^2}^{-1} & -2_{q^2}^{-1} & q^2 2_{q^2}^{-1} & q^3 2_{q^2}^{-1} & -1 \\ 0 & 1 & -q^{-1} & q & -1 & 0 \\ 1 & q & q^2 & q^2 & q^3 & q^4 \end{pmatrix}.$$

The normalizing matrix is

$$D^2 = \operatorname{diag}\left(\frac{q^4}{3_{q^2}}, \frac{q^4}{2_{q^2}^2}, \frac{2_{q^2}}{3_{q^2}4_{q^2}}, \frac{q^2}{3_{q^2}}, \frac{q^2}{2_{q^2}^2}, \frac{2_{q^2}}{3_{q^2}4_{q^2}}\right).$$

Some of these expressions may be simpler if rewritten directly in terms of q, for example $2_{q^2}/4_{q^2} = 1/(1+q^4)$, but as written here the denominators contain only q-numbers which are factors of $n_{q^2}!$. The unquantized intersection cardinality matrix of the sets $\{ab, ac, ad, bc, bd, cd\}$ is

$$J_\# = \begin{pmatrix} 2 & 1 & 1 & 1 & 1 & 0 \\ 1 & 2 & 1 & 1 & 0 & 1 \\ 1 & 1 & 2 & 0 & 1 & 1 \\ 1 & 1 & 0 & 2 & 1 & 1 \\ 1 & 0 & 1 & 1 & 2 & 1 \\ 0 & 1 & 1 & 1 & 1 & 2 \end{pmatrix}$$

The eigenvalues of $J_\#$ in our ordering of the eigenvectors are $0,0,2,2,2,6$. In addition to being symmetric, the matrix is unaltered if we replace every set in $N(2)$ by its complement, or in monomial terms, interchange x and

y. (When this is done in the quantized case we must also replace q by q^{-1}.) To compress the notation, set $a := 2 + 3q^2 + q^6$, $\tilde{a} := 1 + 3q^4 + 2q^6$, $b := 1 + 3q^2 + q^4 + q^6$, $\tilde{b} := 1 + q^2 + 3q^4 + q^6$, $c := -2_{q^2} q^2 (1 - q^2)^2$. With this, the quantized matrix is

$$J_{\#}(q) = 2I + \tfrac{2}{3_{q^2} 4_{q^2}} \times$$

$$\begin{pmatrix} 2_{q^2}(1-q^4)(2+q^4) & qa & q^2 a & q^2 a & q^3 a & c \\ qa & (q^2 - q^4)b & q^3 a & q^3 a & c & q\tilde{a} \\ q^2 a & q^3 a & c & c & q\tilde{a} & q^2\tilde{a} \\ q^2 a & q^3 a & c & c & q\tilde{a} & q^2\tilde{a} \\ q^3 a & c & q\tilde{a} & q\tilde{a} & (q^4 - q^2)\tilde{b} & q^3\tilde{a} \\ c & q\tilde{a} & q^2\tilde{a} & q^2\tilde{a} & q^3\tilde{a} & 2_{q^2}(q^4 - 1)(1 + 2q^4) \end{pmatrix}$$

Notice that wherever the unquantized matrix has 0 we now have $c = -2_{q^2}(q^2 - q^4)^2$, which is negative for real q. Relative to the basis $yyxx$, $yxyx$, $yxxy$, $xyyx$, $xyxy$, $xxyy$, the operation of T_1 is given by the matrix

$$\begin{pmatrix} q & 0 & 0 & 0 & 0 & 0 \\ 0 & 0 & 0 & 1 & 0 & 0 \\ 0 & 0 & 0 & 0 & 1 & 0 \\ 0 & 1 & 0 & q - q^{-1} & 0 & 0 \\ 0 & 0 & 1 & 0 & q - q^{-1} & 0 \\ 0 & 0 & 0 & 0 & 0 & q \end{pmatrix}.$$

It is easy to check that this commutes with $J_{\#}(q)$, as do the other generators of the Hecke algebra. It is an easy matter (which we omit) to give the quantization of the inverse-square law, either directly or by observing that the matrix must be a polynomial in $J_{\#}(q)$, since that already has the maximum number of distinct eigenvalues.

This ends the examples. One may note that in all cases $J_{\#}(q)$ is unchanged if one passes to the opposite of the tensor algebra and replaces q by q^{-1}. This is not accidental. In fact, one can show inductively from the construction of the canonical bases that $J_{\#}(q)$ must commute with all generators of the Hecke algebra but the first, while if one uses the reverse canonical basis the matrix one gets must commute with all but the inverse of the last, and hence with the last also. But one can also check that ultimately one gets the same matrix, which therefore commutes with the operation of all T_i and has the indicated symmetry. (However, there seem to be some symmetries in the matrix which we can not explain.) Moreover, in the three computed examples the matrices remain finite as $q \to \infty$, or equivalently, in view of the foregoing, as $q \to 0$.

Replacing q by, say, $1 + \hbar$ makes clear that what we have solved is really a deformation problem, but the last example also shows the computational difficulties a direct attack would encounter.

References

1. V.G. Drinfel'd, Quantum groups *Proc. ICM 1986*, Ed. A.M. Gleason. Amer. Math. Soc. (Providence, 1987). Pp. 798–820.
2. L.D. Faddeev, N. Yu. Reshetikhin and L. A. Takhtajan, Quantization of Lie groups and Lie algebras. *Leningrad Math. J.* **1** 1990 193–225.
3. M. Gerstenhaber, *Intersection cardinality matrices*, submitted to Amer. Math. Monthly.
4. M. Gerstenhaber, A. Giaquinto and S. D. Schack, Quantum symmetry. *Quantum Groups*, Proceedings, Leningrad 1990. Lecture Notes in Mathematics **1510** Ed. P. P. Kulish. Springer (Berlin, New York, Paris).
5. M. Gerstenhaber and S.D. Schack, Hecke algebras, $U_q sl_n$, and the Donald–Flanigan conjecture for S_n *Trans. Amer. Math. Soc.*. To appear.
6. M. Jimbo, Introduction to the Yang–Baxter equation. *Braid Group, Knot Theory and Statistical Mechanics*, Eds. C. N. Yang and M. L. Ge. World Scientific (Singapore, New Jersey, 1989). Pp. 111–134.
7. H. Wenzl, Hecke algebras of type A_n and subfactors. *Invent. Math.* **92** 1988 349–383.

DEFORMATION QUANTIZATION AND Mp^c STRUCTURES

SIMONE GUTT
Département de Mathématiques, Université Libre de Bruxelles, Campus Plaine, CP 218, 1050 Brussels, Belgium
and
Département de Mathématiques, Université de Metz, Ile du Saulcy, F-57045 Metz Cedex 01, France

AND

JOHN RAWNSLEY
Mathematics Institute, University of Warwick, Coventry CV4 7AL, United Kingdom

Abstract. We relate the condition "$[\beta/2\pi i] - \frac{1}{2}c_1(\omega)^{\mathbb{R}}$ integral" on a symplectic manifold (M, ω) to the existence of Mp^c connections with central curvature β and discuss the appearance of this condition in several contexts relating to quantization where $\beta = i\omega/\hbar$.

1. Introduction

We give a proof of an integrality condition for the existence of Mp^c connections. This supplements the results in [7]. Proofs are provided except for standard facts concerning reduction of principal bundles to maximal compact subgroups, and properties of the Mp^c group such as Theorem 2.7 of [7] which are explicit calculations.

We give a geometrical argument to show that the existence of an Mp^c connection with a prescribed central curvature is a natural condition to impose in the context of deformation quantization when one tries to build approximate representations of a star product. This gives a geometrical interpretation of a condition which has appeared in various contexts of quantization; let us cite amongst others Czyż [2] where it first appeared in the context of Kähler manifolds, Hess [5], Robinson and Rawnsley [7] who introduced the use of Mp^c structures in geometric quantization using

D. Sternheimer et al. (eds.), Deformation Theory and Symplectic Geometry, 103–115.

arbitrary polarizations, Karasev and Maslov [6], Dazord and Patissier [3] in the context of asymptotic operator representation and Fedosov [4] who obtains this condition through a Čech cohomology argument in order for a given star product to have an asymptotic operator representation.

Acknowledgements

We would like to thank the ETH and the organizers of the meeting for the opportunity to present our results in the stimulating environment of Monte Verità. One of us (JR) would like to thank the University of Metz for its hospitality during part of this research. This work was partly supported by a European Commission research contract ERBCHRXCT940701.

2. The Mp^c group and its Lie algebra

Let (V,Ω) be a (real) symplectic vector space and $\mathrm{Sp}(V,\Omega)$ be the Lie group of all linear transformations of V preserving Ω. Let $N(V,\Omega)$ be the Lie group whose underlying manifold is $V \times \mathbb{R}$ with multiplication

$$(v_1,t_1)(v_2,t_2) = (v_1+v_2, t_1+t_2-\tfrac{1}{2}\Omega(v_1,v_2)).$$

$N(V,\Omega)$ is the *Heisenberg group* and is a two-step nilpotent Lie group. $\mathrm{Sp}(V,\Omega)$ acts as a group of automorphisms of $N(V,\Omega)$ by

$$g\cdot(v,t) = (g(v),t).$$

The center of $N(V,\Omega)$ is 1-dimensional, $\{(0,t) \mid t \in \mathbb{R}\}$, and acts by scalars, $(0,t) \mapsto e^{it/\hbar}\,\mathrm{Id}_{\mathcal{H}}$, in any irreducible representation on a Hilbert space $\mathcal{H}$ for some real constant $\hbar$. The Stone–von Neumann uniqueness theorem asserts that, up to equivalence, there is one and only one such irreducible representation for each finite value of $\hbar$. Choose and fix a positive value of $\hbar$ and a Hilbert space $\mathcal{H}$ on which the representation is realised.

By composing with an automorphism g from $\mathrm{Sp}(V,\Omega)$ we get a second representation of $N(V,\Omega)$ on $\mathcal{H}$:

$$(v,t)._g\phi = (g(v),t).\phi$$

and evidently the representation is still irreducible and has the same central parameter $\hbar$. By the uniqueness theorem there is a unitary transformation U of $\mathcal{H}$ such that

$$(g(v),t).\phi = U\left((v,t).(U^{-1}\phi)\right), \quad \forall v\in V,\ t\in\mathbb{R},\ \phi\in\mathcal{H}. \tag{1}$$

The operators U are only determined up to a scalar multiple by the corresponding elements g of $\mathrm{Sp}(V,\Omega)$, and it is not possible to make a consistent choice which respects the group multiplication. Instead we define a new group $\mathrm{Mp}^c(V,\Omega)$ whose elements are pairs (U,g) of unitary transformations U of $\mathcal{H}$ and elements g of $\mathrm{Sp}(V,\Omega)$ satisfying (1). The multiplication in $\mathrm{Mp}^c(V,\Omega)$ is $(U_1,g_1)(U_2,g_2)=(U_1U_2,g_1g_2)$ and the map

$$\sigma(U,g)=g$$

is a surjective homomorphism to $\mathrm{Sp}(V,\Omega)$. The kernel consists of all unitary multiples of the identity, so we have a central extension

$$1\longrightarrow \mathrm{U}(1)\longrightarrow \mathrm{Mp}^c(V,\Omega)\xrightarrow{\sigma}\mathrm{Sp}(V,\Omega)\longrightarrow 1 \tag{2}$$

which does not split.

Although the sequence above does not split, it is known to split on a double covering $\mathrm{Mp}(V,\Omega)$, the *metaplectic group*, and this is the kernel of a homomorphism

$$\eta\colon \mathrm{Mp}^c(V,\Omega)\longrightarrow \mathrm{U}(1)$$

whose restriction to the central $\mathrm{U}(1)$ is the squaring map.

A *positive complex structure* on (V,Ω) is a real endomorphism J with $J^2=-I$ such that

(a) $\Omega(Ju,Jv)=\Omega(u,v)$, $\forall u,\ v\in V$;
(b) $\Omega(Jv,v)>0$, $\forall v\in V,\ v\neq 0$.

If we regard V as a complex vector space via J then it has a Hermitian inner product given by

$$\langle u,v\rangle=\Omega(Ju,v)+i\Omega(u,v)$$

and its unitary group, which we denote by $\mathrm{U}(V,\Omega,J)$, is a maximal compact subgroup of $\mathrm{Sp}(V,\Omega)$. We let $\mathrm{MU}^c(V,\Omega,J)$ be the inverse image of $\mathrm{U}(V,\Omega,J)$ under σ so that (2) induces a corresponding short exact sequence

$$1\longrightarrow \mathrm{U}(1)\longrightarrow \mathrm{MU}^c(V,\Omega,J)\xrightarrow{\sigma}\mathrm{U}(V,\Omega,J)\longrightarrow 1. \tag{3}$$

$\mathrm{MU}^c(V,\Omega,J)$ is, of course, a maximal compact subgroup of $\mathrm{Mp}^c(V,\Omega)$.

Unlike the exact sequence (2), (3) does split canonically by means of a homomorphism

$$\lambda\colon \mathrm{MU}^c(V,\Omega,J)\longrightarrow \mathrm{U}(1).$$

This gives an isomorphism

$$\mathrm{MU}^c(V,\Omega,J)\xrightarrow{\sigma\times\lambda}\mathrm{U}(V,\Omega,J)\times \mathrm{U}(1).$$

In addition we have the determinant character $\mathrm{Det}\colon \mathrm{U}(V,\Omega,J) \longrightarrow \mathrm{U}(1)$ which can be composed with σ to give a character $\mathrm{Det}\circ\sigma$ of $\mathrm{MU}^c(V,\Omega,J)$. The three characters η, λ and $\mathrm{Det}\circ\sigma$ are related by

$$\eta = \lambda^2\,\mathrm{Det}\circ\sigma. \tag{4}$$

Let $\mathfrak{mp}^c(V,\Omega)$ be the Lie algebra of $\mathrm{Mp}^c(V,\Omega)$. Differentiating the exact sequence of Lie groups (2) gives an exact sequence of Lie algebras

$$0 \longrightarrow \mathfrak{u}(1) \longrightarrow \mathfrak{mp}^c(V,\Omega) \xrightarrow{\sigma} \mathfrak{sp}(V,\Omega) \longrightarrow 0. \tag{5}$$

We denote by $\eta\colon \mathfrak{mp}^c(V,\Omega) \longrightarrow \mathfrak{u}(1)$ also the differential of the group homomorphism η, and observe that $\frac{1}{2}\eta$ is a map to $\mathfrak{u}(1)$ which is the identity on the central $\mathfrak{u}(1)$ of $\mathfrak{mp}^c(V,\Omega)$. Hence (5) splits as a sequence of Lie algebras. We shall refer to the component in $\mathfrak{u}(1)$ of an element ξ of $\mathfrak{mp}^c(V,\Omega)$ as its *central component* ξ^c.

3. Mp^c structures

Fix a symplectic vector bundle (E,ω) of rank $2n$ and let (V,Ω) be a symplectic vector space of dimension $2n$. The *symplectic frame bundle* $\pi\colon \mathrm{Sp}(E,\omega)\longrightarrow M$ of (E,ω) is the bundle whose fibre at $x \in M$ consists of all symplectic isomorphisms $b\colon V\longrightarrow E_x$. By composition on the right it becomes a principal $\mathrm{Sp}(V,\Omega)$ bundle.

By an Mp^c *structure* on (E,ω) we mean a principal $\mathrm{Mp}^c(V,\Omega)$ bundle $\pi^c\colon P \longrightarrow M$ with a fibre-preserving map $\phi\colon P \longrightarrow \mathrm{Sp}(E,\omega)$ such that the group actions are compatible:

$$\phi(p\cdot g) = \phi(p)\cdot\sigma(g), \quad \forall p \in P,\ g \in \mathrm{Mp}^c(V,\Omega).$$

If P is an Mp^c structure on (E,ω) then $P(\eta)$ denotes the line bundle associated to P by the representation η of $\mathrm{Mp}^c(V,\Omega)$.

Since any principal bundle can be reduced to the maximal compact subgroup of the structure group, and since $\mathrm{U}(V,\Omega,J)$ is a subgroup of $\mathrm{MU}^c(V,\Omega,J) \subset \mathrm{Mp}^c(V,\Omega)$ it follows that every symplectic vector bundle (E,ω) admits an Mp^c structure, and that the isomorphism classes of Mp^c structures are parameterized by line bundles via the character λ after choosing a positive complex structure.

This parameterization can be made as follows: Choose and fix a fibrewise positive complex structure J on E. This can always be done and any two such positive complex structures are homotopic. Those symplectic frames which are also complex linear form a principal $\mathrm{U}(V,\Omega,J)$ bundle called the unitary frame bundle. The first Chern class of this unitary structure depends only on (E,ω) and we denote it by $c_1(E,\omega)$ or $c_1(\omega)$ when the bundle E is clear from the context. It is clearly the first Chern class of the line bundle associated to the determinant character Det of $\mathrm{U}(V, \Omega, J)$.

If P is an Mp^c structure, we can look at the subset P_J of P lying over the unitary frames. This will be a principal $\mathrm{MU}^c(V, \Omega)$ bundle. The line bundle associated to P_J by the character λ is independent of J, up to isomorphism, and we denote it by $P_J(\lambda)$. The map sending the isomorphism class of P to the isomorphism class of $P_J(\lambda)$ is the required parameterization of Mp^c structures by line bundles. From the relationship (4) between the three characters we have

$$c_1(P(\eta)) = 2c_1(P_J(\lambda)) + c_1(E, \omega).$$

4. Mp^c connections

Let (E, ω) be a symplectic vector bundle and $\pi^c\colon P \longrightarrow M$ an Mp^c structure on it with a fibre-preserving map $\phi\colon P \longrightarrow \mathrm{Sp}(E, \omega)$. Then E is associated to P via the representation σ of $\mathrm{Mp}^c(V, \Omega)$ on V. So any connection in P determines a symplectic connection in E. If $\alpha \in \Omega^1(P) \otimes \mathfrak{mp}^c(V, \Omega)$ denotes the connection form then α^c denotes its central component. This is a $\mathfrak{u}(1)$-valued 1-form, invariant under $\mathrm{Mp}^c(V, \Omega)$. Its derivative $d(\alpha^c) = (d\alpha)^c = (d\alpha + \frac{1}{2}[\alpha \wedge \alpha])^c$ is the central component of the curvature of the connection. It is thus both invariant and horizontal, and hence is the pull-back of a $\mathfrak{u}(1)$-valued 2-form β on M. We call β the *central curvature* of the connection.

Let $H^p(M, \mathbb{Z})^{\mathbb{R}}$ be the image in $H^p(M, \mathbb{R})$ of $H^p(M, \mathbb{Z})$ induced by the inclusion of coefficients. If $c \in H^p(M, \mathbb{Z})$ is an integer cohomology class we let $c^{\mathbb{R}}$ denote its image. We say a class $b \in H^p(M, \mathbb{R})$ is *integral* if it is in $H^p(M, \mathbb{Z})^{\mathbb{R}}$.

Theorem 4.1 *If P is an Mp^c structure in (E, ω) and β the central curvature of an Mp^c connection in P then*

$$[\beta/2\pi i] - \tfrac{1}{2}c_1(E, \omega)^{\mathbb{R}} \in H^p(M, \mathbb{Z})^{\mathbb{R}}.$$

PROOF Recall that if a $\mathrm{U}(1)$-connection has curvature 2-form γ then $\gamma/2\pi i$ represents the first Chern class as a real cohomology class and is also integral, $[\gamma/2\pi i] \in H^p(M, \mathbb{Z})^{\mathbb{R}}$.

Pick a positive complex structure J in (E, ω) and take the reduction P_J of P. Taking the $\mathfrak{mu}^c(V, \Omega, J)$ component of the Mp^c connection and restricting it to P_J gives an $\mathrm{MU}^c(V, \Omega, J)$ connection in P_J such that the central component of the connection form is unchanged. Hence the central curvature is also unchanged. Let γ be the curvature form of the $\mathrm{U}(1)$ connection induced in $P_J(\lambda)$. Then $[\gamma/2\pi i]$ is integral. On the other hand, η is the squaring map on the centre of $\mathrm{Mp}^c(V,\Omega)$ so the connection induced in $P(\eta)$ has curvature 2β. Thus

$$[2\beta/2\pi i] = 2[\gamma/2\pi i] + c_1(E,\omega)^{\mathbb{R}}.$$

Hence

$$[\beta/2\pi i] - \tfrac{1}{2}c_1(E,\omega)^{\mathbb{R}} = [\gamma/2\pi i]$$

is an integral class as claimed. □

Note that for a converse, we see from the proof that we have to choose the Mp^c structure so that the Chern class of $P_J(\lambda)$ is $[\beta/2\pi i] - \frac{1}{2}c_1(E,\omega)^{\mathbb{R}}$.

Theorem 4.2 *If β is a $\mathfrak{u}(1)$-valued 2-form on M such that $[\beta/2\pi i] - \frac{1}{2}c_1(E,\omega)^{\mathbb{R}}$ is integral then there is an Mp^c structure P in (E,ω) and a connection in P whose central curvature is β.*

PROOF [c.f. Theorem 7.2 in [7]] Choose a U(1) bundle L with real Chern class $[\beta/2\pi i]-\frac{1}{2}c_1(E,\omega)^{\mathbb{R}}$ and a unitary reduction of $\mathrm{Sp}(E,\omega)$. The Whitney product of these gives a $\mathrm{U}(V,\Omega,J) \times \mathrm{U}(1)$ bundle, and by extension, an $\mathrm{Mp}^c(V,\Omega)$ bundle P. P gives (E,ω) an Mp^c structure such that $P_J(\lambda)$ is L. We take a unitary connection in E and extend it to an Mp^c connection by picking the central part of the connection form so that the curvature of $P(\eta)$ is 2β. The central curvature is then β. □

5. The formal Weyl algebra

The formal Weyl algebra is a completion, in a suitable grading, of the universal enveloping algebra of the Heisenberg Lie algebra.

Let $\mathfrak{h} = V \oplus \mathbb{R}$ be the Lie algebra of $N(V,\Omega)$ with Lie bracket

$$[(y,h),(y',h')] = (0, -\Omega(y,y'))\,.$$

The dual $\mathfrak{h}^*$ can be identified with $\mathfrak{h}$ by the linear isomorphism $\alpha\colon \mathfrak{h} \to \mathfrak{h}^*$ such that $\langle \alpha(y,h),(y',h')\rangle = \Omega(y,y') + hh'$. We use this isomorphism to transfer the Lie structure to $\mathfrak{h}^*$ to yield $[(x,t),(x',t')] = (0, \Lambda(x,x'))$ where $\Lambda = (\Omega^{-1})$ is the dual form.

Relative to a basis of V, view the coordinates y^i and h as elements of $\mathfrak{h}^*$. Assign the degrees 1 to y^i and 2 to h. This puts a grading on the symmetric algebra $S(\mathfrak{h}^*)$ and so also on the enveloping algebra $U(\mathfrak{h}^*)$ when we identify them by total symmetrization

$$U_m = \left\{ \sum_{2k+l=m} a_{k,i_1,\ldots,i_l} h^k y^{i_1} \ldots y^{i_l} \,\middle|\, a_{k,i_1,\ldots,i_l} \text{ symmetric in } i_1,\ldots,i_l \right\}.$$

The formal Weyl algebra is then the algebra of formal series

$$W = \prod_{m=0}^{\infty} U_m = \left\{ a(y,h) = \sum_{m=0}^{\infty} \left(\sum_{2k+l=m} a_{k,i_1,\ldots,i_l} h^k y^{i_1} \ldots y^{i_l} \right) \right\}.$$

The product in $U(\mathfrak{h}^*)$ is given by (c.f. Moyal star product)

$$(a \circ b)(y,h) = \left(\exp\left(\frac{h}{2} \Lambda^{ij} \frac{\partial}{\partial y^i} \frac{\partial}{\partial z^j} \right) a(y,h) b(z,h) \right) \Big|_{y=z}.$$

The same formula also defines the product in W; W is a graded associative algebra (we denote by W_m the subspace U_m of elements of degree m and by $W_{(m)}$ the ideal of elements of degree $\geq m$).

One defines the Lie algebra structure on W by

$$[a,b]_W = \frac{1}{h}(a \circ b - b \circ a).$$

Note the division by h; W is a graded Lie algebra for a new grading obtained from the one defined above by a shift of -2.

The symplectic group $\mathrm{Sp}(V,\Omega)$ acts as automorphisms of $\mathfrak{h}$ by $A(y,h) = (Ay,h)$ and so has a contragredient action on $\mathfrak{h}^*$ given by $A(x,t) = (x \circ A^{-1}, t)$. This action extends to both $U(\mathfrak{h}^*)$ and W and on the latter is denoted by ρ. It respects the multiplication $\rho(A)(a \circ b) = \rho(A)(a) \circ \rho(A)(b)$, hence acts by automorphisms of the Lie algebra structure of W.

If $B \in \mathfrak{sp}(V,\Omega)$, we associate to it the element $\overline{B} = \frac{1}{2} \sum_{ijr} \Omega_{ir} B^r_j y^i y^j$ of degree two. An easy calculation shows that this correspondence $B \mapsto \overline{B}$ is a Lie algebra embedding of $sp(V,\Omega)$ into W. The natural action $\rho_*(B)$ of $B \in \mathfrak{sp}(V,\Omega)$ on V is given by $\rho_*(B) y^l = [\overline{B}, y^l]_W$. Since both sides act as derivations this extends to all of W as

$$\rho_*(B) a = [\overline{B}, a]_W.$$

6. The bundle of Weyl algebras on a symplectic manifold

If (M,ω) is a symplectic manifold, its tangent bundle TM is a symplectic vector bundle and we denote by $F(M)$ the symplectic frame bundle of its tangent bundle. The associated bundle $\mathcal{W} = F(M) \times_{\mathrm{Sp}(V,\Omega),\rho} W$ is a bundle of algebras on M called the bundle of (formal) Weyl algebras, or, more simply, the Weyl bundle. The topology is the inductive one defined by the grading in W. One can similarly define $\mathcal{W}_m = F(M) \times_{\mathrm{Sp}(V,\Omega),\rho} W_m$ and $\mathcal{W}_{(m)} = F(M) \times_{\mathrm{Sp}(V,\Omega),\rho} W_{(m)}$ since ρ preserves the grading in W.

Locally, sections of the Weyl bundle have the form of formal series

$$a(x,y,h) = \sum_{2k+l\geq 0} h^k a_{k,i_1,\ldots,i_l}(x) y^{i_1} \cdots y^{i_l}$$

where the coefficients $a_{k,i_1,\ldots,i_l}$ are symmetric covariant tensor fields in the indices $i_1, \ldots, i_l$ on M.

Since ρ acts by automorphisms of the associative algebra W the product of two sections taken pointwise makes the space of sections into an algebra, and in terms of the above representation of sections the multiplication has the form

$$(a \circ b)(x,y,h) = \left(\exp\left(\frac{h}{2}\Lambda^{ij}\frac{\partial}{\partial y^i}\frac{\partial}{\partial z^j}\right) a(x,y,h)b(x,z,h)\right)\Big|_{y=z}.$$

Similarly one defines the bracket $[a,b]$ of two sections a and b by taking the bracket pointwise, i.e. $[a,b](x,y,h) = \frac{1}{h}(a \circ b - b \circ a)(x,y,h)$.

Deformation quantization of a symplectic manifold (M,ω) was introduced by Bayen et al. [1]. In this approach, quantization arises as a formal deformation, called a star product, of the associative algebra N of smooth functions on M (with the usual product of functions) which yields by skewsymmetrization a deformation of the Lie algebra structure of N defined by the Poisson bracket. The idea behind Fedosov's construction [4] of a deformation quantization of (M,ω) is to build a subalgebra of the associative algebra $\Gamma(\mathcal{W})$ of sections of $\mathcal{W}$ which is in bijection with the space $N[[\nu]]$ of formal series of smooth functions on M. This he does by considering the space of flat sections for a generalized covariant derivative D with no curvature in $\Gamma(\mathcal{W})$.

7. Generalized connections

Let $p\colon P \to M$ be a principal G bundle. Let W be a Lie algebra "containing" the Lie algebra $\mathfrak{g}$ of G (we denote by $j\colon \mathfrak{g} \to W$ the Lie algebra embedding) and let ρ be a representation of G on W such that each $\rho(g)$ is an automorphism of W and $\rho_*(X)a = [j(X), a]$ $\forall X \in \mathfrak{g}$.

A (G,W)-connection on P is a W-valued 1-form $\tilde{\alpha}$ on P such that

(a) $R_g^*\tilde{\alpha} = \rho(g^{-1})\tilde{\alpha} \quad \forall g \in G$;

(b) $\tilde{\alpha}(X^*) = X \quad \forall X \in \mathfrak{g}$ where $X_q^* = \frac{d}{dt} q.\exp tX|_0$.

The curvature of $\tilde{\alpha}$ is the W-valued 2-form K on P

$$K(X,Y) = d\tilde{\alpha}(X,Y) + [\tilde{\alpha}(X), \tilde{\alpha}(Y)].$$

Since K vanishes on a vertical vector and $R_g^*K = \rho(g^{-1})K$, the curvature can be viewed as a section κ of $\mathcal{W} \otimes \Lambda^2(M)$ with $\mathcal{W} = P \times_{(G,\rho)} W$.

Let E be a vector space carrying representations μ of G and τ of W which are compatible, i.e.

(a) $\mu_*(X) = \tau(j(X))$, $\forall X \in \mathfrak{g} \subset W$;
(b) $\tau(\rho(g)a) = \mu(g)\tau(a)\mu(g^{-1})$, $\forall g \in G$, $\forall a \in W$.

Consider the associated bundle $\mathcal{E} = P \times_{(G,\mu)} E$. The (G, W)-connection $\tilde{\alpha}$ on P defines a connection on $\mathcal{E}$, i.e. a covariant derivative $D^{\mathcal{E}}$ of the sections of $\mathcal{E}$. If $\psi: U \to P$ is a local section of P, it defines a local chart of $\mathcal{E}$; to a section s of $\mathcal{E}$ one associates the local function $\hat{s}: U \to E$ so that $s(x) = [\psi(x), \hat{s}(x)]$. Then $D^{\mathcal{E}}$ is defined by

$$\widehat{D^{\mathcal{E}}_X s}(x) = X_x\hat{s} + \tau(\psi^*\tilde{\alpha}(X))\hat{s}(x).$$

Notice that if α is a $\mathfrak{g}$-valued connection 1-form on P and if the representation ρ "coincides" with the adjoint representation on $\mathfrak{g} \subset W$ (i.e. $\rho(g)j(X) = j(Ad(g)X)\forall g \in G, \forall X \in \mathfrak{g}$)then $\tilde{\alpha} - j\alpha$ can be viewed as a section γ of $\mathcal{W} \otimes \Lambda^1(M)$. If ∂ denotes the covariant derivative of the sections of $\mathcal{E}$ associated to α, then

$$D^{\mathcal{E}}_X s = \partial_X s + \tau(\gamma(X))s$$

where the action τ of a section of $\mathcal{W}$ on a section of $\mathcal{E}$ is defined pointwise.

For any (G, W)-connection, the curvature of $D^{\mathcal{E}}$ is given locally by multiplication of $\hat{s}$ by $\tau(\psi^*K(X,Y))$ or, equivalently

$$(D^{\mathcal{E}}_X D^{\mathcal{E}}_Y - D^{\mathcal{E}}_Y D^{\mathcal{E}}_X - D^{\mathcal{E}}_{[X,Y]})s = \tau(\kappa(X,Y))s.$$

Notice, in particular, that for $E = W$, $\mu = \rho$ and $\tau(a)b = [a, b]$, a (G, W)-connection induces a covariant derivative $D^{\mathcal{W}}$ of the sections of $\mathcal{W} = P \times_{(G,\rho)} W$. $D^{\mathcal{E}}$ has zero curvature if the curvature 2-form κ on P has its values in the center of the Lie algebra W.

8. Symplectic-Weyl connections

Let (M, ω) be a symplectic manifold. A symplectic connection on M is a connection ∇ on TM which is torsion-free and satisfies $\nabla_X \omega = 0$. Such connections always exist but, unlike the Riemannian case, are not unique. We denote the curvature of ∇ by R. A symplectic connection defines a connection in the symplectic frame bundle, i.e. a $\mathfrak{sp}(V, \Omega)$-valued 1-form α on $F(M)$, hence in all associated bundles. In particular we obtain a connection in $\mathcal{W}$ which we denote by ∂. The idea in Fedosov's construction is to modify ∂ to have zero curvature in $\mathcal{W}$.

In order to do this one considers generalized (G, W)-connections as above where $G = \mathrm{Sp}(V, \Omega)$, W is the Weyl Lie algebra, the embedding

j of $\mathfrak{sp}(V,\Omega)$ into W is defined by $j(B) = \overline{B}$, ρ is the representation of $\mathrm{Sp}(V,\Omega)$ on W defined before and $p\colon P \to M$ is the symplectic frame bundle. Remark that the restriction of the action of $\rho(g)$ on the Lie algebra $\mathfrak{sp}(V,\Omega)$ embedded in W coincides with the adjoint action. Hence one can take a generalized connection $\tilde{\alpha}$ such that $\tilde{\alpha} - j\alpha$ corresponds to a section γ of $\mathcal{W} \otimes \Lambda^1(M)$ and we get

$$D^{\mathcal{W}}_X a = \partial_X a + [\gamma(X), a], \quad \forall a \in \Gamma(\mathcal{W})$$

where $[\ ,\ ]$ denotes the bracket of sections of $\mathcal{W}$. The section γ is chosen to be

$$\gamma = \gamma_1 + r$$

where $\gamma_1 = \sum_{ij} \omega_{ij} y^i dx^j$ and r is a section of $\mathcal{W}_{(3)} \otimes \Lambda^1(M)$. The curvature of such a connection is viewed as a section κ of $\mathcal{W} \otimes \Lambda^2(M)$ and we have

$$\begin{aligned}\kappa = \ & \overline{R(X,Y)} + \partial r(X,Y) + [\gamma_1(X), \gamma_1(Y)] \\ & + [\gamma_1(X), r(Y)] + [r(X), \gamma_1(Y)] + [r(X), r(Y)]\end{aligned}$$

where ∂ is extended to sections of $\mathcal{W} \otimes \Lambda^q(M)$ in the usual way $\partial(a \otimes \omega) = \partial(a) \wedge \omega + a \otimes d\omega$.

Proposition 8.1 (Fedosov [4]) *The equation*

$$\overline{R(X,Y)} + \partial r(X,Y) + [\gamma_1(X), r(Y)] + [r(X), \gamma_1(Y)] + [r(X), r(Y)] = 0$$

has a solution $r \in \Gamma(\mathcal{W}_{(3)} \otimes \Lambda^1)$ *(and this solution is unique if it satisfies some normalization condition).*

Notice that for such a choice of r,

$$\kappa(X,Y) = [\gamma_1(X), \gamma_1(Y)] = -\omega(X,Y)$$

and the curvature of $D^{\mathcal{W}}$ vanishes. Notice also that for the given choice of γ_1, it is not possible to choose r so that we get rid of the curvature κ completely.

Since $D = D^{\mathcal{W}}$ acts as a derivation of the pointwise multiplication of sections the space $\mathcal{W}_D$ of flat sections will be a subalgebra of the space of sections of $\mathcal{W}$:

$$\mathcal{W}_D = \{a \in \Gamma(\mathcal{W}) | Da = 0\}.$$

Proposition 8.2 (Fedosov [4]) *For any formal series* $a_0 \in N[[\nu]]$ *where* N *is the space of smooth functions on* M, *there is a unique* $a \in \mathcal{W}_D$ *such that* $a(x, 0, h) = a_0(x, \nu = h/2)$.

The symbol map $\sigma: \Gamma(\mathcal{W}) = N[[\nu]]$, is defined by

$$\sigma(a) = a(x, 0, h), \quad (\nu = \frac{h}{2}).$$

The above proposition tells us that σ is a linear isomorphism when restricted to $\mathcal{W}_D$. It is used to transport the algebra structure of $\mathcal{W}_D$ to $N[[\nu]]$; this defines a star product on N

$$a * b = \sigma\left(\sigma^{-1}(a) \circ \sigma^{-1}(b)\right), \quad a, b \in N[[\nu]].$$

The explicit form of γ_1 ensures that the term C_1 in the expression $u * v = \sum_{r \geq 0} \nu^r C_r(u, v)$ is given by the Poisson bracket

$$C_1(u, v) = \{u, v\}.$$

9. Mp^c-Weyl connections

To build a representation (or a family of approximate representations up to order ν^k for some k) of the associative algebra $(N[[\nu]], *)$ – at least for some given value of ν – a natural way to proceed in our framework is to try to build a representation (or approximate representations) of the associative algebra of flat sections of $\mathcal{W}$ on a space of flat (or approximately flat) sections of a bundle $\mathcal{E}$. A starting point is to get a vector space E which carries a representation μ of the structure group G and a representation τ' of the associative algebra W (or of any truncated series, i.e. of the enveloping algebra $U(\mathfrak{h}^*)$) in a compatible way

$$\tau'(\rho(g)a) = \mu(g)\tau'(a)\mu(g^{-1}),$$

$\forall g \in G$, $\forall a \in W$. Then one has an action of (possibly truncated) sections of $\mathcal{W}$ on sections of $\mathcal{E} = P \times_{G,\mu} E$.

The next point is to build a (G, W)-connection so that the corresponding curvature in $\mathcal{W}$ vanishes; it induces a (possibly approximate) covariant derivative of the sections of $\mathcal{E}$ and one would like the corresponding curvature to also vanish.

A natural candidate for the space E is the space of smooth vectors in $\mathcal{H}$, the Hilbert space carrying an irreducible representation of the Heisenberg group for a given value of $\hbar$. Then the group G acting on that space can not be the symplectic group and we have naturally to consider $G = \mathrm{Mp}^c(V, \Omega)$.

So pick an Mp^c-structure on TM (i.e. a principal $\mathrm{Mp}^c(V, \Omega)$ bundle $\pi^c: P \longrightarrow M$ with a fibre-preserving map $\phi: P \longrightarrow \mathrm{Sp}(E, \omega)$). Observe that the Weyl bundle is also associated to P via the action $\rho \circ \sigma$ of $\mathrm{Mp}^c(V, \Omega)$.

Choose a symplectic connection on (M,ω) and let α be the corresponding $\mathfrak{sp}(V,\Omega)$-valued 1-form on $F(M)$. Pick an Mp^c connection α' on P such that

$$\sigma_*\alpha' = \phi^*\alpha.$$

Denote by α'^c the central component of α', and by β the $\mathfrak{u}(1)$-valued 2-form on M which is the central curvature of α'.

Let W be the complexified Weyl algebra and ρ' the representation of $\mathrm{Mp}^c(V,\Omega)$ on W defined by $\rho' = \rho\circ\sigma$; clearly ρ' acts by automorphisms of W (for the associative and the Lie algebra structures of W) and we have $\rho'_*(X)a = [\overline{\sigma_*(X)}, a]_W$.

Define an embedding j' of $\mathfrak{mp}^c(V,\Omega)$ into W by

$$j'(X) = \overline{\sigma_*(X)} + \frac{1}{2}\eta(X)h.$$

Then $\rho'_*(X)a = [j'(X), a]_W$, $\forall X \in \mathfrak{mp}^c(V,\Omega)$, $\forall a \in W$ and

$$j'(\mathrm{ad}\, g'(X)) = \rho'(g')j'(X)\rho'(g'^{-1})$$

$\forall X \in \mathfrak{mp}^c(V,\Omega)$, $\forall g' \in \mathrm{Mp}^c(V,\Omega)$.

Define a generalized $(\mathrm{Mp}^c(V,\Omega), W)$-connection $\tilde{\alpha}'$ such that $\tilde{\alpha}' - j'\alpha'$ corresponds to the section γ of $\mathcal{W}\otimes\Lambda^1(M)$ defined in the previous paragraph. The curvature of $\tilde{\alpha}'$ is given by the section κ' of $\mathcal{W}\otimes\Lambda^2(M)$

$$\kappa'(X,Y) = j(\beta(X,Y)) - \omega(X,Y) = \beta(X,Y)h - \omega(X,Y).$$

Let E be the space of smooth vectors of the representation χ of $N(V,\Omega)$ on $\mathcal{H}$ and let χ_* be the corresponding representation of $\mathfrak{h}$ on E. Notice that E is also the space of smooth vectors for the natural representation of $\mathrm{Mp}^c(V,\Omega)$ on $\mathcal{H}$; let μ be the representation of $\mathrm{Mp}^c(V,\Omega)$ on E given by $\mu(U,g) = U|_E$.

Let τ' be the representation of the associative algebra $U(\mathfrak{h}^*)$ (i.e. of any truncated element of W) extending

$$\tau'(y) = d\chi_*(\alpha^{-1}(y)), \quad \tau'(h) = d^2\chi_*(\alpha^{-1}(h)) = \frac{id^2}{\hbar}\,\mathrm{Id}$$

where d is a constant. Clearly

$$\tau'(\rho'(g')a) = \mu(g')\tau'(a)\mu(g'^{-1}),$$

$\forall g \in \mathrm{Mp}^c(V,\Omega)'$, $\forall a \in W$.

The representation τ of any truncated element of the Lie algebra W on E is defined to be

$$\tau(a) = (\tau'(h))^{-1}\tau'(a)$$

hence $\tau(h) = 1$ and $\tau(1) = -\frac{i\hbar}{d^2}$. One checks that

$$\tau(j'(X)) = \mu_*(X), \quad \forall X \in \mathsf{mp}^c(V,\Omega).$$

Notice that this embedding implies our expression for j'; for an embedding of $\mathsf{mp}^c(V,\Omega)$ into W, one could have chosen $j'(X) = \overline{\sigma_*(X)} + 1/2\eta(X)f(h)$ with $f(h)$ any element in the center of W but the compatibility between τ and μ_* implies that $\tau(f(h)) = 1$.

The "curvature" in $\mathcal{E}$ is then equal to

$$\tau(\beta(X,Y)h - \omega(X,Y)1) = \beta(X,Y) + \frac{i\hbar}{d^2}\omega(X,Y). \tag{6}$$

Notice that if one defines a family of covariant derivatives in $\mathcal{E}$ by looking at the truncations of γ at a certain order then the corresponding curvature only involves the terms in equation (6) above plus terms of order higher than the level at which we truncated. One then looks at the action of truncated flat sections of $\mathcal{W}$ on sections of $\mathcal{E}$ which are flat at a given order. For $d = \hbar$, one gets in this way a family of approximate representations of the Fedosov star product for $\nu = i\hbar/2$.

In order for the curvature in (6) to vanish, one needs

$$\beta(X,Y) = -\frac{i\hbar}{\hbar^2}\omega(X,Y) = -\frac{i}{\hbar}\omega(X,Y).$$

In view of §4, one can pick an Mp^c structure and build an Mp^c connection with central curvature $-i\omega/\hbar$ if and only if

$$[\omega/2\pi\hbar] - \tfrac{1}{2}c_1(TM,\omega)^{\mathbb{R}} \in H^2(M,\mathbb{Z})^{\mathbb{R}}.$$

This is the integrality condition which can be found in the literature [2, 3, 4, 6, 7].

References

1. F. Bayen, M. Flato, C. Frønsdal, A. Licnerowicz, D. Sternheimer, Deformation theory and quantization. *Lett. in Math. Phys.* **1** (1977) 521–530.
2. J. Czyż, *On some appraoch to geometric quantization.* Springer-Verlag Lecture Notes in Physics **676** (1978) 315–328.
3. P. Dazord and G. Patissier, *La première classe de Chern comme obstruction à la quantification asymptotique.* Springer-Verlag MSRI Publications **20** (1991) 73–97.
4. B. Fedosov, *Deformation quantization and index theory.* Akademie Verlag, Berlin 1996.
5. H. Hess, *On a geometric quantization scheme generalizing those of Kostant-Souriau and Czyż.* Springer-Verlag Lecture Notes in Physics **139** (1981) 1–36.
6. M.V. Karasev and V.P. Maslov, Asymptotic and geometric quantization. *Russian Math. surveys* **39:6** (1984) 133–205.
7. P.L. Robinson and J.H. Rawnsley, The metaplectic representation, Mp^c structures and geometric quantization. *Memoirs of the A.M.S.* vol. **81**, no. **410**. (American Mathematical Society, Providence RI, 1989).

MASSLESS XXZ MODEL AND DEGENERATION OF THE ELLIPTIC ALGEBRA $\mathcal{A}_{q,p}\left(\widehat{sl}_2\right)$

MICHIO JIMBO*
Department of Mathematics, Faculty of Science
Kyoto University, Kyoto 606, Japan

HITOSHI KONNO
The Division of Mathematical and Information Sciences
Faculty of Integrated Arts and Sciences, Hiroshima University
Higashi-Hiroshima 739, Japan

AND

TETSUJI MIWA*
Research Institute for Mathematical Sciences
Kyoto University, Sakyo-ku, Kyoto 606-01, Japan

Abstract. Following Lukyanov's bosonization formula for the sine-Gordon model, we present the bosonization of the massless XXZ model by using the free boson on the real line. We discuss its connection to the degeneration of the elliptic algebra $\mathcal{A}_{q,p}(\widehat{sl}_2)$.

1. Introduction

The one-dimensional spin 1/2 XXZ chain is a system defined by the Hamiltonian

$$H = -\frac{1}{2}\sum_{n=-\infty}^{\infty}\left(\sigma_n^x\sigma_{n+1}^x + \sigma_n^y\sigma_{n+1}^y + \Delta\sigma_n^z\sigma_{n+1}^z\right). \tag{1.1}$$

Here the σ_i^α ($\alpha = x, y, z$) stand for the usual Pauli matrices acting on the i-th lattice site. We are concerned with the case $|\Delta| < 1$, where the spectrum

* The work reported here is based on the talks presented by Jimbo and Miwa, and is partly supported by Grant-in-Aid for Scientific Research on Priority Areas 231, the Ministry of Education, Science and Culture.

D. Sternheimer et al. (eds.), Deformation Theory and Symplectic Geometry, 117–138.

of (1.1) is gapless. In a recent work [1], the problem of describing correlation functions of this model was discussed. Let us first quote the result below.

By a correlation function we mean the ground-state average $\langle \mathcal{O} \rangle$ of a local operator

$$\mathcal{O} = E^{(i_1)}_{\varepsilon_1 \varepsilon'_1} \cdots E^{(i_k)}_{\varepsilon_k \varepsilon'_k}.$$

Here $E_{\varepsilon\varepsilon'} = (\delta_{a\varepsilon}\delta_{b\varepsilon'})_{a,b=\pm}$ is the matrix unit, and $E^{(i)}_{\varepsilon\varepsilon'}$ denotes the operator acting as $E_{\varepsilon\varepsilon'}$ on the i-th site and as identity elsewhere. It suffices to consider $\mathcal{O} = E^{(1)}_{\varepsilon'_1\varepsilon_1} \cdots E^{(n)}_{\varepsilon'_n\varepsilon_n}$, since the general case is reduced to this case by translation and taking linear combinations. Set

$$\Delta = -\cos \pi\nu \qquad (0 < \nu < 1). \tag{1.2}$$

Define the indices $\bar{1}, \cdots, \bar{n}$ $(1 \le \bar{1} < \cdots < \bar{n} \le 2n)$ from $\varepsilon_i, \varepsilon'_i$ by the rule

$$\{\bar{1}, \cdots, \bar{n}\} = \{j \mid 1 \le j \le n, \, \varepsilon'_j = -\} \cup \{j^* \mid 1 \le j \le n, \, \varepsilon_j = +\}$$

where $j^* = 2n + 1 - j$. With these notations, the formula written down in [1] reads as follows:

$$\begin{aligned}
&\langle E^{(1)}_{\varepsilon'_1\varepsilon_1} \cdots E^{(n)}_{\varepsilon'_n\varepsilon_n} \rangle \\
&= \frac{1}{\nu^{n(n-1)/2}} \int_{-\infty}^{\infty} \cdots \int_{-\infty}^{\infty} \prod_{l=1}^{n} \frac{d\alpha_l}{2\pi} \prod_{1 \le l < l' \le n} \frac{\sinh(\alpha_l - \alpha_{l'})}{\sinh \nu(\alpha_l - \alpha_{l'} - \pi i)} \\
&\times \prod_{1 \le \bar{l} \le n} \Big[\prod_{j=1}^{n} \frac{i}{\sinh(\alpha_l + i0)} \prod_{1 \le j < \bar{l}} \sinh \nu(\alpha_l) \prod_{\bar{l} < j \le n} \sinh \nu(-\alpha_l + \pi i) \Big] \\
&\times \prod_{n+1 \le \bar{l} \le 2n} \Big[\prod_{j=1}^{n} \frac{-i}{\sinh(\alpha_l - i0)} \prod_{1 \le j < \bar{l}^*} \sinh \nu(-\alpha_l) \prod_{\bar{l}^* < j \le n} \sinh \nu(\alpha_l + \pi i) \Big].
\end{aligned} \tag{1.3}$$

This is a rather unwieldy, yet completely explicit formula for correlation functions of an arbitrary local operator.

In the earlier works [2, 3], the case of the anti-ferromagnetic regime $\Delta < -1$ was treated. Being written solely in terms of elementary functions, the formula (1.3) is simpler than the corresponding one for the massive case $\Delta < -1$. Nevertheless the situation in the massless case $|\Delta| < 1$ has remained unclear for quite some time. The main reason is that the corner transfer matrix argument is not directly applicable in the massless regime. In the work [1], (1.3) was derived through the following indirect argument. The massless XXZ model is a limiting case of the XYZ model. The latter is massive, and the corner transfer matrix method applies. This leads to the following system of difference equations.

$$G_n(\cdots,\beta_{j+1},\beta_j,\cdots)_{\cdots,\varepsilon_{j+1},\varepsilon_j,\cdots} = \sum_{\varepsilon'_j,\varepsilon'_{j+1}} R^{\varepsilon'_j,\varepsilon'_{j+1}}_{\varepsilon_j,\varepsilon_{j+1}}(\beta_j-\beta_{j+1})G_n(\cdots,\beta_j,\beta_{j+1},\cdots)_{\cdots,\varepsilon'_j,\varepsilon'_{j+1},\cdots} \tag{1.4}$$

$$G_n(\beta_1,\cdots,\beta_{2n-1},\beta_{2n}-i\lambda)_{\varepsilon_1,\cdots,\varepsilon_{2n}} = G_n(\beta_{2n},\beta_1,\cdots,\beta_{2n-1})_{\varepsilon_{2n},\varepsilon_1,\cdots,\varepsilon_{2n-1}}, \tag{1.5}$$

$$G_n(\beta_1,\cdots,\beta_{2n})_{\varepsilon_1,\cdots,\varepsilon_{2n}}\Big|_{\beta_{2n}=\beta_{2n-1}+\pi i} = \delta_{\varepsilon_{2n-1}+\varepsilon_{2n},0}G_{n-1}(\beta_1,\cdots,\beta_{2n-2})_{\varepsilon_1,\cdots,\varepsilon_{2n-2}}. \tag{1.6}$$

Here G_n is a function with values in $(\mathbb{C}^2)^{\otimes 2n}$, and R denotes the elliptic R matrix (see (2.3) below). The correlation functions are obtained by taking $\lambda = 2\pi$ and specializing the variables as

$$(\beta_1,\cdots,\beta_{2n}) = (\overbrace{\beta+\pi i,\cdots,\beta+\pi i}^{n},\overbrace{\beta,\cdots,\beta}^{n}).$$

We note that these equations imply in particular the quantum Knizhnik-Zamolodchikov (qKZ) equation. In the elliptic case, however, construction of solutions of the above system is still an open problem. The formula (1.3) was obtained by solving the corresponding difference equations directly in the XXZ limit. The difficulty in this approach is that the solutions of difference equations are determined only up to arbitrary periodic functions, so one has to single out in some way the unique solution which corresponds to the correlation functions. [1] For this reason, most part of [1] is devoted to the verification that (1.3) reproduces the correct results in all known cases: the XXX limit ($\nu = 0$), Ising limit ($\nu = 1/2$), and the nearest neighbor two point function (ν general). Thus it is fair to call (1.3) a conjecture, supported by these evidences.

For the massive XXZ model $\Delta < -1$, the ambiguity of solutions is resolved since the canonical solution is provided by a trace of vertex operators. The latter has a clear meaning in the framework of representation theory of the quantum affine algebra $U_q(\widehat{sl}_2)$, and via bosonization, enables us to write down an explicit integral formula for the trace similar to (1.3). The situation in the massless case would become much clearer if a similar bosonization is found for the full elliptic algebra $\mathcal{A}_{q,p}\left(\widehat{sl}_2\right)$ [5] corresponding to the XYZ model. In this connection, we should mention that

[1]Compare with the form factor bootstrap approach [4], in which the abundance of solutions corresponds to that of local operators. In our case there is only one physically meaningful solution.

Lukyanov [6] had previously constructed a bosonization of the vertex operators for the sine-Gordon model. Although his formulas look quite similar to (1.3), the precise relation has not been worked out. This note is intended to make explicit the connections between the elliptic algebra, Lukyanov's bosonization, and the correlation functions (1.3).

Lukyanov's construction starts with bosonized vertex operators with a cutoff, and the form factors of the sine-Gordon model are derived after removing the cutoff parameter at the final stage. Here we prefer to work directly with operators with the cutoff parameter removed. The regularization is introduced in the integrals appearing in the contractions of these operators. The vertex operators are written in terms of bosonic oscillators alone, without involving 'zero mode' operators P, Q (see the remark at the end of section 3). We verify that the formula (1.3) for the correlations coincides with the trace of these vertex operators (of type I in the terminology of [3]). The vertex operators considered here can be regarded as a degeneration of those of the elliptic algebra $\mathcal{A}_{q,p}\left(\widehat{sl}_2\right)$ in the limit $p, q \to 1$. The precise relation is given in the text. In particular we show that the symmetry property of the L operators which play a key role in the definition of $\mathcal{A}_{q,p}\left(\widehat{sl}_2\right)$ is indeed satisfied in this limit.

2. Degeneration of the elliptic algebra

In this section, we discuss the degeneration of the elliptic algebra $\mathcal{A}_{q,p}\left(\widehat{sl}_2\right)$ and the vertex operators associated with it.

2.1. R MATRIX

In order to fix the notation, let us begin by recalling the elliptic R matrix from Baxter's book [7]. The Boltzmann weights of the eight vertex model are given by (see eq.(10.4.21) in [7])

$$a(u) = \frac{\mathrm{snh}(\lambda - u)}{\mathrm{snh}(\lambda)}, \quad b(u) = \frac{\mathrm{snh}(u)}{\mathrm{snh}(\lambda)}, \quad c(u) = 1, \quad d(u) = k\,\mathrm{snh}(\lambda - u)\mathrm{snh}(u). \tag{2.1}$$

Here $\mathrm{snh}(u) = -i\mathrm{sn}(iu)$, and $\mathrm{sn}(u)$ denotes Jacobi's elliptic function with modulus k. Let K, K' be the corresponding complete elliptic integrals. We shall use the variables

$$p = e^{-\frac{\pi K'}{K}}, \qquad q = -e^{-\frac{\pi\lambda}{2K}}, \qquad \zeta = e^{\frac{\pi u}{2K}}, \tag{2.2}$$

and regard (2.1) as functions of ζ, p, q.

The R matrix is defined to be

$$R(\zeta)=R(\zeta;p^{1/2},q^{1/2})=\frac{1}{\mu(\zeta)}\begin{pmatrix} a(u) & & & d(u) \\ & b(u) & c(u) & \\ & c(u) & b(u) & \\ d(u) & & & a(u) \end{pmatrix}. \tag{2.3}$$

We choose the overall scalar factor $\mu(\zeta)$ as follows. [2]

$$\frac{1}{\mu(\zeta)}=\frac{1}{\overline{\kappa}(\zeta^2)}\frac{(p^2;p^2)_\infty}{(p;p)_\infty^2}\frac{\Theta_{p^2}(q^2)\Theta_{p^2}(p\zeta^2)}{\Theta_{p^2}(q^2\zeta^2)}, \tag{2.4}$$

$$\frac{1}{\overline{\kappa}(z)}=\frac{(q^4z^{-1};p,q^4)_\infty(q^2z;p,q^4)_\infty(pz^{-1};p,q^4)_\infty(pq^2z;p,q^4)_\infty}{(q^4z;p,q^4)_\infty(q^2z^{-1};p,q^4)_\infty(pz;p,q^4)_\infty(pq^2z^{-1};p,q^4)_\infty} \tag{2.5}$$

where

$$(z;p_1,\cdots,p_m)_\infty = \prod_{n_1,\cdots,n_m\ge 0}(1-zp_1^{n_1}\cdots p_m^{n_m}), \tag{2.6}$$

$$\Theta_q(z) = (z;q)_\infty(qz^{-1};q)_\infty(q;q)_\infty. \tag{2.7}$$

With this choice of $\mu(\zeta)$, (2.3) satisfies the unitarity and crossing relations as well as the Yang-Baxter equation. We skip the details since we do not need them.

As usual, we shall regard (2.3) as a linear map acting on $V\otimes V$ with $V=\mathbb{C}^2$. In the standard basis v_+,v_- of V we write

$$R(\zeta)v_{\varepsilon_1}\otimes v_{\varepsilon_2}=\sum_{\varepsilon_1',\varepsilon_2'=\pm} v_{\varepsilon_1'}\otimes v_{\varepsilon_2'}\, R_{\varepsilon_1\varepsilon_2}^{\varepsilon_1'\varepsilon_2'}(\zeta).$$

2.2. ELLIPTIC ALGEBRA $\mathcal{A}_{Q,P}\left(\widehat{SL}_2\right)$

The algebra $\mathcal{A}_{q,p}\left(\widehat{sl}_2\right)$ [5, 9] was proposed as an elliptic extension of the quantized affine algebra $U_q\left(\widehat{sl}_2\right)$. It is presented in terms of the formal generating series

$$L^\pm(\zeta)=\sum_{n=-\infty}^{\infty}L_n^\pm\zeta^{-n}, \qquad L_n^\pm=\left(L_{\varepsilon\varepsilon',n}^\pm\right)_{\varepsilon,\varepsilon'=\pm}, \tag{2.8}$$

$$L_{\varepsilon\varepsilon',n}^\pm=0 \quad \text{if } \varepsilon\varepsilon'\neq(-1)^n. \tag{2.9}$$

[2]In this occasion let us correct some misprints in [8]. On page 434, the second line from the bottom, the formula for $d(\zeta)$ should read $k\,\mathrm{snh}(\lambda-u)\mathrm{snh}(u)/\mu(\zeta)$. In the first line of eq.(2.18), the factor $(p^2;p^2)_\infty^2$ should read $(p^2;p^2)_\infty$.

By definition, $\mathcal{A}_{q,p}\left(\widehat{sl}_2\right)$ is the algebra generated by the symbols $L^{\pm}_{\varepsilon\varepsilon',n}$ ($n \in \mathbb{Z}, \varepsilon, \varepsilon' = \pm, \varepsilon\varepsilon' = (-1)^n$) and a central element c, through the following defining relations.

$$R^{\pm}_{12}(\zeta_1/\zeta_2) \overset{1}{L^{\pm}}(\zeta_1) \overset{2}{L^{\pm}}(\zeta_2) = \overset{2}{L^{\pm}}(\zeta_2) \overset{1}{L^{\pm}}(\zeta_1) R^{*\pm}_{12}(\zeta_1/\zeta_2), \tag{2.10}$$

$$R^{+}_{12}(q^{c/2}\zeta_1/\zeta_2) \overset{1}{L^{+}}(\zeta_1) \overset{2}{L^{-}}(\zeta_2) = \overset{2}{L^{-}}(\zeta_2) \overset{1}{L^{+}}(\zeta_1) R^{*+}_{12}(q^{-c/2}\zeta_1/\zeta_2), \tag{2.11}$$

$$q\text{-det}L^{+}(\zeta) \equiv L^{+}_{++}(q^{-1}\zeta)L^{+}_{--}(\zeta) - L^{+}_{-+}(q^{-1}\zeta)L^{+}_{+-}(\zeta) = q^{c/2}, \tag{2.12}$$

$$L^{-}_{\varepsilon\varepsilon'}(\zeta) = \varepsilon\varepsilon' L^{+}_{-\varepsilon,-\varepsilon'}(p^{1/2}q^{-c/2}\zeta). \tag{2.13}$$

Let us explain the notation. We set

$$R^{+}(\zeta) = \tau(q^{1/2}\zeta^{-1})R(\zeta), \qquad R^{-}(\zeta) = \tau(q^{1/2}\zeta)^{-1}R(\zeta),$$

where

$$\tau(\zeta) = \zeta^{-1}\frac{(q\zeta^2;q^4)_\infty(q^3\zeta^{-2};q^4)_\infty}{(q^3\zeta^2;q^4)_\infty(q\zeta^{-2};q^4)_\infty}. \tag{2.14}$$

We use also the R matrices in which the parameter p is changed:

$$R^{*\pm}(\zeta) = R^{\pm}(\zeta; p^{*1/2}, q^{1/2}), \qquad p^* = pq^{-2c}.$$

Our generating series (2.8) contain both positive and negative powers of ζ. If we set

$$L^{+}_{\varepsilon\varepsilon',n} = \left(-p^{1/2}\right)^{\max(n,0)} \overline{L}^{+}_{\varepsilon\varepsilon',n}$$

and let formally $p \to 0$, then $L^{+}(\zeta)$ and $L^{-}(\zeta)$ become power series in ζ and ζ^{-1} respectively. In this limit, the relations (2.10)-(2.13) reduce to the presentation of the quantum affine algebra $U_q(\widehat{sl}_2)$ due to Reshetikhin and Semenov-Tian-Shanskii [10]. As it is argued in [5], the symmetry condition (2.13) means that for generic p we have the same number of generators as we have for $p = 0$.

2.3. VERTEX OPERATORS

In the application of the representation theory of $U_q(\widehat{sl}_2)$, an important role is played by the integrable highest weight modules and intertwiners between them (vertex operators). From now on, we shall focus attention to the level one case. In this case, there exist two highest weight modules $\mathcal{H}^{(i)}$ ($i = 0, 1$). The vertex operators are intertwiners of the form

$$\Phi^{(1-i,i)}(\zeta) : \mathcal{H}^{(i)} \longrightarrow \mathcal{H}^{(1-i)} \otimes V_\zeta, \quad \Phi^{(1-i,i)}(\zeta) = \sum \Phi^{(1-i,i)}_{\varepsilon}(\zeta) \otimes v_\varepsilon,$$

$$\Psi^{*(1-i,i)}(\zeta) : V_\zeta \otimes \mathcal{H}^{(i)} \longrightarrow \mathcal{H}^{(1-i)}, \quad \Psi^{*(1-i,i)}_{\varepsilon}(\zeta) = \Psi^{*(1-i,i)}(\zeta)\left(v_\varepsilon \otimes \cdot\right).$$

Here V_ζ denotes the evaluation module based on $V = \mathbb{C}^2 = \mathbb{C}v_+ \oplus \mathbb{C}v_-$ (see e.g.[3] for the details).

It was conjectured in [5] that in the elliptic case $\mathcal{A}_{q,p}\left(\widehat{sl}_2\right)$ there exist natural analogs of the level one modules and vertex operators. (Here we say that a representation of $\mathcal{A}_{q,p}\left(\widehat{sl}_2\right)$ has level k if the central element c acts as k times the identity.) Let us use the same letters $\mathcal{H}^{(i)}$, $\Phi_\varepsilon^{(1-i,i)}(\zeta)$, $\Psi_\varepsilon^{*(1-i,i)}(\zeta)$ for the elliptic counterpart. We shall often omit writing the upper indices i when there is no fear of confusion. The conjecture says in particular that they obey the following commutation relations.

$$\Phi_{\varepsilon_2}(\zeta_2)\Phi_{\varepsilon_1}(\zeta_1) = \sum_{\varepsilon_1',\varepsilon_2'=\pm} R_{\varepsilon_1\varepsilon_2}^{\varepsilon_1'\varepsilon_2'}(\zeta_1/\zeta_2)\Phi_{\varepsilon_1'}(\zeta_1)\Phi_{\varepsilon_2'}(\zeta_2), \tag{2.15}$$

$$\Phi_{\varepsilon_1}(\zeta_1)\Psi^*_{\varepsilon_2}(\zeta_2) = \tau(\zeta_1/\zeta_2)\Psi^*_{\varepsilon_2}(\zeta_2)\Phi_{\varepsilon_1}(\zeta_1), \tag{2.16}$$

$$\Psi^*_{\varepsilon_1}(\zeta_1)\Psi^*_{\varepsilon_2}(\zeta_2) = -\sum_{\varepsilon_1',\varepsilon_2'=\pm} R^{*\varepsilon_1'\varepsilon_2'}_{\varepsilon_1\varepsilon_2}(\zeta_1/\zeta_2)\Psi^*_{\varepsilon_2'}(\zeta_2)\Psi^*_{\varepsilon_1'}(\zeta_1). \tag{2.17}$$

In addition, they should satisfy the normalization conditions

$$\begin{aligned}&\Phi_{\varepsilon_1}^{(i,1-i)}(\zeta_1)\Phi_{\varepsilon_2}^{(1-i,i)}(\zeta_2)\\&\quad= (-1)^{1-i}\varepsilon_2 g^{-1}\delta_{\varepsilon_1+\varepsilon_2,0} + O(\zeta_1 - q\zeta_2) \quad (\zeta_1 \to q\zeta_2),\end{aligned} \tag{2.18}$$

$$\begin{aligned}&\Psi_{\varepsilon_1}^{*(i,1-i)}(\zeta_1)\Psi_{\varepsilon_2}^{*(1-i,i)}(\zeta_2)\\&\quad= \frac{(-1)^{1-i}\varepsilon_1}{1-q^{-2}\zeta_2^2/\zeta_1^2}\left(\frac{\zeta_2}{q\zeta_1}\right)^{i+(1+\varepsilon_1)/2} g\delta_{\varepsilon_1+\varepsilon_2,0} + O(1) \quad (\zeta_1 \to q^{-1}\zeta_2),\end{aligned} \tag{2.19}$$

with some constant g.

In terms of the vertex operators, the $L^\pm$ operators acting on $\mathcal{H}^{(i)}$ can be expressed as

$$L^+_{\varepsilon\varepsilon'}(\zeta) = \kappa\Psi^*_{\varepsilon'}(\zeta)\Phi_\varepsilon(q^{1/2}\zeta), \tag{2.20}$$

$$L^-_{\varepsilon\varepsilon'}(\zeta) = \kappa\Phi_\varepsilon(\zeta)\Psi^*_{\varepsilon'}(q^{1/2}\zeta). \tag{2.21}$$

Here κ is a normalization constant. The defining relations (2.10),(2.11) are immediate consequences of the commutation relations (2.15) -(2.17). The condition (2.12) for the quantum determinant also follows from (2.18),(2.19) with an appropriate choice of κ. The symmetry (2.13) of the L operators entails the following relation for the vertex operators.

$$\Phi_\varepsilon(\zeta)\Psi^*_{\varepsilon'}(q^{1/2}\zeta) = \varepsilon\varepsilon'\Psi^*_{-\varepsilon'}(p^{1/2}q^{-1/2}\zeta)\Phi_{-\varepsilon}(p^{1/2}\zeta). \tag{2.22}$$

2.4. DEGENERATION

As we mentioned already, the elliptic algebra degenerates in the limit $p \to 0$ to the quantum affine algebra $U_q(\widehat{sl}_2)$. We now study the degeneration in another limit $p, q \to 1$. More precisely, we let

$$p = q^{2(\xi+1)}, \quad \zeta = q^{i\beta/\pi}, \qquad q \to 1, \tag{2.23}$$

where ξ and β are kept fixed.

Remark. In [11, 2, 3], the anti-ferroelectric regime of the six vertex model is discussed. This is a limiting case $p \to 0$ of the eight vertex model in the principal regime $0 < p^{1/2} < -q < \zeta^{-1} < 1$. On the other hand, the limit (2.23) is *not* a boundary of the principal regime.

In the limit (2.23), the elliptic R matrix (2.3) degenerates to a trigonometric one. However the limit has the form

$$R = \begin{pmatrix} a & & & d \\ & b & c & \\ & c & b & \\ d & & & a \end{pmatrix}$$

with $d \neq 0$, as opposed to the standard trigonometric R matrix coming from the universal R matrix of $U_q(\widehat{sl}_2)$. In order to bring it to the usual form, we need to introduce a 'gauge' transformation. Define

$$U = \begin{pmatrix} 1 & -i \\ 1 & i \end{pmatrix}, \qquad U_0 = U\,\sigma^z, \quad U_1 = \sigma^z\,U\,\sigma^z.$$

Then we find

$$\lim\,(U_1 \otimes U_0)\,R(\zeta; p^{1/2}, q^{1/2})\,(U_0 \otimes U_1)^{-1} = R(-\beta), \tag{2.24}$$

$$\lim\,(U_1 \otimes U_0)\left(-R(\zeta; p^{*1/2}, q^{1/2})\right)(U_0 \otimes U_1)^{-1} = S(\beta), \tag{2.25}$$

where

$$S(\beta) = \frac{S_0(\beta)}{\sinh\frac{i\pi-\beta}{\xi}} \begin{pmatrix} \sinh\frac{i\pi-\beta}{\xi} & & & \\ & \sinh\frac{\beta}{\xi} & \sinh\frac{i\pi}{\xi} & \\ & \sinh\frac{i\pi}{\xi} & \sinh\frac{\beta}{\xi} & \\ & & & \sinh\frac{i\pi-\beta}{\xi} \end{pmatrix} \tag{2.26}$$

$$R(\beta) = -S(-\beta)|_{\xi\to\xi+1}. \tag{2.27}$$

Here $S_0(\beta)$ is given in Appendix, (A.11). The matrices $\left(S_{ab}^{cd}(\beta)\right)$, $\left(R_{ab}^{cd}(\beta)\right)$ coincide with those given respectively in eq.(2.33), (9.15) of [6].

Note that

$$\begin{aligned}(\sigma^z \otimes \sigma^z)(U_1 \otimes U_0) &= U_0 \otimes U_1,\\ (U_0 \otimes U_1)^{-1}(\sigma^z \otimes \sigma^z) &= (U_1 \otimes U_0)^{-1}.\end{aligned}$$

Since $R(-\beta)$ and $S(\beta)$ commute with $\sigma^z \otimes \sigma^z$, (2.24),(2.25) can also be written as

$$\lim (U_0 \otimes U_1)\, R(\zeta; p^{1/2}, q^{1/2})\, (U_1 \otimes U_0)^{-1} = R(-\beta), \tag{2.28}$$

$$\lim (U_0 \otimes U_1) \left(-R(\zeta; p^{*1/2}, q^{1/2})\right) (U_1 \otimes U_0)^{-1} = S(\beta). \tag{2.29}$$

The transformations (2.24), (2.25), (2.28), (2.29) are not similarity transformations. This may seem strange at first glance. However, this is understood to be natural because in (2.15) and (2.17) the order of ζ_1 and ζ_2 is opposite in the left and right hand sides.

Now we will rewrite (2.15) and (2.17) in the limit. Write $\Phi_a(\beta), \Psi^*_a(\beta)$ for the limit of $\Phi_a(\zeta), \Psi^*_a(\zeta)$. In accordance with (2.24),(2.25), (2.28), (2.29) we set

$$Z_a^{(1,0)}(\beta) = 2\sum_b (U_0^{-1})_{ba} \Psi_b^{*(1,0)}(\beta) = \Psi_+^{*(1,0)}(\beta) - ia\Psi_-^{*(1,0)}(\beta) \tag{2.30}$$

$$Z_a^{(0,1)}(\beta) = 2\sum_b (U_1^{-1})_{ba} \Psi_b^{*(0,1)}(\beta) = a\Psi_+^{*(0,1)}(\beta) - i\Psi_-^{*(0,1)}(\beta) \tag{2.31}$$

$$Z_a^{\prime(1,0)}(\beta) = \sum_b (U_0)_{ab} \Phi_b^{(1,0)}(\beta) = \Phi_+^{(1,0)}(\beta) + ia\Phi_-^{(1,0)}(\beta), \tag{2.32}$$

$$Z_a^{\prime(0,1)}(\beta) = \sum_b (U_1)_{ab} \Phi_b^{(0,1)}(\beta) = a\Phi_+^{(0,1)}(\beta) + i\Phi_-^{(0,1)}(\beta). \tag{2.33}$$

Their commutation relations can be determined using (2.37),(2.24) and (2.25). Dropping the upper indices, we find

$$Z_a(\beta_1) Z_b(\beta_2) = \sum_{c,d} S_{ab}^{cd}(\beta_1 - \beta_2) Z_d(\beta_2) Z_c(\beta_1), \tag{2.34}$$

$$Z'_a(\beta_1) Z'_b(\beta_2) = \sum_{c,d} R_{ab}^{cd}(\beta_1 - \beta_2) Z'_d(\beta_2) Z'_c(\beta_1), \tag{2.35}$$

$$Z_a(\beta_1) Z'_b(\beta_2) = ab \tan\left(\frac{\pi}{4} + i\frac{\beta_1 - \beta_2}{2}\right) Z'_b(\beta_2) Z_a(\beta_1). \tag{2.36}$$

Here we have used

$$\lim \tau(q^{i\beta/\pi}) = \tan\left(\frac{\pi}{4} + \frac{i\beta}{2}\right). \tag{2.37}$$

The conditions (2.18),(2.19) become

$$Z_a(\beta_1)Z_b(\beta_2) = \frac{C}{\beta_1-\beta_2-\pi i}\delta_{a+b,0}+O(1) \qquad (\beta_1\to\beta_2+\pi i), \tag{2.38}$$

$$Z'_a(\beta)Z'_b(\beta+\pi i) = C'\delta_{a+b,0}. \tag{2.39}$$

Here C, C' are constants depending on the normalization of $Z_a(\beta), Z'_a(\beta)$. In addition, the symmetry relation (2.22) reduces to the following.

$$Z'_a(\beta)Z_b(\beta-\frac{\pi i}{2}) = Z_b(\beta^*)Z'_a(\beta^*-\frac{\pi i}{2}), \qquad \beta^*=\beta-\pi i(\xi+\frac{1}{2}). \tag{2.40}$$

In the next section we shall discuss the bosonization of these relations (2.34)-(2.40).

3. Bosonization of the vertex operators

In this section we give a bosonization of the operators $Z_\pm(\alpha)$ and $Z'_\pm(\alpha)$.

Let us consider free bosons $a(t)$ $(t\in\mathbb{R})$ which satisfy

$$[a(t),a(t')] = \frac{\sinh\frac{\pi t}{2}\sinh\pi t\sinh\frac{\pi t(\xi+1)}{2}}{t\sinh\frac{\pi t\xi}{2}}\delta(t+t'). \tag{3.1}$$

We also use $a'(t)$ defined by

$$a'(t)\sinh\frac{\pi t(\xi+1)}{2} = a(t)\sinh\frac{\pi t\xi}{2}.$$

We consider the Fock space $\mathcal{H}$ generated by $|\mathrm{vac}\rangle$ which satisfies

$$a(t)|\mathrm{vac}\rangle = 0 \quad \text{if} \quad t>0.$$

We set

$$V(\alpha) = :e^{i\phi(\alpha)}:, \qquad i\phi(\alpha) = \int_{-\infty}^{\infty}\frac{a(t)}{\sinh\pi t}e^{i\alpha t}dt,$$

$$\overline{V}(\alpha) = :e^{-i\bar\phi(\alpha)}:, \qquad i\bar\phi(\alpha) = \int_{-\infty}^{\infty}\frac{a(t)}{\sinh\frac{\pi t}{2}}e^{i\alpha t}dt,$$

$$V'(\alpha) = :e^{i\phi'(\alpha)}:, \qquad i\phi'(\alpha) = -\int_{-\infty}^{\infty}\frac{a'(t)}{\sinh\pi t}e^{i\alpha t}dt,$$

$$\overline{V}'(\alpha) = :e^{-i\bar\phi'(\alpha)}:, \qquad i\bar\phi'(\alpha) = -\int_{-\infty}^{\infty}\frac{a'(t)}{\sinh\frac{\pi t}{2}}e^{i\alpha t}dt.$$

In Appendix 1, we list the operator products of these operators.

A comment is in order here. When we compute the contractions of operators, we often encounter an integral

$$\int_0^\infty F(t)dt$$

which is divergent at $t = 0$. Here we adopt the following prescription for the regularization: it should be understood as the contour integral

$$\int_C F(t)\frac{\log(-t)}{2\pi i}dt, \tag{3.2}$$

where C is the contour

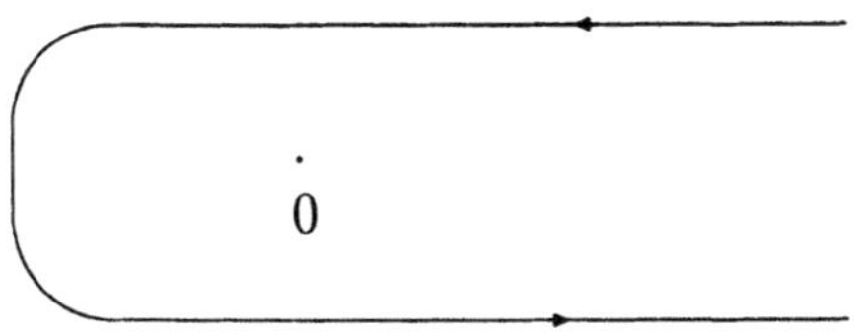

Now let us define

$$\begin{aligned} Z_+(\beta) &= V(\beta), && (3.3)\\ Z_-(\beta) &= \int_{C_1}\frac{d\alpha}{2\pi} : e^{i\phi(\beta)-i\bar{\phi}(\alpha)} : f(\alpha-\beta), && (3.4)\\ Z'_+(\beta) &= V'(\beta), && (3.5)\\ Z'_-(\beta) &= \int_{C_2}\frac{d\alpha}{2\pi} : e^{i\phi'(\beta)-i\bar{\phi}'(\alpha)} : f'(\alpha-\beta), && (3.6) \end{aligned}$$

where

$$\begin{aligned} f(\alpha) &= \Gamma\left(\frac{i\alpha}{\pi\xi}-\frac{1}{2\xi}\right)\Gamma\left(-\frac{i\alpha}{\pi\xi}-\frac{1}{2\xi}\right),\\ f'(\alpha) &= \Gamma\left(\frac{i\alpha}{\pi(\xi+1)}+\frac{1}{2(\xi+1)}\right)\Gamma\left(-\frac{i\alpha}{\pi(\xi+1)}+\frac{1}{2(\xi+1)}\right). \end{aligned}$$

Here the integration contours are chosen as follows. The contour C_1 is $(-\infty,\infty)$ except that the poles

$$\alpha-\beta = -\frac{\pi i}{2}+n\pi\xi i \quad (n\in\mathbb{Z}_{\geq 0})$$

of $\Gamma(\frac{i(\alpha-\beta)}{\pi\xi}-\frac{1}{2\xi})$ are above C_1 and the poles

$$\alpha-\beta = \frac{\pi i}{2}-n\pi\xi i \quad (n\in\mathbb{Z}_{\geq 0})$$

of $\Gamma(-\frac{i(\alpha-\beta)}{\pi\xi}-\frac{1}{2\xi})$ are below C_1. The contour C_2 is $(-\infty,\infty)$. The poles

$$\alpha-\beta=\frac{\pi i}{2}+n\pi(\xi+1)i \quad (n\in Z_{\geq 0})$$

of $\Gamma(\frac{i(\alpha-\beta)}{\pi(\xi+1)}+\frac{1}{2(\xi+1)})$ are above C_2 and the poles

$$\alpha-\beta=-\frac{\pi i}{2}-n\pi(\xi+1)i \quad (n\in Z_{\geq 0})$$

of $\Gamma(-\frac{i(\alpha-\beta)}{\pi(\xi+1)}+\frac{1}{2(\xi+1)})$ are below C_2. We are going to show that the operators (3.3)-(3.5) satisfy the relations for the degeneration of the elliptic algebra.

Proposition 3.1 *The operators $Z_\pm(\beta)$ and $Z'_\pm(\beta)$ satisfy the commutation relations (2.34), (2.35) and (2.36).*

Proof. First we consider (2.34). The equality

$$Z_+(\beta_1)Z_+(\beta_2)=S_0(\beta_1-\beta_2)Z_+(\beta_2)Z_+(\beta_1)$$

follows immediately from (A.1).

Note that

$$w(\alpha)=c_1 f(\alpha)\sinh\frac{\alpha-\frac{\pi i}{2}}{\xi},\qquad c_1=\frac{e^{-(\gamma+\log(\pi\xi))\frac{\xi+1}{\xi}}}{\pi i}.$$

We have

$$\begin{aligned} Z_+(\beta_1)Z_-(\beta_2) &= c_1 g(\beta_2-\beta_1)\int_{C_1}\frac{d\alpha}{2\pi} :e^{i\phi(\beta_1)+i\phi(\beta_2)-i\bar{\phi}(\alpha)}: \\ &\times\ f(\alpha-\beta_1)f(\alpha-\beta_2)\sinh\frac{\alpha-\beta_1-\frac{\pi i}{2}}{\xi}. \end{aligned} \tag{3.7}$$

The contour C_1 must be properly chosen with respect to the location of the poles of $f(\alpha-\beta_1)$ and $f(\alpha-\beta_2)$. As for $f(\alpha-\beta_2)$ this is already discussed. Since the factor $f(\alpha-\beta_1)\sinh\frac{\alpha-\beta_1-\frac{\pi i}{2}}{\xi}$ is from the contraction of the kind (A.2), the choice of C_1 is such that $\mathrm{Im}(\alpha-\beta_1)<-\frac{\pi}{2}$. Note that the zeros of $\sinh\frac{\alpha-\beta_1-\frac{\pi i}{2}}{\xi}$ cancel the poles of $\Gamma(-\frac{i\alpha}{\pi\xi}-\frac{1}{2\xi})$. Therefore, we can keep the same condition on C_1 with respect to the poles of $f(\alpha-\beta_1)$ as $f(\alpha-\beta_2)$. Similarly, we have

$$Z_+(\beta_2)Z_-(\beta_1) = c_1 g(\beta_1-\beta_2)\int_{C_1}\frac{d\alpha}{2\pi} :e^{i\phi(\beta_1)+i\phi(\beta_2)-i\bar{\phi}(\alpha)}:$$

$$\times f(\alpha-\beta_1)f(\alpha-\beta_2)\sinh\frac{\alpha-\beta_2-\frac{\pi i}{2}}{\xi}, \tag{3.8}$$

$$\begin{aligned} Z_-(\beta_2)Z_+(\beta_1) &= -c_1 g(\beta_1-\beta_2)\int_{C_1}\frac{d\alpha}{2\pi} : e^{i\phi(\beta_1)+i\phi(\beta_2)-i\bar{\phi}(\alpha)} : \\ &\quad\times f(\alpha-\beta_1)f(\alpha-\beta_2)\sinh\frac{\alpha-\beta_1+\frac{\pi i}{2}}{\xi}, \end{aligned} \tag{3.9}$$

We used (A.12) in the latter equality. The condition on the contour C_1 is the same for all (3.7), (3.8), (3.9). Therefore, the equality

$$\begin{aligned} Z_+(\beta_1)Z_-(\beta_2) &= S_0(\beta_1-\beta_2)\{\frac{\sinh\frac{\pi i}{\xi}}{\sinh\frac{\pi i-\beta_1+\beta_2}{\xi}}Z_+(\beta_2)Z_-(\beta_1) \\ &+ \frac{\sinh\frac{\beta_1-\beta_2}{\xi}}{\sinh\frac{\pi_i-\beta_1+\beta_2}{\xi}}Z_-(\beta_2)Z_+(\beta_1)\} \end{aligned}$$

follows from the identity

$$\begin{aligned} &\sinh\frac{\alpha-\beta_1-\frac{\pi i}{2}}{\xi}\sinh\frac{\pi_i-\beta_1+\beta_2}{\xi} \\ =\ &\sinh\frac{\alpha-\beta_2-\frac{\pi i}{2}}{\xi}\sinh\frac{\pi i}{\xi}-\sinh\frac{\alpha-\beta_1+\frac{\pi i}{2}}{\xi}\sinh\frac{\beta_1-\beta_2}{\xi}. \end{aligned}$$

The proof of the equality

$$\begin{aligned} Z_-(\beta_1)Z_+(\beta_2) &= S_0(\beta_1-\beta_2)\{\frac{\sinh\frac{\pi i}{\xi}}{\sinh\frac{\pi_i-\beta_1+\beta_2}{\xi}}Z_-(\beta_2)Z_+(\beta_1) \\ &+\frac{\sinh\frac{\beta_1-\beta_2}{\xi}}{\sinh\frac{\pi_i-\beta_1+\beta_2}{\xi}}Z_+(\beta_2)Z_-(\beta_1)\} \end{aligned}$$

is similar. Now, we proceed to the equality

$$Z_-(\beta_1)Z_-(\beta_2) = S_0(\beta_1-\beta_2)Z_-(\beta_2)Z_-(\beta_1). \tag{3.10}$$

We have

$$\begin{aligned} Z_-(\beta_1)Z_-(\beta_2) = g(\beta_2-\beta_1)\int_{C_1}\frac{d\alpha_1}{2\pi}\int_{C_1}\frac{d\alpha_2}{2\pi} : e^{i\phi(\beta_1)+i\phi(\beta_2)-i\bar{\phi}(\alpha_1)-i\bar{\phi}(\alpha_2)} : \\ \times f(\alpha_1-\beta_1)f(\alpha_2-\beta_2)\bar{g}(\alpha_2-\alpha_1)w(\alpha_2-\beta_1)w(\beta_2-\alpha_1). \end{aligned} \tag{3.11}$$

The factors $\bar{g}(\alpha_2-\alpha_1), w(\alpha_2-\beta_1), w(\beta_2-\alpha_1)$ arise with the restriction

$$\mathrm{Im}(\alpha_2-\alpha_1) < -\pi,\ \mathrm{Im}(\alpha_2-\beta_1) < -\frac{\pi}{2},\ \mathrm{Im}(\beta_2-\alpha_1) < -\frac{\pi}{2}$$

respectively. However, the location of the poles of these functions is consistent with the modification of the contours for α_1 and α_2 to C_1. Therefore, the equality (3.10) follows from the identity

$$\begin{aligned} & S(f(\alpha_1-\beta_1)f(\alpha_2-\beta_2)\bar{g}(\alpha_2-\alpha_1)w(\alpha_2-\beta_1)w(\beta_2-\alpha_1)) \\ = \; & S(f(\alpha_1-\beta_1)f(\alpha_2-\beta_2)\bar{g}(\alpha_1-\alpha_2)w(\alpha_1-\beta_2)w(\beta_1-\alpha_2)) \end{aligned}$$

where S means the symmetrization of (α_1, α_2).

The proof of (2.35) is similar. In fact, it is even simpler because the contour C_2 is simpler than C_1. We omit the detail.

Let us consider (2.36). The case $a = b = 1$ follows immediately from (A.7). The case $a = -b = 1$ follows from (A.7), (A.8), (A.9). Note that the operator products (A.8), (A.9) produce no pole on the line $\mathrm{Im}\,(\beta_2-\beta_1) = 0$. Therefore, the contours in $Z'_-(\alpha)$ of the left and right hand sides are equal. The case $a = -b = -1$ is similar. Finally, consider the case $a = b = -1$. We have

$$\begin{aligned} Z_-(\beta_1)Z'_-(\beta_2) = -ie^{-\gamma} \int_{C_1} \frac{d\alpha_1}{2\pi} \int_{C_2} \frac{d\alpha_2}{2\pi} :e^{i\phi(\beta_1)-i\bar{\phi}(\alpha_1)+i\phi'(\beta_2)-i\bar{\phi}'(\alpha_2)}: \\ \times f(\alpha_1-\beta_1)f'(\alpha_2-\beta_2)h(\beta_2-\beta_1)\frac{(\beta_2-\alpha_1)(\alpha_2-\beta_1)}{(\alpha_1-\alpha_2)^2+\frac{\pi^2}{4}}. \end{aligned} \tag{3.12}$$

The contour for α_1 must go below $\beta_1 - \frac{\pi i}{2}$, while it must also avoid the pole at $\alpha_1 = \alpha_2 - \frac{\pi i}{2}$. Apparently there is a chance of pinching at $\mathrm{Im}\beta_1 = 0$. However the contour for α_2 can be deformed because $f'(\alpha_1 - \beta_2)$ has no pole if $\mathrm{Im}\,(\alpha_2 - \beta_1) < \frac{\pi}{2}$. Therefore, the pinching does not occur. With a similar consideration for $Z'_-(\beta_2)Z_-(\beta_1)$ we obtain (2.36). □

Proposition 3.2 *The operators $Z_\pm(\beta)$ and $Z'_\pm(\beta)$ satisfy (2.40).*

Proof. Note that in (A.7) the restriction to $\beta_2 - \beta_1 = -\frac{\pi i}{2}$ is regular. The case $a = b = +$ follows immediately from the identity

$$\begin{aligned} & -\sinh\frac{\pi t\xi}{2}e^{i\beta t} + \sinh\frac{\pi t(\xi+1)}{2}e^{i(\beta-\frac{\pi i}{2})t} \\ & = \sinh\frac{\pi t(\xi+1)}{2}e^{i(\beta-\pi i(\xi+\frac{1}{2}))t} - \sinh\frac{\pi t\xi}{2}e^{i(\beta-\pi i(\xi+1))t}. \end{aligned} \tag{3.13}$$

Next consider the case $a = -b = +$. We have an integration in Z_-. Call the integration variable α. The equality of the integrands in the left and right hand sides follows from (3.13) and the identity

$$(\alpha-\beta)f(\alpha-\beta+\frac{\pi i}{2}) = f(\alpha-\beta+\pi i(\xi+\frac{1}{2}))(\beta-\alpha-\pi i(\xi+1)).$$

The equality of the contours also follows because the irrelevant poles are canceled by the factors $\alpha - \beta$ and $\beta - \alpha - \pi i(\xi + 1)$. The case $a = -b = -$ is similar.

In the case $a = b = -$, the only additional care we should take is to choose the integration contour for the variable α_1 in the factors $Z'_-(\beta)$, $Z'_-(\beta^* - \frac{\pi i}{2})$, and C_2 for α_2 in $Z_-(\beta - \frac{\pi i}{2})$, $Z_-(\beta^*)$ in such a way that C_1 is above $\alpha_2 + \frac{\pi i}{2}$ in the left hand side, and below $\alpha_2 - \frac{\pi i}{2}$ in the right hand side. This is possible because the pole of $f(\alpha_2 - \beta + \frac{\pi i}{2})$ at $\alpha_2 = \beta$ and the pole of $f(\alpha_2 - \beta + \pi i(\xi + \frac{1}{2}))$ at $\alpha_2 = \beta - \pi i(\xi + 1)$ are canceled. □

Proposition 3.3 *The operators $Z_\pm(\beta)$ satisfy (2.38). Likewise $Z'_\pm(\beta)$ satisfy (2.39). The constants C, C' are given by*

$$C = c_1 \sin\frac{\pi}{\xi}\left(\pi\xi\Gamma(-\frac{1}{\xi})\right)^2 g(-\pi i), \qquad c_1 = \frac{1}{\pi i}e^{-(\gamma+\log(\pi\xi))(\xi+1)/\xi},$$
$$C' = c'_1 \sin\frac{\pi}{\xi+1}\left(\pi(\xi+1)\Gamma(\frac{1}{\xi+1})\right)^2 \lim_{\beta\to 0}\frac{g'(\beta+\pi i)}{\beta},$$
$$c'_1 = \frac{1}{\pi i}e^{-(\gamma+\log(\pi(\xi+1)))\xi/(\xi+1)}.$$

Proof. We shall show the case of $Z_\pm(\beta)$.

Since $g(\beta_1 - \beta_2)$ is regular at $\beta_1 = \beta_2 + \pi i$, $Z_+(\beta_1)Z_+(\beta_2)$ is also regular there. Consider $Z_+(\beta_1)Z_-(\beta_2)$ given in (3.7). As β_1 tends to $\beta_2 + \pi i$, the contour is pinched. Taking the residue at $\alpha = \beta_1 - \pi i/2$ we find that modulo regular terms the right hand side is written as

$$c_1 g(\beta_2 - \beta_1) f(\beta_1 - \beta_2 - \frac{\pi i}{2})\frac{\pi\xi}{i}\sin\frac{\pi}{\xi}\Gamma\left(-\frac{1}{\xi}\right) : e^{-i\phi(\beta_1-\pi i)+i\phi(\beta_2)} : .$$

Here we have used

$$\bar{\phi}(\beta) = \phi(\beta + \frac{\pi i}{2}) + \phi(\beta - \frac{\pi i}{2}).$$

The behavior at $\beta_1 = \beta_2 + \pi i$ is easily calculated from this. The case of $Z_-(\beta_1)Z_+(\beta_2)$ is similar.

Finally consider $Z_-(\beta_1)Z_-(\beta_2)$. In the right hand side of (3.11), we rewrite the integral into three parts as

$$\begin{aligned} I_1 + I_2 + I_3 \;=\; & \int_{C_{\beta_1-\pi i/2}}\frac{d\alpha_1}{2\pi}\int_{C_1}\frac{d\alpha_2}{2\pi} + \\ & + \int_{\tilde{C}_1}\frac{d\alpha_1}{2\pi}\int_{C_{\beta_1-\pi i/2}}\frac{d\alpha_2}{2\pi} + \int_{\tilde{C}_1}\frac{d\alpha_1}{2\pi}\int_{\tilde{C}_1}\frac{d\alpha_2}{2\pi}, \end{aligned}$$

where $C_{\beta_1-\pi i/2}$ is a small circle around $\beta_1-\pi i/2$, and $\tilde{C}_1$ is the same as C_1 except that $\beta_1-\pi i/2$ is below $\tilde{C}_1$. Clearly I_3 is regular. Taking the residues we find

$$\begin{aligned} I_1 &= Af(\beta_1-\beta_2-\pi i/2)\sinh\frac{\beta_2-\beta_1}{\xi} \\ &\times \int_{C_1}\frac{d\alpha_2}{2\pi}\frac{f(\alpha_2-\beta_2)}{w(\alpha_2-\beta_1+\pi)} : e^{-i\phi(\beta_1-\pi i)+i\phi(\beta_2)-i\bar{\phi}(\alpha_2)} :, \\ I_2 &= Af(\beta_1-\beta_2-\pi i/2)\sinh\frac{-\pi i}{\xi} \\ &\times \int_{\tilde{C}_1}\frac{d\alpha_1}{2\pi}\frac{f(\alpha_1-\beta_1)w(\beta_2-\alpha_1)}{w(\alpha_1-\beta_1-\pi)w(\beta_1-\alpha_1)} : e^{-i\phi(\beta_1-\pi i)+i\phi(\beta_2)-i\bar{\phi}(\alpha_1)} :, \end{aligned}$$

where A is a constant. In the above, the integral part for I_1 is now regular because of the zero of the integrand. The pole at $\beta_1=\beta_2+\pi i$ comes from $f(\beta_1-\beta_2-\pi i/2)$. Taking the residue we find that the residues for I_1 and I_2 cancel each other. This completes the proof. □

A similar calculation shows the following.

Proposition 3.4

$$\begin{aligned} &\mathrm{Res}_{\beta_1=\beta_2+\pi i-\pi i\xi}Z_a(\beta_1)Z_b(\beta_2) = const.\delta_{a+b,0}T(\beta), \\ &T(\beta) =: e^{\lambda(\beta+\pi i/2)} : + : e^{-\lambda(\beta-\pi i/2)} :, \end{aligned}$$

where

$$\lambda(\beta) = i\phi(\beta-\frac{\pi i}{2}) - i\phi(\beta-\frac{\pi i}{2}+\pi i\xi) = \int\frac{2\sinh(\pi\xi t/2)}{\sinh\pi t}e^{it\beta+\pi(\xi+1)t/2}a(t)dt.$$

The operator $T(\beta)$ is nothing but the limit of the current of the deformed Virasoro algebra introduced by Shiraishi et al. [12]. This agrees with Lukyanov's observation [13] that the deformed Virasoro algebra (in the limit) and the creation operators of breathers in sine-Gordon theory satisfy the same commutation relations.

Remark The bosonization given above is basically obtained from Lukyanov's [6] by setting $t=m\varepsilon$, $a(t)=a_m/\varepsilon$ and letting $\varepsilon\to 0$. This procedure gives rise to the following expressions for the vertex operators

$$Z_+(\beta) = e^{-2\beta/\xi} : e^{i\varphi(\beta)} :, \tag{3.14}$$

$$Z_-(\beta) = e^{2\beta/\xi}\int\frac{d\alpha}{2\pi}e^{(\alpha-\beta)/\xi}f(\alpha-\beta) : e^{i\varphi(\beta)-i\bar{\varphi}(\alpha)} :, \tag{3.15}$$

and similar formulas for $Z'_\pm(\beta)$. Here

$$i\varphi(\beta) = \sqrt{\frac{\xi+1}{2\xi}}iQ + i\phi(\beta), \qquad i\bar{\varphi}(\beta) = i\varphi(\beta-\frac{\pi i}{2}) + i\varphi(\beta+\frac{\pi i}{2}).$$

They differ from the formulas (3.3),(3.4) by the presence of a piece Q of the 'zero mode' operators, and an exponential factor. They satisfy the same commutation relations (2.34)-(2.36). However we have not been able to recover the formulas for the correlation functions (1.3) using (3.14),(3.15) and their analogs. For this reason we modified the operators as given in (3.3)-(3.6).

4. Massless correlation functions

In this section, we derive a solution of the difference equations (1.4)-(1.6) algebraically, and obtain an integral formula for it.

Let us define the boost operator H as

$$H = \int_0^\infty dt \frac{t^2 \sinh\frac{\pi t(\xi+1)}{2}}{\sinh\frac{\pi t}{2}\sinh\pi t\sinh\frac{\pi t\xi}{2}} a'(-t)a'(t) \tag{4.1}$$

which enjoys the property

$$e^{\lambda H}a'(t)e^{-\lambda H} = e^{-\lambda t}a'(t). \tag{4.2}$$

Hence we have

$$e^{\lambda H}X(\beta)e^{-\lambda H} = X(\beta+i\lambda), \tag{4.3}$$

for $X = V, \bar{V}, V', \bar{V}'$. Now let us consider the following trace function.

$$G(\beta_1, \cdots, \beta_{2n})_{\varepsilon_1,\cdots,\varepsilon_{2n}} = \frac{\mathrm{tr}_{\mathcal{H}}(e^{-\lambda H} Z'_{\varepsilon_1}(\beta_1)\cdots Z'_{\varepsilon_{2n}}(\beta_{2n}))}{\mathrm{tr}_{\mathcal{H}}(e^{-\lambda H})}. \tag{4.4}$$

By using the relations (2.35) and (4.3), one can show that the function $G(\beta_1, \cdots, \beta_{2n})_{\varepsilon_1,\cdots,\varepsilon_{2n}}$ satisfies the desired difference equations (1.4)-(1.6).

The bosonization formulas (3.3)-(3.5) for the vertex operators make the evaluation of the trace in (4.4) possible. Here we consider the basic function $G(\beta_1, \cdots, \beta_n, \beta_{n+1} \cdots, \beta_{2n})_{+\cdots+-\cdots-}$. The general components are obtainable from this by the relation (1.4).

From (3.5) and (3.6), we have

$$\begin{aligned} &G(\beta_1, \cdots, \beta_n, \beta_{n+1} \cdots, \beta_{2n})_{+\cdots+-\cdots-} \\ &= \frac{1}{\mathrm{tr}_{\mathcal{H}}(e^{-\lambda H})} \int \prod_{a=n+1}^{2n} \left(\frac{d\alpha_a}{2\pi} f'(\beta_a - \alpha_a)\right) \\ &\times \mathrm{tr}_{\mathcal{H}}(e^{-\lambda H} V'(\beta_1)\cdots V'(\beta_n) : e^{i\phi(\beta_{n+1})-i\bar{\phi}(\alpha_{n+1})} : \cdots : e^{i\phi(\beta_{2n})-i\bar{\phi}(\alpha_{2n})} :). \end{aligned} \tag{4.5}$$

Carrying out a normal ordering procedure, we obtain

$$\begin{aligned} & V'(\beta_1)\cdots V'(\beta_n) : e^{i\phi(\beta_{n+1})-i\bar{\phi}(\alpha_{n+1})} : \cdots : e^{i\phi(\beta_{2n})-i\bar{\phi}(\alpha_{2n})} : \\ & = \prod_{1\le j<k\le 2n} g'(\beta_k-\beta_j) \prod_{1\le j\le 2n}\prod_{n+1\le a\le 2n} w'(\alpha_a-\beta_j) \\ & \times \prod_{n+1\le a<b\le 2n} w'(\alpha_b-\beta_a)w'(\beta_b-\alpha_a)\bar{g}'(\alpha_b-\alpha_a) \\ & \quad \times e^{\int_0^\infty dt\mathcal{F}(-t)a'(-t)} e^{\int_0^\infty dt\mathcal{F}(t)a'(t)}, \end{aligned} \tag{4.6}$$

where

$$\mathcal{F}(t) = -\sum_{j=1}^{2n} \frac{e^{i\beta_j t}}{\sinh\pi t} + \sum_{a=n+1}^{2n} \frac{e^{i\alpha_a t}}{\sinh\frac{\pi t}{2}}. \tag{4.7}$$

Then the trace is evaluated by using the following formula.

$$\begin{aligned} & \mathrm{tr}_{\mathcal{H}}(e^{-\lambda H} e^{\int_0^\infty dt g(t)a'(-t)} e^{\int_0^\infty dt h(t)a'(t)})/\mathrm{tr}_{\mathcal{H}}(e^{-\lambda H}) \\ & = \exp\Big\{\int_0^\infty dt \frac{\sinh\frac{\pi t}{2}\sinh\pi t\sinh\frac{\pi t\xi}{2}}{t\sinh\frac{\pi t(\xi+1)}{2}(e^{\lambda t}-1)} g(t)h(t)\Big\} \end{aligned} \tag{4.8}$$

for some functions $g(t)$ and $h(t)$.

Set $\nu = 1/(\xi+1)$. Using this formula in (4.5), we have the following result [1].

$$\begin{aligned} & G(\beta_1,\cdots,\beta_n,\beta_{n+1}\cdots,\beta_{2n})_{+\cdots+-\cdots-} \\ & = \prod_{1\le j<k\le 2n} \rho(\beta_j-\beta_k) \prod_{a=n+1}^{2n} \int_{C_a} \frac{d\alpha_a}{2\pi} \prod_{j=1}^{2n}\prod_{a=n+1}^{2n} \varphi(\beta_j-\alpha_a) \\ & \times \prod_{n+1\le a<b\le 2n} \frac{\psi(\alpha_a-\alpha_b)}{\sinh\nu(\alpha_a-\alpha_b-\pi i)} \\ & \times \prod_{\substack{j<a \\ 1\le j\le 2n \\ n+1\le a\le 2n}} \sinh\nu(\alpha_a-\beta_j+\frac{\pi i}{2}) \prod_{n+1\le a<j\le 2n} \sinh\nu(\beta_j-\alpha_a+\frac{\pi i}{2}) \end{aligned} \tag{4.9}$$

Here we omitted an irrelevant constant factor. The functions $\rho(\beta), \varphi(\beta)$ and $\psi(\alpha)$ are defined by

$$\rho(\beta) = \sinh\frac{\pi\beta}{\lambda} \frac{S_3(\pi-i\beta|\lambda,2\pi,\frac{\pi}{\nu})S_3(\pi+\lambda+i\beta|\lambda,2\pi,\frac{\pi}{\nu})}{S_3(-i\beta|\lambda,2\pi,\frac{\pi}{\nu})S_3(\lambda+i\beta|\lambda,2\pi,\frac{\pi}{\nu})}, \tag{4.10}$$

$$\varphi(\beta) = \frac{2}{S_2(\frac{\pi}{2}+i\beta|\lambda,\frac{\pi}{\nu})S_2(\frac{\pi}{2}-i\beta|\lambda,\frac{\pi}{\nu})}, \tag{4.11}$$

$$\psi(\alpha) = \sinh\frac{\pi\beta}{\lambda} S_2(\pi+i\beta|\lambda,\frac{\pi}{\nu})S_2(\pi-i\beta|\lambda,\frac{\pi}{\nu}). \tag{4.12}$$

For the definition of the multiple sine function $S_r(x|\omega_1, \cdots, \omega_r)$, see Appendix A of [1].

In the derivation of (4.9), we made the following identification.

$$g'(\beta_k - \beta_j)\exp\Big\{-\int_0^\infty dt \frac{\sinh\frac{\pi t}{2}\sinh\frac{\pi t\xi}{2}}{t\sinh\pi t\sinh\frac{\pi t(\xi+1)}{2}} \frac{e^{i(\beta_j-\beta_k)t}+e^{-i(\beta_j-\beta_k)t}}{e^{\lambda t}-1}\Big\}$$

$$= \text{const.}\rho(\beta_j - \beta_k), \tag{4.13}$$

$$\bar{g}'(\alpha_b - \alpha_a)\exp\Big\{-\int_0^\infty dt \frac{\sinh\pi t\sinh\frac{\pi t\xi}{2}}{t\sinh\frac{\pi t}{2}\sinh\frac{\pi t(\xi+1)}{2}} \frac{e^{i(\alpha_a-\alpha_b)t}+e^{-i(\alpha_a-\alpha_b)t}}{e^{\lambda t}-1}\Big\}$$

$$= \text{const.}\frac{\psi(\alpha_a - \alpha_b)}{\sinh\nu(\alpha_a - \alpha_b - \pi i)}, \tag{4.14}$$

$$w'(\alpha - \beta)\exp\Big\{\int_0^\infty dt \frac{\sinh\frac{\pi t\xi}{2}}{t\sinh\frac{\pi t(\xi+1)}{2}} \frac{e^{i(\beta-\alpha)t}+e^{-i(\beta-\alpha)t}}{e^{\lambda t}-1}\Big\}$$

$$= \text{const.}\varphi(\beta - \alpha)\sinh\nu(\alpha - \beta + \frac{\pi i}{2}). \tag{4.15}$$

The resultant expression (4.9) coincides with the one obtained in [1] previously.

Acknowledgements

Two of us (M.J. and T.M.) are indebted to the organizers of the workshop, in particular to Moshe Flato and Daniel Sternheimer, for kind invitation and hospitality. We thank S. Khoroshkin, D. Lebedev, S. Lukyanov, S. Pakuliak and Y. Pugai for stimulating discussions.

A. Appendix

Here we list the formulas of the form

$$X(\beta_1)Y(\beta_2) = C_{X,Y}(\beta_2 - \beta_1) : X(\beta_1)Y(\beta_2) :$$

where $X, Y = V, \bar{V}, V', \bar{V}'$ and $C_{X,Y}(\beta)$ is a meromorphic function on $\mathbb{C}$. The equality $C_{X,Y} = C_{Y,X}$ is valid in all cases. Therefore, we omit redundant cases.

These formulas follow from the definition of $V(\alpha)$, $\bar{V}(\alpha)$, $V'(\alpha)$, $\bar{V}'(\alpha)$ and the commutation relation (3.1) . We first obtain $\log C_{X,Y}$ as an integral of the form:

$$\int_0^\infty f(t)dt = \int_C f(t)\frac{\log(-t)dt}{2\pi i}$$

where the regularization prescription (3.2) is implied. The resulting integrals can be written by using the gamma function $\Gamma(x)$ or the double gamma function $\Gamma_2(x|\omega_1,\omega_2)$. For a summary of definitions and properties of the latter, see Appendix A in [1]. In each formula, we have shown the region of the absolute convergence of the integral by $\mathrm{Im}(\beta_2-\beta_1)<0$, etc. This information is useful when we fix the integration contours in the formulas (3.4) and (3.6).

$$V(\beta_1)V(\beta_2)=g(\beta_2-\beta_1):e^{i\phi(\beta_1)+i\phi(\beta_2)}:\quad (\mathrm{Im}(\beta_2-\beta_1)<0) \tag{A.1}$$

$$g(\beta)=\frac{\Gamma_2(i\beta+\pi|2\pi,\pi\xi)\Gamma_2(i\beta+\pi(\xi+1)|2\pi,\pi\xi)}{\Gamma_2(i\beta|2\pi,\pi\xi)\Gamma_2(i\beta+\pi(\xi+2)|2\pi,\pi\xi)}e^{\frac{\gamma(\xi+1)}{2\xi}}$$

$$V(\beta_1)\bar{V}(\beta_2)=w(\beta_2-\beta_1):e^{i\phi(\beta_1)-i\bar{\phi}(\beta_2)}:\quad (\mathrm{Im}(\beta_2-\beta_1)<-\frac{\pi}{2}) \tag{A.2}$$

$$w(\beta)=\frac{\Gamma(\frac{i\beta}{\pi\xi}-\frac{1}{2\xi})}{\Gamma(\frac{i\beta}{\pi\xi}+1+\frac{1}{2\xi})}e^{-(\gamma+\log(\pi\xi))\frac{\xi+1}{\xi}}=\frac{1}{g(\beta+\frac{\pi i}{2})g(\beta-\frac{\pi i}{2})}$$

$$\bar{V}(\beta_1)\bar{V}(\beta_2)=\bar{g}(\beta_2-\beta_1):e^{-i\bar{\phi}(\beta_1)-i\bar{\phi}(\beta_2)}:\quad (\mathrm{Im}(\beta_2-\beta_1)<-\pi) \tag{A.3}$$

$$\bar{g}(\beta)=\frac{i\beta}{\pi\xi}\frac{\Gamma(\frac{i\beta}{\pi\xi}+1+\frac{1}{\xi})}{\Gamma(\frac{i\beta}{\pi\xi}-\frac{1}{\xi})}e^{2(\gamma+\log(\pi\xi))\frac{\xi+1}{\xi}}=\frac{1}{w(\beta+\frac{\pi i}{2})w(\beta-\frac{\pi i}{2})}$$

$$V'(\beta_1)V'(\beta_2)=g'(\beta_2-\beta_1):e^{i\phi'(\beta_1)+i\phi'(\beta_2)}:\quad (\mathrm{Im}(\beta_2-\beta_1)<\pi) \tag{A.4}$$

$$g'(\beta)=\frac{\Gamma_2(i\beta+2\pi|2\pi,\pi(\xi+1))\Gamma_2(i\beta+\pi(\xi+1)|2\pi,\pi(\xi+1))}{\Gamma_2(i\beta+\pi|2\pi,\pi(\xi+1))\Gamma_2(i\beta+\pi(\xi+2)|2\pi,\pi(\xi+1))}e^{\frac{\gamma\xi}{2(\xi+1)}}$$

$$V'(\beta_1)\bar{V}'(\beta_2)=w'(\beta_2-\beta_1):e^{i\phi'(\beta_1)-i\bar{\phi}'(\beta_2)}:\quad (\mathrm{Im}(\beta_2-\beta_1)<\frac{\pi}{2}) \tag{A.5}$$

$$w'(\beta)=\frac{\Gamma(\frac{i\beta}{\pi(\xi+1)}+\frac{1}{2(\xi+1)})}{\Gamma(\frac{i\beta}{\pi(\xi+1)}+1-\frac{1}{2(\xi+1)})}e^{-(\gamma+\log\pi(\xi+1))\frac{\xi}{\xi+1}}$$

$$=\frac{1}{g'(\beta+\frac{\pi i}{2})g'(\beta-\frac{\pi i}{2})}$$

$$\bar{V}'(\beta_1)\bar{V}'(\beta_2)=\bar{g}'(\beta_2-\beta_1):e^{-i\bar{\phi}'(\beta_1)-i\bar{\phi}'(\beta_2)}:\quad (\mathrm{Im}(\beta_2-\beta_1)<0) \tag{A.6}$$

$$\bar{g}'(\beta)=\frac{i\beta}{\pi(\xi+1)}\frac{\Gamma(\frac{i\beta}{\pi(\xi+1)}+1-\frac{1}{\xi+1})}{\Gamma(\frac{i\beta}{\pi(\xi+1)}+\frac{1}{\xi+1})}e^{2(\gamma+\log\pi(\xi+1))\frac{\xi}{\xi+1}}$$

$$=\frac{1}{w'(\beta+\frac{\pi i}{2})w'(\beta-\frac{\pi i}{2})}$$

$$V(\beta_1)V'(\beta_2)=h(\beta_2-\beta_1):e^{i\phi(\beta_1)+i\phi'(\beta_2)}:\quad (\mathrm{Im}(\beta_2-\beta_1)<\frac{\pi}{2}) \tag{A.7}$$

$$h(\beta)=\frac{\Gamma(\frac{i\beta}{2\pi}+\frac{1}{4})}{\Gamma(\frac{i\beta}{2\pi}+\frac{3}{4})}e^{-\frac{1}{2}(\gamma+\log(2\pi))}$$

$$V(\beta_1)\bar{V}'(\beta_2)=i(\beta_2-\beta_1)e^{\gamma}:e^{i\phi(\beta_1)-i\bar{\phi}'(\beta_2)}:\quad (\mathrm{Im}(\beta_2-\beta_1)<0) \tag{A.8}$$

$$\bar{V}(\beta_1)V'(\beta_2) = i(\beta_2-\beta_1)e^{\gamma} : e^{-i\bar{\phi}(\beta_1)+i\phi'(\beta_2)} : \quad (\mathrm{Im}(\beta_2-\beta_1)<0) \tag{A.9}$$

$$\bar{V}(\beta_1)\bar{V}'(\beta_2) = -\frac{e^{-2\gamma}}{(\beta_2-\beta_1)^2+\frac{\pi^2}{4}} : e^{-i\bar{\phi}(\beta_1)-i\bar{\phi}'(\beta_2)} : \quad (\mathrm{Im}(\beta_2-\beta_1)<-\frac{\pi}{2}) \tag{A.10}$$

We set

$$S_0(\beta) = \frac{g(-\beta)}{g(\beta)}, \qquad R_0(\beta) = \frac{g'(-\beta)}{g'(\beta)}. \tag{A.11}$$

The following relations are sometimes useful.

$$\frac{w(\beta)}{w(-\beta)} = -\frac{\sinh(\beta-\frac{\pi i}{2})/\xi}{\sinh(\beta+\frac{\pi i}{2})/\xi}, \tag{A.12}$$

$$\frac{\bar{g}(\beta)}{\bar{g}(-\beta)} = \frac{\sinh(\beta+\pi i)/\xi}{\sinh(\beta-\pi i)/\xi}, \tag{A.13}$$

$$\frac{w'(\beta)}{w'(-\beta)} = -\frac{\sinh\frac{\beta+\frac{\pi i}{2}}{\xi+1}}{\sinh\frac{\beta-\frac{\pi i}{2}}{\xi+1}}, \tag{A.14}$$

$$\frac{\bar{g}'(\beta)}{\bar{g}'(-\beta)} = \frac{\sinh\frac{\beta-\pi i}{\xi+1}}{\sinh\frac{\beta+\pi i}{\xi+1}}, \tag{A.15}$$

$$\frac{h(\beta)}{h(-\beta)} = -\frac{\sinh(\frac{\beta}{2}-\frac{\pi i}{4})}{\sinh(\frac{\beta}{2}+\frac{\pi i}{4})}. \tag{A.16}$$

References

1. M. Jimbo and T. Miwa. Quantum KZ equation with $|q|=1$ and correlation functions of the XXZ model in the gapless regime. *J. Phys. A*, 29:2923–2958, 1996.
2. M. Jimbo, K. Miki, T. Miwa, and A. Nakayashiki. Correlation functions of the XXZ model for $\Delta < -1$. *Phys. Lett. A*, 168:256–263, 1992.
3. M. Jimbo and T. Miwa. *Algebraic Analysis of Solvable Lattice Models.* CBMS Regional Conference Series in Mathematics vol. 85, AMS, 1994.
4. F. A. Smirnov. *Form Factors in Completely Integrable Models of Quantum Field Theory.* World Scientific, Singapore, 1992.
5. O. Foda, K. Iohara, M. Jimbo, R. Kedem, T. Miwa, and H. Yan. An elliptic quantum algebra for $\widehat{sl}_2$. *Lett. Math. Phys.*, 32:259–268, 1994.
6. S. Lukyanov. Free field representation for massive integrable models. *Comm. Math. Phys.*, 167:183–226, 1995.
7. R. J. Baxter. *Exactly Solved Models in Statistical Mechanics.* Academic, London, 1982.
8. M. Jimbo, R. Kedem, H. Konno, T. Miwa, and R. Weston. Difference equations in spin chains with a boundary. *Nucl. Phys.*, pages 429–456, 1995.
9. O. Foda, K. Iohara, M. Jimbo, R. Kedem, T. Miwa, and H. Yan. Notes on highest weight modules of the elliptic algebra $\mathcal{A}_{q,p}(\widehat{sl}_2)$. *Prog. Theoret. Phys., Supplement*, 118:1–34, 1995.

10. N. Yu Reshetikhin and M. Semenov-Tian-Shansky. Central extensions of quantum current groups. *Lett. Math. Phys.*, 19:133–142, 1990.
11. B. Davies, O. Foda, M. Jimbo, T. Miwa, and A. Nakayashiki. Diagonalization of the XXZ Hamiltonian by vertex operators. *Comm. Math. Phys.*, 151:89–153, 1993.
12. J. Shiraishi, H. Kubo, H. Awata, and S. Odake. A quantum deformation of the Virasoro algebra and the Macdonald symmetric functions, 1995. YITP/U-95-30, q-alg/9507034.
13. S. Lukyanov. A note on the deformed Virasoro algebra. *Phys. Lett. B*, pages 121–125, 1996.

FORMALITY CONJECTURE

MAXIM KONTSEVICH
I. H. E. S.,
35 Route de Chartres, F-91440 Bures-sur-Yvette, France
maxim@ihes.fr

Abstract. We propose a conjecture of a homological nature which implies that every Poisson manifold admits a deformation quantization. Brute force calculations confirms the conjecture up to 6-th order in the perturbation theory. Also, we construct a strange flow on the space of equivalence classes of germs of Poisson manifolds.

1. Introduction

This paper is devoted to a conjecture concerning deformation quantization. This conjecture implies that an arbitrary smooth Poisson manifold can be formally quantized, and the equivalence class of the resulting algebra is canonically defined. In other terms, it means that non-commutative geometry, in the formal approximation to the commutative geometry of smooth spaces, is described by the semi-classical approximation.

Recently an article by A. Voronov (see [15]) inluding the exposition of the formality conjecture appeared on the net. The present paper can be seen as a companion to [15]. Here I present the conjecture in a slightly different form (in Section 4). In order to do it I include some preparatory material on deformation theory and homotopy theory of differential graded Lie algebras (Sections 2,3). Direct calculations of possible obstructions to the conjecture involve a version of Graph Cohomology (Section 5). An immediate consequence is that the conjecture holds up to the 6-th order. As a by-product, I obtained a formula for a new flow on the space of germs of Poisson manifolds. Also I propose a reformulation of the conjecture and further evidence (Section 6).

D. Sternheimer et al. (eds.), Deformation Theory and Symplectic Geometry, 139–156.

1.1. DEFORMATION QUANTIZATION

Let A be the algebra of smooth functions on a C^∞-manifold X. We are interested in star-products on A (see [2]), i.e. associative $\mathbf{R}[[\hbar]]$-linear products on $A[[\hbar]]$ given by formulas

$$(f, g) \mapsto f * g = fg + \hbar B_1(f, g) + \hbar^2 B_2(f, g) + \ldots \in A[[\hbar]]$$

where $\hbar$ is the formal variable, and B_i are bi-differential operators. There is a natural gauge group acting on star-products. It acts via linear transformations $A \longrightarrow A$ parameterized by $\hbar$:

$$f \mapsto f + \hbar D_1(f) + \hbar^2 D_2(f) + \ldots$$

where D_i are differential operators. The formality conjecture will give a clear picture of star-products. In particular, for any Poisson bracket B_1, there should be a canonical gauge equivalence class of star-products with B_1 as the first term.

2. Deformation theory via differential graded Lie algebras

This part is essentially standard (see [8, 9, 13],...).

Let $\mathbf{g}$ be a differential graded Lie algebra over a field of characteristic zero:

$$\mathbf{g} = \bigoplus_{k \in \mathbf{Z}} \mathbf{g}^k, \quad [\, , \,] : \mathbf{g}^k \otimes \mathbf{g}^l \longrightarrow \mathbf{g}^{k+l}, \quad d : \mathbf{g}^k \longrightarrow \mathbf{g}^{k+1}, \quad d^2 = 0\,.$$

In other words, $\mathbf{g}$ is a Lie algebra in the tensor category of complexes of vector spaces.

We associate with it a functor Def on finite-dimensional commutative associative algebras over the same ground field with values in the category of sets.

First of all, let us assume that $\mathbf{g}$ is a nilpotent Lie superalgebra. We define set $\mathcal{M}(\mathbf{g})$ (the set of solutions of the Maurer-Cartan equation) by the formula

$$\mathcal{M}(\mathbf{g}) := \left\{ \gamma \in \mathbf{g}^1 | \ d\gamma + \frac{1}{2}[\gamma, \gamma] = 0 \right\} / \Gamma^0$$

where Γ^0 is the nilpotent group associated with the nilpotent Lie algebra $\mathbf{g}^0$. The action of Γ^0 can be defined by the exponentiation of the infinitesimal action of its Lie algebra:

$$\alpha \in \mathbf{g}^0 \mapsto (\dot{\gamma} = d\alpha + [\alpha, \gamma])\ .$$

Now we are ready to define the functor Def. Technically, it is convenient to define it on the category of finite-dimensional nilpotent commutative associative algebras without unit. Let $\mathbf{m}$ be such an algebra, $\mathbf{m}^{dim\,\mathbf{m}+1} = 0$. The functor is given (on objects) by the formula

$$Def(\mathbf{m}) = \mathcal{M}(\mathbf{g} \otimes \mathbf{m}).$$

In the conventional approach $\mathbf{m}$ is the maximal ideal in a finite-dimensional Artin algebra with unit

$$\mathbf{m}' := \mathbf{m} \oplus (\text{ground field}) \cdot \mathbf{1}.$$

In general, it is convenient to think about a commutative associative algebras without unit as about objects dual to spaces with base points. Algebra corresponding to a space with a base point is the algebra of functions vanishing at the base point.

2.1. EXAMPLES

There are many standard examples of differential graded Lie algebras and related moduli problems. We recall two cases:

1) Let X be a complex manifold. Define $\mathbf{g}^k$ for $k \geq 0$ as

$$\Gamma(X, \Omega^{0,k} \otimes T^{1,0})$$

with the differential equal to $\overline{\partial}$ and the Lie bracket coming from the cup-product on $\overline{\partial}$-forms and the usual Lie bracket on holomorphic vector fields. Then the deformation functor related with $\mathbf{g}$ is canonically equivalent to the usual deformation functor for complex structures on X.

2) Let A be an associative algebra. We define $\mathbf{g}^k$ for $k \geq -1$ as

$$Hom(A^{\otimes(k+1)}, A).$$

The differential is the usual differential in the Hochschild complex, and the Lie bracket is the Gerstenhaber bracket. We would like to recall here the definition of these structures. Let F denote the free coassociative graded super coalgebra with counit cogenerated by the graded vector space $A[1]$, i.e. A endowed with the pure grading in degree -1. Then $\mathbf{g}$ is the Lie algebra of coderivations of F in the tensor category of $\mathbf{Z}$-graded super vector spaces. The associative product on A defines an element $m \in \mathbf{g}^1$ satisfying the equation $[m, m] = 0$. The differential d in $\mathbf{g}$ is defined as $ad(m)$. Again, the deformation functor related with g is equivalent to the usual deformation functor for algebraic structures.

3. L_∞-morphisms, L_∞-algebras and quasi-isomorphisms

Let $\mathbf{g}_1$ and $\mathbf{g}_2$ be two differential graded Lie algebras.

Definition. *A pre-L_∞-morphism f from $\mathbf{g}_1$ to $\mathbf{g}_2$ is an infinite sequence of linear maps $f = (f_1, f_2, \ldots)$ between $\mathbf{Z}$-graded vector spaces:*

$$\begin{aligned} f_1 &: \mathbf{g}_1 \longrightarrow \mathbf{g}_2 \\ f_2 &: \wedge^2(\mathbf{g}_1) \longrightarrow \mathbf{g}_2[-1] \\ f_3 &: \wedge^3(\mathbf{g}_1) \longrightarrow \mathbf{g}_2[-2] \\ &\ldots \end{aligned}$$

In formulas above the exterior power of $\mathbf{g}_1$ is taken in the tensor category of graded super vector spaces. The suffix $[n]$, $n \in \mathbf{Z}$ denotes the tensor product with the standard one-dimensional space endowed with the grading in degree $-n$. Thus, in plain terms we have a collection of linear maps between graded components

$$f_{(k_1,\ldots,k_n)} : \mathbf{g}_1^{k_1} \otimes \ldots \otimes \mathbf{g}_1^{k_n} \longrightarrow \mathbf{g}_2^{k_1+\ldots+k_n+(1-n)}$$

with the symmetry property

$$\begin{aligned} &f_{(k_1,\ldots,k_n)}(\gamma_1 \otimes \ldots \otimes \gamma_n) = \\ &\quad = -(-1)^{k_i k_{i+1}} f_{(k_1,\ldots,k_{i+1},k_i,\ldots,k_n)}(\gamma_1 \otimes \ldots \otimes \gamma_{i+1} \otimes \gamma_i \otimes \ldots \otimes \gamma_n). \end{aligned}$$

As we see, in this definition we use only the structure of graded vector spaces on $\mathbf{g}_1, \mathbf{g}_2$.

With any graded vector space $\mathbf{g}$ we associate the cofree coassociative cocommutative graded superalgebra without counit cogenerated by $\mathbf{g}[1]$:

$$C(\mathbf{g}) := \bigoplus_{k=1}^{\infty} Sym^k(\mathbf{g}[1]) = \bigoplus_{k=1}^{\infty} \wedge^k(\mathbf{g})[k].$$

Notice that in this definition we use $\mathbf{g}$ with the *reversed* parity.

Intuitively, we think of $C(\mathbf{g})$ as of an object corresponding to a formal supermanifold with a based point, possibly infinite-dimensional:

$$(\text{Formal neighborhood of zero in } \mathbf{g}[1],\, 0)\ .$$

The reason for this is that if $\mathbf{g}$ is finite-dimensional then the dual space to $C(\mathbf{g})$ is the algebra of formal power series vanishing at the origin on the super vector space $\mathbf{g}[1]$. Pre-L_∞-morphisms from $\mathbf{g}_1$ to $\mathbf{g}_2$ correspond to graded morphisms of coalgebras

$$C(\mathbf{g}_1) \longrightarrow C(\mathbf{g}_2)$$

and (intuitively) to maps between formal supermanifolds with based points.

Suppose now that $\mathbf{g}$ is a differential graded Lie algebra. Then it is easy to see that the differential and the Lie bracket in $\mathbf{g}$ define a differential Q, $Q^2 = 0$ of degree $+1$ of the coalgebra $C(\mathbf{g})$. In the case of ordinary Lie algebras (in degree 0) we get the standard reduced chain complex. In geometrical terms, we have an action of $(0|1)$-dimensional graded abelian Lie supergroup $\mathbf{A}^1[-1]$ on a formal supermanifold preserving the base point.

The chain complex of a differential graded Lie algebra gives a formal supermanifold with a flat structure in which the odd vector field Q generating $\mathbf{A}^1[-1]$-action has terms of degree 1 and 2 only in the Taylor expansion at the origin.

In paper [1] supermanifolds with the action of $\mathbf{A}^1[-1]$ are called Q-manifolds. By analogy, we introduce here the following

Definition. *A formal Q-manifold is a differential graded coalgebra $\mathcal{C}$ which is isomorphic as a graded coalgebra to $C(\mathbf{g})$ for some graded vector space $\mathbf{g}$.*

We want to stress that the specific isomorphism between $\mathcal{C}$ and $C(\mathbf{g})$ is *not* considered as a part of data. If we fix it, then we will obtain so called L_∞-algebra (or homotopy Lie algebras, see [10]).

Definition. *An L_∞-algebra is a graded vector space $\mathbf{g}$ and a differential Q of degree $+1$ on the graded coalgebra $C(\mathbf{g})$.*

The structure of an L_∞-algebra is given by the infinite sequence of Taylor coefficients Q_i of the odd vector field Q (coderivation of $C(\mathbf{g})$):

$$
\begin{aligned}
Q_1 &: \mathbf{g} \longrightarrow \mathbf{g}[-1] \\
Q_2 &: \wedge^2(\mathbf{g}) \longrightarrow \mathbf{g} \\
Q_3 &: \wedge^3(\mathbf{g}) \longrightarrow \mathbf{g}[1] \\
&\dots
\end{aligned}
$$

The condition $Q^2 = 0$ can be translated into an infinite sequence of quadratic constraints on polylinear maps Q_i. First of these constraints means that Q_1 is the differential of the graded space $\mathbf{g}$. The second constraint means that Q_2 is a skew-symmetric bilinear product on $\mathbf{g}$ compatible with Q_1 by the Leibnitz rule. The third constraint means that Q_2 satisfies the Jacobi identity up to homotopy given by Q_3, etc. As we have seen, a differential graded Lie algebra is the same as an L_∞-algebra with $Q_3 = Q_4 = \ldots = 0$.

Nevertheless, I recommend to return to the geometric point of view and think in terms of formal Q-manifolds. This naturally leads to the following

Definition. *An L_∞-morphism between two L_∞-algebras is a morphism between corresponding differential graded cocommutative coalgebras.*

For the case of differential graded Lie algebras an L_∞-morphism can be identified with a pre-L_∞-morphism satisfying the following equation for any $n = 1, 2, \ldots$:

$$df_n(\gamma_1 \wedge \gamma_2 \wedge \ldots \wedge \gamma_n) - \sum_{i=1}^{n} \pm f_n(\gamma_1 \wedge \ldots \wedge d\gamma_i \wedge \ldots \wedge \gamma_n) =$$
$$= \frac{1}{2} \sum_{k,l \geq 1, k+l=n} \frac{1}{k!l!} \sum_{\sigma \in \Sigma_n} \pm [f_k(\gamma_{\sigma_1} \wedge \ldots \wedge \gamma_{\sigma_k}), f_l(\gamma_{\sigma_{k+1}} \wedge \ldots \wedge \gamma_{\sigma_n})] +$$
$$+ \sum_{i<j} \pm f_{n-1}([\gamma_i, \gamma_j] \wedge \gamma_1 \wedge \ldots \wedge \gamma_n).$$

Here are first two equations in the explicit form:

$$df_1(\gamma_1) = f_1(d\gamma_1),$$

$$df_2(\gamma_1 \wedge \gamma_2) - f_2(d\gamma_1 \wedge \gamma_2) - (-1)^{deg(\gamma_1)} f_2(\gamma_1 \wedge d\gamma_2) =$$
$$= f_1([\gamma_1, \gamma_2]) - [f_1(\gamma_1), f_1(\gamma_2)].$$

We see that f_1 is a morphism of complexes.

L_∞-morphisms generalize usual morphisms of differential graded Lie algebras. We call a L_∞-morphism f a quasi-isomorphism if and only if the first component f_1 induces isomorphism between cohomology groups of complexes $\mathbf{g}_1$ and $\mathbf{g}_2$.

The essence of the homotopy/deformation theory is contained in the following

Theorem. *Let f be an L_∞-morphism form $\mathbf{g}_1$ to $\mathbf{g}_2$. Assume that f is a quasi-isomorphism. Then there exists an L_∞-morphism form $\mathbf{g}_1$ to $\mathbf{g}_2$ inducing the inverse isomorphism between cohomology of complexes g_i, $i = 1, 2$. Also, for the case of differential graded algebras, L_∞-morphism f induces an equivalence between deformation functors associated with $\mathbf{g}_i$.*

The first part of this theorem shows that if $\mathbf{g}_1$ is quasi-isomorphic to $\mathbf{g}_2$ then $\mathbf{g}_2$ is quasi-isomorphic to $\mathbf{g}_1$, i.e. we get an equivalence relation.

The second part of the theorem works as follows. Any solution of the Maurer-Cartan equation in $\mathbf{g}_1$ depending formally on a parameter $\hbar$:

$$\gamma(\hbar) = \gamma_1 \hbar + \gamma_2 \hbar^2 + \ldots \in \mathbf{g}_1^1[[\hbar]], \quad d\gamma(\hbar) + \frac{1}{2}[\gamma(\hbar), \gamma(\hbar)] = 0$$

produces a solution of the Maurer-Cartan equation in $\mathbf{g}_2$:

$$\tilde{\gamma}(\hbar) = \sum_{n=1}^{\infty} \frac{1}{n!} f_n(\gamma(\hbar) \wedge \ldots \wedge \gamma(\hbar)) \in \mathbf{g}_2^1[[\hbar]], \quad d\tilde{\gamma}(\hbar) + \frac{1}{2}[\tilde{\gamma}(\hbar), \tilde{\gamma}(\hbar)] = 0.$$

This theorem is essentially standard (see related results in [8, 9]). My approach consists in the translation of all relevant notions to the geometric language of formal Q-manifolds. The main technical result in this approach is a (dual) version of Sullivan's theory of minimal models (see [14]).

In order to formulate it we introduce two notions:

Definition. *An L_∞-algebra $\mathbf{g}$ is called* **minimal** *if the first Taylor coefficient Q_1 of the differential Q vanishes.*

Definition. *An L_∞-algebra $\mathbf{g}$ is called* **linear contractible** *if all higher Taylor coefficients Q_k, $k \geq 2$ of the differential Q vanish, and the cohomology of the differential Q_1 on the graded space $\mathbf{g}$ vanish.*

Lemma (Minimal Model Theory). *Any L_∞-algebra $\mathbf{g}$ is L_∞-isomorphic to the product of a minimal L_∞-algebra $\mathbf{g}_{min}$ and a linear contractible L_∞-algebra $\mathbf{g}_{contr}$.*

This lemma can be proved by induction in degrees of terms of coefficients of Taylor expansions (see [12]). From this lemma follows that the set of quasi-isomorphisms classes of L_∞-algebras is in one-to-one correspondence with the set of L_∞-equivalence classes of minimal algebras.

4. The main conjecture

Let X be a smooth manifold. We associate with it two differential graded Lie algebras. The first Lie superalgebra $D^*(X)$ is a subalgebra of the Hochschild complex of the algebra A of functions on X (see the last paragraph of Section 2). The space $D^n(X)$, $n \geq -1$ consists of local Hochschild cochains $A^{\otimes(n+1)} \longrightarrow A$ given by polydifferential operators. In local coordinates (x_i) any element of D^n can be written as

$$\phi_0 \otimes \ldots \otimes \phi_n \mapsto \sum_{(I_0,\ldots,I_n)} C_{I_0,\ldots,I_n}(x) \cdot \partial^{I_0}(\phi_0) \ldots \partial^{I_n}(\phi_n)$$

where the sum is finite, I_k denote multi-indices, ∂^{I_k} denote corresponding partial derivatives, and ϕ_k and $C_{I_0,\ldots,I_n}$ are functions in (x_i).

The second differential graded Lie algebra $T^*(X)$ is the Lie superalgebra of polyvector fields on X:

$$T^n(X) = \Gamma(X, \wedge^{n+1} T_X), \quad n \geq -1$$

endowed with the standard Schouten-Nijenhuis bracket and with the differential $d := 0$.

We have an evident map $f_1 : T^*(X) \longrightarrow D^*(X)$:

$$f_1 : (\xi_0 \wedge \ldots \wedge \xi_n) \mapsto \left(\phi_0 \otimes \ldots \otimes \phi_n \mapsto \frac{1}{n!} \sum_{\sigma \in \Sigma_n} sgn(\sigma) \prod_{i=1}^{n} \xi_{\sigma_i}(\phi_i) \right) .$$

Here ξ_i are vector fields on X and ϕ_i are functions. By a version of Kostant-Hochschild-Rosenberg theorem this map is a quasi-isomorphism of complexes.

Formality Conjecture. *There exists an L_∞-morphism f from $T^*(X)$ to $D^*(X)$ with the first term f_1 fixed as above.*

In other words, this conjecture claims that $T^*(X)$ and $D^*(X)$ are quasi-isomorphic differential graded Lie algebras. In analogous situation in rational homotopy theory (see [14]), a differential graded commutative algebra is called formal if it is quasi-isomorphic to its cohomology algebra endowed with zero differential. This explains the name of my conjecture.

Solutions of the Maurer-Cartan equation in $T^*(X)$ are exactly Poisson structures on X. The gauge group action is the action of the diffeomorphism group by conjugation. Thus, the conjecture implies that every Poisson manifold can be canonically quantized.

4.1. REFORMULATION

Here I present an attempt to reformulate the formality conjecture in some vague terms. This part is not essential for the rest of the paper and can be regarded as a deviation.

The Lie superalgebra $T^*(X)$ is the Lie superalgebra of functions on supermanifold ΠT^*X (the odd cotangent bundle to X) with respect to the odd Poisson bracket. Its quotient modulo the center (constant functions) coincides with the Lie superalgebra of hamiltonian vector fields on ΠT^*X. In order to get a more neat formulation we can add one extra odd coordinate to ΠT^*X and get a supermanifold Y with an odd contact 1-form α. Lie superalgebra $T^*(X)$ coincides with the Lie superalgebra of infinitesimal automorphisms of (Y, α).

I propose to consider the differential graded Lie algebra $D^*(X)$ as the Lie algebra of infinitesimal automorphisms in homotopy sense of the triangulated category $D^b(A-mod)$ (or, better, A_∞-category, see [11]), where A is the algebra of functions on X. First of all, the truncated algebra $D^{\geq 0}(X)$ is responsible for automorphisms/deformations of A in the homotopy sense. When we add to it the term $D^{-1}(X) \simeq A$ then it means that we consider inner automorphisms of A as trivial. Morally, it is analogous to the passage from groups to groupoids, i.e. to categories. I am sorry, but I can't be more precise there.

Thus, the formality conjecture means that there exists a natural construction of an A_∞-category from a supermanifold with an odd contact 1-form (and vice versa).

For me it looks as an odd version of the usual star-quantization of even symplectic manifolds. Constructions by De Wilde - Lecomte and Fedosov (see [5, 7]) of the star-quantization of symplectic manifolds show that with every symplectic manifold one can associate in a canonical way an equivalence class of an associative algebra $\mathcal{A}$ over $\mathbf{R}[[\hbar]]$. Unfortunately, the algebra $\mathcal{A}$ itself can not be constructed canonically. Nevertheless, after an additional work one can construct canonically an abelian category which is equivalent to the category of $\mathcal{A}$-modules (J. Bernstein, M.K., unpublished).

In the usual even situation the quantization of the *algebraic* symplectic manifold $\mathbf{R}^{2n}$ with the standard symplectic structure gives the algebra of polynomial differential operators on $\mathbf{R}^n$. The symplectic group acts by automorphisms of this algebra. In particular, the Fourier transform identifies algebras of polynomial differential operators on dual vector spaces. It seems that in the odd symplectic geometry it corresponds to the result of Beilinson and Bernstein-Gelfand-Gelfand on the equivalence (after small modifications) of bounded derived categories of modules over symmetric and exterior algebras (see [3] and [4]). Morally, triangulated categories for algebras of functions on $\mathbf{R}^{n|0}$ and on $\mathbf{R}^{0|n}$ are equivalent.

In paper [1] we proposed a Lagrangian for a topological two-dimensional sigma-model with the target space being an odd symplectic manifold. There are some heuristic arguments (see [11]) showing that boundary conditions for topological field theories in two dimensions coupled with gravity form an A_∞-category. Thus, one can hope that the string theory can provide a proof of the formality conjecture.

5. Obstructions to the formality

In this section we will show that if the formality conjecture is wrong then there exists a non-zero cohomology class in certain universal complex, the odd Graph Complex (various versions of the Graph Complex were introduced in [12]). Direct calculations with this complex show that the appropriate cohomology groups vanish up to degree 6 in perturbation theory. It seems quite possible that all these cohomology groups of the odd Graph Complex vanish.

5.1. GENERAL OBSTRUCTION THEORY

Let $\mathbf{g}$ be a differential graded Lie algebra and $\mathbf{g}_1$ be its algebra of cohomology considered as a differential graded Lie algebra (with zero differential). We are looking for a quasi-isomorphism between $\mathbf{g}$ and $\mathbf{g}_1$, i.e. asking whether $\mathbf{g}$ is formal, or not. Let us start to construct components f_i of an L_∞-morphism from $\mathbf{g}_1$ to $\mathbf{g}$ by induction. The first component can be constructed as a quasi-isomorphism of complexes $f_1 : \mathbf{g}_1 \longrightarrow \mathbf{g}$ which identifies graded Lie algebras $\mathbf{g}_1$ and $H^*(\mathbf{g})$.

We assume that we already constructed components $f_1, \dots, f_N$ satisfying first N equations on an L_∞-morphism:

$$\forall n \leq N \qquad df_n(\gamma_1 \wedge \ldots \wedge \gamma_n) = \sum_{i<j} \pm f_{n-1}([\gamma_i, \gamma_j] \wedge \gamma_1 \wedge \ldots \wedge \gamma_n) +$$

$$+\frac{1}{2} \sum_{k,l \geq 1,\ k+l=n} \frac{1}{k!l!} \sum_{\sigma \in \Sigma_n} \pm [f_k(\gamma_{\sigma_1} \wedge \ldots \wedge \gamma_{\sigma_k}), f_l(\gamma_{\sigma_{k+1}} \wedge \ldots \wedge \gamma_{\sigma_n})] =$$

$$= \sum_{i<j} \pm f_{n-1}([\gamma_i, \gamma_j] \wedge \gamma_1 \wedge \ldots \wedge \gamma_n) +$$

$$+ \sum_i \pm [f_1(\gamma_i), f_{n-1}(\gamma_1 \wedge \ldots \wedge \gamma_{i-1} \wedge \gamma_{i+1} \wedge \ldots \wedge \gamma_n)] +$$

$$+\frac{1}{2} \sum_{k,l \geq 2,\ k+l=n} \frac{1}{k!l!} \sum_{\sigma \in \Sigma_n} \pm [f_k(\gamma_{\sigma_1} \wedge \ldots \wedge \gamma_{\sigma_k}), f_l(\gamma_{\sigma_{k+1}} \wedge \ldots \wedge \gamma_{\sigma_n})].$$

Here we use the fact that the differential in $\mathbf{g}_1$ is equal to 0.

We want to solve analogous equation for $n = N+1$. It is easy to see that the right hand side of this equation is an element of $Ker(d) \subset \mathbf{g}$. This follows directly from all previous equations. If for all γ_i the right hand side belongs to $Im(d) \subset Ker(d)$, we are ready to construct f_{N+1}. The obstruction is a polylinear map:

$$\Phi : \wedge^{N+1}(\mathbf{g}_1) \longrightarrow \mathbf{g}_1[1-N].$$

I claim that this map is cocycle of the graded Lie superalgebra $\mathbf{g}_1$ with the coefficients in the (twisted by $[1-N]$) adjoint representation. It means that

$$\sum_{i<j} \pm \Phi([\gamma_i, \gamma_j] \wedge \gamma_0 \wedge \ldots \wedge \gamma_{N+1}) + \sum_i \pm [\gamma_i, \Phi(\gamma_0 \wedge \ldots \wedge \gamma_{i-1} \wedge \gamma_{i+1} \wedge \ldots)] = 0.$$

Again, it follows directly from all previous equations.

Let us assume that the cohomology class of this cocycle is zero:

$$\Phi(\gamma_0 \wedge \ldots \wedge \gamma_N) = \sum_{i<j} \pm \Psi([\gamma_i, \gamma_j] \wedge \gamma_0 \wedge \ldots \wedge \gamma_N) +$$

$$+ \sum_i \pm [\gamma_i, \Psi(\gamma_0 \wedge \ldots \wedge \gamma_{i-1} \wedge \gamma_{i+1} \wedge \ldots \wedge \gamma_N)]$$

for some linear map $\Psi : \wedge^N(\mathbf{g}_1) \longrightarrow \mathbf{g}_1[1-N]$.

Then, if we replace f_N by

$$\tilde{f}_N : \gamma_1 \wedge \ldots \wedge \gamma_N \mapsto f_N(\gamma_1 \wedge \ldots \wedge \gamma_N) + f_1(\Psi(f_N))$$

we obtain a new collection of maps $(f_1, \ldots, \tilde{f}_N)$ with vanishing obstruction to the existence of f_{N+1}.

We see that the obstruction to the formality lies in cohomology groups which can be loosely denoted by $H^N(\mathbf{g}_1, \mathbf{g}_1[1-N])$. These cohomology groups are cohomology of complexes consisting in skew-symmetric polylinear maps (in the tensor category of graded vector spaces) from $\mathbf{g}_1$ to $\mathbf{g}_1[1-N]$ with the differential imitating the usual Chevalley-Eilenberg differential.

5.2. POLYVECTOR FIELDS AND THE GRAPH COMPLEX

Let us apply the general result to the Lie superalgebra of polyvector fields in the flat space $\mathbf{R}^d$. As a graded vector space this superalgebra $T^*(\mathbf{R}^d)$ can be identified with the space of functions on supermanifold $\mathbf{R}^{d|d}$ with coordinates x_i, ξ^i, $i = 1, \ldots, d$. The first group of coordinates, (x_i) is pure even, in degree 0, the second group (ξ^i) is purely odd, in degree +1. Loosely speaking, ξ^i corresponds to $\partial/\partial x_i$.

One can show that the first non-vanishing obstruction class, if it exists, should be in certain sense stable (independent of the dimension). There is a natural class of stable cochains of $T^*(\mathbf{R}^d)$ associated with finite graphs. I claim that the obstruction class to the formality conjecture comes from a cohomology class in the (odd) Graph Complex. The proof of this statement is too long and technical to include there. Thus, the reader should consider the rest of this section only as an announcement.

Here follows the description of cochains associated with graphs. Let Γ be a finite non-oriented graph (= finite 1-dimensional CW-complex) without multiple edges. If we enumerate vertices of this graph by $\{1, \ldots, n\}$ then this graph is defined by its incidence matrix $M = (M_{kl})$ which is symmetric with entries in $\{0, 1\}$.

Let us also enumerate the set of edges of Γ by $\{1, \ldots, m\}$. The corresponding cochain of $T^*(\mathbf{R}^d)$ is the obtained by antisymmetrization from the following polylinear map $\mathbf{g}_1^{\otimes n} \longrightarrow \mathbf{g}_1[n-m-1]$:

$$\phi_1 \otimes \ldots \otimes \phi_n \mapsto \left(\left(\prod_{k \le l,\ M_{kl}=1} \Omega_{kl} \right) (\phi_1 \otimes \ldots \otimes \phi_n) \right)_{diag} .$$

Here on the right hand side the tensor product of functions ϕ_i on $\mathbf{R}^{d|d}$ is considered as a function on $\left(\mathbf{R}^{d|d}\right)^n$. The ste of factors in the product of

Ω_{kl} is in one-to-one correspondence with the ste of edges of Γ. Differential operator Ω acts on functions on $\mathbf{R}^{d|d} \times \mathbf{R}^{d|d}$ by the formula

$$\Omega = \sum_i \left(\frac{\partial}{\partial x_i} \otimes \frac{\partial}{\partial \xi^i} + \frac{\partial}{\partial \xi^i} \otimes \frac{\partial}{\partial x_i} \right) .$$

Operator Ω_{kl} is a copy of Ω acting on the product of the k-th and the l-th factor of $\left(\mathbf{R}^{d|d}\right)^n$ for $k \neq l$. For $k = l$ we define Ω_{kk} as the differential operator on $\mathbf{R}^{d|d}$

$$\sum_i \frac{\partial^2}{\partial x_i \partial \xi^i}$$

acting along the k-th factor of $\left(\mathbf{R}^{d|d}\right)^n$.

Operators Ω_{kl} are odd anti-commuting differential operators acting on functions on $\left(\mathbf{R}^{d|d}\right)^n$. In order to specify the meaning of their composition we use the enumeration of the set of edges of Γ (up to an even permutation). Then we restrict the result of the application of the composite operator to the tensor product of functions ϕ_k to the diagonal, and obtain again a function on $\mathbf{R}^{d|d}$.

Polylinear maps $\mathbf{g}_1^{\otimes n} \longrightarrow \mathbf{g}_1$ obtained in this way are not antisymmetric in general. After antisymmetrization some graphs give automatically zero cochain. The result of the antisymmetrization does not depend on the enumeration of vertices. Also, a boundary of the cochain given by a graph is obtained from a universal linear combination of graphs.

Now we are going to describe the part of the graph complex relevant for the formality conjecture.

Definition. *A finite graph is called* **good** *if it is nonempty, connected, nonseparable, has no multiple edges, and the valency (degree) of each vertex is greater than or equal to 3. The condition of nonseparability means that the complement to any vertex is connected.*

It follows from the definition that any good graph has no simple loops. In terms of the incidence matrix it means that all diagonal entries are 0. The simplest good graph has 4 vertices, it is the complete graph (the skeleton of a tetrahedron). For a graph Γ we define a one-dimensional vector space $Or(\Gamma)$ over $\mathbf{Q}$. It is the top exterior power of the vector space $\mathbf{Q}^{edges(\Gamma)}$ spanned by the set of edges of Γ. The automorphism group of Γ acts on the space $Or(\Gamma)$. If this action is non-trivial, i.e. there exists an automorphism acting by (-1) on $Or(\Gamma)$, then the corresponding cochain of the Lie superalgebra of polyvector fields vanishes.

The Graph complex will be bi-graded, with the differential of degree $(+1, +1)$.

The graded component $G^{n,m}$ is defined as

$$\bigoplus_{\text{classes of } \Gamma} (Or(\Gamma))_{Aut(\Gamma)} .$$

Here the sum is taken over all isomorphism classes of good graphs with n vertices and m edges. For each isomorphism class we choose a representative and take the space of coinvariants. This space does not depend on the choice of a representative.

It follows form the definition of the Graph complex that every good graph Γ with enumerated edges defines an element in $G^{n,m}$. We denote this element by $[\Gamma, enum]$ where $enum$ denotes the chosen enumeration of edges. We are going to write a formula for $d[\Gamma, enum]$.

Let v be a vertex of graph Γ of degree ≥ 4. Denote by S_v the set of edges connected to v. Let us represent S_v as a disjoint union of two sets of cardinality ≥ 2: $S_v = S_v^1 \sqcup S_v^2$. With such a structure we associate a new good graph $\Gamma' = \Gamma'(v, S_v^1, S_v^2)$. This graph has $n+1$ vertices and $m+1$ edges.

First of all, we remove vertex v and all edges from S_v from Γ. Then we add two new vertices v_1 and v_2. Edges from the set S_v^1 we connect to v_1 (and to same vertices in $\Gamma \setminus \{v\}$ as it was in Γ), edges from S_v^2 connect to v_2, both new vertices v_1 and v_2 connect by a new edge e. We obtain graph Γ'. The set of edges of Γ' is obtained from the set of edges of Γ by adding a new element e. Thus, we get an enumeration $enum'$ of edges of Γ' by attaching number $m+1$ to the edge e.

The formula for the differential in the Graph Complex is

$$d[\Gamma, enum] = \sum_{(v, S_v^1, S_v^2)} [\Gamma'(v, S_v^1, S_v^2), enum'(v, S_v^1, S_v^2)] .$$

The square of this operator is equal to zero.

Claim. *Let us assume that the formality conjecture for flat spaces* $X = \mathbf{R}^d$, $d = 1, 2, \ldots$ *holds up to the* $(n-1)$*-st term in the perturbation theory. Then the possible obstruction on to the formality on the* n*-th step comes from a cohomology class of the graph complex in degree* $(n, 2n-3)$.

Elementary counting shows that $G^{n,2n-3} = 0$ for $n \leq 6$. Thus, we get a confirmation of the formality conjecture with a great precision.

In fact, any good 3-valent graph gives a cohomology class of the Graph Complex. Thus one can expect plenty of non-trivial cohomology classes in degrees (k, l) where $l/k = 3/2$. It seems quite plausible that degrees in which cohomology is non-trivial are close to the line $l/k = 3/2$ and there is no cohomology in degrees $(n, 2n-3)$ (close to the line $l/k = 2$) at all!

5.3. A FLOW ON THE SPACE OF POISSON MANIFOLDS

If we believe in the formality conjecture, we can ask ourselves whether the desired quasi-isomorphism is unique up to homotopy, or not. The corresponding cohomology groups of the graph complex have degrees $(n, 2n-2)$. I am aware of only one such a class, it corresponds to simplest good graph, the complete graph with 4 vertices (and 6 edges). This class gives a remarkable vector field on the space of bi-vector fields on $\mathbf{R}^d$. The evolution with respect to the time t is described by the following non-linear partial differential equation:

$$\frac{d\alpha}{dt} := \sum_{i,j,k,l,m,k',l',m'} \frac{\partial^3 \alpha_{ij}}{\partial x_k \partial x_l \partial x_m} \frac{\partial \alpha_{kk'}}{\partial x_{l'}} \frac{\partial \alpha_{ll'}}{\partial x_{m'}} \frac{\partial \alpha_{mm'}}{\partial x_{k'}} \left(\frac{\partial}{\partial x_i} \wedge \frac{\partial}{\partial x_j} \right)$$

where $\alpha = \sum_{i,j} \alpha_{ij}(x) \frac{\partial}{\partial x_i} \wedge \frac{\partial}{\partial x_j}$ is a bi-vector field on $\mathbf{R}^d$.

A priori we can guarantee the existence of a solution of the evolution only for small times and real-analytic initial data.

More generally, this class produces a formal one-parameter group of L_∞-automorphisms of $T^*(\mathbf{R}^d)$.

It follows from general properties of cohomology that 1) this evolution preserves the class of (real-analytic) Poisson structures, 2) if two Poisson structures are conjugate by a real-analytic diffeomorphism then the same will hold after the evolution. Thus, our evolution operator is essentially intrinsic and does not depend on the choice of coordinates.

In fact, I cheated a little bit. In the formula for the vector field on the space of bivector fields which one get from the tetrahedron graph, an additional term is present. This term is equal (up to a numerical factor) to

$$\sum_{i,j,k,l,m,n,p,r} \frac{\partial^2 \alpha_{ij}}{\partial x_k \partial x_l} \frac{\partial^2 \alpha_{km}}{\partial x_n \partial x_p} \frac{\partial \alpha_{nl}}{\partial x_r} \frac{\partial \alpha_{rp}}{\partial x_j} \left(\frac{\partial}{\partial x_i} \wedge \frac{\partial}{\partial x_m} \right).$$

It is possible to prove formally that if α is a Poisson bracket, i.e. if $[\alpha, \alpha] = 0 \in T^2(\mathbf{R}^d)$, then the additional term shown above vanishes.

I tried in vain to find any example where my evolution operator really changes the conjugacy class of a Poisson bracket. The problem is that principal examples of Poisson brackets have coefficients which are either linear (Kirillov bracket), or quadratic polynomials (classical Sklyanin algebras). In the evolution I use third derivatives. Also, in dimension $d = 2$ the direct calculation shows that the evolution operator gives a conjugation of bivector field α by a vector field whose coefficients are differential polynomials in coefficients of α. Nevertheless, it is easy to check that in higher dimensions it is impossible to obtain a formal proof of triviality of the evolution operator. I don't know whether one can construct the evolution operator on arbitrary Poisson manifolds, not necessarily embeddable into the Euclidean space of the same dimension.

The formality conjecture implies that there should be analogous (formal) evolution operator on equivalence classes of non-commutative algebras close to algebras of functions on smooth manifolds.

As a side remark, I would like to mention that the cohomology of the odd Graph Complex form a Lie superalgebra which acts on the space of equivalence classes of germs of QP-manifolds (see [1] for the definition).

6. Comments and further evidence

In the formulation of the formality conjecture I didn't express all my wishes. First of all, graded components $f_{(k_1,\dots,k_n)}$ of the desired L_∞-morphism f form $T^*(X)$ to $D^*(X)$ should be given by local expressions, i.e. by polydifferential operators in coefficients of polyvector fields.

Also, it would be nice to formulate the conjecture using the language of sheaves:

6.1. SHEAVES

Differential graded Lie algebras $D^*(X)$ and $T^*(X)$ are global sections of natural sheaves D^* and T^*. By "naturalness" I mean that sheaves are defined in a coordinate-independent way. Sheaves D^* and T^* can be defined not only for C^∞-manifolds, but also for complex-analytic manifolds and for algebraic manifolds defined over a field of characteristic zero. One can just hope that there is a natural sheaf $\mathcal{X}^*$ of differential graded Lie algebras on manifolds (in a wide sense) with natural morphisms which are quasi-isomorphisms, to both sheaves D^* and T^*.

6.2. CUP-PRODUCTS ON THE TANGENT COHOMOLOGY

If $\mathbf{g}$ is a differential graded Lie algebra and $\gamma \in \mathbf{g} \otimes \mathbf{m}$ is a solution of the Maurer-Cartan equation, where $\mathbf{m}$ is an Artin algebra, then one can define the tangent cohomology $T^*(\gamma)$ responsible for the deformation of γ. This is the cohomology group of the complex $\mathbf{g} \otimes \mathbf{m}$ endowed with the differential $d' := d + ad(\gamma)$. Cohomology groups are modules over $\mathbf{m}$. A more delicate invariant is the isomorphism class of this complex in the derived category of $\mathbf{m}$-modules. Again, any quasi-isomorphism induces the isomorphism of the tangent cohomology.

Let (X, α) be a Poisson manifold, i.e. $\alpha \in T^1(X)$ $[\alpha, \alpha] = 0$. Thus we get a solution of the Maurer-Cartan equation in $T^*(X)$ depending on formal variable $\hbar$:

$$\alpha(\hbar) = \alpha \cdot \hbar \in T^1[[\hbar]] .$$

Assuming the formality conjecture we get a solution of the Maurer-Cartan equation in $D^*(X)$, i.e. an associative algebra $\mathcal{A}$ over $\mathbf{R}[[\hbar]]$ which is isomorphic to $C^\infty(X)[[\hbar]]$ as $\mathbf{R}[[\hbar]]$-module.

On both tangent cohomology for $T^*(X)$ and for $D^*(X)$ there is a natural cup-product of degree $+1$.

For the the Lie superalgebra $T^*(X)$ we use just the usual cup-product of polyvector fields. For the case of $D^*(X)$ we use the interpretation of tangent cohomology (Hochschild cohomology) as the Ext-groups shifted by [1]:

$$Ext^*_{\mathcal{A}\otimes_{\mathbf{R}[[\hbar]]}\mathcal{A}^o-mod}(\mathcal{A},\mathcal{A})[1]\,.$$

Then the Yoneda product gives a cup-product in tangent cohomology.

The tangent cohomology for the graded Lie algebra $T^*(X)$ coincides (after tensoring $\otimes_{\mathbf{R}[[\hbar]]}\mathbf{R}((\hbar))$) with the tensor product of $\mathbf{R}((\hbar))$ with the Poisson cohomology of (X,α) shifted by [1]. The Poisson cohomology $HP^*(X,\alpha)$ is defined as the cohomology of the complex $T^*(X)[-1]$ endowed with the differential $ad(\alpha)$.

Conjecture about cup-products. *Assuming the formality conjecture, the cup product on tangent cohomology for a solution of the Maurer-Cartan equation in $T^*(X)$ coincides with the cup-product on the tangent cohomology for the corresponding solution in $D^*(X)$.*

I am aware about one non-trivial example, confirming this conjecture. Namely, let X be the vector space dual to a Lie algebra $\mathbf{g}$ and α will be the usual Kirillov-Poisson structure on $X=\mathbf{g}^*$. The space X is considered here as an algebraic manifold, not a C^∞-manifold. I claim that the quantization of this Poisson manifold should be the universal enveloping algebra $U_\hbar(\mathbf{g})$ where the variable $\hbar$ is used as the deformation parameter:

$$U_\hbar(\mathbf{g}) = \mathbf{R}[[\hbar]]\text{-algebra generated by } \mathbf{g}/\{\text{relations } X\cdot Y - Y\cdot X = \hbar[X,Y]\}.$$

The reason for this claim is based on certain considerations in invariant theory. Namely, the deformation $U_\hbar(\mathbf{g})$ should correspond to a 1-parameter family of Poisson structures on $X=\mathbf{g}^*$. Using $GL(\mathbf{g})$-invariant expressions in coefficients of the structure tensor of the Lie algebra $\mathbf{g}$ one can get only the standard linear Kirillov bracket. There is no invariant way to construct a polylinear map $\wedge^2(\mathbf{g})\longrightarrow Sym^k(\mathbf{g})$ for $k\geq 2$, i.e. no way to get higher coefficients in the Taylor expansion for a Poisson bracket.

The Poisson cohomology in degree 0 of (X,α) is equal to the algebra of functions on X invariant under the coadjoint action. The Hochshild cohomology of $U(\mathbf{g})$ in degree 0 is the center $\mathcal{Z}(U(\mathbf{g}))$ of the universal enveloping algebra. Of course, in the definition of tangent cohomology we should include variable $\hbar$ into the game.

It is well-known that vector spaces $HP^0(X,\alpha)$ and $Z(U(\mathbf{g}))$ can be canonically identified:

$$HP^0(X,\alpha) \simeq \bigoplus_{k=0}^{\infty} \left(Sym^k(\mathbf{g})\right)^{\mathbf{g}} \simeq (U(\mathbf{g}))^{\mathbf{g}} \simeq \mathcal{Z}(U(\mathbf{g}))\,.$$

Nevertheless, the most naive identification does not preserve the algebra structure.

Fortunately, a long time ago M. Duflo (see [6]) based on Kirillov's "universal formula" for characters of representations of Lie groups, constructed a different map $HP^0(X,\alpha)\longrightarrow\mathcal{Z}(U(\mathbf{g}))$ which gives an isomorphism of algebras. For the case of semi-simple Lie algebras it is the Harish-Chandra isomorphism. Strictly speaking, the construction of Duflo was done only for solvable Lie algebras. Nevertheless, if one include the formal parameter $\hbar$, the formula became universal and didn't require the classification theory of Lie algebras. It seems that it works for finite-dimensional Lie algebras in arbitrary tensor categories in characteristic zero, although I am not aware of a general proof. Coefficients in Duflo-Kirillov formulas are quite non-trivial, they are equal essentially to the Bernoulli numbers.

References

1. M. Alexandrov, M. Kontsevich, A. Schwarz, O. Zaboronsky, *The Geometry of the Master Equation and Toplogical Quantum Field Theory*, hep-th/9502010.
2. F. Bayen, M. Flato, C. Frønsdal, A. Lichnerowicz, D. Sternheimer, *Deformation theory and quantization. I. Deformations of symplectic structures*, Ann. Physics **111** (1978), no. 1, 61 - 110.
3. A. Beilinson, *Coherent sheaves on P^N and problems of linear algebra*, Funct. Anal. Appl., **12** (1978), 214 - 216.
4. I. Bernstein, I. Gelfand, S. Gelfand, *Algebraic bundles on P^N and problems of linear algebra*, Funct. Anal. Appl., **12** (1978), 212 - 214.
5. M. De Wilde, P. B. A. Lecomte, *Existence of star-products and of formal deformations in Poisson Lie algebra of arbitrary symplectic manifolds*, Let. math. Phys., **7** (1983), 487 - 496.
6. M. Duflo, *Caractéres des algèbres de Lie résolubles*, C. R. .Acad. Sci., **269** (1969), série a, p. 437 - 438.
7. B. Fedosov, *A simple geometric construction of deformation quantization*, J. Diff. Geom., bf 40 (1994) 2, 213 - 238.
8. W. Goldman, J. Millson, *The homotopy invariance of the Kuranishi space*, Ill. J. Math., **34** (1990), no. 2, 337 - 367.
9. V. Hinich, V. Schechtman, *Deformation theory and Lie algebra homology*, alg-geom/9405013.
10. V. Hinich, V. Schechtman, *Homotopy Lie algebras*, I. M. Gelfand Seminar, Adv. Sov. Math., bf 16 (1993), part 2,, 1- 28.
11. M. Kontsevich, *Homological algebra of mirror symmetry*, Proceedings of ICM, Zürich 1994, vol. I, Birkhäuser (1995), 120 - 139.
12. M. Kontsevich, *Formal (non)-commutative symplectic geometry*, The Gelfand Mathematical Seminars, 1990 - 1992, Birkhäuser (1993), 173 - 187.
13. M. Schlessinger, J. Stasheff, *The Lie algebra structure on tangent cohomology and deformation theory*, J. Pure Appl. Algebra **89** (1993), 231 - 235.

14. D. Sullivan, *Infinitesimal computations in topology*, I. H. E. S. Publ. Math., no. 47 (1977), 269 - 331.
15. A. Voronov, *Quantizing Poisson Manifolds*, q-alg/9701017.

QUANTUM COHOMOLOGY OF THE FLAG MANIFOLD AS AN ALGEBRA OF RATIONAL FUNCTIONS ON A UNIPOTENT ALGEBRAIC GROUP

BERTRAM KOSTANT*
Department of Mathematics, MIT
Cambridge, MA 02139, USA
kostant@math.mit.edu

Abstract. Let G be a simply connected complex semisimple Lie group, $\mathfrak{g} = Lie\, G$ and $X^\vee$ the flag manifold associated to the Langlands dual to G. Let $CH(X^\vee)$ be the complexified quantum cohomology of $X^\vee$. Let $f \in \mathfrak{g}$ be a principal nilpotent element, G_o^f the identity component of the centralizer of f in $\mathfrak{g}$, and $\mathfrak{g}^f = Lie\, G_o^f$. Let F_f be the field of rational functions on G_o^f. Using results of Dale Peterson and earlier results of ours, particularly on the Toda lattice, we show that $CH(X^\vee)$ naturally injects into F_f. Among other things it follows that $CH(X^\vee)$ has the structure of $U(\mathfrak{g}^f)$ module where $U(\mathfrak{g}^f)$ is the universal enveloping algebra of $\mathfrak{g}^f$. Furthermore generators of $CH(X^\vee)$ are explicitly given in terms of the matrix entries of $\pi_\rho|G_o^f$ where π_ρ is the irreducible representation of G whose highest weight is one-half the sum of the positive roots. A particularly important role here is played by Newton's equation of motion for the Toda lattice. For example the quantum parameters q_i appear as "forces".

1. Introduction

1.1. Let $\mathfrak{g}$ be a complex semisimple Lie algebra and let, respectively, $\mathfrak{h} \subset \mathfrak{b}$ be a Cartan subalgebra and Borel subalgebra of $\mathfrak{g}$. Let $H \subset B \subset G$ be Lie groups corresponding, respectively, to $\mathfrak{h} \subset \mathfrak{b} \subset \mathfrak{g}$ where G is taken to be simply-connected. The flag manifold X for G is the projective variety of all Borel subgroups of G. Let $G^\vee$ be the Langlands dual to G and let $X^\vee$ be the flag manifold for $G^\vee$.

*Research supported in part by NSF grant DMS-9307460

D. Sternheimer et al. (eds.), Deformation Theory and Symplectic Geometry, 157–175.

Let $\mathfrak{h}'$ be the dual space to $\mathfrak{h}$ and let $\Delta_+ \subset \Delta \subset \mathfrak{h}'$ be the set of roots for the pairs $(\mathfrak{h}, \mathfrak{b})$ and $(\mathfrak{h}, \mathfrak{g})$, respectively. Regard Δ_+ as a choice of a set of positive roots in Δ. Let $l = rank\, \mathfrak{g}$, $r = dim\, X$ and let (x, y), for $x, y \in \mathfrak{g}$, be a nonsingular invariant symmetric bilinear form on $\mathfrak{g}$. Let $\Pi = \{\alpha_1, \ldots, \alpha_l\} \subset \Delta_+$ be the set of simple positive roots. For $i = 1, \ldots, l$, let $e_i, e_{-i} \in \mathfrak{g}$ be root vectors for α_i and $-\alpha_i$, respectively, and normalized so that $(e_i, e_{-i}) = 1$. Let $\Lambda \subset \mathfrak{h}'$ be the set of dominant integral linear forms on $\mathfrak{h}$. For any $\lambda \in \Lambda$ let $\pi_\lambda : G \to Aut\, V_\lambda$ be an irreducible representation with highest weight λ. Let $\rho_i \in \Lambda$ be the highest weight of the fundamental representation of G corresponding to α_i. We will be particularly concerned in this paper with the 2^r-dimensional representation π_ρ where $\rho = \sum_{i=1}^l \rho_i$.

1.2. Let $f = -\sum_{i=1}^l e_{-i}$ so that f is a principal nilpotent element in $\mathfrak{g}$ and let G_o^f be the identity component of the centralizer G^f of f in G. Our determination, in [8], of the action angle coordinates on a special Lagrangian leaf, Y_0^*, of the Toda lattice corresponding to G, has lead to G_o^f. It also leads to the projective subvariety P_f (which we refer to as the Peterson variety) of X defined as follows: Let $o \in X$ be the unique B fixed point in X and let $Q = G^f \cdot o$ so that $Q \simeq G_o^f$. Then P_f is the closure of Q in X.

Let U be the unique open orbit of B in X. One proves $R = P_f \cap U$ is not empty so that Q and R are two open affine subvarieties of P_f. Let $A(Q)$ and $A(R)$ be their respective affine rings. Let $CH(X^\vee)$ be the complexified quantum cohomology ring of $X^\vee$. If $G = SL(l+1, \mathbb{C})$ then using results of Givental, Kim and Ciocan-Fontanine we proved in [3] that $CH(X^\vee)$ is an integral domain and that as algebras

$$CH(X^\vee) \simeq A(R) \tag{a}$$

so that in fact we can regard

$$R = Spec\; CH(X^\vee) \tag{b}$$

Subsequently Dale Peterson announced a proof of (a) and hence (b) for all G. We assume in this paper the validity of Peterson's results and identify $CH(X^\vee)$ with $A(R)$. But since $A(R)$ and $A(Q)$ have the same quotient field this defines an injection

$$CH(X^\vee) \to F(Q) \tag{c}$$

where $F(Q)$ is the quotient field of $A(Q)$. That is, quantum cohomology classes in $CH(X^\vee)$ define rational functions on Q or equivalently on G_o^f. One of the main points of the present paper is to explicitly determine the rational functions corresponding to certain key quantum cohomology classes.

1.3. The functions in $A(Q)$ arise from matrix entries of representations of G_o^f. In [7] we expressed the key quantum cohomology classes as quotients of matrix entries of the representations $\pi_{\rho_i}|G_o^f$. For computational purposes the use of π_{ρ_i} is less than satisfactory since π_{ρ_i} is not explicitly given in general. In this paper we show, among other things, that we may use π_ρ instead. The point is that π_ρ can be explicitly given. Doing so is one of the main points in [5]. In more detail, $CH(X^\vee)$ is a graded algebra. The gradation (see Theorem 35 in [7]) arises from the action, on R, of the one dimensional subgroup of H which normalizes $\mathfrak{g}^f$. One has $dim\, CH^2(X^\vee) = l$ and a natural linear isomorphism

$$\mathfrak{h}' \to CH^2(X^\vee), \quad \nu \mapsto x_\nu \tag{d}$$

On the other hand the quantum parameters $q_1, \ldots, q_l \in CH^2(X^\vee)$ are algebraically independent and if $C = \mathbb{C}[q_1, \ldots, q_l]$ then $CH^2(X^\vee)$ is generated as a C-algebra by the x_{ρ_i}. That is,

$$CH^2(X^\vee) = C[x_{\rho_1}, \ldots, x_{\rho_l}] \tag{e}$$

Let $v_\rho \in V_\rho$ be a highest weight vector and (since π_ρ is self contragredient) let (u, v), $u, v \in V_\rho$ be a nonsingular G-invariant bilinear form on V_ρ. Let $G_\rho^f = \{g \in G_o^f | (v_\rho, g^{-1} v_\rho) \neq 0\}$. Then if $g \in G_o^f$ one proves that $g \in G_\rho^f$ if and only if $g \cdot o \in Q \cap R$. This establishes isomorphisms

$$G_\rho^f \simeq Q \cap R \simeq Y_0^* \tag{e}$$

But then the generator x_{ρ_i} on $Q \cap R$ is given by

$$x_{\rho_i}(g \cdot o) = (v_\rho, g^{-1} v_i)/(v_\rho, g^{-1} v_\rho) \tag{f}$$

for $g \in G_\rho^f$ where $v_i \in V_\rho$ is a (unique up to scalar multiplication) weight vector for the weight $\rho - \alpha_i$.

1.4. For $z \in \mathfrak{g}^f$ let ξ^z be the corresponding vector field on the G^f-orbit Q. The restrictions $\xi | Q \cap R$ exactly correspond to the Hamiltonian vector fields on the Lagrangian leaf Y_0^* given by the system in involution for the completely integrable Toda lattice. Under this bijection ξ^f corresponds to the Hamiltonian vector field for the Toda Hamiltonian function itself (the "energy" function). On the other hand if $U(\mathfrak{g}^f)$ is the enveloping algebra of $\mathfrak{g}^f$ the action by the vector fields ξ^z defines the structure of a $U(\mathfrak{g}^f)$-module on the quantum cohomology $CH(X^\vee)$. (See Remark 18 for possible ramifications of this action.) A key result is the equation

$$\xi^f x_{\rho_i} = \frac{-2}{(\alpha_i, \alpha_i)} q_i \tag{g}$$

From the point of view of the Toda lattice (g) is Newton's equation of motion, $F = ma$, since the x_{ρ_i} correspond to momenta. However for our purposes here the equation (g) applied to (f) yields an explicit expression (see (52)) for the q_i in terms of the representation $\pi_\rho|G_o^f$. Even more explicitly it involves matrix entries for the natural action of G_o^f on $\wedge^r\mathfrak{g}$. See Remark 22.

1.5. Another consequence of (g) is an analogue of a result (see [2]) of J. Carrell. It was indicated but not explicitly stated in [7]. Let $(\xi^f\, CH(X^\vee))$ be the ideal in $CH(X^\vee)$ generated by the image, $\xi^f\, CH(X^\vee)$ of ξ^f on $CH(X^\vee)$. Then where $H^*(X^\vee)$ is ordinary cohomology

$$CH(X^\vee)/(\xi^f\, CH(X^\vee)) \simeq H^*(X^\vee) \tag{h}$$

1.6. The paper also contains a number of structure results concerning the finite morphism $\mu : R \to \mathbb{C}^l$ induced by the embedding of C in $CH(X^\vee)$. That is μ is the map induced by taking all possible specializations of the quantum parameters. In particular it is shown that if the scheme theoretic fiber of μ at $c \in \mathbb{C}^l$ is reduced then $c \in (\mathbb{C}^*)^l$. Also a quantum Vandermonde condition is given for μ to be étale at $p \in Q \cap R$.

2. Connections with the (nonperiodic) Toda lattice

2.1. Let $\mathfrak{g}$ be a complex semisimple Lie algebra and let, respectively, $\mathfrak{h} \subset \mathfrak{b}$ be a Cartan subalgebra and Borel subalgebra of $\mathfrak{g}$. Let $H \subset B \subset G$ be Lie groups corresponding, respectively, to $\mathfrak{h} \subset \mathfrak{b} \subset \mathfrak{g}$ where G is taken to be simply-connected. The flag manifold $X = G/B$ will be regarded as the variety of all Borel subgroups of G and, with the usual tautology, for any $p \in X$, let $B_p \subset G$ be the corresponding Borel subgroup. The point $o \in X$ is such that $B_o = B$.

Let $\mathfrak{h}'$ be the dual space to $\mathfrak{h}$ and let $\Delta_+ \subset \Delta \subset \mathfrak{h}'$ be the set of roots for the pairs $(\mathfrak{h}, \mathfrak{b})$ and $(\mathfrak{h}, \mathfrak{g})$, respectively. Regard Δ_+ as a choice of a set of positive roots in Δ. Let $l = rank\, \mathfrak{g}$, $r = dim_{\mathbb{C}} X$ and let $B_{\mathfrak{g}}$ be a nonsingular, invariant, symmetric bilinear form on $\mathfrak{g}$. Its value on $x, y \in \mathfrak{g}$ is denoted by (x, y). Let $\Pi = \{\alpha_1, \ldots, \alpha_l\} \subset \Delta_+$ be the set of simple positive roots. For $i = 1, \ldots, l$ let $e_i, e_{-i} \in \mathfrak{g}$ be root vectors for α_i and $-\alpha_i$, respectively, and normalized so that $(e_i, e_{-i}) = 1$.

2.2. Let $f = -\sum_{i=1}^{l} e_{-i}$ so that f is a principal nilpotent element in $\mathfrak{g}$. Let $\mathfrak{e} = \sum_{i=1}^{l} \mathbb{C}e_i$ and translating the $2l$-dimensional subspace $\mathfrak{h} + \mathfrak{e} \subset \mathfrak{b}$ by f let Y be the "Jacobi" subvariety of $\mathfrak{g}$ given by putting

$$Y = f + \mathfrak{h} + \mathfrak{e} \tag{1}$$

In this paper $A(Z)$ will denote the algebra of regular functions on any complex affine variety Z. The adjoint action induces a G-module structure on the graded algebra $A(\mathfrak{g})$. Then (Chevalley) if $m_i \in \mathbb{Z}_+$, $i = 1, \ldots, l$, say in increasing order, are the exponents of $\mathfrak{g}$ then the algebra $A(\mathfrak{g})^G$ of G-invariants is a polynomial ring $\mathbb{C}[\Phi_1, \ldots, \Phi_l]$ where $\Phi_i \in A^{m_i+1}(\mathfrak{g})$.

Let $F_i = \Phi_i|Y$ so that $F_i \in A(Y)$. Let $\mathcal{I}$ be the ideal in $A(Y)$ defined by putting

$$\mathcal{I} = (F_1, \ldots F_l)$$

Let $\mathcal{N}$ be the nilcone of $\mathfrak{g}$, i.e. the variety of all nilpotent elements in $\mathfrak{g}$ and put $Y_0 = Y \cap \mathcal{N}$. It is clear that $Y_0 \subset Y$ is the zero variety for the ideal $\mathcal{I}$. On the other hand if $CH(X)$ denotes the complexified quantum cohomology ring for the flag manifold X then it is a result of Givental, Kim and Ciocan-Fontanine (see [3]) that, in case $G = Sl(l+1, \mathbb{C})$, there is an isomorphism

$$CH(X) \simeq A(Y)/\mathcal{I}$$

Using results about the generalized (nonperiodic) Toda lattice established in [8] we have proved for all G (see Theorem 6 in [7]), among other things, that $\mathcal{I}$ is a prime ideal so that in all cases $A(Y)/\mathcal{I}$ is an integral domain and in fact

$$A(Y)/\mathcal{I} = A(Y_0)$$

In particular then for $G = Sl(l+1, \mathbb{C})$, the quantum cohomology ring $CH(X)$ is an integral domain and one has an isomorphism

$$CH(X) \simeq A(Y_0) \tag{2}$$

Using more appropriate notation than that introduced at the end of the introduction of [7] let $G^\vee$ be the Langlands dual to G and let $X^\vee$ be the flag manifold for $G^\vee$. For $G = Sl(l+1, \mathbb{C})$ we may identify X with $X^\vee$ so that (1) can be written

$$CH(X^\vee) \simeq A(Y_0) \tag{3}$$

Dale Peterson has announced results which among other things establishes (3) for all complex semisimple Lie groups G. We shall put this statement in another form where Y_0 is replaced by an isomorphic (l-dimensional) subvariety R of X.

2.3. Let G^f be the centralizer of f in G. Since f is principal nilpotent the identity component of G^f is a commutative l-dimensional unipotent subgroup of G. The isomorphism of R with Y_0 will arise from the action of G^f on X. See (7) below. The group G^f enters the picture, as will now be described, because of a connection with the Toda lattice and results in [8] involving the determination of the action angle coordinates.

Let $\pi_\mathfrak{e} : Y \to \mathfrak{e}$ be the projection defined by the decomposition (1). If $e \in \mathfrak{e}$ write

$$e = \sum_{i=1}^{l} c_i e_i \tag{4}$$

for $c_i \in \mathbb{C}$. In the notation of (4) let $\mathfrak{e}^* = \{e \in \mathfrak{e} | c_i \neq 0,\ i = 1, \ldots, l\}$ and let $Y^* = \pi_{\mathfrak{e}}^{-1}(\mathfrak{e}^*)$. Then Y^* has the structure of a symplectic manifold (in the complex sense). See e.g. Proposition 4 in [7]. Furthermore if $\Psi, \Psi' \in A(\mathfrak{g})^G$ then the functions $\Psi | Y^*$, $\Psi' | Y^*$ Poisson commute. See Theorem 1.5 in [8]. Fix Ψ so that $\Psi(x) = \frac{1}{2}(x, x)$. We may adjust $B_{\mathfrak{g}}$ so that $\Psi | Y^*$ is the Toda Hamiltonian (see (7.1.11) in [8]) and $\Omega_i | Y^* = F_i | Y^*$, $i = 1, \ldots, l$, defines a corresponding completely integrable system. But now if $Y_0^* = Y^* \cap Y_0$ then Y_0^* is the Lagrangian leaf corresponding to zero values of the integrable system. Theorem 6 in [7] asserts that Y_0^* is a nonsingular, irreducible, Zariski open and dense subvariety of Y_0. The Hamiltonian vector fields corresponding to $\Psi | Y^*$ and all $\Omega_i | Y^*$ are tangent to Y_0^*. Denote their restrictions to the leaf Y_0^* by ξ_Ψ and ξ_{Ω_i} respectively.

2.4. One of the main ideas of [8], applied to the case at hand, leads to a natural isomorphism (of commutative Lie algebras) between $\mathfrak{g}^f = Lie\, G^f$ and the span of the ξ_{Ω_i}. This may be accomplished as follows. Consider the G^f-orbit $Q = G^f \cdot o$ in X. As affine varieties obviously

$$Q \simeq G_o^f \tag{5}$$

where G_o^f is the identity component of G^f. Let P_f be the closure of Q in X. We refer to P_f as the (Dale) Peterson variety since he is, as far as I know, the first person to intensely study this rather remarkable unipotent version of a toric variety. Since f is principal nilpotent it lies in a unique Borel subalgebra $\bar{\mathfrak{b}}$. Of course $\bar{\mathfrak{b}}$ is an opposed Borel subalgebra to $\mathfrak{b}$ and is uniquely characterized by the condition that $\mathfrak{b} \cap \bar{\mathfrak{b}} = \mathfrak{h}$. Let $\overline{B}$ be the Borel subgroup corresponding to $\bar{\mathfrak{b}}$ and let $p_f \in X$ be defined so that $B_{p_f} = \overline{B}$. As established in §4.4 of [7] one has $p_f \in P_f$ and if $U = B \cdot p_f$ then U is the "big" Schubert cell in X defined by the action of B. Let

$$R = P_f \cap U$$

We have thus isolated 2 affine subvarieties, Q and R, of the projective variety P_f.

2.5. Let N be the unipotent radical of B. Then Corollary 18 in [7] ($\mathfrak{b}$ is misprinted as $\mathfrak{g}$ in (57) of [7]) asserts that

$$\eta : (f + \mathfrak{b}) \cap \mathcal{N} \to U \tag{6}$$

is an isomorphism where if $y \in (f + \mathfrak{b}) \cap \mathcal{N}$ there exists a unique $u \in N$ such that $Ad\, u^{-1}(f) = y$ and one puts $\eta(y) = u \cdot p_f$. Note that $Y_0 \subset (f + \mathfrak{b}) \cap \mathcal{N}$. The following result is a strengthening of part of the statement of Theorem 19 in [7]. It also depends on part of Theorem 6 in [7].

Theorem 1. *One has* $\eta(Y_0) = R$ *and*

$$\eta : Y_0 \to R \tag{7}$$

is an isomorphism of affine varieties. Furthermore $\eta(Y_0^*) = Q \cap R$ *and*

$$\eta : Y_0^* \to Q \cap R \tag{8}$$

is an isomorphism of nonsingular affine varieties.

Remark 2. The use of the word strengthening above is the recognition here that Y_0^* is affine and not just quasi-affine. This is immediate since it is given by the simultaneous nonvanishing of the functions on Y_0 defined by the c_i in (4). Actually this was already recognized in [7]. See Remark 22 in [7].

The isomorphism (7) of course defines an isomorphism $A(R) \simeq A(Y_0)$ so that the announced result of Peterson is an isomorphism

$$CH(X^\vee) \simeq A(R) \tag{9}$$

The affine variety R, outside of trivial examples involving $Sl(2,\mathbb{C})$, is always singular but it admits a stratification as a union of 2^l nonsingular affine subvarieties. See Theorems 6 and 9 in [7] noting the strengthening of the latter using the argument in Remark 2. On the other hand Q is always a nonsingular affine variety. In fact clearly the map

$$\gamma : G_o^f \to Q \tag{10}$$

where $\gamma(g) = g \cdot o$ is an isomorphism of nonsingular affine varieties.

2.6. Henceforth we will assume the validity of (9) and, by abuse of notation, identify $CH(X^\vee)$ with $A(R)$. In particular then

$$R = Spec\, CH(X^\vee)$$

In any case, strictly on the basis of published results, all the remaining statements in this paper are valid where $A(R)$ replaces $CH(X^\vee)$.

Let $FH(X^\vee)$ be the quotient field of the quantum cohomology algebra $CH(X^\vee)$.

Theorem 3. *The field* $FH(X^\vee)$ *is the quotient field of* $A(Q)$ *and consequently* $FH(X^\vee)$ *is isomorphic to the field of rational functions in* l *indeterminates.*

Proof. Since Q and R are open affine subvarieties of the same irreducible variety P_f the quotient fields of $A(Q)$ and $A(R)$ are the same. But the isomorphism (10) implies that Q is an l-dimensional rational variety. QED

We have considered the Toda leaf $Q \cap R \simeq Y_0^*$ (see (8)) from the "point of view" of $R \simeq Y_0$. We now describe it from the "point of view" of Q.

2.7. Let $\rho \in \mathfrak{h}^*$, as usual, be one half the sum of the positive roots. Consider an irreducible representation

$$\pi_\rho : G \to Aut\, V_\rho$$

with highest weight ρ. This representation was the central theme in [5]. The representation π_ρ is self-contragredient. Let (u, v), $u, v \in V_\rho$ be a non-singular G-invariant bilinear form B_ρ on V_ρ. Let $0 \neq v_\rho$ be a highest weight vector. For reference purposes normalize the choice of v_ρ to be consistent with its definition in §5.2 in [7] and normalize B_ρ so that in the notation of §5.2 of [7] one has $\langle v'_\rho, v \rangle = (v_\rho, v)$ for any $v \in V_\rho$. If $0 \neq v \in V_\rho$ is a weight vector with weight ν note that $(v_\rho, v) = 0$ unless $\nu = -\rho$. In particular $(v_\rho, v_\rho) = 0$. We are concerned here with the G_o^f-module structure on V_ρ defined by the restriction $\pi_\rho | G_o^f$. Let $\phi_\rho \in A(G_o^f)$ be defined by putting

$$\phi_\rho(g) = (v_\rho, g^{-1} v_\rho) \tag{11}$$

and let

$$G_\rho^f = \{ g \in G_o^f | \phi_\rho(g) \neq 0 \}$$

It is not immediately obvious (but not difficult to directly prove) that G_ρ^f is not empty. The nontriviality of G_ρ^f of course implies that G_ρ^f is a Zariski open, dense, affine subvariety of G_o^f. The nontriviality is implicitly implied in

Theorem 4. *One has $\gamma(G_\rho^f) = Q \cap R$ (see (10)) so that*

$$\gamma : G_\rho^f \to Q \cap R \tag{12}$$

is an isomorphism of affine varieties.

Proof. Recalling the notation of (79) in [7] one has an element $\psi_\rho \in A(Q)$ where $\psi_\rho(\gamma(g)) = (v_\rho, g^{-1} v_\rho)$. Using the notation of (80) note that ψ_ρ is the product of ψ_i, $i = 1, \ldots, l$, since, as one knows, all $n_i = 1$ for $\lambda = \rho$. The result then follows from Proposition 21 in [7]. QED

Remark 5. It may be noted that Theorem 4 is just a restatement of (60) in [7]. Indeed, using the Bruhat decomposition, note that $G_\rho^f = G_*^f$ where, as in §3.4 of [7], G_*^f is defined to be the intersection of G_o^f with the big double B-coset in G.

2.8. For any $z \in \mathfrak{g}^f$ let ξ^z be the G^f-invariant vector field on Q defined so that if $g \in G^f_o$ and $\gamma(g) = p \in Q$ then ξ^z_p is the tangent vector to the curve $t \mapsto \gamma(g\, exp(-tz))$ at $t = 0$. To relate the completely integrable Toda lattice to $\mathfrak{g}^f$ (see Theorem 6 below) one first notes that there is a surjection

$$A(\mathfrak{g})^G \to \mathfrak{g}^f, \qquad \Phi \mapsto z_\Phi \tag{13}$$

defined so that for any $x \in \mathfrak{g}$,

$$(z_\Phi, x) = \frac{d}{dt}\Phi(f + tx)|_{t=0} \tag{14}$$

See (24),(25) and (26) in [7]. Furthermore if $z_i = z_{\Phi_i}$ then $\{z_i\}$, $i = 1, \ldots, l$ is a basis of $\mathfrak{g}^f$. In addition

$$z_\Psi = f \tag{15}$$

See (108) in [7].

Theorem 6. *For the isomorphism on vector fields induced by (8) one has*

$$\eta_*(\xi_\Phi) = \xi^{z_\Phi}|Q \cap R \tag{16}$$

for any $\Phi \in A(\mathfrak{g})^G$. In particular for the hamiltonian vector field ξ_Ψ on the leaf Y_0^* corresponding to the Toda hamiltonian function $\Psi|Y^*$, one has

$$\eta_*(\xi_\Psi) = \xi^f|Q \cap R \tag{17}$$

Proof. The map β referred to in Theorem 29 of [7] is an isomorphism $G^f_\rho \to Y_0^*$. See Remark 5 above and (16) in [7]. Let $I : G^f \to G^f$ be the map of group inversion. By (64) in [7] it follows that G^f_ρ is stable under I. (Taking the transpose this is also clear from the definition above of G^f_ρ.) But then one has $\gamma = \eta \circ \beta \circ I$ on G^f_ρ by (94) in [7]. (Note that η_0 in this reference is just $\eta|Y_0$. See (58) in [7].) Thus $\eta = \gamma \circ I \circ \beta^{-1}$ on Y_0^*. But clearly $(\gamma \circ I \circ \beta^{-1})_*(\xi_\Phi) = \xi^{z_\Phi}|Q \cap R$ by (98) in [7]. Note that ξ_Φ here is $\xi_\Phi|Y_0^*$ in [7]. This proves (16). The equation (17) follows from (15). QED

3. Specializing the quantum parameters; the finite morphism, $\mu : R \to \mathbb{C}^l$

3.1. For $i = 1, \ldots, l$, let $q_i \in CH(X^\vee)$ be defined so that for $p \in R$ one has $q_i(p) = c_i$, using the notation of (4), where $e = \pi_\mathfrak{e}(\eta^{-1}(p))$. The q_i are called the quantum parameters. By Theorem 20 in [7] the q_i are algebraically independent. Furthermore, by Theorem 20 in [7], if $C = \mathbb{C}[q_1, \ldots, q_l]$ is the polynomial subalgebra of $CH(X^\vee)$ which they generate then $CH(X^\vee)$ is a free C-module of dimension $|W|$ (= $card\, W$). That is, there exists a subspace $\mathcal{W} \subset CH(X^\vee)$ of complex dimension $|W|$ such that

$$CH(X^\vee) = C \otimes \mathcal{W} \tag{18}$$

Remark 7. In [6] we proved that $A(\mathfrak{g})$ is a free $A(\mathfrak{g})^G$ module and exhibited a subspace $\mathcal{H}$ such that $A(\mathfrak{g}) = A(\mathfrak{g})^G \otimes \mathcal{H}$. Another subspace $\mathcal{H}_1$ satisfying this condition was constructed by Bernstein-Lunts in [1]. Our proof of (18) in [7] was based upon the decomposition $A(\mathfrak{g}) = A(\mathfrak{g})^G \otimes \mathcal{H}_1$. By modifying $\mathcal{H}$ the decomposition $A(\mathfrak{g}) = A(\mathfrak{g})^G \otimes \mathcal{H}_1$ was discovered earlier by Wallach. See §11.2.4 in [10].

3.2. Changing notation from [7] let F, here, be the quotient field of C in $FH(X^\vee)$. Classical ring theory implies

Theorem 8. *The integral domain $CH(X^\vee)$ is integral and free of dimension $|W|$ over C. Also*

$$FH(X^\vee) = F \otimes \mathcal{W} \tag{19}$$

so that $FH(X^\vee)$ is an algebraic extension of dimension $|W|$ over F.

Remark 9. Note that, by Corollary 25 and Example (1) in §5.2 in [7], if $\mathfrak{g}$ is simple then $CH(X^\vee)$ is not integrally closed in $FH(X^\vee)$ (i.e. R is not a normal variety) unless $G \simeq SL(2,\mathbb{C})$ or $G \simeq SL(3,\mathbb{C})$. On the other hand if $q = q_1 \cdots q_l$ and we localize with respect to q then by (8)

$$Spec\; CH(X^\vee)_q = Q \cap R$$

and $Q \cap R$ is nonsingular. See Theorem 1.

3.3. We may regard $\mathbb{C}^l = Spec\, C$ and $(\mathbb{C}^*)^l = Spec\, C_q$. The injections $C \to CH(X^\vee)$ and $C_q \to CH(X^\vee)_q$ define finite morphisms of degree $|W|$ (see (18) and (19))

$$\mu : R \to \mathbb{C}^l \tag{20}$$

and (by restriction)

$$\mu : Q \cap R \to (\mathbb{C}^*)^l \tag{21}$$

For any $c = (c_1, \ldots, c_l) \in \mathbb{C}^l$ let $\mathbb{C}_c = \mathbb{C}$ regarded as a C-module by specialization at the point c. The scheme theoretic fiber of (20) at c is $CH(X^\vee) \otimes_C \mathbb{C}_c$ and hence setwise

$$\mu^{-1}(c) = Spec\; CH(X^\vee) \otimes_C \mathbb{C}_c \tag{22}$$

By (18) one has

$$dim\, CH(X^\vee) \otimes_C \mathbb{C}_c = |W| \tag{23}$$

Theorem 10. *For any evaluation $c \in \mathbb{C}^l$ of the quantum parameters one has*

$$Card\,\mu^{-1}(c) \leq |W| \tag{24}$$

Furthermore if equality occurs in (24) (i.e. the ring $CH(X^\vee) \otimes_C \mathbb{C}_c$ is reduced) then $c \in (\mathbb{C}^*)^l$.

Proof. The first statement follows immediately from (23). Now assume $c \notin (\mathbb{C}^*)^l$. Then, using the notation of §3.5 in [7], it follows from (21) in [7], that there exists $\gamma \in \Gamma$ such that Π_γ is a proper subset of Π, and such that $\mu^{-1}(c) \subset (Y_\gamma)_0^*$. By definition $(Y_\gamma)_0^* = f_\gamma^o + (X_\gamma)_0^*$. But then by (24) applied to the case where $\mathfrak{g}_\gamma$ replaces $\mathfrak{g}$ one has $card\,\mu^{-1}(c) \leq |W_\gamma|$ where W_γ is the Weyl group of $(\mathfrak{h}_\gamma, \mathfrak{g}_\gamma)$. But clearly W_γ identifies with a proper subgroup of W so that $|W_\gamma| < |W|$. QED

3.4. Let $H^*(X)$ be the complexified ordinary cohomology algebra of X. Given its usual decription as $A(\mathfrak{h})/I$ where I is the ideal in $A(\mathfrak{h})$ generated by the augmentation ideal in $A(\mathfrak{h})^W$ one readily has an isomorphism

$$H^*(X) \simeq H^*(X^\vee) \tag{25}$$

(Indeed the Weyl groups are isomorphic and there is an isomorphism of Cartan subalgebras which respects the isomorphism of Weyl groups.)

Let $\mathcal{Q}$ be the ideal $(q_1, \ldots, q_l)$ in $CH(X^\vee)$ and let 0 here denote the origin of $\mathbb{C}^l$. It is a special point for the map μ. As noted in the last sentence preceding Theorem 9 in [7], setwise one has

$$\mu^{-1}(0) = \{p_f\} \tag{26}$$

On the other hand the fiber at 0, scheme theoretically, is

$$\begin{aligned} CH(X^\vee) \otimes_C \mathbb{C}_0 &= CH(X^\vee)/\mathcal{Q} \\ &\simeq H^*(X^\vee) \end{aligned} \tag{27}$$

The isomorphism in (27) follows from (26) and (71) in [7]. (In (71) of [7] the quotient $A(R)/I(R)$ has been misprinted as $A(R)I(R)$.)

Remark 11. Note that (27) justifies regarding the quantum cohomology $CH(X^\vee)$ as a deformation of the ordinary cohomology $H^*(X^\vee)$.

Let

$$\epsilon_C : CH(X^\vee) \to H^*(X^\vee) \tag{28}$$

be the epimorphism with kernel $\mathcal{Q}$ defined by (27). Note that for any choice of $\mathcal{W} \subset CH(X^\vee)$ satisfying (18) the restriction

$$\epsilon_C : \mathcal{W} \to H^*(X^\vee) \tag{29}$$

is a linear isomorphism.

4. Elements of $CH^2(X^\vee)$ as quotients of matrix entries of $\pi_\rho|G_o^f$

4.1. By Theorem 1 all the elements in $CH(X^\vee)$ are rational functions on Q. Hence, using (5), they can be expressed as rational functions on G_o^f. Since the elements of $A(G_o^f)$ arise as matrix entries of representations of G_o^f one of the main goals of [7] was to locate the representations and vectors in the corresponding module which can be used to write down generators of $CH(X^\vee)$. Let $L \subset \mathfrak{h}^*$ be the weight lattice and let $\Lambda \subset L$ be the set of dominant elements in L. For each $\lambda \in \Lambda$ let $\pi_\lambda : G \to Aut\, V_\lambda$ be an irreducible representation with highest weight λ. The representations of G_o^f, which are needed, are the restrictions $\pi_\lambda|G_o^f$. Here π_ρ plays a special role. For any $v \in V_\rho$ let $x_\rho^{(v)} \in FH(X^\vee)$ (see Theorem 3) such that for $g \in G_\rho^f$ one has

$$x_\rho^{(v)}(g \cdot o) = (v_\rho, g^{-1}v)/\phi_\rho(g) \tag{30}$$

using the notation of (11). Let $U_\rho = \{v \in V_\rho | x_\rho^{(v)} \in CH(X^\vee)\}$ and let $CH(X^\vee)^\rho = \{x_\rho^{(u)} | u \in U_\rho\}$. Recall $r = dim_{\mathbb{C}} X$. One knows $dim\, V_\rho = 2^r$ (see [Cliff]) and hence

$$dim\, CH(X^\vee)^\rho \leq 2^r \tag{31}$$

On the other hand one has

$$dim\, CH(X^\vee)^\rho \geq |W| \tag{32}$$

as a consequence of

Theorem 12. *The restriction of ϵ_C (see (29)) to $CH(X^\vee)^\rho$ defines a surjective map*

$$\epsilon_C : CH(X^\vee)^\rho \to H^*(X^\vee) \tag{33}$$

Proof. Let $\mathcal{W}_\rho$ be defined as in Theorem 43 of [7]. Then by the latter result we may choose $\mathcal{W}_\rho = \mathcal{W}$. Hence $\epsilon_C|\mathcal{W}_\rho$ is surjective by (29). However $\mathcal{W}_\rho \subset CH(X^\vee)^\rho$ by definition of $\mathcal{W}_\rho$. See §9.2 in [7]. QED

4.2. As one knows, the Weyl group W naturally operates on $H^*(X^\vee)$ (using (25)) and this representation is equivalent to the regular representation of W. In particular $dim\, Hom_W(\mathfrak{h}', H^*(X^\vee)) = l$. The reflection representation primary component, $H^*(X^\vee)_{rflt}$, is clearly the span of $\sigma(\mathfrak{h}')$ over all $\sigma \in Hom_W(\mathfrak{h}', H^*(X^\vee))$. One thus has $dim\, H^*(X^\vee)_{rflt} = l^2$.

By Theorem 32, (106), (107) in [7] for any $z \in \mathfrak{g}^f$ and $\nu \in L$ there exists an element $x^z_\nu \in CH(X^\vee)$ such that (1) x^z_ν is bilinear in ν and z and (2) is such that if $\lambda \in \Lambda$ and $g \in G^f_\rho$ then using π_λ,

$$x^z_\lambda(g \cdot o) = \frac{\langle v'_\lambda, g^{-1} z v_\lambda \rangle}{\langle v'_\lambda, g^{-1} v_\lambda \rangle} \tag{34}$$

where $0 \neq v_\lambda \in V_\lambda$ is a highest weight vector and $0 \neq v'_\lambda$ is a highest weight vector of the contragredient representation. Note that (34) is well-defined using (80) and (82) in [7]. For $i = 1, \ldots, l$, let $\rho_i \in \Lambda$ be defined by $\frac{2(\rho_i, \alpha_j)}{(\alpha_j, \alpha_j)} = \delta_{ij}$ so that the ρ_i are the highest weights of the fundamental representations of G.

Theorem 13. *For any $\nu \in L$ and $z \in \mathfrak{g}^f$ one has $x^z_\nu \in CH(X^\vee)^\rho$.*

Proof. By linearity to prove that $x^z_{\rho_i} \in CH(X^\vee)^\rho$ for any $z \in \mathfrak{g}^f$ and $i = 1, \ldots, l$. Note $x^z_{\rho_i} = x^{(w_i)}_{\rho_i}$, by (139) in [7], where $w_i = z v_{\rho_i}$. But then if, in the notation of (144) in [7], we put $w_j = v_{\rho_j}$ for $j \neq i$ one has $x^w_\rho = x^{(w_i)}_{\rho_i}$ since, clearly, $x^{(w_j)}_{\rho_j} = 1$ for $j \neq i$. But since $x^z_{\rho_i} \in CH(X^\vee)$ this proves that $w \in U_\rho$ and $x^z_{\rho_i} \in CH(X^\vee)^\rho$. QED

By linear extension define $x^z_\nu \in CH(X^\vee)$ for any $z \in \mathfrak{g}^f$ and $\nu \in \mathfrak{h}'$ (and not just for $\nu \in L$). Let $CH(X^\vee)_{rflt}$ be the $\mathbb{C}$ span of x^z_ν over all $z \in \mathfrak{g}^f$ and $\nu \in \mathfrak{h}'$ so that

$$CH(X^\vee)_{rflt} \subset CH(X^\vee)^\rho \tag{35}$$

by Theorem 13. The following result says that we have isolated a distinguished l^2-dimensional subspace of $CH(X^\vee)$.

Theorem 14. *The restriction of (33) to $CH(X^\vee)_{rflt}$ defines an isomorphism*

$$\epsilon_C : CH(X^\vee)_{rflt} \to H^*(X^\vee)_{rflt} \tag{36}$$

In fact there exists an isomorphism

$$\mathfrak{g}^f \to Hom_W(\mathfrak{h}', H^*(X^\vee)), \quad z \mapsto \sigma_z$$

such that

$$\sigma_z(\nu) = \epsilon_C(x^z_\nu) \tag{37}$$

for $\nu \in L$.

Proof. Let $d_i \in \mathfrak{h}$ be defined as in (5) of [7]. Let $z \in \mathfrak{g}^f$ and let $\Phi \in A(\mathfrak{g})^G$ be the unique element in the span of the Φ_i such that $z_\Phi = z$. It is clear that there exists a unique element $\tau_z \in Hom_W(\mathfrak{h}', H^*(X^\vee))$ such that

$$\tau_z(\alpha_i) = \epsilon((\iota(d_i)\Phi)|\mathfrak{h}) \tag{38}$$

where ϵ in (38), as in (1) of [7], is the Borel-Cartan-Leray homomorphism

$$\epsilon : A(\mathfrak{h}) \to H^*(X^\vee) \tag{39}$$

(see also (25)). Furthermore well-known facts about the generalized Vandermonde determinant and the top class in $H^*(X^\vee)$ imply that the map

$$\mathfrak{g}^f \to Hom_{W'}(\mathfrak{h}', H^*(X^\vee)), \qquad z \mapsto \tau_z \tag{40}$$

is an isomorphism. But, by (16), it follows from the equalities (101a) and (112) that

$$x^z_{\alpha_i} \circ \eta = \iota(d_i)\Phi \quad on\ Y_0^* \tag{41}$$

But since $x^z_{\alpha_i} \in CH(X^\vee)$ the equality (41) extends to Y_0. But then by putting $\sigma_z = \tau_z$ it follows from (70) in [7] that (37) is satisfied. Theorem 14 then follows from the isomorphism (40). QED

4.4. As we will see the case in Theorem 14 where $z = f$ deserves special attention. For notational convenience write $x_\nu = x^f_\nu$ for any $\nu \in \mathfrak{h}'$. The normalizer, D, of $\mathfrak{g}^f$ in the maximal torus H of G is a 1-dimensional subgroup of H. Recalling §4.3 and §4.4 in [7] the action of D on P_f induces a $\mathbb{Z}$-graded structure on both $A(Q)$ and $CH(X^\vee)$. We used subscripts to indicate degree. However in the light of (9), for the case of $CH(X^\vee)$ we will change to superscript. By Proposition 17 in [7] the algebra $A(Q)$ is graded by $-2\mathbb{Z}_+$ whereas, in contrast, by Theorem 35, the quantum cohomology algebra, $CH(X^\vee)$, is graded by $2\mathbb{Z}_+$. For the quantum parameters, q_i, one has

$$q_i \in CH^4(X^\vee) \tag{42}$$

by (104) in [7]. Since $CH(X^\vee)$ is a graded subspace of $FH(X^\vee)$ note that the subspace $CH(X^\vee)^\rho$ is graded by (138) in [7]. In particular $x_\nu \in CH^2(X^\vee)$ for any $\nu \in \mathfrak{h}'$. In fact

$$CH^2(X^\vee) = \{x_\nu | \nu \in \mathfrak{h}'\} \tag{43}$$

so that

$$dim\, CH^2(X^\vee) = l \tag{44}$$

See the first statement of Theorem 35 in [7].

4.5. Let $C[x_{\rho_1}, \ldots, x_{\rho_l}]$ be the subalgebra of $CH^2(X^\vee)$ generated by all x_ν, $\nu \in \mathfrak{h}'$ over the polynomial ring $C = \mathbb{C}[q_1, \ldots, q_l]$.

Theorem 15. *One has*

$$CH(X^\vee) = C[x_{\rho_1}, \ldots, x_{\rho_l}] \tag{45}$$

Furthermore the map ϵ_C (see (28)) is a homomorphism of graded algebras. Moreover for any $\nu \in \mathfrak{h}'$ one has

$$\epsilon_C(x_\nu) = \epsilon(\nu) \tag{46}$$

so that

$$H^2(X^\vee) = \{\epsilon_C(x_\nu) | \nu \in \mathfrak{h}'\} \tag{47}$$

Proof. By (15), (37) and (38) one has

$$\epsilon_C(x_{\alpha_i}) = \epsilon((\iota(d_i)\Psi)|\mathfrak{h})$$

But, by (5), (110) in [7], one has $(\iota(d_i)\Psi)|\mathfrak{h} = \alpha_i$. This proves (46), by linearity. But then (47) follows also, since, as one knows, $\epsilon(\mathfrak{h}') = H^2(X^\vee)$. Let $\mathcal{A}$ be the (obviously graded) subalgebra of $CH(X^\vee)$ generated, over $\mathbb{C}$, by all x_ν, $\nu \in \mathfrak{h}'$. Since, as one knows, $H^*(X^\vee)$ is generated by $H^2(X^\vee)$ it follows from (47) that $\epsilon_C|\mathcal{A}$ is an epimorphism of graded algebras. But then, clearly, $\mathcal{A}$ contains a subspace $\mathcal{W}$ satisfying (18). But then (18) implies (45). Furthermore since $\epsilon_C(q_i) = 0$ it is obviously from (42) that ϵ_C itself is an epimorphism of graded algebras. QED

4.6. We return now to a major theme of this paper and of [7]–namely writing elements of $CH(X^\vee)$ as specific rational functions on Q. In [7] we expressed the generators x_{ρ_i} of (45) as a quotient of matrix entries using the fundamental representation π_{ρ_i} of G. See (119) in [7]. On the other hand, by (35), in this paper, we could use π_ρ instead, thereby achieving a uniformity for all i. We have not as yet named which vector $v_i \in U_\rho \subset V_\rho$ yields x_{ρ_i}. We do so now and note that the answer is quite elegant.

First note that for $i = 1, \ldots, l$ the elements $\rho - \alpha_i \in L$ are extremal weights for π_ρ. In fact $\rho - \alpha_i$ is the image of ρ with respect to the fundamental reflection in W corresponding to α_i. Thus a weight vector v_i in V_ρ with weight $\rho - \alpha_i$ is unique up to scalar multiplication.

Theorem 16. *Let ϕ_ρ be defined as in (11). Then for $i = 1, \ldots, l$, the weight vector $v_i \in V_\rho$ (of weight $\rho - \alpha_i$) can be normalized so that for any $g \in G^f_\rho$,*

$$x_{\rho_i}(g \cdot o) = (v_\rho, g^{-1} v_i)/\phi_\rho(g) \tag{48}$$

Proof. In (146) of [7] let $n_i = 1$ and $n_j = 0$ for $j \neq i$. But then the right side of (146) equals x_{ρ_i}. Using the notation of §8.4 in [7] note then that (146) implies that $x_\rho^{(w)} = x_{\rho_i}$ where $w = w(n_1, \ldots, n_l)$. Put $v_i = w$. We have only to observe that w has weight $\rho - \alpha_i$. Obviously the map (143) carries weight vectors to weight vectors having the same weight. But w is the image of $w_1 \otimes \cdots \otimes w_l$ where $w_j = f^{n_j} v_{\rho_j}$. Thus $w_j = v_{\rho_j}$ for $j \neq i$ and $w_i = f v_{\rho_i} = -e_{-i} v_{\rho_i}$. But then it is immediate that w is a weight vector of weight $\rho - \alpha_i$. QED

5. The $U(\mathfrak{g}^f)$-module structure on $CH(X^\vee)$ and the q_i as quadratic quotients of matrix elements of $\pi_\rho|G_o^f$

5.1. Let $U(\mathfrak{g}^f)$ be the universal enveloping algebra of $\mathfrak{g}^f$. We now observe that the quantum cohomology algebra $CH(X^\vee)$ has the structure of a $U(\mathfrak{g}^f)$- module. For any $z \in \mathfrak{g}^f$ the G^f-invariant vector field ξ^z on Q of course defines a derivation of $A(Q)$ and, consequently by extension of domain, a derivation of the quotient field $FH(X^\vee)$ of $A(Q)$ (see Theorem 3). This defines a representation of $\mathfrak{g}^f$ on $FH(X^\vee)$. By extension, therefore, $FH(X^\vee)$ has the structure of a $U(\mathfrak{g}^f)$-module. The obvious question as to whether $CH(X^\vee)$ is stabilized by $U(\mathfrak{g}^f)$ is answered in the affirmative.

Theorem 17. *The quantum cohomology algebra $CH(X^\vee)$ is stable under the action of $U(\mathfrak{g}^f)$ on $FH(X^\vee)$. In particular then, $CH(X^\vee)$ inherits the structure of a $U(\mathfrak{g}^f)$-module. Furthermore if $U_+(\mathfrak{g}^f)$ is the augmentation ideal in $U(\mathfrak{g}^f)$ then (see (27))*

$$U_+(\mathfrak{g}^f)\,CH(X^\vee) \subset \mathcal{Q} \tag{49}$$

so that the action of $U(\mathfrak{g}^f)$ on $CH(X^\vee)$ induces the trivial action of $U(\mathfrak{g}^f)$ on $H^*(X^\vee)$.

Proof. Given Peterson's result (9), Theorem 17 is just the first 2 statements of Theorem 38 in [7]. See also Theorem 30 in [7] for a relevant definition. QED

Remark 18. Let $K \subset G$ be a maximal compact subgroup and let $\Omega(K)$ be the group of based loops in K. Also let $K^\vee$ be a maximal compact subgroup of $G^\vee$. V. Ginzburg (see Proposition 1.7.2 and Theorem 1.7.6 in [4]) and, independently, Dale Peterson ([9]) have shown that $U(\mathfrak{g}^f)$ is naturally isomorphic to the ordinary cohomology $H^*(\Omega(K^\vee))$. On the other hand Peterson obtains (9) by relating the Gromov-Witten invariants defining $CH(X^\vee)$ to the structure constants defining Pontrjagin product in the homology group $H_*(\Omega(K))$. It is therefore suggestive that the action in Theorem 17 may arise from some general topological considerations.

5.2. Recall that the action of $\mathfrak{g}^f$ on $CH(X^\vee)$ is via the Lie algebra of vector fields $\{\xi^z\}$, $z \in g^f$, on Q. The case where $z = f$ is distinguished. Recalling (17) the restriction ξ^f to $Q \cap R$ corresponds, using η, to the Hamiltonian vector field ξ_Ψ for the Toda Hamiltonian function, on the

Lagrangian leaf Y_0^*. A key result in [7] (see (125) in [7]) is the equation $\xi^f x_{\nu_i} = -q_i$, $i = 1, \ldots, l$ where $\{\nu_i\}$ is the basis of $\mathfrak{h}'$ which is dual to the basis (see (5)) $\{d_i\}$ of $\mathfrak{h}$. But by (5) in [7] it follows that $\nu_i = \frac{2}{(\alpha_i,\alpha_i)}\rho_i$, using the bilinear form on $\mathfrak{h}'$, induced by $B_{\mathfrak{g}}$. It follows then that (125) in [7] can be rewritten

$$\xi^f x_{\rho_i} = \frac{-2}{(\alpha_i, \alpha_i)} q_i \tag{50}$$

Remark 19. As noted in [7] the equation (125) in [7] or its slight modification, (50) above, has a number of interpretations. From the point of view of the Toda lattice the functions x_{ρ_i} correspond to momenta and ξ^f corresponds to time differentiation. Thus (50) can be understood as Newton's equation of motion, $F = ma$, for the Toda lattice on the Lagrangian leaf Y_0^*. It says the "force" is $\frac{-2}{(\alpha_i,\alpha_i)}q_i$. On the other hand (50) plays a major role in proving that we need go no further than taking matrix entries from $\pi_\rho | G_o^f$ to construct a subspace $\mathcal{W}$ which satisfies (18) above. The critical result for doing this is Theorem 42 in [7].

5.3. We will draw 2 other conclusions from (50). As mentioned in [7], the idea that something like (50) might be true seemed suggestive to us by a result of J. Carrell (see [2]) on a construction of $H^*(X)$ using a vector field on X which corresponds to a principal nilpotent element in $\mathfrak{g}$. In fact the analogue of Carrell's theorem is true.

Theorem 20. *Let $(\xi^f\, CH(X^\vee))$ be the ideal in $CH(X^\vee)$ generated by the image, $\xi^f\, CH(X^\vee)$ of ξ^f on $CH(X^\vee)$. Then*

$$CH(X^\vee)/(\xi^f\, CH(X^\vee)) \simeq H^*(X^\vee) \tag{51}$$

Proof. By (27) we have only to show that $(\xi^f\, CH(X^\vee)) = \mathcal{Q}$. But this follows immediately from (50) since $(\xi^f\, CH(X^\vee)) \subset \mathcal{Q}$, by (49). QED

5.4. An expression for the quantum parameters q_i as rational functions on Q was given in Theorem 27 in [7] using powers of matrix entries arising from all the fundamental representations π_{ρ_i} of $\mathfrak{g}$. But now using (48) together with (50) all the q_i can be written using matrix entries just from π_ρ.

Theorem 21. *Let $g \in G_\rho^f$. Then, using the notation of (48), for $i = 1, \ldots, l$,*

$$q_i(g \cdot o) = \frac{(\alpha_i, \alpha_i)}{2}((v_\rho, g^{-1}v_i)(v_\rho, g^{-1}fv_\rho) - (v_\rho, g^{-1}v_\rho)(v_\rho, g^{-1}fv_i))/\phi_\rho(g)^2 \tag{52}$$

Proof. We have just carried out the differentiation in (50) to the expression for x_{ρ_i} given in (48). QED

Remark 22. Note that both numerator and denominator in (52) are quadratic expressions in the matrix entries of $\pi_\rho|G^f_\rho$. Moreover, in all the terms, note that v_ρ appears by itself on the left side of each inner product. It follows then that both numerator and denominator are matrix entries of $\pi_{2\rho}|G^f_\rho$. Moreover $\pi_{2\rho}$ may be explicitly constructed using the adjoint action of $\mathfrak{g}$ on $\wedge^r\mathfrak{g}$. See [5] for the Cartan product role played by $\pi_{2\rho}$ in $\wedge\mathfrak{g}$.

6. The quantum Vandermonde determinant E_ρ and the condition that μ is étale at $p \in Q \cap R$

6.1 Now recall (see (34), (35) and (36)) the elements $x^z_\nu \in CH(X^\vee)$, $z \in \mathfrak{g}^f$, $\nu \in \mathfrak{h}'$ and the subspace, $CH(X^\vee)_{rflt}$ that they span. The $l \times l$ matrix M with entries in the integral domain $CH(X^\vee)$, where $M_{ij} = x^{z_i}_{\rho_j}$, is invertible since $dim\, CH(X^\vee)_{rflt} = l^2$. Thus if $E_\rho = det\, x^{z_i}_{\rho_j}$ then $E_\rho \neq 0$. Let $O = \{p \in Q \cap R | E_\rho(p) \neq 0\}$ so that O is a nonsingular, open, dense, affine subvariety of $Q \cap R$. The following result is just a restatement of part of Proposition 44 and Theorem 45 in [7].

Theorem 23. *One has*

$$E_\rho \in CH^{2r}(X^\vee)^\rho \tag{53}$$

and $\epsilon_C(E_\rho)$ spans $H^{2r}(X^\vee)$, the top cohomology of $X^\vee$. Furthermore, recalling the finite morphism μ, (see (21)),

$$O = \{p \in Q \cap R | \mu \textit{ is étale at } p\} \tag{54}$$

6.2. By Theorem 23 we may use the quantum parameters q_i as coordinate functions on O and consequently introduce corresponding vector fields $\frac{\partial}{\partial q_i}$ on O. Our final result will be an expression for the vector fields, $\xi^z|O$, $z \in \mathfrak{g}^f$, in terms of the frame $\{\frac{\partial}{\partial q_i}\}$.

Theorem 24. *Let $z \in \mathfrak{g}^f$. Then, on O,*

$$\xi^z = \sum_{i=1}^{l} q_i x^z_{-\alpha_i} \frac{\partial}{\partial q_i} \tag{55}$$

In particular, the "Toda" vector field, ξ^f, on O, is given by

$$\xi^f = \sum_{i=1}^{l} q_i x_{-\alpha_i} \frac{\partial}{\partial q_i} \tag{56}$$

Proof. Theorem 24 is an immediate consequence of (101a) in [7]. QED

References

1. J. Bernstein and V. Lunts, A simple proof of Kostant's theorem that $U(\mathfrak{g})$ is free over its center, Technical report 5-94, 1994, The School of Math. Sciences, Tel Aviv Univ.
2. J. Carrell, Vector fields, flag manifolds and Schubert calculus, Proc. 1989 Hyderabad Conference on Algebraic Groups and Applications, 23-55.
3. I. Ciocan-Fontanine, Quantum cohomology of flag manifolds, *IMNR Notices*, 1995, No. 6, 263-277.
4. V. Ginzburg, Perverse sheaves on a loop group and Langlands duality, preprint 1995.
5. B. Kostant, Clifford algebra analogue of the Hopf-Koszul-Samelson theorem, the ρ-decomposition $C(\mathfrak{g}) = End\, V_\rho \otimes C(P)$, and the $\mathfrak{g}$-module structure of $\wedge\mathfrak{g}$, *Adv. in Math.* **125**, No. 2 (1997) 275-350.
6. B. Kostant, Lie group representations on polynomial rings, *Amer. J. of Math.* **85** (1963), 327-404.
7. B. Kostant, Flag manifold quantum cohomology, the Toda lattice, and the representation with highest weight ρ, *Selecta Mathematica, New Series* **2**:1 (1996), 43-91.
8. B. Kostant, The solution to a generalized Toda lattice and representation theory, *Adv. in Math.* **34** (1979), 195-338.
9. Personal communication.
10. N. Wallach, *Real Reductive Groups II*, Academic Press, 1992.

HOMOLOGIES ASSOCIATED WITH POISSON STRUCTURES

OLIVIER MATHIEU
IRMA, Université Louis Pasteur
7 rue René Descartes, 67000 Strasbourg, France

Abstract. We investigate various homologies attached to Poisson manifolds and to symplectic manifolds. For any Poisson manifold X, we introduce a one parameter family of degree -1 differentials $(\delta_\tau)_{\tau \in \mathbf{CP}^1}$ defined on the space of forms Ω_X^*. The Koszul differential δ [11, 2], which computes the canonical homology of X, is δ_∞. There is a canonical map from $H_*(C^\infty(X))$ (the homology of the Lie algebra $C^\infty(X)$) to $H_*^0(X)[1]$ (the homology of δ_0). Next we consider the canonical operations associated with symplectic structures [13]. We use them to define some maps $\Theta^n : H_*(C^\infty(X)) \to H_*^0(X)[2n+1]$ for X symplectic. We also describe a test which could give a negative answer to the question of the quantizability of Poisson manifolds.

Introduction

Let X be a Poisson manifold. In [11], Koszul introduced a degree -1 differential δ on the space Ω_X^* of forms on X. The homology of δ, denoted by $H_*^{can}(X)$ is called the canonical homology of X. In [2], Brylinski showed that the canonical homology of X is connected with the adjoint homology of the Lie algebra $C^\infty(X)$, namely there is a natural map $Br : H_*(C^\infty(X), C^\infty(X)) \to H_*^{can}(X)$.

We generalize the Koszul differential δ and we show that there is a one parameter family of degree -1 differentials $(\delta_\tau)_{\tau \in \mathbf{CP}^1}$ such that $\delta = \delta_\infty$. This allows us to define a canonical homology $H_*^\tau(X)$ with parameter τ (see Section 1). Then we show that the canonical homology of X with parameter 0 is connected with the homology of the Lie algebra $C^\infty(X)$ with trivial coefficients. More precisely, there is a canonical map $\Theta^0 : H_*(C^\infty(X)) \to H_*^0(X)[1]$ (see Section 2; in what follows, we set $H_*[m] = H_{*+m}$ for any graded vector space H_*).

D. Sternheimer et al. (eds.), Deformation Theory and Symplectic Geometry, 177–199.

For general Poisson manifolds, it seems that there is almost no relation between the canonical homologies $H_*^\tau(X)$ for various values of τ. However, for symplectic manifolds X, we can use that Ω_X^* is a $\mathfrak{sl}(2)$-module [17] to prove that $H_*^\tau(X)$ is isomorphic to the usual canonical homology for all τ outside a finite set of special values (see Section 3). For the special values of τ, $H_*^\tau(X)$ is very different and seems quite close to the cyclic or Hochschild homology of X. As 0 is not a special value, there is a well-defined map $\Theta : H_*(C^\infty(X)) \to H_*^{can}(X)[1]$. When X satisfy the strong Lefschetz theorem, we show that Θ is a refinement of Brylinski map Br (see Section 5).

Then we recall the definition of some differential operators canonically attached to connected symplectic manifolds X (see [13] and Section 7). These operators, called canonical operations, are all n-ary differential operators which are invariant by all diffeomorphisms preserving the symplectic structure. The space $Can_n(2m)$ of all these operators depends only on the dimension $2m$ of X. It contains a subspace $Quant_n(2m)$ called the space of quantizable canonical operators (see Section 9). We show that these operators play a role in the computation of $H_*(C^\infty(X))$ quite similar to Connes operator S in the homology of the Lie algebra of matrices [12, 16]. Indeed we use them to define maps $\Theta^n : H_*(C^\infty(X)) \to H_*^0(X)[2n+1]$, for infinitely many n (see Section 11).

Thus, we show that the canonical operations can be used to test the conjectural claim that any smooth Poisson manifold can be quantized (see Section 10). Indeed, we show that if a Poisson manifold X contains an open leaf U of dimension $2m$, then any $\theta \in Quant_n(2m)$ defines canonically an n-ary differential operator $\overline{\theta}$ on U. Moreover, if X can be quantized, then $\overline{\theta}$ can be extended to the whole of X. In other words, if one of these differential operators $\overline{\theta}$ is not extendable, then the manifold is not quantizable. The space $Quant_n(2m)$ can be defined purely combinatorially. Because of the combinatorial complexity of this definition, it is difficult to perform the test.

Acknowledgements

We thank I.M. Gelfand who introduced us to the question of computing the Lie homology of Poisson algebras. We also thank J. Barge, M. Duflo and M. Flato for helpful discussions.

Conventions

In what follows, tensor products should be understood in the topological sense. For example, for any manifold X, $C^\infty(X) \otimes C^\infty(X)$ denotes

$C^\infty(X \times X)$. The same conventions apply to the exterior algebra $\bigwedge^* V$ and the symmetric algebra $Sym^* V$ of a topological vector space V.

1. The differential δ_κ

In this section, we will generalize the canonical homology of Poisson manifolds.

Let X be a Poisson manifold. Denote by ν the two-vector defining the Poisson bracket, namely we have $\{f, g\} = i_\nu(df \wedge dg)$ for any smooth functions f and g. Also denote by $\Omega_X^* = \oplus_{n \geq 0} \Omega_X^n$ the space of complex-valued forms on X. In [11], Koszul introduced a degree -1 operator $\delta : \Omega_X^* \to \Omega_X^*$ satisfying $\delta^2 = 0$. The homology of δ, denoted by $H_*^{can}(X)$, is called the canonical homology of X. More precisely, what is defined here is the canonical homology with complex coefficients, i.e. the complexification of Koszul's definition.

Since Ω_X^* is a graded vector space, the space of **C**-linear endomorphisms of Ω_X^* has a natural structure of Lie super-algebra. The bracket $[A, B]$ of two homogeneous endomorphisms A, B is defined by $AB + BA$ if both A and B have odd degree, and by $AB - BA$ otherwise. The operators d, δ and i_ν satisfy the following Koszul relations (see [11, 2]):

$$d^2 = \delta^2 = 0,$$

$$[d, \delta] = 0,$$

$$[i_\nu, d] = \delta,$$

$$[i_\nu, \delta] = 0.$$

Define two degree -1 operators $\delta^\pm : \Omega_X^* \to \Omega_X^*$ by the formulas $\delta^+ = d \circ i_\nu$ and $\delta^- = i_\nu \circ d$. Also set $\Theta = -\delta \circ d \circ i_\nu$. Note that $[i_\nu, \delta \circ d] = [i_\nu, \delta] \circ d + \delta \circ [i_\nu, d] = \delta^2 = 0$. Hence we also have $\Theta = -i_\nu \circ \delta \circ d$.

Lemma 1.1 *We have*
(i) $\delta^+ \circ \delta^+ = \Theta$,
(ii) $\delta^- \circ \delta^- = \Theta$,
(iii) $\delta^+ \circ \delta^- = 2\Theta$,
(iv) $\delta^- \circ \delta^+ = 0$.

Proof Proof of (i): we have

$$\begin{aligned}
(\delta^+)^2 &= d \circ i_\nu \circ d \circ i_\nu \\
&= [d, i_\nu] \circ d \circ i_\nu + i_\nu \circ d \circ d \circ i_\nu \\
&= -\delta \circ d \circ i_\nu \\
&= \Theta.
\end{aligned}$$

Proof of (ii): we have

$$\begin{aligned}(\delta^-)^2 &= i_\nu \circ d \circ i_\nu \circ d \\ &= i_\nu \circ d \circ [i_\nu, d] + i_\nu \circ d \circ d \circ i_\nu \\ &= i_\nu \circ d \circ \delta \\ &= -i_\nu \circ \delta \circ d \\ &= \Theta.\end{aligned}$$

Proof of (iii): we have

$$\begin{aligned}\delta^+ \circ \delta^- &= d \circ i_\nu \circ i_\nu \circ d \\ &= [d, i_\nu] \circ i_\nu \circ d + i_\nu \circ [d \circ i_\nu] \circ d + i_\nu \circ i_\nu \circ d \circ d \\ &= -\delta \circ i_\nu \circ d - i_\nu \circ \delta \circ d \\ &= 2\Theta.\end{aligned}$$

Proof of (iv): it follows from $d^2 = 0$. Q.E.D.

For $\tau \in \mathbf{C}$ and $\alpha \in \Omega_X^n$, set $\delta_\tau \alpha = (n+\tau)\delta^- \alpha - (n+1+\tau)\delta^+ \alpha$.

Lemma 1.2 *We have* $\delta_\tau^2 = 0$.

Proof Let $\alpha \in \Omega_X^n$. Using Lemma 1.1, we get:

$$\begin{aligned}\delta_\tau^2.\alpha &= ((n+\tau)\delta^+ + (n+\tau-1)\delta^-) \circ ((n+\tau+1)\delta^+ + (n+\tau)\delta^-)\alpha \\ &= ((n+\tau)(n+\tau+1) + (n+\tau-1)(n+\tau) - 2(n+\tau)^2)\Theta(\alpha) \\ &= ((n+\tau)(2n+2\tau) - 2(n+\tau)^2)\Theta(\alpha) \\ &= 0. \quad \text{Q.E.D.}\end{aligned}$$

For a convenient function $\lambda : \mathbf{C} \to \mathbf{C}^*$, $\lambda(\tau).d_\tau$ goes to δ when τ goes to ∞. So it will be convenient to set $\delta_\infty = \delta$. Of course, there is a topological obstruction to find such a function λ which is continuous, so it is not possible to renormalize the operator d_τ to find a continuous family of operators on the whole space $\mathbf{C} \cup \{\infty\}$. Define the canonical homology with parameter τ as the homology of δ_τ. These homology groups will be denoted by $H_*^\tau(X)$ (so $H_*^\infty(X) = H_*^{can}(X)$). For a finite τ, define $\kappa_\tau : \Omega_X^* \to \Omega_X^*$ as the multiplication by $n+\tau$ on the component Ω_X^n. For any degree k endomorphism of Ω_X^*, we have $A \circ \kappa_\tau = \kappa_{\tau-k} \circ A$.

Example 1.3 The differentials δ_τ are also defined in the context of algebraic Poisson manifolds. For example, let X be the dual of a Lie algebra $\mathfrak{g}$. The complex $(Sym^*\mathfrak{g} \otimes \bigwedge^* g, \delta_\tau)$ splits as $\oplus_{n\geq 0}(Sym^n\mathfrak{g} \otimes \bigwedge^* g, \delta_\tau)$. Hence we can define the homology groups $H_*^\tau(\mathfrak{g}, Sym^n\mathfrak{g})$. For $\tau = \infty$, these homology groups are the usual homology of $\mathfrak{g}$ with value in $Sym^n\mathfrak{g}$.

Lemma 1.4 *We have $\delta_\tau(Ker\, d) \subset Ker\, d$ and $\delta_\tau(Im\, d) \subset Im\, d$. In particular, δ_τ induces a differential on $\Omega^*_X/d\Omega^*_X$.*

Proof For τ infinite, it follows from Koszul's relation. For τ finite, we have:

$$\begin{aligned} \delta_\tau \circ d &= -d \circ \kappa_{\tau-1} \circ \delta^-, \\ d \circ \delta_\tau &= \delta^+ \circ \kappa_{\tau-1} \circ d. \end{aligned}$$

Hence δ_τ stabilizes $Ker\, d$ and $Im\, d$. Q.E.D.

Lemma 1.5 *We have $i_\nu \circ \delta_\tau = \delta_{\tau+1} \circ i_\nu$. Hence i_ν induces a degree -2 operator $i_\nu : H^\tau_*(X) \to H^{\tau+1}_*(X)$.*

Proof Using that $[i_\nu, \delta_\infty] = 0$, we get:

$$\begin{aligned} i_\nu \circ \delta_\tau &= i_\nu \circ \delta_\infty \circ \kappa_\tau - i_\nu \circ d \circ i_\nu \\ &= \delta_\infty \circ (\kappa_\tau + 2) \circ i_\nu - \delta^- \circ i_\nu \\ &= (\delta^- \circ (\kappa_\tau + 1) - \delta^+ \circ (\kappa_\tau + 2)) \circ i_\nu \\ &= \delta_{\tau+1} \circ i_\nu. \quad \text{Q.E.D.} \end{aligned}$$

For general Poisson manifolds X, the canonical homologies $H^\tau_*(X)$ are strictly different from the usual canonical homology $H^\infty_*(X)$. We will show it on three examples:

Example 1.6 First let us show that the homology of a product is usually difficult to compute except for $\tau = \infty$. Let us consider $X = M \times N$ be a product of two Poisson manifolds where N is endowed with a trivial Poisson product. In such a case, we have $H^\infty_*(M \times N) = H^\infty_*(M) \otimes \Omega^*_N$.

Now let $\tau \neq \infty$. Set $\Omega^{m,n} = \Omega^m_M \otimes \Omega^n_N$. We have

$$\delta_\tau \Omega^{m,n} \subset \Omega^{m-1,n} \oplus \Omega^{m-2,n+1}.$$

Set $\mathcal{K}_m = \oplus_{p \le m, q} \Omega^{p,q}$. The filtration $\mathcal{K}_*$ of the complex $\Omega^*_{M\times N}, \delta_\tau$ gives rise to a spectral sequence E_* and we have isomorphisms of complexes:

$$\begin{aligned} (E_0, d_0) &= (\Omega^*_{M\times N}, 0) \\ (E_1, d_1) &= \oplus_{n\ge 0}(\Omega^{*,n}, \delta_{\tau+n}) \end{aligned}$$

Then we have

$$E_2 = \oplus H^{\tau+n}_*(M) \otimes \Omega^n_N$$

and its differential is

$$i_\nu \otimes d : H^{\tau+n}_*(M) \otimes \Omega^n_N \to H^{\tau+n+1}_*(M) \otimes \Omega^{n+1}_N,$$

where the operator i_ν is defined in the previous lemma. So there are examples with $d_2 \neq 0$. When we write down the same spectral sequence at $\tau = \infty$, all differentials d_i for $i \geq 2$ vanish. Hence the canonical homology $H_*^\tau(X)$ behaves differently from the usual canonical homology. Indeed, this example raises the following question of the existence of a Kunneth formula for $\tau \neq \infty$. For two Poisson manifolds, is it possible to compute the groups $H_*^\tau(X \times Y)$ in terms of the canonical homologies of X and Y?

Example 1.7 Let X be a symplectic manifold of dimension two. We have $\delta_\tau(\alpha) = (\tau+1)\delta_\infty\alpha$ for $\alpha \in \Omega_X^1$ and $\delta_\tau(\alpha) = (\tau+3)\delta_\infty\alpha$ for $\alpha \in \Omega_X^2$. Hence for $\tau \neq -1, -3$, we have $H_*^\tau(X) \simeq H_*^\infty(X)$. However we have $H_*^{-1}(X) = \Omega_X^0 \oplus (\Omega_X^1/\delta_\infty\Omega_X^2) \oplus H_2^\infty(X)$ and $H_*^{-3}(X) = H_0^\infty(X) \oplus Z_1^\infty \oplus \Omega_X^2$, where $Z_1^\infty = \{\alpha \in \Omega_X^1 | \delta_\infty(\alpha) = 0\}$. So at the special values $\tau = -1, -3$, the canonical homology is close to cyclic and Hochschild homology of X. For other values of τ, the canonical homology is isomorphic to the usual canonical homology.

Example 1.8 Let X be a compact oriented Poisson manifold of dimension N. Let $\tau, \tau' \in \mathbf{C}$ with $\tau + \tau' + N + 2 = 0$. The usual pairing $\Omega_X^* \times \Omega_X^* \to \mathbf{C}$, defined by $(\alpha, \beta) = \int_X \alpha \wedge \beta$ induces a natural pairing $H_*^\tau(X) \times H_*^{\tau'}(X) \to \mathbf{C}$. Thus only $H_*^{-1-dim(X)/2}(X)$ and $H_*^\infty(X)$ carry an inner pairing.

2. Lie homology of Poisson algebras

Let X be a Poisson manifold. In this section, we will define a map Θ^0 from the homology of the Lie algebra $C^\infty(X)$ to $H_*^0(X)[1]$.

Recall that the homology of the Lie algebra of all vector fields is computed by Gelfand-Fuks theory, see e.g. [8]. However, little is known about the homology of the Lie algebra of the underlying Poisson structure (or about the Lie algebra of all hamiltonian vector fields for a symplectic manifold). See e.g. [7, 9, 10].

We will denote by $\bigwedge^*(C^\infty(X))$ and by $C^\infty(X) \otimes \bigwedge^*(C^\infty(X))$ the Koszul complex of the Lie algebra $C^\infty(X)$ with trivial and adjoint coefficients. (By convention, $\bigwedge^n(C^\infty(X))$ is the space of skew-symmetric smooth functions on X^n.) The Koszul differentials will be denoted by d_{triv} and by d_{ad}. The homology of these groups are the trivial and the adjoint Lie homology of $C^\infty(X)$ and are denoted by $H_*(C^\infty(X))$ and by $H_*(C^\infty(X), C^\infty(X))$.

Let V be a vector space, and let M be a graded left $\bigwedge^* V$-module. For any $v \in V$, we denote by $e(v)$ the left multiplication by v on $\bigwedge^* V$ and on M. An endomorphism $\psi : \bigwedge^* V \to M$ is called a differential operator of order $\leq n$ if and only if $[e(v_1)[\ldots[e(v_{n+1})(\psi)]\ldots] = 0$ for any $v_1, \ldots, v_{n+1} \in V$ (here the brackets are defined with the sign convention of the previous section). A differential operator of order n is uniquely determined by its restriction to $\oplus_{i\leq n} \bigwedge^i V$.

For example, d_{triv} is the unique order two differential operator from $\bigwedge^* C^\infty(X)$ to itself such that $d_{triv}\, 1 = 0$, $d_{triv}\, f = 0$ and $d_{triv}(x \wedge y) = -[x,y]$, for any $x, y \in C^\infty(X)$. The element $1 \in C^\infty(X)$ defines the multiplication operator $e(1)$ which should not be confused with the unit 1 in $\bigwedge^* C^\infty(X)$. Denote by I_ν the unique order 3 differential operator from $\bigwedge^* C^\infty(X)$ to itself such that $I_\nu(1) = 0$, $I_\nu(f) = 0$, $I_\nu(f \wedge g) = 0$, and $I_\nu(f \wedge g \wedge k) = \{f,g\}.k + \{g,k\}.f + \{k,f\}.g$, for any $f, g, k \in C^\infty(X)$. For $n \geq 3$, we have the explicit formula

$$I_\nu(f_1 \wedge \ldots \wedge f_n) = \sum_{1\leq i<j<k\leq n} (-1)^{i+j+k} I_\nu(f_i, f_j, f_k) df_1 \wedge \ldots \widehat{df_i} \ldots \widehat{df_j} \ldots \widehat{df_k} \ldots \wedge df_n,$$

with the usual convention that the symbol $\widehat{}$ denotes a missing factor.

Lemma 2.1 *We have* $d_{triv} = [e(1), I_\nu]$.

Proof The endomorphisms d and $[e(1), I_\nu]$ are differential of order two, and they are zero on $\mathbf{C}$ and $\bigwedge^1 C^\infty(X)$. Hence it suffices to check that they are equal on components of degree ≤ 2. We have: $[e(1), I_\nu](f \wedge g) = -I_\nu(1 \wedge f \wedge g) = -\{f, g\}$. Q.E.D.

Define the map $\pi_0, \pi : \bigwedge^* C^\infty(X) \to \Omega_X^*$ by

$$\pi_0(f_1 \wedge \ldots \wedge f_n) = \sum_{1\leq i\leq n} (-1)^i f_i df_1 \wedge \ldots \wedge \widehat{df_i} \wedge \ldots \wedge df_n$$

and

$$\pi(f_1 \wedge \ldots \wedge f_n) = (1/n!)\pi_0(f_1 \wedge \ldots \wedge f_n).$$

For simplicity, set $\kappa = \kappa_0$.

Lemma 2.2 *We have*
(i) $\pi_0 \circ I_\nu = i_\nu \circ (\kappa - 1) \circ \pi_0$;
(ii) $\pi_0 \circ e(1) = d \circ 1/(\kappa + 1) \circ \pi_0$;
(iii) $\pi_0 \circ d_{triv} = (\delta^- \circ (\kappa/(\kappa+1)) - \delta^+) \circ \pi_0$.

Proof Let $f_1, \ldots, f_n \in C^\infty(X)$. For any indices i, j, k with $1 \leq i < j < k \leq n$, set $\omega_{i,j,k} = df_1 \wedge \ldots \wedge \hat{d} f_i \ldots \wedge \hat{d} f_j \ldots \wedge \hat{d} f_k \ldots df_n$. More generally, for any three distinct indices i, j, k, set $\omega_{i,j,k} = \epsilon(\sigma)\omega_{\sigma(i),\sigma(j),\sigma(k)}$, where $\sigma : \{i,j,k\} \to \{i,j,k\}$ is the unique bijective map with $\sigma(i) < \sigma(j) < \sigma(k)$ and $\epsilon(\sigma)$ is its signature. For any four distinct indices i, j, k, l, define in a similar way the $n-4$-form $\omega_{i,j,k,l}$.

Set $\alpha = \pi_0 \circ I_\nu(f_1 \wedge \ldots \wedge f_n)$ and $\beta = i_\nu \circ \pi_0(f_1 \wedge \ldots \wedge f_n)$. Clearly, α is a linear combination of the forms $f_i\{f_j, f_k\}\omega_{i,j,k}$ and $f_i df_l\{f_j, f_k\} \wedge \omega_{i,j,k,l}$. Using that $df_l\{f_j, f_k\} = f_l d\{f_j, f_k\} + \{f_j, f_k\}df_l$, we can express α as a linear combinations of $f_i\{f_j, f_k\}\omega_{i,j,k}$ and of $f_i f_l d\{f_j, f_k\} \wedge \omega_{i,j,k,l}$. However the expression $f_i f_l d\{f_j, f_k\}$ is symmetric into f_i and f_l, so α is a linear combination of the forms $f_i\{f_j, f_k\}\omega_{i,j,k}$. The corresponding coefficient is $(-1)^{i+j+k}(n-2)$.

It is easy to compute β and we have

$$\beta = \sum_i \sum_{j<k} (-1)^{i+j+k} f_i\{f_j, f_k\}\omega_{i,j,k}.$$

So we get $\alpha = (n-2)\beta$. As $\pi_0(f_1 \wedge \ldots \wedge f_n)$ is a $n-1$-form, we get formula (i). Formula (ii) is obvious, and formula (iii) follows from Formulas (i) and (ii). Q.E.D.

Proposition 2.3 *The map* $\pi : (\bigwedge^* C^\infty(X), d_{triv}) \to (\Omega_X^*[1], \delta_0)$ *commutes with the differential. Hence* π *induces a map* $\Theta^0 : H_*(C^\infty(X)) \to H_*^0(X)[1]$.

Proof The formula $\pi \circ d_{triv} = \delta_0 \circ \pi$ follows easily from the previous lemma.

Example 2.4 The map π is onto, and its component in degree 1 is an isomorphism. Hence the map $\Theta^0 : H_2(C^\infty(X)) \to H_1^0(X)$ is surjective.

3. Canonical homologies for symplectic manifolds

In this section, we compare $H_*^\tau(X)$ with the usual canonical homology for X being a symplectic manifold. Indeed theses homology groups are isomorphic, except for integral values of τ such that $-dim(X) \leq \tau + 1 \leq 0$ and $\tau + 1 \neq (-1/2)dim(X)$.

Let $\mathfrak{g} = \mathfrak{g}_0 \oplus \mathfrak{g}_1$ be the Lie super-algebra over $\mathbf{C}$ defined as follows. Its degree 0 part has basis $\{e, f, h\}$ and Lie brackets $[h, e] = 2e, [h, f] = -2f, [e, f] = h$. As Lie algebra, $\mathfrak{g}_0$ is isomorphic with $\mathfrak{sl}(2)$. The degree one part $\mathfrak{g}_1$ has basis $\{d, \delta\}$ and is an abelian. The remaining brackets $[e, d] = 0, [h, d] = d, [f, d] = \delta, [e, \delta] = d, [h, \delta] = -\delta, [f, \delta] = 0$ correspond with the natural action of $\mathfrak{sl}(2)$ over the two-dimensional space. Thus $\mathfrak{g}$ is the Lie super-algebra $\mathfrak{sl}(2) \times \mathbf{C}^2$. Set $\delta^- = f.d$ and $\delta^+ = d.f$.

For $N \geq O$, let $\mathcal{C}_N$ be the category of all $\mathfrak{g}$-modules M on which h acts diagonally with integral eigenvalues, all of them between $-N$ and N (the multiplicity of each eigenvalue can be infinite). Set $\mathcal{C} = \cup_{N \geq 0}\mathcal{C}_N$. In order to simplify the statements, we do not require that the $\mathfrak{g}$-modules are $\mathbf{Z}/2\mathbf{Z}$-graded. Denote by $L(n)$ the unique irreducible $\mathfrak{sl}(2)$-module of dimension $n+1$, and let $P(n)$ be the $\mathfrak{g}$-module induced from $L(n)$ (it is isomorphic to the module co-induced from $L(n)$). As $\mathfrak{sl}(2)$-module, any $\mathfrak{g}$-module $M \in \mathcal{C}$

can be decomposed as a direct sum of $L(n)$. Moreover, it is easy to show that the modules $P(n)$ are indecomposable (see the explicit model given below). Hence the family $(P(n))_{n\geq 0}$ is the set of all indecomposable projective modules in $\mathcal{C}$.

Let $\mathcal{A}$ be the centralizer of h in $U(\mathfrak{sl}(2))$. As it is well known, $\mathcal{A}$ is a commutative algebra freely generated by h and the Casimir operator $\Omega = 2ef + 2fe + h^2$. So it is also generated by ef and fe. An explicit model for the $\mathfrak{g}$-module $\oplus_{n\geq 0}L(n)$ is $\mathcal{M} = \mathbf{C}[x, y]$ on which e acts as $x\partial/\partial y$ and f by $y\partial/\partial x$. For any $a \in \mathcal{A}$, $a.x^n y^m$ is proportional to $x^n y^m$. Hence for any integers $n, m \geq 0$, we can define the character $\chi_{n,m}$ of $\mathcal{A}$ by the requirement $a.x^n y^m = \chi_{n,m}(a).x^n y^m$, for all $a \in \mathcal{A}$. Let $M \in \mathcal{C}$. We have $M = \oplus_{n\geq 0}M[n]$, where $M[n]$ is the $\mathfrak{sl}(2)$ isotypical component of type $L(n)$. We can also uniquely decompose M as $M = \oplus_{n,m\geq 0}M_{n,m}$ in a such way that any $a \in \mathcal{A}$ acts as $\chi_{n,m}(a)$ on the component $M_{n,m}$.

Recall that $L(1) \otimes L(n) = L(n+1) \oplus L(n-1)$ for any $n \geq 0$ (with the convention that $L(-1) = 0$). As δ is isotypical of type $L(1)$, we have $\delta(M[n]) \subset M[n-1] \oplus M[n+1]$. Denote by $\overline{U}$ the completion of $U(\mathfrak{g})$ with respect to the family of finite codimension ideals. Define the raising and lowering operators $R, L \in U(\mathfrak{g})$ by the requirements: $\delta = R - L$, $R(M[n]) \subset M[n+1]$, $L(M[n]) \subset M[n-1]$ for any $M \in \mathcal{C}$. Also define $\underline{m}, \underline{n} \in \overline{U}$ by the requirements $\underline{n}.x = nx$, $\underline{m}.x = mx$ for any $x \in M_{n,m}$ and any $M \in \mathcal{C}$.

For $\tau \in \mathbf{C}$, set $\delta'_\tau = \delta^- \circ (h + \tau) - \delta^+ \circ (h + \tau + 1)$.

Lemma 3.1 *Let $M \in \mathcal{C}$ and let $m, n \geq 0$ be integers.*

(i) We have $\delta^{\pm}(M_{n,m}) \subset M_{n,m+1} \oplus M_{n-1,m}$.

(ii) We have $R(M_{n,m}) \subset M_{n,m+1}$ and $L(M_{n,m}) \subset M_{n-1,m}$.

(iii) We have $\delta'_\tau = R \circ (\tau - \underline{m}) - L \circ (\tau + \underline{n} + 1)$.

Proof We have $h = \underline{n} - \underline{m}$. Hence we have $\delta(M_{n,m}) \subset \oplus_{k-l=n-m-1}M_{k,l}$. Moreover $M[p] = \oplus_{k+l=p}M_{k,l}$. As $\delta(M[l]) \subset M[l-1] \oplus M[l+1]$, the assertions (i) and (ii) follow.

To prove (iii), first note that it is enough to prove the assertion for all projective modules $P(n)$. So we will use an explicit model $\mathcal{P}$ for $\oplus_{n\geq 0}P(n)$. Set $\mathcal{P} = \mathbf{C}[x, y, \xi, \eta]$, where ξ, η are odd variables. We let act e by $x\partial/\partial y$, f by $y\partial/\partial x$, d by $\xi.x + \eta\partial/\partial y$ and δ by $\xi.y - \eta\partial/\partial x$. In this model, we have $R = \xi.y$, $L = \eta\partial/\partial x$, $\underline{n} = x\partial/\partial x$, and $\underline{m} = y\partial/\partial y$.

Hence, we have:

$$\begin{aligned}\delta^+ &= d \circ f \\ &= (\xi.x + \eta\partial/\partial y) \circ y\partial/\partial x \\ &= \xi.yx\partial/\partial x + \eta\partial/\partial x(1 + y\partial/\partial y) \\ &= R \circ \underline{n} + L \circ (1 + \underline{m}).\end{aligned}$$

Similarly, we have

$$\begin{aligned}\delta^- &= f \circ d \circ (h+\tau)\\ &= y\partial/\partial x \circ (\xi.x + \eta\partial/\partial y)\\ &= \xi.y + \xi.yx\partial/\partial x + \eta\partial/\partial xy\partial/\partial y\\ &= R \circ (1+\underline{n}) + L \circ \underline{m}.\end{aligned}$$

Hence, we have

$$\begin{aligned}\delta'_\tau &= \delta^- \circ (h+\tau) - \delta^+ \circ (h+\tau+1)\\ &= (R\circ(1+\underline{n}) + L\circ\underline{m})\circ(\underline{n}-\underline{m}+\tau)\\ &\qquad -(R\circ\underline{n} + L\circ(1+\underline{m}))\circ(\underline{n}-\underline{m}+\tau+1)\\ &= R\circ((1+\underline{n})(\underline{n}-\underline{m}+\tau) - \underline{n}(\underline{n}-\underline{m}+\tau+1))\\ &\qquad +L\circ(\underline{m}(\underline{n}-\underline{m}+\tau) - (1+\underline{m})(\underline{n}-\underline{m}+\tau+1))\\ &= R\circ(\tau-\underline{m}) - L\circ(\tau+1+\underline{n}).\quad \text{Q.E.D.}\end{aligned}$$

Let $\mathcal{B}'_N$ the set of all integers $p \neq -1$ with $-N-1 \leq p \leq N-1$.

Lemma 3.2 *Let $N \geq 0$, and let $M \in \mathcal{C}_N$. Assume that $\tau \notin \mathcal{B}'_N$. Then there exists $u \in \mathcal{A}$ such that $u|_M$ is invertible and $u \circ \delta \circ u^{-1}|_M = \delta'_\tau$.*

Proof Set $\Gamma = \{(n,m) \in \mathbf{Z}^2_{\geq 0} | n+m \leq N\}$, and let $\phi : \Gamma \to \mathbf{C}^*$. Denote by $\overline{\phi}$ the endomorphism of M define by $\overline{\phi}(v) = \phi(n,m)$ for any $v \in M_{n,m}$. It is clear that $\overline{\phi}$ is the action of some element of $\mathcal{A}$. Now fix two functions $a, b : \{0,1,\ldots,N\} \to \mathbf{C}^*$ such that

$$\begin{aligned}a(n-1) &= -(n+\tau+1).a(n) \quad \text{for any } 1 \leq n \leq N,\\ b(m+1) &= (\tau-m).b(m) \qquad\quad \text{for any } 0 \leq m \leq N-1.\end{aligned}$$

The hypothesis $\tau \notin \mathcal{B}_N$ is the precise condition for the existence of such a and b. Set $\phi(n,m) = a(n)b(m)$. Then for $v \in M_{n,m}$, we have

$$\begin{aligned}\overline{\phi}\circ\delta\circ\overline{\phi}^{-1}(v) &= \phi(n,m+1)/\phi(n,m)R(v) + \phi(n-1,m)/\phi(n,m)L(v)\\ &= b(m+1)/b(m)R(v) + a(n)/a(n-1)L(v)\\ &= (\tau-m)R(v) - (n+\tau+1)L(v)\\ &= \delta'_\tau(v).\end{aligned}$$

In the previous computation, we should take into account the special cases $m = N$ or $n = 0$. In such a case, the identity also holds because we have $R(v) = 0$ if $m = N$ and $L(v) = 0$ if $n = 0$. Q.E.D.

Remark Indeed it is possible to choose

$$\phi(n,m) = \gamma(\tau)\Gamma(-(n+\tau+1))/\Gamma(\tau+1-m),$$

where $\gamma(\tau)$ is some meromorphic function in a such way that the functions $\tau \mapsto \phi(n,m)$ have no poles or zeroes outside $\mathcal{B}'_N$.

Let X be a connected symplectic manifold. The two-form defining the symplectic structure will be denoted by ω. Let $\mathcal{B}_X$ the set of all integers p such that $p+1 \neq -(dimX)/2$ and $-dimX \leq p+1 \leq 0$.

Proposition 3.3 *Let $\tau \notin \mathcal{B}_X$.*

(i) There exists a $C^\infty(X)$-linear degree 0 isomorphism $u : \Omega^*_X \to \Omega^*_X$ such that $u \circ \delta_\infty \circ u^{-1} = \delta_\tau$. Moreover such u is unique up to scalar multiple.

(ii) There is a polynomial in two variables P such that $u = P(e(\omega) \circ i_\nu, i_\nu \circ e(\omega))$.

(iii) The isomorphism u commutes with all diffeomorphisms and all vector fields preserving ω.

Proof Recall from [14] that Ω^*_X is indeed a $\mathfrak{g}$-module. The operators e and f are the multiplication by the two-form ω and the contraction by ν. This module belongs to $\mathcal{C}_N$, where $N = (dim\, X)/2$ and we have $\kappa_0 = h + N$. Hence we have $\delta_\tau = \delta'_{\tau+dim(X)/2}$. Hence the existence of such u follows from Lemma 3.2. Since u is in the action of $\mathcal{A}$, we can write $u = P(ef, fe)$ for some polynomial P, which proves Assertion (ii).

Let u, u' be two such isomorphisms. Then $u.u'^{-1}$ is $C^\infty(X)$-linear, commutes with δ_∞ and preserves the grading. Set $\Phi = *u.u'^{-1}(*)^{-1}$, where $*$ is the Hodge operator. Then Φ is $C^\infty(X)$-linear, commutes with $d = *\delta(*)^{-1}$ and preserves the grading. We have $d.\Phi(1) = \Phi(d.1) = 0$. Hence $\Phi(1)$ is a scalar, and the restriction of Φ to Ω^0_X is the scalar multiplication by $\Phi(1)$. For any $n \geq 1$, Ω^n_X is the $C^\infty(X)$-module generated by $d.\Omega^{n-1}_X$. Hence by induction on n, Φ is the scalar multiplication by $\Phi(1)$ on Ω^n_X. So u is unique up to scalar multiplications. Q.E.D.

Corollary 3.4 *For any $\tau \notin \mathcal{B}_X$, there is a canonical isomorphism (up to scalar multiple) $H^\tau_*(X) \simeq H^{can}_*(X)$.*

Remark It is easy to show that for any $\tau \in \mathcal{B}_X$, the canonical homologies $H^\tau_*(X)$ are infinite dimensional. Hence for these values, the canonical homologies are different. See Example 1.7 for symplectic manifolds of dimension 2.

Recall that $H_*(C^\infty(X))$ denotes the homology of $C^\infty(X)$ as a Lie algebra. The composition of the natural map $\Theta^0 : H_*(C^\infty(X)) \to H^0_*(X)[1]$ and the previous corollary shows the following:

Corollary 3.5 *There is a canonical map (up to scalar multiple)*

$$\Theta : H_*(C^\infty(X)) \to H^{can}_*(X)[1].$$

Example 3.6 Let X be the sphere $x^2 + y^2 + z^2 = 1$. For its natural symplectic structure $\omega = xdy \wedge dz + ydz \wedge dx + zdx \wedge dy$, we have $\{x, y\} = z$, $\{y, z\} = x$ and $\{z, x\} = y$. We have $d_{triv}(x \wedge y \wedge z) = 0$, and $\pi(x \wedge y \wedge z) = 1/2\omega$. Hence the map $\Theta : H_3(C^\infty(X)) \to H_2^{can}(X)$ is onto.

Example 3.7 Let X be $\mathbf{R}^{2n}$ with its usual symplectic structure $\omega = \sum_{1 \leq i \leq n} dx_i \wedge dy_i$. By the Poincaré lemma, we have $H_i^{can}(X) = 0$ for $i \neq 2n$ and $H^{2n}(X) = \mathbf{C}[\omega^n]$. Set $\alpha = 1 \wedge x_1 \wedge y_1 \ldots \wedge x_n \wedge y_n$. We have $d_{triv}(\alpha) = 0$ and $\pi(\alpha) = 2^n/(2n+1).\omega^n$. Hence the map $\Theta : H_*(C^\infty(X)) \to H_*^{can}(X)$ is onto. Indeed, in an unpublished work, G. Papadopoulo showed that $H_i(C^\infty(X)) = 0$ for $0 < i \leq 2n$ and $H_{2n+1}(C^\infty(X)) = \mathbf{C}[\alpha]$.

4. A difference between the cohomology of Lie algebras of vector fields and the cohomology of groups of diffeomorphisms

At first, it seems that the map Θ defined in the previous section is very similar to the method of finding Euler classes in group cohomology. However, it is not the same, as we show in this section.

First remember that there is a analogous map for the cohomology of groups. Indeed let G be a connected topological group acting by diffeomorphisms on a connected manifold X. Assume that X is $(n-1)$ homologically connected, that is $H^i(X) = 0$ for $0 < i < n$. Then there is an exact sequence of G-modules

$$0 \to \mathbf{C} \to \Omega_X^0 \to \ldots \to Z_X^n \to H^n(X) \to 0.$$

Since G acts trivially on $H^n(X)$, the exact sequence is an $n+1$-extension of the trivial module $H^n(X)$ by $\mathbf{C}$, and it gives rise to map $\mu : H^n(X) \to H^{n+1}(G, \mathbf{C})$. This well-known construction gives rise to many non-trivial cohomology classes. For example the Euler class is the image of the canonical generator of $H^1(S^1)$ in $H^2(Diff^+(S^1))$, and more generally there are Euler classes in $H^{n+1}(Diff(S^n))$.

So in general we are expecting a map $H^*(X)[1] \to H^*(G)$. Indeed, there is no such map, but there is a spectral sequence E_* with $E_2 = H^*(G) \otimes H^*(X)$ and which computes the hypercohomology group $H^*(G, \Omega_X^*)$. So there is a well-defined map from some subquotient of $H^*(X)[1]$ to $H^*(G)$, whose one component is the map μ. More precisely, for X being $(n-1)$ homologically connected, $H^n(X) = H^0(G, H^n(X))$ and $H^{n+1}(G, \mathbf{C}) = H^{n+1}(G, H^0(X))$ are components of E_{n+1} and μ is the corresponding component of d_{n+1}. The non-triviality of the map μ is due to the fact that in general the complex of G-modules Ω_X^* is not quasi-isomorphic to its cohomology.

It is natural to ask if we can get the maps Θ and Θ^0 by the same kind of arguments. So let X be a manifold, let W_X the Lie algebra of all vector fields on X, let $\mathfrak{g} \to W_X$ be a morphism of Lie algebras. Similarly, there is a spectral sequence E with $E_2 = H^*(\mathfrak{g}) \otimes H^*(X)$ and which computes the hypercohomology group $H^*(\mathfrak{g}, \Omega_X^*)$. However the corresponding "map" $H^*(X)[1] \to H^*(\mathfrak{g})$ is zero, that is all differentials d_i of the spectral sequence are zero for $i \geq 2$. This is due to the following fact, usually attributed to Deligne.

Proposition 4.1 (Deligne) *The complex of W_X-modules Ω_X^* is quasi-isomorphic to its cohomology.*

Proof For the convenience of the reader, we give a proof (a similar trick is used in [4]). Choose an open covering $(U_i)_{i \in I}$ of X such that for any $n \geq 1$ and any n-tuple $\mathbf{i} = (i_1, \dots, i_n)$ the open set $U_{\mathbf{i}} = U_{i_1} \cap \dots \cap U_{i_n}$ is empty or contractible. Let C^{**} the double complex with components $C^{p,q} = \prod_{\mathbf{i} \in I^{p+1}} \Omega^q_{U_{\mathbf{i}}}$, and let C^* the simple complex whose components are $\prod_{\mathbf{i} \in I^{p+1}} H^0(U_{\mathbf{i}})$. There are canonical quasi-isomorphisms of complexes of W_X-modules:

$$\Omega_X^* \leftarrow C^{**} \to C^*.$$

Hence the complex of W_X-modules Ω_X^* is quasi-isomorphic to the complex C^*. Since W_X acts trivially on C^*, it is quasi-isomorphic to its cohomology as a complex of W_X-modules. Q.E.D.

Corollary 4.2 *Let X be a symplectic manifold, and let $\tau \notin \mathcal{B}_X$. Under the action of the Lie algebra $C^\infty(X)$, the complex $(\Omega_X^*, \delta_\tau)$ is quasi-isomorphic to $H^\tau(X)$.*

Proof The operator $*$ is $C^\infty(X)$-invariant and conjugates δ_∞ and d. Hence the proposition follows Propositions 3.3 and 4.1. Q.E.D.

Hence, the map θ cannot be obtained by the same methods as the Euler classes in group cohomology. However it is natural to ask if one can use the double complex structure of Ω_X to write down explicit extensions.

5. Another construction of Θ^0 in some special cases

Let (X, ω) be a connected symplectic manifold. In this section, we will show another construction of Θ^0 when X satisfies the strong Lefschetz theorem. Then we will compare Θ^0 with a Brylinski map connecting the homology of $C^\infty(X)$ with adjoint coefficients and the canonical homology. Indeed we will show that Θ^0 is a refinement of Brylinski map.

Let $\mathfrak{g}$ be the Lie super-algebra introduced in Section 3, and let $M \in \mathcal{C}$. We have $M = \oplus_{n \in \mathbf{Z}} M_n$. Hence (M, d) and (M, δ) are two complexes, and we will denote by $h_*(M, d)$ and by $h_*(M, \delta)$ their homologies. The harmonic elements in M are the elements $m \in M$ such that $dm = \delta m = 0$. We denote by $h_{har}(M,d)$ the image of harmonic elements in $h_*(M,d)$. Since $[d,e]=0$ and $[f,\delta]=0$, e acts on $h_*(M,d)$ and f acts on $h_*(M,\delta)$. Moreover we have $e.h_i(M,d) \subset h_{i+2}(M,d)$ and $f.h_i(M,\delta) \subset h_{i-2}(M,\delta)$. For any $n \geq 0$ denote by $L(n)^e$ the unique simple $\mathfrak{g}$-module of dimension $n+1$. As $\mathfrak{sl}(2)$-module, it is simple and $\mathfrak{g}_1$ acts trivially.

Proposition 5.1 *The following conditions are equivalent:*

(i) We have $h_{har}(M,d) = h_*(M,d)$

(ii) For any $i \geq 0$, *the map* $e^i : h_{-i}(M,d) \to h_i(M,d)$ *is an isomorphism.*

(iii) We have $M = S \oplus P$ *as* $\mathfrak{g}$*-modules, where* S *is a direct sum of simple modules* $L(n)^e$ *and* P *is projective.*

(iv) The morphism of complexes $(M,\delta) \to (M/dM,\delta)$ *is a quasi-isomorphism.*

Proof In [14], we proved that any $M \in C$ is a direct sum indecomposable modules of finite dimension, and we describe explicitly all indecomposable. So it is enough to check the result on each indecomposable module, what can be done easily as in [14].

Remark The $\mathfrak{sl}(2)$-action on M comes from an action of the group $SL(2)$, and we denote by $*$ the action of the matrix $\left(\begin{smallmatrix}0 & 1\\ -1 & 0\end{smallmatrix}\right)$. Indeed $*$ is the Hodge operator. As $*$ conjugates d and δ and e and f, the conditions (i), (ii) and (iv) can be expressed equivalently in terms of δ and f instead.

Let $2n$ be the dimension of X. We say that X satisfy the strong Lefschetz theorem if for any i with $0 \leq i \leq n$, the map $e([\omega^i]) : H^{n-i}(X) \to H^{n+i}(X)$ is an isomorphism. Such a manifold is necessarily compact, because $H^{2n}(X) = \mathbf{C}[\omega^n]$. Apply Proposition 5.1 to the $\mathfrak{g}$-module Ω_X^* to get:

Theorem 5.2 *Let* X *be a symplectic manifold of dimension* $2n$. *The following conditions are equivalent:*

(i) The strong Lefschetz theorem holds for X.

(ii) The morphism of complexes $(\Omega_X^*,\delta) \to (\Omega_X^*/d\Omega_X^*,\delta)$ *induces an isomorphism in homology.*

(iii) Any class in $H_*^{can}(X)$ *contains an harmonic representative.*

The equivalence of (i) and (iii) was obtained in [14] (for a proof simpler than the author's original proof, see D. Yan [19]). The homologies of δ and $\delta \circ (\kappa+1)$ are obviously the same. Hence the morphism of complexes $(\Omega_X^*,\delta_0) \to (\Omega_X^*/d\Omega_X^*, \delta\circ(\kappa+1))$ induces the map $H_*^0(X) \to h_*(\Omega_X^*/d\Omega_X^*,\delta)$ (see Lemma 1.4).

Corollary 5.3 *Assume that the strong Lefschetz theorem holds for* X, *then the natural maps*

$$H_*^0(X) \to h_*(\Omega_X^*/d\Omega_X^*,\delta) \quad \textit{and} \quad H_*^{can}(X) \to h_*(\Omega_X^*/d\Omega_X^*,\delta)$$

are both isomorphisms.

In [2], Brylinski defined a morphism of complexes

$$(C^\infty(X) \otimes \bigwedge^* C^\infty(X), d_{ad}) \to (\Omega_X^*, \delta)$$

by the formula $f_0 \otimes f_1 \wedge \dots f_n \mapsto f_0 df_1 \wedge \dots df_n$. Denote by

$$Br : H_*(C^\infty(X), C^\infty(X)) \to H_*^{can}(X)$$

the induced morphism. As with any Lie algebra, the natural map $C^\infty(X) \otimes \bigwedge^* C^\infty(X) \to \bigwedge^* C^\infty(X)$ defined by $f \otimes \alpha \mapsto f \wedge \alpha$ is a morphism of complexes and gives rise to a natural map $Nat : H_*(C^\infty(X), C^\infty(X))[1] \to H_*(C^\infty(X))$.

Corollary 5.4 *Assume that the strong Lefschetz theorem holds for X, then the Brylinski map Br factors through $H_*(C^\infty(X))$. More precisely, we have $Br([\alpha]) = n!\,\Theta^0 \circ Nat([\alpha])$ for any $[\alpha] \in H_n(C^\infty(X), C^\infty(X))$.*

Proof By Corollary 5.3, we can use the complex $(\Omega_X^*/d\Omega_X^*, \delta)$ instead of (Ω_X^*, δ_0) to define Br and Θ^0. At the chain level, Brylinsky map is defined by $f_0 \otimes f_1 \wedge \dots f_n \mapsto f_0 df_1 \wedge \dots df_n$ modulo $d\Omega_X^*$. The map $\Theta^0 \circ Nat$ is defined by $f_0 \otimes f_1 \wedge \dots f_n \mapsto 1/(n+1)! \sum_{0 \le i \le n} (-1)^i f_i df_0 \wedge \dots \wedge \hat{df_i} \dots \wedge df_n$ modulo $d\Omega_X^*$. Up to a d-boundary, $f_0 df_1 \wedge \dots df_n$ and $(-1)^i df_i \dots \wedge df_n$ are equal, which proves the corollary.

6. Canonical operations for symplectic manifolds

In this section, we will recall the results of [13] that we will use in the next section. Let X be a connected symplectic manifold. For each $f \in C^\infty(X)$, denote by H_f the hamiltonian vector field associated with f, and let $\mathcal{H}_X$ the Lie algebra of all hamiltonian vector fields. By definition, the Lie derivative of H_f is the linear map $g \in C^\infty(X) \mapsto \{f, g\} \in C^\infty(X)$.

By definition, a *canonical operation* is a multilinear differential operator on X which is invariant by $\mathcal{H}_X$ (or equivalently by the group of all diffeomorphisms preserving the symplectic structure). For any $n \ge 1$ we will denote by $Inv_n(X)$ the space of all n-multilinear canonical operations. Of course, the Poisson bracket and multiplication of functions are binary canonical operations. The map $f, g, k \in C^\infty(X) \mapsto f.\{g, h\}$ is an example of trinary canonical operation. More generally, the canonical operations which are "compositions" (in a suitable sense) of Poisson brackets and multiplications will be called *semi-classical*. The subspace of n-ary semi-classical operations will be denoted by $SC_n(X)$.

The main observations are the following:

(i) the space $Inv_n(X)$ depends only on the dimension of X;

(ii) there are canonical operations which are not semi-classical.

Let us state more precisely the statement (1.1). Let m be an integer. Denote by $\mathfrak{so}(n)$ the space of square skew-matrices of size n and by $\mathfrak{so}(n)[2m]$ the subspace of matrices of rank $\leq 2m$. Set $A_n[2m] = \mathbf{C}[\mathfrak{so}(n)[2m]]$. By removing the first row and column, one gets a map $\pi : \mathfrak{so}(n+1)[2m] \to \mathfrak{so}(n)[2m]$ which induces a structure of $A_n[2m]$-module on $A_{n+1}[2m]$.

There is a differential operator $K : A_n[2m] \to A_{n+1}[2m]$ defined in [13]. Denote by $Can_n(2m)$ its kernel. A precise version of (1.1) is the following.

Theorem 6.1 *There is a canonical isomorphism $Inv_n(X) \simeq Can_n(2m)$, where $2m$ is the dimension of X.*

Note that the spaces $\mathfrak{so}(n)[2m]$ and $\mathfrak{so}(n+1)[2m]$ are cones. Hence there is a natural grading of the algebra $A_n[2m]$ and of its module $A_{n+1}[2m]$. Moreover, the differential operator K is homogeneous. Hence, $Inv_n(X)$ bear a natural grading. Thus, denote by $L : Inv_n(X) \to Inv_n(X)$ the endomorphism defined by the multiplication by k on the degree k component of $Inv_n(X)$. When the symplectic two-form ω is exact, there is another way to describe L. In such a case, there is a vector field ξ such that $\xi.\omega = -\omega$ (ξ is a Liouville vector field). For $\theta \in Inv_n(X)$, let $[\xi, \theta] : C^\infty(X)^{\otimes n} \to C^\infty(X)$ be the map $f_1 \otimes \ldots \otimes f_n \mapsto \xi.\theta(f_1, \ldots, f_n) - \sum_{1 \leq i \leq n} \theta(f_1, \ldots, \xi.f_i, \ldots, f_n)$. We have $\xi\{f, g\} = \{\xi.f, g\} + \{f, \xi.g\}$ for any $f, g \in C^\infty(X)$. Hence $[\xi, \theta]$ belongs to $Inv_n(X)$ and we have $L(\theta) = [\xi, \theta]$. See [13] (Section 3.5) for a precise definition of L and for a proof of the next statement.

Lemma 6.2 *Let $\theta \in Inv_n(X)$ be non zero and let $k \in \mathbf{Z}_{\geq 0}$. If $L(\theta) = k\theta$, then the order of the differential operator θ is $2k$.*

7. Canonical operations for Poisson manifolds

This section contains technical statements used in Section 10. Let X be a Poisson manifold. As in the previous section, we will denote by H_f the hamiltonian vector field associated with any $f \in C^\infty(X)$, and let $\mathcal{H}_X$ the Lie algebra of all hamiltonian vector fields.

Recall that X has a natural foliation by symplectic leaves. An n-ary multilinear differential operator $\theta : C^\infty(X)^n \mapsto C^\infty(X)$ will be called tangential if it is tangent along each leaf. Stated otherwise, it means that for any leaf $Y \subset X$, $\theta(f_1, \ldots, f_n)$ vanishes on Y whenever at least one of the functions f_i vanishes on Y. Denote by $Inv_n(X)$ the space of all multilinear differential operators $\theta : C^\infty(X)^n \mapsto C^\infty(X)$ which are $\mathcal{H}_X$-invariant, i.e. such that

$$\{f, \theta(f_1, \ldots, f_n\} = \sum_{1 \leq 1 \leq n} \theta(f_1, \ldots, f_{i-1}, \{f, f_i\}, \ldots, f_n),$$

for any $f, f_1, f_2, \ldots \in C^\infty(X)$. Also denote by $Inv_n^{tang}(X)$ the subspace of $Inv_n(X)$ of all $\mathcal{H}_X$-invariant tangential differential operators. The elements in $Inv_n(X)$ (respectively: $Inv_n^{tang}(X)$) will be called invariant operations (respectively: tangential invariant operations). When the Poisson structure comes from a symplectic structure, invariant operations are the canonical operations.

Lemma 7.1 *Let U be an open subset of X, and let $\theta \in Inv_n(X)$. Then its restriction to U is in $Inv_n(U)$.*

Proof We want to show that θ is H_f-invariant for any $f \in C^\infty(X)$. However f can be written as a locally finite sum $\sum f_\alpha$, where the $f_\alpha \in C^\infty(U)$ are compactly supported. Each f_α extends to X, so θ is H_{f_α}-invariant. Hence θ is H_f-invariant. Q.E.D.

Let $\Omega(X)$ the set of all open subsets of X. By the previous lemma, there are restriction maps $Inv_n(U) \to Inv_n(V)$, $Inv_n^{tang}(U) \to Inv_n^{tang}(V)$ for any $U, V \in \Omega(X)$ with $U \supset V$ of X. The proof of the next statement is obvious.

Lemma 7.2 *The two maps $U \in \Omega(X) \mapsto Inv_n(U), Inv_n^{tang}(U)$ are sheaves.*

Let X be a Poisson manifold. Denote by $C^{\infty,tang}(X)$ the space of smooth functions on X which are constant along all symplectic leaves. The Poisson manifold is called regular if all symplectic leaves have the same dimension. The following statement follows from Section 6 and Lemma 7.2:

Lemma 7.3 *Let X be a regular Poisson manifold, and let $2m$ be the dimension of its symplectic leaves. Then there is a canonical isomorphism*

$$Inv_n^{tang}(X) \simeq C^{\infty,tang}(X) \otimes Can_n(2m).$$

8. Quantization of Poisson algebras

In the section, we will recall the well-known conjecture about the quantizability of all Poisson manifolds. The formality conjecture of Kontsevitch, which implies the quantizability of all Poisson manifolds, will be not involved.

Let k be a field. By Poisson algebra, we mean a commutative k-algebra A endowed with a Lie bracket $\{,\}$ satisfying Leibnitz identity: $\{fg, h\} = f\{g, h\} + g\{f, h\}$. Denote by h a formal variable. Let $* : A[[h]] \times A[[h]] \to A[[h]]$ be a bilinear $k[[h]]$-map. Define $\mu_n : A \times A \to A$ by the requirement: $f * g = \sum_{n \geq 0} h^n.\mu_n(f, g)$, for any $f, g \in A$. Following the work of Bayen, Flato, Frønsdal, Lichnerowicz and Sternheimer [1], the pair $(A, *)$ is called a quantization of the Poisson structure (and $*$ is called the star product of the quantization) if and only if the following requirements hold:

(i) $\mu_0(f,g) = f.g$,

(ii) $\mu_1(f,g) = \{f,g\}$,

(iii) the product $*$ is associative,

(iv) each μ_n is a differential operator.

Often there are additional requirements for the star product (like some symmetry conditions or a symbol condition on μ_n or that 1 is a unit), but it is not necessary here. For any Poisson manifold X, $C^\infty(X)$ is a Poisson algebra. Thus a quantization of X means a quantization of its Poisson algebra $C^\infty(X)$. If Y is a symplectic leaf of X, we say that the star product $*$ is tangential to Y if all differential operators μ_i are tangential to Y.

Conjecture: Any Poisson manifold is quantizable, i.e. it admits a $*$-product.

Recall that any symplectic manifold and more generally any regular Poisson manifold is quantizable: this follows from the work of Fedosov [6], de Wilde and Lecomte [5], Omori, Maeda and Yoshioka [15] (see also [18] for an account of the subject). It is natural to ask about the possibility to quantize any Poisson algebra, not only the Poisson algebras coming from Poisson manifolds. However, there are very singular examples of Poisson algebras and the general answer is negative as proved by the next easy example. Let $\mathfrak{g}$ be a Lie algebra over a field of characteristic zero. The ideal I of $S\mathfrak{g}$ generated by $S^2\mathfrak{g}$ is a Poisson ideal. Set $B = S\mathfrak{g}/I$.

Proposition 8.1 *Assume that $\mathfrak{g}$ is simple and is not of type A. Then the Poisson algebra B is not quantizable.*

Proof Consider a star product on B, and set $B((h)) = B \otimes k((h))$. Denote by B_L and by $B((h))_L$ the Lie algebra structures on B and $B((h))$ (the latter is given by the $*$-bracket). We have $H^2(B_L, B_L) = 0$, hence the Lie algebra B_L is rigid. It follows that $B((h))_L$ is isomorphic to $B_L \otimes k((h))$ as a Lie algebra. In particular, any Levi factor of $B((h))_L$ is simple of the same type as $\mathfrak{g}$. However, $B((h))$ is a finite dimensional associative $k((h))$-algebra. Hence $B((h))_L$ is isomorphic to the semi-direct product $\mathfrak{s} \times \mathfrak{r}$, where $\mathfrak{s}$ is a direct sum of $\mathfrak{gl}(n)$ and $\mathfrak{r}$ is nilpotent. Hence its Levi factor contains only type A-factor. Thus we get a contradiction.

9. The subspace $Quant_n(2m)$ of quantizable canonical operations

Let n, m be ≥ 1 integers. We will recall the construction of the subspace $Quant_n(2m)$ in $Can_n(2m)$. Choose any connected symplectic manifold X of dimension $2m$, and let $*$ be any star product. Denote by Σ_n the symmetric group of degree n. For any element $a = \sum_{\sigma\in\Sigma_n} a_\sigma.\sigma \in \mathbf{C}[[h]][\Sigma_n]$, the map $f_1, \ldots, f_n \mapsto \sum_{\sigma\in\Sigma_n} a_\sigma f_{\sigma(1)} * f_{\sigma(2)} * \ldots * f_{\sigma(n)}$ is a formal series $\tau_a = \sum_{i\geq 0} h^i \tau_i$ of differential operators. Let i be the smallest integer with $\tau_i \neq 0$. It is easy to prove that the differential operator τ_i (the first non-zero derivative of the formal series τ_a) is canonical. We define $Quant_n(X, *)$ as the set of all first non-zero derivatives of series τ_a when a runs over $\mathbf{C}[[h]][\Sigma_n]$.

Theorem 9.1 [13]

(i) The vector space $Quant_n(X, *)$ is a regular representation of Σ_n (in particular it has dimension $n!$).

(ii) As a subspace of $Can_n(2m)$, it does not depend on the choice of the symplectic manifold X or on the choice of the star-product $*$.

In brief, the space $Quant_n(X, *)$ depends only on the dimension of X. So we will set $Quant_n(2m) = Quant_n(X, *)$. The elements in $Quant_n(2m)$ will be called quantizable canonical operations. For $n > 2m + 1$, some elements in $Quant_n(X, 2m)$ cannot be written in terms of products and Poisson brackets. To compute the space $Quant_n(2m)$, it is enough to choose one star product on one symplectic manifold of dimension $2m$. By using the Moyal product on $\mathbf{R}^{2m}$, one gets a purely combinatorial description of $Quant_n(2m)$. However, its combinatorial analysis is very complicated.

Example 9.2 Let X be a symplectic manifold of dimension $2m$, and let n be an integer. For $f_1, \ldots, f_n \in C^\infty(X)$, set

$$\tau_n(f_1, \ldots, f_n) = \sum_{\sigma\in\Sigma_n} \epsilon(\sigma) f_{\sigma(1)} * \ldots * f_{\sigma(n)},$$

where $*$ is a star-product on X. Then τ_n is a formal series $\sum_{i\geq 0} h^i \tau_{n,i}$ of differential operators. Moreover τ_n is not zero [13]. The full formal series τ_n depends upon $*$, but its first non-zero derivative θ_n and the order of the formal series τ_n are independent of it. When $n \leq 2m + 1$, it is very easy to compute θ_n. Indeed for $n = 2a \leq 2m + 1$, we have

$$\theta_{2a}(f_1, \ldots, f_{2a}) = (1/2^a) \sum_{\sigma\in\Sigma_{2a}} \epsilon(\sigma)\{f_{\sigma(1)}, f_{\sigma(2)}\}\{f_{\sigma(3)}, f_{\sigma(4)}\}\cdots$$

and for $n = 2a + 1 \leq 2m + 1$, we have

$$\theta_{2a+1}(f_1, \ldots, f_{2a+1}) = (1/2^a) \sum_{\sigma\in\Sigma_{2a+1}} \epsilon(\sigma) f_{\sigma(1)}\{f_{\sigma(2)}, f_{\sigma(3)}\}\{f_{\sigma(4)}, f_{\sigma(5)}\}\cdots$$

In both cases the order of the formal series τ_n is a, and the order of the differential operators θ_{2a} and θ_{2a+1} is $2a$. However for $n > 2m + 1$, θ_n is not semi-classical and the explicit formula is very complicated.

10. A possible obstruction for quantizability of Poisson manifolds

In this section, we will show that the quantizable differential operators can be used to test the conjecture of quantizability of Poisson manifolds which contains an open leaf. A negative answer to this test would be a counter-example of the conjecture of Section 8. Theorem 6.1 reduces the computation to a difficult combinatorial problem, and we were still not able to perform this test on interesting examples.

Let X be a connected Poisson manifold, let $2m$ the maximal dimension of its symplectic leaves and let U be the open subset of all points $x \in X$ such that ν_x has maximal rank $2m$. A star product $f * g = \sum_{n \geq 0} h^n \mu_n(f, g)$ on X is called generically tangential if all operators μ_n are tangential to all symplectic leaves of maximal dimension.

Lemma 10.1 *(i) Any $\theta \in Quant_n(2m)$ defines a canonical n-ary differential operator $\overline{\theta}$ on U which is tangential to any symplectic leaf of U.*

(ii) Moreover assume that X admits a generically tangential star product. Then $\overline{\theta}$ can be extended to X.

Proof Point (i) follows from Lemma 7.3. Since $Quant_n(2m)$ does not depend on the choice of a star-product, there exists $a = \sum_{\sigma \Sigma_N} a_\sigma . \sigma \in \mathbf{C}[\Sigma_n][[h]]$ and an integer N such that for any symplectic manifold Y of dimension $2m$ and any star-product $*$ on Y, we have

$$\theta(f_1, \ldots, f_n) = d^N/dh^N \left(\sum_{\sigma \Sigma_N} a_\sigma f_{\sigma(1)} * \ldots * f_{\sigma(n)}\right) \qquad (1)$$

and

$$d^j/dh^j \left(\sum_{\sigma \Sigma_N} a_\sigma f_{\sigma(1)} * \ldots * f_{\sigma(n)}\right) = 0, \qquad (2)$$

for any $j < N$, for any $f_1, \ldots, f_n \in C^\infty(Y)$.

By choosing any tangential star-product on U, formula (1) defines a n-ary differential operator $\overline{\theta}$ on U which is tangential to any symplectic leaf of U and $\overline{\theta}$ is independent of the choice of the star product. Moreover, if X admits a generically tangential star-product, then formula (1) shows that $\overline{\theta}$ extends to the whole X. Q.E.D.

When the manifold X contains an open leaf, then all star products are generically transversal. So we get:

Proposition 10.2 (Test for quantizability) *Let X be a connected Poisson manifold which contains an open leaf of dimension $2m$. Assume that X admits a star-product. Then for any $n \geq 0$ and any $\theta \in Quant_n(2m)$, the n-ary differential operator $\overline{\theta}$ can be extended from U to the whole X.*

Conversely, if some $\overline{\theta}$ is not extendable, then X is not quantizable. It would be interesting to perform the test for the operations θ_n (defined in example 9.2) for $n \gg 0$, but explicit computations seem very complicated.

11. An extension of Θ^0 for symplectic manifolds.

In [10], the authors investigated the Lie algebra homology of $C^\infty(T)$, where T is an even dimensional torus endowed with an invariant symplectic structure. The main point of the paper was to find a non-canonical map $\Psi : H_*(C^\infty(T)) \to H^*(T) \otimes HC(\mathbf{C})[1]$. Non-canonical means here that Ψ is canonically defined on the graded vector space associated to some filtration of $H_*(C^\infty(T))$. The cogebra structure of the Lie homology induces a map $Sym(\Psi) : H_*(C^\infty(T)) \to Sym^*(H^*(T) \otimes HC(\mathbf{C})[1])$. In view of [10], there is a strong evidence that the map $Sym(\Psi)$ is onto. In [10], it has been conjectured that it is indeed an isomorphism.

We have $H^*(T) \simeq H_*^0(T)$. Hence, the map Ψ can be decomposed into a collection of maps $\Psi^n : H_*(C^\infty(T)) \to H_*^0(T)[2n+1]$. In this section, we will define some natural maps $\Theta^n : H_*(C^\infty(X)) \to H_*^0(X)[2n+1]$, for any symplectic manifolds X and for infinitely many n. The map Θ^n is very similar to the map S^n in cyclic homology which appears in the computation of the homology of matrix algebras (see [12, 16] where S is Connes map [3]).

Let m, n be integers. The symmetric group Σ_n acts on $Can_n(2m)$. Denote by $Can_n^\epsilon(2m)$ the subspace of skew symmetric canonical operations in $Can_n(2m)$. Let X be a symplectic manifold of dimension $2m$. As in Section 2, any $\theta \in Can_n^\epsilon(2m)$ extends uniquely to a differential operator $\overline{\theta} : \bigwedge^* C^\infty(X) \to C^\infty(X)[n-1]$ of order n. By the map $\theta \mapsto \overline{\theta}$, we identify the space $Can_*^\epsilon(2m) = \oplus_{n\geq 0} Can_n^\epsilon(2m)$ with a Lie super-subalgebra of $End(\bigwedge^* C^\infty(X))$.

Lemma 11.1 *Let m be an integer.*

(i) For any $n \geq 0$, the canonical operation $\theta_n \in Can_n^\epsilon(2m)$ is an eigenvector of L. Denote by λ_n the corresponding eigenvalue.

(ii) Let $Z(2m)$ the set of all integers $n \geq 0$ such that $\lambda_{2n+2} \neq \lambda_{2n+1} + 1$. The set $Z(2m)$ is infinite.

(iii) For any $n \in Z(2m)$, we have $[d_{triv}, \overline{\theta}_{2n+1}] = 0$.

Proof By Theorem 9.1, we can choose any symplectic manifold X of dimension $2m$ and any quantization to define the operations θ_n. We can assume that $X = \mathbf{R}^{2m}$, the symplectic form is $\omega = \sum_{1\leq i\leq m} dx_i \wedge dy_i$, the quantization is the Moyal product and the Liouville field is

$$\xi = (-1/2m) \sum_{1\leq i\leq m} x_i\partial/\partial x_i + y_i\partial/\partial y_i.$$

Set $f * g = \sum_{i\geq 0} h^n R_n(f,g)$, where $f, g \in C^\infty(\mathbf{R}^{2m})$. For this very particular example, we have $[\xi, R_i] = i.R_i$ for any $i \geq 0$. For $f_1, \ldots, f_n \in C^\infty(\mathbf{R}^{2m})$, set $\tau_n(f_1, \ldots, f_n) = \sum_{\sigma\in\Sigma_n} \epsilon(\sigma) f_{\sigma(1)} * \ldots * f_{\sigma(n)}$. Then τ_n is a formal series $\sum_{i\geq 0} h^i \tau_{n,i}$ of differential operators and we have $[\xi, \tau_{n,i}] = i.\tau_{n,i}$. Hence the canonical operations θ_n are eigenvectors of L.

First prove that $\lambda_{2n+1} = \lambda_{2n}$. We have

$$\tau_{2n} = h^{\lambda_{2n}} \theta_{2n} + \text{ terms of higher order},$$

and

$$\tau_{2n+1}(f_1, \ldots, f_{2n+1}) = \sum_{1\leq i\leq 2n+1} (-1)^i f_i * \tau_{2n}(f_1, \ldots, \hat{f_i}, \ldots, f_{2n+1}).$$

Thus we have: $\tau_{2n} = h^{\lambda_{2n}} \theta' +$ terms of higher order, where θ' is defined by:

$$\theta'(f_1, \ldots, f_{2n+1}) = \sum_{1\leq i\leq 2n+1} (-1)^{i+1} f_i \theta_{2n}(f_1, \ldots, \hat{f_i}, \ldots, f_{2n+1}).$$

We have $\tau_{2n+1}(1, f_2, \ldots, f_{2n+1}) = \tau_{2n}(f_2, \ldots, f_{2n+1})$. Hence θ' is not zero, and the order of the series θ_{2n} and θ_{2n+1} are the same. Thus the claim $\lambda_{2n+1} = \lambda_{2n}$ is proved.

Next prove that $\lambda_{2n+2} \geq \lambda_{2n+1} + 1$. We have

$$\tau_{2n+2}(f_1, \ldots, f_{2n+2}) =$$

$$\sum_{1\leq i<j\leq 2n+2} (-1)^{i+j+1} (f_i * f_j - f_j * f_i) * \tau_{2n}(f_1, \ldots, \hat{f_i}, \ldots, \hat{f_j}, \ldots, f_{2n+2}).$$

We get: $\tau_{2n+2} = h^{\lambda_{2n}+1} \theta'' +$ terms of higher order, where θ'' is defined by:

$$\theta''(f_1, \ldots, f_{2n+2}) =$$

$$\sum_{1\leq i<j\leq 2n+2} (-1)^{i+j+1} \{f_i, f_j\} \theta_{2n}(f_1, \ldots, \hat{f_i}, \ldots, \hat{f_j}, \ldots, f_{2n+2}).$$

The inequality $\lambda_{2n+2} \geq \lambda_{2n+1} + 1$ follows.

The proof of Corollary 8.8 of [13] shows that λ_n/n goes to ∞ as n goes to ∞. Hence the set $Z(2m)$ is infinite. Moreover n belongs to $Z(2m)$ if and only if $\theta" = 0$. Hence Assertion (iii) follows from the easy identity $\theta" = [d_{ad}, \theta_{2n+1}]$.

Q.E.D.

Theorem 11.2 *Let X be a symplectic manifold of dimension $2m$. For any $n \in Z(2m) \cup \{0\}$, the map $\pi_0 \circ \theta_{2n+1}$ induces a natural map*

$$\Theta^n : H_*(C^\infty(X)) \to H^0_*(X)[2n+1].$$

Proof It follows from Proposition 2.3 and Lemma 11.1 that the maps θ_{2n+1} and π_0 commute with the differentials. Q.E.D.

References

1. F. Bayen, M. Flato, C. Frønsdal, A. Lichnerowicz and D. Sternheimer, Deformation theory and quantization, *Ann. of Phys.* 111 (1978) 61–151.
2. J.L. Brylinski, A differential complex for Poisson manifold, *J. Diff. Geom.* 28 (1988) 93–114.
3. A. Connes, Cohomologie cyclique et foncteurs Ext^n, *CRAS* 296 (1983) 953–958.
4. P. Deligne and C. Illusie, Revêtement modulo p^2 et scindage du complexe de De Rham, *Invent. Math.* 89 (1987)247–270.
5. M. de Wilde and P. Lecomte, Existence of star-product and formal deformations in Poisson Lie algebra of arbitrary symplectic manifold, *Lett. Math. Phys.* 7 (1983) 487–496.
6. B. Fedosov, A simple geometrical construction of deformation quantization, *J. Diff. Geom.* 40 (1994) 213–238.
7. J. Ferchik *Cohomology of hamiltonian and related formal vector fields algebras*, Thesis, Harvard (1975).
8. I.M. Gelfand, D.B Fuks, Cohomology of the Lie algebra of tangential vector fields, *Funct. An. Appl.* 4 (1970) 23–31.
9. I.M. Gelfand, D.B. Fuks, D.I. Kalinin, Cohomology of the Lie algebra of hamiltonian vector fields, *Fuct. An. Appl.* 6 (1972) 193–196.
10. I.M. Gelfand and O. Mathieu, On the cohomology of hamiltonian vector fields, *J. Funct. Anal.* 115 (1992) 121–132.
11. J.L. Koszul, *Crochet de Shouten-Nijenjuis et cohomologie* in "Elie Cartan et les mathèmatique d'aujourd'huis", Asterisque (1985) 251–271.
12. J.L. Loday, D. Quillen, Cyclic homology and the homology of matrices, *Comment. Math. Helv.* 4 (1984) 23–31.
13. O. Mathieu, *The symplectic operad*, in Functional analysis on the eve of the 21th Century, (ed. S. Gindikin, J. Lepowsky, R.L. Wilson), Birkhauser. Prog. in Mat. 131, (1995), 223–243.
14. O. Mathieu, Harmonic cohomology classes of symplectic manifold, *Comment. Math. Helv.* 70 (1995) 1–9.
15. H. Omori, Y. Maeda, A. Yoshioka, Weyl manifolds and deformation quantization, *Adv. Math.* 85 (1991) 224–255.
16. B.L. Tsygan, Homology of matrix algebras over rings and the Hochschild homology, *Usp. Math. Nauk* 38 (1983) 217–218.
17. A. Weil, *Varietes Kahleriennes*, Hermann, Paris. Publ. Inst. Math. Univ. Nancago, 6 (1958).
18. A. Weinstein, Deformation quantization, Bourbaki Seminar (June 1994), to appear in Astérisque.
19. D. Yan, *Hodge Theory on Symplectic manifolds*, Preprint.

MAXIMAL TORI OF SOME SYMPLECTOMORPHISM GROUPS AND APPLICATIONS TO CONVEXITY

ANTHONY BLOCH[1]
Department of Mathematics, University of Michigan,
Ann Arbor MI 48109, USA
MOHAMED OULD EL HADRAMI[2]
Department of Mathematics, University of Arizona,
Tucson, AZ 85721, USA
HERMANN FLASCHKA
Department of Mathematics, University of Arizona,
Tucson, AZ 85721, USA
AND
TUDOR S. RATIU[3]
Department of Mathematics, University of California,
Santa Cruz, CA 95064, USA

Abstract. In this paper we study some analogies between symplectomorpism groups and compact Lie groups. In particular we discuss the approximation of the Lie algebra of Hamiltonian vector fields by matrix Lie algebras, the existence of maximal tori in a suitable geometric setting, and an infinite-dimensional version of the Kostant convexity theorem for a nonsmooth completion of the symplectomorphism group.

1. Introduction

The goal of the work reported in this paper is to find analogies between symplectomorphism groups and compact Lie groups. For the reader's con-

[1] Research partially supported by the National Science Foundation PYI grant DMS-91-57566, AFOSR grant F49620-96-1-0100, a Guggenheim Fellowship, and The Institute for Advanced Study, Princeton.

[2] Research supported by USAID through the Afgrad Program of the African American Institute. Present address: B.P. 2871, Nouakchott, Mauritania.

[3] Research partially supported by NSF Grant DMS-9503273 and DOE contract DE-FG03-95ER25245-A000.

D. Sternheimer et al. (eds.), Deformation Theory and Symplectic Geometry, 201–222.

venience, we summarize material from our earlier papers, but we also describe recent developments which, we hope, will eventually shed new light on infinite dimensional Lie theory.

It might be useful to begin with the two papers that stimulated our research in this area. The first one (Bao and Ratiu [1992]) identifies a maximal torus in the group of area–preserving Sobolev diffeomorphisms of an annulus. This torus consists of the twist maps: if the annulus is parametrized by $(z, \theta) \in [0,1] \times [0, 2\pi]$, a twist map sends (z, θ) to $(z, \theta + f(z))$. The set of such maps turns out to be a submanifold of the group of area–preserving diffeomorphisms, it is maximal abelian, and in a natural weak Riemannian metric, it is flat and totally geodesic. (We will state the precise theorem in Section 2.) These are the most important properties of maximal tori in compact Lie groups. We generalize this result to projective toric varieties. Instead of twist maps, one now has the canonical transformations that preserve a Lagrangean foliation by tori. These form a maximal torus in the group of symplectomorphisms with respect to the Kähler symplectic form on the toric variety (it is unfortunate that the word "torus" is used in two different ways!). The mathematical point here is a technical one: one set of diffeomorphisms — our maximal torus — is a Hilbert submanifold of another manifold of Sobolev diffeomorphisms.

Our second motivation comes from Brockett and Bloch [1990]. They consider the dispersionless Toda lattice, and show that some solutions exhibit the so–called "sorting property". It would take too long to elaborate on the details, and we shall not need them in any case. Basically, the "sorting property" can be explained as a gradient flow from the top vertex to the bottom vertex in a certain momentum polytope. This was well known in the finite–dimensional setting, and that led us to the infinite–dimensional convexity theorem described in Section 5.

Let us return now to analogies between symplectomorphism groups and compact Lie groups. On the global level, we have the maximal tori referred to above. The infinitesimal version (Lie algebras) is much better developed (Saveliev and Vershik [1989], [1990]). When the underlying symplectic manifold is two–dimensional, the Lie algebra of Hamiltonian vector fields is $\mathbb{Z}$–graded and shares many properties with semisimple Lie algebras. Our generalization to toric varieties appears to offer a natural extension of the Saveliev–Vershik ideas; this is one reason for our interest in this problem, but we have nothing definitive to report as yet. The convexity theorem described below lies halfway between global and infinitesimal: it involves the action of a symplectomorphism group on its Lie algebra of Hamiltonian vector fields. As will become clear, this result is pure functional analysis — and that is the current dilemma of the infinite–dimensional Lie theory we envision. One thinks of compact Lie groups (at the very basic level which we can currently hope to adapt to infinite dimensions) as geometric objects, yet the transition to symplectomorphism groups requires analysis, measure theory, completions, and even the introduction of semigroups. We shall attempt to clarify these rather vague remarks later in the paper.

Perhaps the best way to understand the analogies we have in mind is by "approximating" the Lie algebra of Hamiltonian vector fields on a symplectic manifold by finite-dimensional matrix Lie algebras. For our purposes, the notion of *Toeplitz quantization* is particularly convenient. We now summarize some explicit calculations that might make the rest of our paper more understandable. We emphasize that Toeplitz quantization, a variant of the more familiar "geometric quantization", should only be thought of as a guide for intuition. Eventually, it might become a technical tool as well, but we have not succeeded in pushing through certain (apparently delicate) estimates.

Consider the Riemann sphere $\mathbb{CP}(1)$ with the Fubini–Study symplectic form

$$\Omega = \frac{i}{(1+w\bar{w})^2}\, dw \wedge d\bar{w}, \tag{1.1}$$

and the standard hyperplane bundle L. The tensor power $L^{\otimes m}$ has $m+1$ linearly independent sections, which in the local coordinate w are just $1, w, \ldots, w^m$. The bundle $L^{\otimes m}$ comes equipped with the Hermitean metric

$$< s_1, s_2 > (w) = \frac{1}{(1+|w|^2)^m} s_1(w)\overline{s_2(w)}.$$

Now let Γ_2^m be the space of square-integrable sections of $L^{\otimes m}$ and Γ_{hol}^m the space of holomorphic sections (the span of $1, w, \ldots, w^m$). The orthogonal projection $\Gamma_2^m \to \Gamma_{\text{hol}}^m$ is denoted by Π^m.

Definition 1.1 *Let $f : \mathbb{CP}(1) \to \mathbb{R}$ be a smooth function. The* Toeplitz quantization *of f is the map $T_f^m : \Gamma_{\text{hol}}^m \to \Gamma_{\text{hol}}^m$ defined by*

$$T_f^m = \Pi^m \circ M_f \circ \Pi^m,$$

where M_f is multiplication by f. Note that T_f^m is a Hermitean matrix.

Guillemin [1979] and Bordemann, Meinrenken, Schlichenmaier [1994] proved the following results, in the much more general case where $\mathbb{CP}(1)$ is replaced by a projective Kähler variety ($\|\cdot\|$ is the usual operator norm on matrices):

Theorem 1.2 *(i)* $\lim_{m\to\infty} \|im[T_f^m, T_g^m] - T_{\{f,g\}}^m\| = 0$;
(ii) $\lim_{m\to\infty}$ Trace $(T_f^m)^p/m = \int_{\mathbb{CP}(1)} f^p\,\Omega$ *for* $p = 0, 1, 2, \ldots$.

This theorem says that, after suitable renormalization with $1/m$ playing the role of Planck's constant $\hbar$, the matrix commutator $[T_f^m, T_g^m]$ "approximates" the Poisson bracket $\{f, g\}$, and that the eigenvalues of T_f^m distribute themselves like the values of the function f.

Now to the point. We compute the limit of T_f^m in the case of the Riemann sphere, and thereby understand the correspondence between matrices (maps from Γ_{hol}^m to itself) and functions on $\mathbb{CP}(1)$ (Hamiltonians that generate the Lie algebra of the symplectomorphism group of the sphere). We will think of functions on $\mathbb{CP}(1)$ as "continuum matrices". This interpretation, as we shall see, is surprisingly fruitful.

Proposition 1.3 *(i) Let* $\gamma_{m,k} = \sqrt{\frac{2\pi}{(m+1)\binom{m}{k}}}$. *The sections of* Γ_{hol}^m *given (locally) by* $\gamma_{m,k} w^k$, $k = 0, 1, \ldots, m$ *are orthonormal.*
(ii) The (k, ℓ) *entry of the matrix* T_f^m *is*

$$\gamma_{m,k}\gamma_{m,\ell} \int_0^{2\pi} \int_0^{\infty} f(r, \theta) \frac{r^{k+\ell+1} \exp(i(k-\ell)\theta)}{(1+r^2)^{m+2}} \, dr \, d\theta,$$

where $w = r\exp(i\theta)$.
(iii) Expand $f(r, \theta)$ *in a Fourier series:*

$$f(r, \theta) = \sum_{s=-\infty}^{\infty} \hat{f}_s(r) \exp(is\theta).$$

Let $m, \ell, k \to \infty$ *so that* $\frac{k+\ell}{m} \to a$ *and* $k - \ell = s \in \mathbb{Z}$. *Then*

$$(T_f^m)_{k,\ell} \to \hat{f}_{-s}(r_0),$$

where $r_0 = \sqrt{a/(2-a)}$.
(iv) f *is independent of* θ *if, and only if,* T_f^m *is diagonal for every* $m = 0, 1, 2, \ldots$.

We will not elaborate on the proof (see El Hadrami [1996]): it requires integral tables, Stirling's formula, and the Laplace approximation. The intuitive meaning of the result is more important. As $m \to \infty$ (meaning that $\hbar \to 0$), the matrices T_f^m of course become bigger. Under the conditions stated in the Proposition, the super– and sub–diagonals of T_f^m are asymptotically indexed by the Fourier exponentials $\exp(is\theta)$. For example, $s = 0$ corresponds to the main diagonal of T_f^m, $s = 1$ to the first super–diagonal, and so on. Asymptotically, all the diagonals are "equally long", and position along them is indexed by the radial parameter r.

Our general philosophy is simple: *what is true for matrices* T_f^m *should have a counterpart for the functions* $f(r, \theta)$. Of course there will be complications, but much of the finite–dimensional theory seems to survive. The challenging aspect of this version of infinite–dimensional Lie theory, we feel, is the generalization to symplectomorphism groups of higher–dimensional manifolds. In the next section, we begin with a few precise results involving toric varieties.

2. Maximal tori

For sake of concreteness, start with the Riemann sphere $\mathbb{CP}(1)$. The group of diffeomorphisms $\mathcal{D}^s$ of Sobolev class $s > 2$, and its subgroup $\mathcal{D}^s_\Omega$ of diffeomorphisms preserving the symplectic form Ω are both infinite–dimensional manifolds (see Ebin and Marsden [1970] for details). We claim that there is a natural "maximal torus" sitting inside $\mathcal{D}^s_\Omega$. Take the sphere and twist it about the axis passing between the north and south poles, rotating each circle of latitude rigidly through an angle depending on the circle. Call the set of these twist maps $\mathcal{T}^s$. Transformations belonging to $\mathcal{T}^s$ obviously preserve area, they commute with each other, and $\mathcal{T}^s$ has differential geometric properties reminiscent of the important features of maximal tori in compact Lie groups. Bao and Ratiu [1992] first drew attention to this analogy:

Theorem 2.1 *Let $\mathcal{A}$ be the annulus (or cylinder) $\{0 \le z \le 1\} \times \{\exp 2\pi i\theta \mid 0 \le \theta < 1\}$. Let* $\mathrm{SDiff}^s(\mathcal{A})$ *be the group of Sobolev diffeomorphisms (of class $s > 2$) that preserve area, and let $\mathcal{T}^s$ be its subgroup of diffeomorphisms that preserve the circles $z =$* constant.
(i) $\mathcal{T}^s$ is a smooth, H^s-closed, path connected submanifold of $\mathrm{SDiff}^s(\mathcal{A})$*;*
(ii) it is a maximal abelian subgroup of $\mathrm{SDiff}^s(\mathcal{A})$*;*
(iii) it is totally geodesic and flat with respect to the hydrodynamic metric

$$\langle X, Y\rangle = \int_{\mathcal{A}} X \cdot Y d(\text{ area}) \tag{2.1}$$

where X and Y are (Sobolev class) divergence free vector fields on $\mathcal{A}$.

We extend this result by replacing $\mathcal{A}$ with $\mathbb{CP}(1)$, $\mathbb{CP}(2)$ (El Hadrami [1996]) and then by projective toric varieties. The subtle part of the problem is the one most easily taken for granted: do the twists form a Hilbert submanifold of $\mathcal{D}^s_\Omega$? In other words, does the "twist" subgroup deserve to be called "Lie subgroup" (or "maximal torus", etc.)?

Let us isolate the technical problems first. To prove that A is a submanifold of B, one can try three strategies.

- Exhibit a chart in which B looks locally like $\mathbf{E} \times \mathbf{F}$ and A looks like $\mathbf{E}$, where $\mathbf{E}$ and $\mathbf{F}$ are Hilbert spaces. This is the approach taken by Ebin and Marsden [1970] and Bao and Ratiu [1992], but it does not seem to work for $\mathbb{CP}(1)$. An explicit chart for a diffeomorphism group is usually based on geodesic flow on the underlying finite–dimensional manifold.

Because the circles of latitude on the sphere, and in particular, the north and south poles, cannot be geodesics, we did not succeed in constructing a chart of this type.

- Use the implicit function theorem: find a submersion F from B to another manifold, and a regular value a of F, so that $F^{-1}(a) = A$.
- Use transversality: A is the transverse intersection of two (or more) submanifolds of B.

Our proof, due to El Hadrami [1996], uses the implicit function theorem and transversality. There are three steps to the argument. We begin with $\mathbb{CP}(1)$.

Step 1: Think of $\mathbb{CP}(1)$ as the sphere in $\mathbb{R}^3$. The one–dimensional torus S^1 acts on it by rigid rotations about the vertical axis. This is a Hamiltonian action (with respect to the area form Ω) and it has a momentum mapping $J : \mathbb{CP}(1) \to \mathbb{R}$ that associates to each point its height above the equatorial plane. Let $G^{\text{fix}} = \{\phi \in \mathcal{D}^s \mid \phi \text{ fixes the poles}\}$. This is a manifold (Ebin and Marsden [1970]). One proves that the map $F : \phi \mapsto J \circ \phi$ is a submersion onto a certain affine subspace of a Sobolev space of real–valued maps. The set $F^{-1}(J)$ is a submanifold of G^{fix}. It consists of the diffeomorphisms that preserve the level sets of J, i. e. the circles of latitude on the sphere. Note that these maps are not necessarily just twists, and so they need not conserve the symplectic form.

Step 2: Define a function from $F^{-1}(J)$ to the space of cohomology classes of two–forms on $\mathbb{CP}(1)$ by

$$\gamma : \phi \to [\phi^* \Omega] .$$

Note that $\gamma^{-1}(\Omega)$ is exactly the set of maps that preserve the symplectic form Ω on $\mathbb{CP}(1)$ *and* the level sets of J. This is the proposed maximal torus in $\mathcal{D}^s_\Omega$. We prove that γ is a submersion at Ω, so that our proposed maximal torus $\gamma^{-1}(\Omega)$ is a submanifold of $F^{-1}(J)$ (this argument follows Ebin and Marsden [1970], but the details are more complicated).

The proof now concludes as follows. The candidate maximal torus $\gamma^{-1}(\Omega)$ is a submanifold of $F^{-1}(J)$, which is known to be a submanifold of $\mathcal{D}^s$. By Ebin and Marsden [1970] we know that $\mathcal{D}^s_\Omega$ is a submanifold of $\mathcal{D}^s$. Since

$$\gamma^{-1}(\Omega) \subset \mathcal{D}^s_\Omega,$$

a technical argument with coordinate charts shows that indeed the torus $\gamma^{-1}(\Omega)$ is a submanifold of $\mathcal{D}^s_\Omega$.

We want to stress that almost all the effort in adapting Theorem 2.1 to $\mathbb{CP}(1)$ goes into proving that the group of twist mappings has a manifold structure, and that it is a submanifold of the larger group $\mathcal{D}^s_\Omega$. Parts (ii) and (iii) of Theorem 2.1 carry over with little change.

The proof outlined above can be adapted to symplectomorphism groups of higher–dimensional manifolds. We sketch the situation for $\mathbb{CP}(2)$. A local coordinate is given by $(w_1, w_2) \in \mathbb{C}^2$. Write $w_j = r_j \exp(i\theta_j)$. The role of twist maps is now taken over by the group of canonical (with respect to the Kähler form) transformations preserving every torus defined by fixing r_1, r_2. The natural action of the torus $S^1 \times S^1$ on $\mathbb{CP}(2)$ has a momentum mapping

$$J : \mathbb{CP}(2) \rightarrow \text{ a triangle } \Delta.$$

(See, for example, Audin [1991]). The group of Sobolev diffeomorphisms fixing Ω is again denoted by $\mathcal{D}^s_\Omega$, but now $s > 3$. The candidate maximal torus, naturally, is the set of maps preserving Ω and leaving the coordinates r_1, r_2 unchanged.

The proof (that this group of symplectomorphisms is a submanifold of $\mathcal{D}^s_\Omega$) requires a modification of G^{fix} in Step 1 above. This time, G^{fix} consists of diffeomorphisms that fix the pre–images of the edges of the momentum polytope Δ. The results of Ebin and Marsden [1970] no longer guarantee the manifold structure of this set of maps, because the maps preserve a singular union of three spheres (the pre–images of the edges of Δ). Ebin and Marsden [1970] prove that the subgroup of diffeomorphisms preserving a smooth submanifold of $\mathbb{CP}(2)$ is itself a manifold, but the maps in G^{fix} preserve a singular variety. We solve this problem by: (a) using the fact that the set of maps preserving any one sphere is a manifold; (b) using transversality of the intersections of the spheres to prove transversality of the sets of maps preserving the individual spheres. The other steps in the "submanifold" proof are more tedious than for $\mathbb{CP}(1)$, but not essentially different.

With some careful book–keeping, this can be generalized to $\mathbb{CP}(n)$, and to projective toric varieties. One needs the momentum polytope, the transversality of the pre–images of its faces, and a little work in certain surjectivity arguments. The details will be published elsewhere.

3. The Schur–Horn–Kostant Theorem

In order to facilitate the transition to infinite dimensions, in this section we will review the classical Schur, Horn, and Kostant convexity theorems. The main tools for the proof of the Schur and Horn theorems will also be presented since these generalize to the infinite dimensional setting to be discussed in the next section. These ideas have been discussed, in more detail, in our earlier papers: Bloch, Flaschka, and Ratiu [1993,1996].

We shall use the following notation:

1. $\sum_n x$ is the orbit of x under the symmetric group on n letters, i.e. the collection of points $(x_{s(1)}, \ldots, x_{s(n)})$, where s ranges over all $n!$ permutations.
2. For $C \subset \mathbb{R}^n$, $\widehat{C}$ denotes the convex hull of C.

Theorem 3.1 (Schur [1923]) *Let A be a Hermitean matrix with eigenvalues λ_j (arranged in non-increasing order). Let $\lambda = (\lambda_1, \ldots, \lambda_n)$, and let $A^0 = (A_{11}, \ldots, A_{nn})$ be the diagonal of A. Then*

$$A^0 \in \widehat{\textstyle\sum_n \lambda}.$$

Theorem 3.2 (Horn [1954]) *Let $\lambda \in \mathbb{R}^n$, with components arranged in non-increasing order. If $A^0 \in \widehat{\sum_n \lambda}$, there is a Hermitean matrix A with eigenvalues λ whose diagonal is A^0.*

These theorems have a Lie theoretic interpretation discovered by Kostant [1973]. Let $K = U(n)$, and let T be the subgroup of diagonal matrices. The Lie algebras of K and T are denoted, respectively, by $\mathfrak{k} = \mathfrak{u}(n)$ and $\mathfrak{t}$; $\mathfrak{u}(n)$ consists of all skew Hermitean matrices and $\mathfrak{t}$ is the set of all purely imaginary diagonal matrices. Let $W := N(T)/T$ denote the Weyl group which in this case coincides with $\sum_n$, the symmetric group. For $\lambda \in \mathfrak{t}$, let $\mathcal{O}_\lambda$ denote the orbit $\{k\lambda k^{-1} \mid k \in K\} \subset \mathfrak{u}(n)$ of skew Hermitean matrices with fixed eigenvalues λ_j. Let $\pi : \mathfrak{u}(n) \to \mathfrak{t}$ be the orthogonal projection relative to the inner product on $\mathfrak{u}(n)$ given by minus the trace of the product of two matrices; the effect of π is to pick up the diagonal part of a matrix. Identify the Lie algebra $\mathfrak{u}(n)$ of skew Hermitean matrices with the vector space of Hermitean matrices via multiplication by $\sqrt{-1}$. The theorems of Schur and Horn state that $\pi(\sqrt{-1}\mathcal{O}_\lambda)$ is a convex polytope which coincides with the convex hull of the Weyl group orbit $\sqrt{-1}\sum_n \lambda$. In this form, Theorems 3.1 and 3.2 generalize to all compact connected Lie groups (Kostant [1973]).

It is well known that the Schur–Horn theorems, and Kostant's generalization, are most naturally viewed in the setting of symplectic geometry, being consequences of the Atiyah–Guillemin–Sternberg theorem about the convexity of the image of the momentum map (Atiyah [1982], Guillemin and Sternberg [1982]). We will not discuss these results here since our infinite dimensional theorems adapt the three classical theorems to a certain completion of the group of symplectomorphisms of toric varieties, but (so far) nothing we do involves infinite-dimensional symplectic geometry in any essential way.

One of the classical proofs of the Schur–Horn theorem is based on the following facts which, as we explain in Section 4, have infinite-dimensional counterparts. Symplectic geometry plays no role in this proof, and Lie theoretic methods are not needed (although much of the argument, including equation (3.1), can be phrased Lie–algebraically; see Atiyah and Bott [1982]).

Definition 3.3 (Hardy, Littlewood, Pólya [1934]). *For $x \in \mathbb{R}^n$, let x^* denote the vector obtained by rearranging the components of x in nonincreasing order. We say that y majorizes x, written $x \prec y$, if*

$$x_1^* + \ldots + x_k^* \leq y_1^* + \ldots y_k^*, \quad \text{for } 1 \leq k \leq n-1 \quad \text{and} \quad \sum_{j=1}^n x_j^* = \sum_{j=1}^n y_j^*. \quad (3.1)$$

Definition 3.4 *An $n \times n$ real matrix P is called doubly stochastic if $P_{ij} \geq 0$, and the sum of the entries on each row and column is* 1.

Theorem 3.5 *(i)* (Marshall and Olkin [1979]) *P is doubly stochastic if, and only if, $Pe = e$ and $e'P = e'$, where e is the column vector all of whose entries are* 1, *and e' is its transpose.*
(ii) (Marshall and Olkin [1979]) *P is doubly stochastic if, and only if, $Px \prec x$ for all $x \in \mathbb{R}^n$.*
(iii) (Hardy, Littlewood, Pólya [1934]) *$x \prec y$ if, and only if, there is a doubly stochastic P such that $x = Py$.*
(iv) (Birkhoff [1946]) *The set of doubly stochastic matrices is the convex hull of its extreme points, which are precisely the permutation matrices.*

Corollary 3.6 $\{x \mid x \prec y\} = \widehat{\Sigma_n y}$.

The Schur theorem is proved by first diagonalizing the Hermitean matrix A, $A = Q\lambda Q^*$, Q unitary. Then $A_{ii} = \sum_j |Q_{ij}|^2 \lambda_j$. If Q is unitary, the matrix $P_{ij} \stackrel{\text{def}}{=} |Q_{ij}|^2$ is doubly stochastic and the Corollary gives the conclusion. Horn proved the converse by a rather intricate argument, deducing that when $x \prec \lambda$, there must be a doubly stochastic P of the form $P_{ij} = |Q_{ij}|^2$, with Q unitary, satisfying $x = P\lambda$; $Q\lambda Q^*$ is then the desired Hermitean A having eigenvalues λ_j and diagonal x.

4. Measure preserving maps of the sphere

This section continues the review of the Lie theoretic properties of the group $K = \mathrm{SDiff}^s(\mathbb{CP}(1))$ of area (but not necessarily orientation) preserving Sobolev diffeomorphisms of the sphere. Using Sections 1 and 2 as motivation, we formulate the Schur–Horn theorems 3.1 and 3.2 in Kostant's Lie–theoretic language in the infinite–dimensional setting. This leads us to introduce a certain completion of K.

To conform to the notation used in our earlier papers, we write w for the local coordinate on $\mathbb{CP}(1)$, set $w = r \exp i\theta$ as in Section 1, and introduce the *real* variable $z \in [0, 1)$ by $z = r^2/(1 + r^2)$. The symplectic form Ω in equation (1.1) becomes $2dz \wedge d\theta$.

The Lie algebra $\mathfrak{k} = \mathrm{sdiff}^s(\mathbb{CP}(1))$ of $\mathrm{SDiff}^s(\mathbb{CP}(1))$ is the algebra of divergence–free Sobolev vector fields on $\mathbb{CP}(1)$. These vector fields are globally Hamiltonian with respect to Ω. We will identify the (formal) Lie algebra $\mathfrak{k}$ with the (formal) Poisson algebra of H^{s+1} functions on $\mathbb{CP}(1)$. Strictly speaking, the latter is a central extension of $\mathfrak{k}$, and it is not closed under Poisson bracket, but that will not matter. The considerations in this section serve to motivate the results in the next section, but are not used in their proof.

As suggested by Proposition 1.3, $\mathfrak{k}$ is analogous to $u(n)$, and thus $\mathrm{SDiff}^s(\mathbb{CP}(1))$ should be analogous to $U(n)$. Furthermore, according to part (iv) of Proposition 1.3, the Poisson subalgebra $\mathfrak{t}^s$ of $\mathfrak{k}$ consisting of functions independent of θ may be thought of as the "diagonal" Cartan subalgebra of $\mathfrak{k}$. An $f \in \mathfrak{t}^s$ generates the Hamiltonian vector field

$$\dot{z} = \frac{\partial f}{\partial \theta} = 0\,, \qquad \dot{\theta} = -\frac{\partial f}{\partial z}\,, \tag{4.1}$$

which integrates to give the symplectic twist map

$$(z,\theta) \mapsto \left(z, \theta - t\frac{\partial f(z)}{\partial z}\right)\,. \tag{4.2}$$

In view of Section 2, the set $\mathcal{T}^s$ of twist maps (4.2) is the maximal torus corresponding to $\mathfrak{t}^s$.

Next, we introduce the adjoint orbits—"continuum matrices with fixed spectrum". The adjoint representation of $K = \mathrm{SDiff}^s(\mathbb{CP}(1))$ on its Lie algebra $\mathfrak{k}$ is given by the map $P_k : x \mapsto x \circ k$ for $k \in \mathrm{SDiff}^s(\mathcal{A})$ and $x \in \mathfrak{k}$. Thus the adjoint orbit through $y \in \mathfrak{k}$ is $\mathcal{O}_y = \{y \circ k \mid k \in K\}$.

Since the function

$$\mathrm{Tr}\, x = \int_{\mathbb{CP}(1)} x\, dz\, d\theta$$

is invariant under the adjoint action, we can define a weakly nondegenerate invariant inner product on $\mathfrak{k}$ by

$$\langle x, y\rangle = \mathrm{Tr}\, xy\,, \qquad x, y \in \mathfrak{k}\,. \tag{4.3}$$

Hence we regard the space $\mathfrak{k}$ as its own (algebraic) dual, and identify the co-adjoint action with the adjoint action.

Finally, the orthogonal projection of $\mathfrak{k}$ onto the diagonal Cartan subalgebra $\mathfrak{t}^s$ is naturally given by

$$\pi : x(z,\theta) \mapsto \int_0^1 x(z,\theta)\, d\theta\,. \tag{4.4}$$

At this point, all the ingredients for the convexity theorem are present and one might hope to prove the following statement:

Conjecture *Let $\lambda \in \mathfrak{t}^s$ be a non-increasing function of z (and independent of θ), and let $\mathcal{O}_\lambda$ be the orbit of* SDiff$^s(\mathbb{CP}(1))$ *through λ. The image $\pi(\mathcal{O}_\lambda)$ is the convex hull of its extreme points, which coincides with the orbit of the Weyl group $N(\mathcal{T}^s)/\mathcal{T}^s$.*

Unfortunately, this conjecture is false. The Weyl group W^s of $\mathcal{T}^s$ is essentially trivial, so that the convex hull of its orbit $W^s \cdot \lambda$ will not be very interesting:

Proposition 4.1 *The normalizer $N(\mathcal{T}^s)$ of $\mathcal{T}^s$ is the group of twist maps (possibly orientation reversing):*

$$(z,\theta) \mapsto (z, j\theta + \phi(z)) \text{ or } (-z+1, j\theta + \phi(z)), \qquad j = +1 \text{ or } -1.$$

The Weyl group $N(\mathcal{T}^s)/\mathcal{T}^s$ is the four-element group

$$\{\text{identity}, (z,\theta) \mapsto (z,-\theta), (z,\theta) \mapsto (-z+1,-\theta), (z,\theta) \mapsto (-z+1,\theta)\},$$

which is isomorphic to $\mathbb{Z}_2 \times \mathbb{Z}_2$.

The resolution of this difficulty just described lies in a completion process. It is simply a fact that diffeomorphism groups are not rich enough for a satisfactory convexity theorem. We must leave the arena of groups and deal with semigroups. It is perhaps surprising that these semigroups, which (as far as we know) are not infinite dimensional smooth manifolds, lead to a convexity theorem analogous to that for compact Lie groups.

Let SMeas$(\mathbb{CP}(1))$ denote the group of invertible measure preserving transformations of the sphere $\mathbb{CP}(1)$. Each $g \in$ SMeas$(\mathbb{CP}(1))$ determines a unitary operator P_g on $L^2(\mathbb{CP}(1))$ by $P_g x = x \circ g$. The strong operator topology induces a topology on SMeas$(\mathbb{CP}(1))$. Halmos [1944] described a basis for this topology and showed that relative to it, SMeas$(\mathbb{CP}(1))$ is a complete topological group (we specialize his result to $\mathbb{CP}(1)$).

Once one passes to this completion, triviality of the Weyl group is no longer an obstacle. Let $\mathcal{T}$ be the subgroup of pure twist maps, $(z,\theta) \mapsto (z, \theta+\phi(z))$, with measurable ϕ. This subgroup will play the role of maximal torus.

Proposition 4.2 *The normalizer $N(\mathcal{T})$ of $\mathcal{T}$ is the group of maps*

$$(z,\theta) \mapsto (a(z), j(z)\theta + k(z)), \tag{4.5}$$

with a an invertible measure preserving transformation of $[0,1]$, j measurable and equal to ± 1 almost everywhere, and k measurable. The "Weyl group" $W = N(\mathcal{T})/\mathcal{T}$ may be identified with the group of maps of the form

$$(z,\theta) \mapsto (a(z), j(z)\theta). \tag{4.6}$$

See Bloch, Flaschka and Ratiu [1993] for the proof.

Now denote by $\overline{\text{SMeas}(\mathbb{CP}(1))}$ the semigroup of all measure preserving (not necessarily invertible) maps of $\mathbb{CP}(1)$. It follows from Brown [1966] that $\overline{\text{SMeas}(\mathbb{CP}(1))}$ is the closure (in the topology induced from the strong operator topology on the P_g) of $\text{SMeas}(\mathbb{CP}(1))$. According to Oxtoby [1973], given an invertible measure preserving map g and an $\epsilon > 0$, there exists a measure preserving homeomorphism ϕ of $\mathbb{CP}(1)$ such that

$$\text{measure of } \{x \mid g(x) \neq \phi(x)\} < \epsilon.$$

In particular, this implies that g and ϕ are close in the topology on $\text{SMeas}(\mathbb{CP}(1))$. Moser [1992] showed that every measure preserving homeomorphism of a $2n$-dimensional *cube* can be approximated arbitrarily closely by a symplectic diffeomorphism. We used this result for the *square* in Bloch, Flaschka, Ratiu [1993], where the convexity theorems of Section 5 below were stated for an annulus $\mathcal{A}$ (as in Theorem 2.1) rather than for $\mathbb{CP}(1)$, to prove:

Theorem 4.3 (Completion Theorem) *The strong closure of* $\{P_k \mid k \in \text{SDiff}^s(\mathcal{A})\}$ *is* $\{P_g \mid g \in \overline{\text{SMeas}(\mathcal{A})}\}$.

Remark. It is possible (Alpern [1978]) to express the topology on $\text{SMeas}(\mathcal{A})$ (or $\text{SMeas}(\mathbb{CP}(1))$) directly in terms of the maps g rather than the induced operators P_g.

We expect that the Completion Theorem remains true when $\mathcal{A}$ is replaced by $\mathbb{CP}(1)$ (or by more general symplectic manifolds), but have not tried to prove this. The convexity theorems in the next section do not rely on the process of completion; at the moment, passage from diffeomorphisms to measure preserving maps serves only to motivate the formulation of the convexity results. It may well turn out, however, that the completion will be important in a more systematic infinite-dimensional Lie theory.

Let us assume, for sake of argument, that $\overline{\text{SMeas}(\mathbb{CP}(1))}$ is indeed the completion of $\text{SDiff}^s(\mathbb{CP}(1))$. Other Lie-theoretic objects associated to $\text{SDiff}^s(\mathbb{CP}(1))$ then also have natural completions. The strong closure $\overline{W}$ of the Weyl group W consists of maps of the form (4.6) in which $a(z)$ is merely a measure preserving map of $[0, 1]$, not necessarily invertible. We replace the Lie algebra $\mathfrak{k}$ by its completion $L^2(\mathbb{CP}(1))$. The closure in $L^2(\mathbb{CP}(1))$ of the Lie algebra $\mathfrak{t}^s$ of the maximal torus is the subspace $\mathfrak{t}$ of functions that are independent of θ; it is convenient to identify it with $L^2([0, 1])$. The Weyl semigroup $\overline{W}$ acts on $\mathfrak{t}$ only through the first component $a(z)$; it is, so to say, the "permutation group" of the interval $0 \leq z \leq 1$. The L^2-orthogonal projection that associates to every function its zero order Fourier coefficient (equation (4.4)) is the same for the measurable case as for the smooth case. We again have all the ingredients for the convexity theorem, but now in the setting of completions.

5. The convexity theorems

In this section, we summarize the main results in Bloch, Flaschka, Ratiu [1993]. Proofs are omitted, with one exception: we take the opportunity to correct a faulty proof of the "Spectral Theorem" (Proposition 5.3).

It should perhaps be stressed again that the theorems below are proved without any reference to Lie theory—they are pure analysis. The material in the earlier sections is an attempt to create a "natural" setting for the convexity result, which might otherwise seem unmotivated and probably less interesting.

The first task is to define the "spectrum" of a continuum matrix $x(z,\theta)$ and to understand the "adjoint orbit" of all $y(z,\theta)$ with that spectrum. Since the traces of powers of a matrix determine its eigenvalues, we try to characterize the orbit by the values of the moments $\int x^p$ of x.

Theorem 5.1 (Spectral Theorem) *Let* $x \in L^2(\mathbb{CP}(1)) \cap L^\infty(\mathbb{CP}(1))$ *(this hypothesis ensures that all moments of* x *exist), and set*

$$I_p = \int_{\mathbb{CP}(1)} x^p \Omega\,, \qquad p \in \mathbb{Z}^+ .$$

There exists a unique, nonincreasing, right-continuous function λ *on* $[0,1]$ *such that*

$$I_p = \int_0^1 \lambda^p(z)dz\,, \qquad p \in \mathbb{Z}^+ .$$

The function λ is the familiar continuous analog of the ordered vector x^* in (3.1):

Definition 5.2 *Let* (M,m) *be a probability measure space, and let* $f \in L^1(M)$. *Set* $\psi(t) = m(\{u \in M \mid f(u) > t\})$ *and, for* $0 \le z < 1$, *set*

$$f^*(z) = \sup\{t \mid \psi(t) > z\}\,.$$

The nonincreasing, right continuous function f^* *is called the* nonincreasing rearrangement *of* f.

Theorem 5.1 is essentially contained in Chong and Rice [1971].

Passing to the semigroup $\overline{\mathrm{SMeas}(\mathbb{CP}(1))}$ creates new difficulties with the notion of "orbit". Namely, even if x and y have the same moments, $\int x^p = \int y^p$ for all $p \in \mathbb{Z}^+$, there may not exist any measure preserving φ such that $y = x \circ \varphi$.

Example. Let $f(z) = z - z^2$, $z \in [0,1]$. Then

$$\psi(t) = \begin{cases} 1\,, & \text{if } t \le 0 \\ \sqrt{1-4t}\,, & \text{if } 0 \le t \le 1/4 \\ 0\,, & \text{if } t \ge 1/4 \end{cases}$$

and the nonincreasing rearrangement of f is $f^*(z) = (1 - z^2)/4$. The moments of f and f^* are the same. Note that the map φ: $[0,1] \to [0,1]$ given by $\varphi(z) = 2\left|z - \frac{1}{2}\right|$ is measure preserving, transforms f^* to f (i.e., $f = f^* \circ \varphi$), but is non–invertible. Ryff [1963] proves that there is no measure preserving ψ such that $f^* = f \circ \psi$. ∎

To get a satisfactory description of an orbit of $\overline{\mathrm{SMeas}(\mathbb{CP}(1))}$, one must appeal to nonconstructive results from measure theory. It is known (Halmos [1950], §41) that a nonatomic, separable, totally finite normalized measure algebra is isomorphic to the measure algebra of the unit interval $[0,1]$. Fix such an isomorphism $\tilde{q}$ between the measure algebras of $\mathbb{CP}(1)$ and $[0,1]$. A theorem of von Neumann (Royden [1963], p. 329) says that $\tilde{q}$ is implemented by a point mapping $q : \mathbb{CP}(1) \to [0,1]$ which is one–to–one off two sets of measure zero (in the domain and range).

Proposition 5.3 *Let $x \in L^2(\mathbb{CP}(1)) \cap L^\infty(\mathbb{CP}(1))$, and let λ be as in the Spectral Theorem. There exists a measure preserving map $\psi : \mathbb{CP}(1) - \mathcal{Z}_1 \to \mathbb{CP}(1) - \mathcal{Z}_2$, with $\mathcal{Z}_i$ of measure zero, such that $x = \lambda \circ q \circ \psi$.*

Proof. Clearly, $x \circ q^{-1} \stackrel{\text{def}}{=} g$ is equimeasurable with x. According to Ryff [1965], there is a measure preserving map ϕ of $[0,1]$ such that $g = \lambda \circ \phi$. Therefore $x = \lambda \circ q \circ q^{-1} \circ \phi \circ q$. The map $\psi = q^{-1} \circ \phi \circ q$ is the required measure preserving transformation of $\mathbb{CP}(1)$. ∎

Recall that functions of z alone are continuum analogs of diagonal matrices. The function $\lambda \circ q$ in Proposition 5.3 is far from "diagonal". It would be more consistent with linear algebra if every x with given moments I_p were obtained by applying a measure preserving transformation to the "diagonal" function $\lambda_2(z,\theta) \stackrel{\text{def}}{=} \lambda(z)$. This is not the case.

Example Let $x : \mathbb{CP}(1) \to [0,1]$ be an invertible measure preserving transformation (e.g., the q in Proposition 5.3). The λ given by the Spectral Theorem is easily seen to be the identity, $\lambda(z) = z$. Let $\sigma : (z,\theta) \mapsto (\mu(z,\theta), \nu(z,\theta))$ be an invertible measure preserving transformation of $\mathbb{CP}(1)$, such that $x = \lambda_2 \circ \sigma$. Then $x = \lambda \circ \mu = \mu$. Because $\nu : \mathbb{CP}(1) \to [0,1]$ is also measure preserving, we find, for all measurable $A, B \subset [0,1]$ (m is normalized Lebesgue measure on the sphere):

$$\begin{aligned} m(\mu^{-1}(A) \cap \nu^{-1}(B)) &= m(\sigma^{-1}(A \times B)) \\ &= m(A \times B) \\ &= |A|\,|B| \\ &= m(\mu^{-1}(A))\, m(\nu^{-1}(B)) \qquad (5.1) \end{aligned}$$

For every B, one can find A so that $\mu^{-1}(A) = \nu^{-1}(B)$ because μ is invertible. Thus (5.1) implies

$$m(\mu^{-1}(A)) = m(\mu^{-1}(A))^2,$$

whence $|B| = m(\mu^{-1}(A)) = |A| = 0$ or 1. This is impossible, because B was arbitrary. ∎

We conclude from this example that

$$\mathcal{O}_{\lambda_2} \subset \mathcal{O}_{\lambda\circ q} \text{ and } \mathcal{O}_{\lambda_2} \neq \mathcal{O}_{\lambda\circ q} \tag{5.2}$$

The orbit $\mathcal{O}_{\lambda\circ q}$ is characterized in the next Proposition as the set of all functions $y : \mathbb{CP}(1) \to \mathbb{R}$ with fixed moments I_p. We do not know a characterization of $\mathcal{O}_{\lambda_2}$, the "orbit through a diagonal matrix". [1]

Theorem 5.4 (Orbit Theorem) *Let $x \in L^2(\mathbb{CP}(1)) \cap L^\infty(\mathbb{CP}(1))$. There is a unique (necessarily bounded) nonincreasing, right continuous function λ of z alone such that $x \in \mathcal{O}_{\lambda\circ q}$. The orbit $\mathcal{O}_{\lambda\circ q}$ consists of all functions y for which*

$$\int_{\mathbb{CP}(1)} y^p = \int_0^1 \lambda(z)^p dz \quad \left(= \int_{\mathbb{CP}(1)} x^p \right), \quad p = 1, 2, \dots .$$

If one thinks of the set of nonincreasing functions of z alone as the positive Weyl chamber in the Cartan subalgebra $\mathfrak{t}$, the first part of this theorem asserts that every maximal orbit intersects *some* Weyl chamber in a unique point. The counterexample above shows that it is *not* in general the *positive* Weyl chamber (this was the mistake in our earlier paper).

Finally, we state the convexity theorems.

Theorem 5.5 (Schur's Theorem) *Let $x \in L^2(\mathbb{CP}(1)) \cap L^\infty(\mathbb{CP}(1))$, let $\pi(x)$ be its zeroth Fourier coefficient,*

$$\pi(x)(z) = \int_0^1 x(z, \theta) d\theta \,, \tag{5.3}$$

and let λ be as in the Spectral Theorem. Then $\pi(x)$ belongs to the closed convex hull of the orbit of the Weyl semigroup $\overline{W}$ through λ,

$$\overline{W} \cdot \lambda = \{\lambda \circ \phi \mid \phi \text{ is a measure preserving transformation of } [0,1]\} .$$

Theorem 5.6 (Horn's Theorem) *Let λ be a bounded, nonincreasing function on $[0, 1]$, and let X lie in the closed convex hull of $\overline{W} \cdot \lambda$. Then there exists an $x \in L^2(\mathbb{CP}(1)) \cap L^\infty(\mathbb{CP}(1))$ such that*

$$(i)\; X(z) = \pi(x)(z) = \int_0^1 x(z, \theta) d\theta \,. \tag{5.4}$$

$$(ii)\; \int_0^1 \int_0^1 x(z, \theta)^p d\theta dz = \int_0^1 \lambda(z)^p dz \,, \qquad p \in \mathbb{Z}^+ . \tag{5.5}$$

[1]Incidentally, this turns out to be an important question in the theory of stability of the beam solution for the Poisson–Vlasov system; the energy–Casimir method for this type of solution cannot be applied because no Casimir functionals characterizing the coadjoint orbit through such a solution are known.

These theorems are summarized and extended in the following result (π is as in (5.3)):

Theorem 5.7 (Convexity Theorem) *Let $\lambda \in L^2(\mathbb{CP}(1))$ be a bounded, nonincreasing, right continuous function of z alone, and let $\mathcal{O}_{\lambda\circ q}$ be the orbit of $\overline{\mathrm{SMeas}(\mathbb{CP}(1))}$ through $\lambda \circ q$. Then $\pi(\mathcal{O}_{\lambda\circ q}) \subset L^2([0,1])$ is a weakly compact, convex set. Its set of extreme points is the orbit $\overline{W} \cdot \lambda$ of the Weyl semigroup through λ, i.e., the set of all measurable rearrangements ("permutations") of λ.*

The new item is weak compactness of $\pi(\mathcal{O}_\lambda)$ which follows from Ryff's Theorem below. This is perhaps surprising because $\mathcal{O}_\lambda$ is itself not weakly compact in $L^2(\mathbb{CP}(1))$. For example, let $\{r_n\}$ be the Rademacher orthogonal system in $L^2([0,1])$, extended to $\mathbb{CP}(1)$ as functions independent of θ. (Recall that the Rademacher system consists of "square waves" that are alternately $+1$ and -1 on on intervals of length 2^{-n}.) The moments $I_p(r_n)$ are 1 (p even) or 0 (p odd). However, $r_n \rightharpoonup 0$. Similarly, one sees that $\mathcal{O}_\lambda$ is not w*–compact in L^∞.

The proof relies on Theorems due to Ryff [1963], [1965], [1967].

Definition 5.8 *(i) Let $f, g \in L^1([0,1])$. We say that f majorizes g (written $g \prec f$) if*

$$\begin{aligned} \textstyle\int_0^s g^*(z)dz &\le \textstyle\int_0^s f^*(z)dz\,, \qquad 0 \le s < 1\,, \\ \textstyle\int_0^1 g^*(z)dz &= \textstyle\int_0^1 f^*(z)dz\,. \end{aligned} \tag{5.6}$$

(ii) A linear operator P on $L^1([0,1])$ is called doubly stochastic if $Pf \prec f$ for all $f \in L^1([0,1])$.

Theorem 5.9 (Ryff [1965], [1967]) *In $L^1([0,1])$,$g \prec f$ if, and only if, there is a doubly stochastic P such that $g = Pf$. The set $\Omega(f) \stackrel{\text{def}}{=} \{g \mid g \prec f\}$ is weakly compact and convex. Its set of extreme points is $\{f^* \circ \phi \mid \phi$ is a measure preserving transformation of $[0,1]\}$.*

The proof of Schur's Theorem follows the finite–dimensional proof quite closely. We show that the "diagonal part" $\pi(x)$ of the "matrix" $x(z,\theta)$ is obtained from λ by the action of a doubly stochastic operator and then use Ryff's theorem. For the Horn part, we assume that $X \prec \lambda$, which, according to Ryff's Theorem, means that $X = P\lambda$ for a doubly stochastic P. From this P, we build an $x(z,\theta)$ with the same moments as λ, for which $\pi(x) = X$.

6. Towards an infinite dimensional Lie theory

We want to present here the beginnings of a dictionary between finite dimensional Lie theory and measure theory that is an indication of the type of results that seem within reach for a genuine infinite dimensional Lie theory for completions of diffeomorphisms groups. The results presented above are already an indication of what is possible for the analogues of compact Lie groups. The situation for noncompact groups is far less satisfactory but even here one can discern the beginnings of such a theory due to some remarkable results of Brenier [1991] that are particularly adapted to the polar decomposition for Lie groups. We will briefly describe these theorems before presenting the aforementioned dictionary. Throughout this section, because Brenier's results have not yet been generalized to compact manifolds, we take the annulus $\mathcal{A}$ to be the basic domain, and work with $\mathrm{SDiff}^s(\mathcal{A})$, $\mathrm{SMeas}(\mathcal{A})$, etc.

There is a straightforward infinite-dimensional version of the Lie algebraic picture. We summarize the picture in a table; examples in Bloch, Flaschka, Ratiu [1996] support the analogies.

Given a function $x(z,\theta)$ on $\mathcal{A}$, expand it in a Fourier series

$$x(z,\theta) = \sum_{n=-\infty}^{\infty} x_n(z) e^{2\pi n i\theta}, \qquad x_n(z) \in \mathbb{C}. \tag{6.1}$$

Matrix Lie Algebras	***Functional Lie Algebras***
$\mathfrak{gl}(N,\mathbb{C})$	$L^2(\mathcal{A},\mathbb{C})$
$\mathfrak{sl}(N,\mathbb{C})$	$\{x \in L^2(\mathcal{A},\mathbb{C}) \mid \int_{\mathcal{A}} x(z,\theta)\,dz\,d\theta = 0\}$
Complex conjugation	$\operatorname{conj} x(z,\theta) = -\overline{x(z,-\theta)}$
Transpose	$x(z,\theta)^T = x(z,-\theta)$
Adjoint (transpose conjugate)	$x(z,\theta) = -\overline{x(z,\theta)}$
$\mathfrak{gl}(N,\mathbb{R})$	$\{x \in L^2(\mathcal{A},\mathbb{C}) \mid x(z,\theta) = -\overline{x(z,-\theta)}\}$
$\mathfrak{sl}(N,\mathbb{R})$	$\{x \in L^2(\mathcal{A},\mathbb{C}) \mid x(z,\theta) = -\overline{x(z,-\theta)}, \int_{\mathcal{A}} x(z,\theta)\,dz\,d\theta = 0\}$
$\mathfrak{so}(N)$	$\{x \in L^2(\mathcal{A},\mathbb{C}) \mid x(z,\theta) = -x(z,-\theta)\}$
$\mathfrak{b}$ (upper triangular matrices with real entries on the diagonal)	functions $x(z,\theta)$ having Fourier expansion only for $n \geq 0$ and first term purely imaginary
symmetric matrices	functions $x(z,\theta)$ having Fourier expansion only in $\cos\theta$

Group elements corresponding to the various Lie algebra elements in the table should be obtained by exponentiating the vector fields generated by the functions $x(z, \theta)$. When x is complex-valued, this is not possible. Indeed, it is known that $\mathrm{SDiff}^s(\mathcal{A})$ does not admit a complexification. The subgroup of $\mathrm{SDiff}^s(\mathcal{A})$ of diffeomorphisms fixing one of the boundaries of $\mathcal{A}$ is a normal subgroup. If there were a complexification of $\mathrm{SDiff}^s(\mathcal{A})$, the quotient group, which is just the diffeomorphism group of the circle, would also admit a complexification. This is, however, known to be false (see Pressley and Segal [1986]). Thus, at a naive level, there is no complexification of $\mathrm{SDiff}^s(\mathcal{A})$. The situation is analogous to what we saw when trying to generalize the convexity theorem: it was false for $\mathrm{SDiff}^s(\mathcal{A})$ and true for $\overline{\mathrm{SMeas}(\mathcal{A})}$. Perhaps, then, one should try to find a "complexification" for $\overline{\mathrm{SMeas}(\mathcal{A})}$ instead of $\mathrm{SDiff}^s(\mathcal{A})$ or another "smooth" object playing the role of the complexification of $\mathrm{SDiff}^s(\mathcal{A})$.

At this point Brenier's [1991] results come in by providing a monoid playing the role of the complexification of $\overline{\mathrm{SMeas}(\mathcal{A})}$. We state Brenier's results in full generality, since we believe that they may be of use in future generalizations of the Lie theoretical results discussed here for measure preserving maps of other manifolds than the annulus $\mathcal{A}$. The notations are those of Brenier [1991], except where there already is a standard Lie theoretical notation.

We fix a probability space (X, μ), an open bounded connected set $\Omega \subset \mathbb{R}^d$ with smooth boundary $\partial\Omega$, and a Lebesgue integrable non-negative function $\beta : \overline{\Omega} \to \mathbb{R}$ which is bounded away from 0 on any compact subset of Ω as well as bounded away from 0 and ∞ on $\overline{\Omega}$. We denote by dm the Lebesgue measure on $\mathbb{R}^d$ and assume that βdm is a probability measure on $\overline{\Omega}$. Thus βdm has the same negligible subsets as dm (in particular, the measure of $\partial\Omega$ is zero) and is absolutely continuous relative to dm and conversely. $L^p(X, \mu, \mathbb{R}^d)$ and $L^p(\Omega, \beta, \mathbb{R}^d)$ denote the $\mathbb{R}^d$-valued L^p-functions on X and Ω respectively, relative to the measures μ and βdm. We shall use the standard notation

$$W^{1,p}(\Omega, \beta) \stackrel{\text{def}}{=} \{f \in L^p(\Omega, \beta, \mathbb{R}) \mid \nabla f \in L^p(\Omega, \beta, \mathbb{R}^d)\}$$

for the Sobolev space of functions that are L^p-integrable together with their first derivatives. In all that follows it is assumed that $1 \leq p < \infty$. Under the conditions on β imposed above $W^{1,p}(\Omega, \beta)$ is compactly embedded in $L^p(\Omega, \beta, \mathbb{R})$.

Definition 6.1 *A* rearrangement *of $u \in L^p(X, \mu, \mathbb{R}^d)$ on Ω is a function $v \in L^p(\Omega, \beta, \mathbb{R}^d)$ satisfying*

$$\int_X f(u(x))d\mu(x) = \int_\Omega f(v(y))\beta(y)dm$$

for all $f \in C^0(\mathbb{R}^d, \mathbb{R})$ such that $|f(\zeta)| \leq Constant(1 + \|\zeta\|^p)$.

Theorem 6.2 (Symmetric Space Theorem) *For each* $u \in L^p(X, \mu, \mathbb{R}^d)$ *there is a unique rearrangement in the set*

$$P = \{\nabla\psi \mid \psi \in W^{1,p}(\Omega, \beta),\ \psi \text{ convex}\} \subset L^p(\Omega, \beta, \mathbb{R}^d)$$

and the mapping associating to u its unique rearrangement in P is continuous from $L^p(X, \mu, \mathbb{R}^d)$ *to* $L^p(\Omega, \beta, \mathbb{R}^d)$.

To say that the function ψ is ***convex*** means that it is convex in the usual sense on the convex hull of Ω. This is theorem 1.1 in Brenier [1991]. This space plays the role of the positive definite Hermitean matrices (or the usual symmetric space $P = \exp \mathfrak{p}$ in Lie theory), as is seen from the following theorem that also is the main reason why Brenier's results seem to identify the complexification of $\overline{\mathrm{SMeas}(\mathcal{A})}$. We need one more definition. A function $u \in L^p(X, \mu, \mathbb{R}^d)$ is called ***non-degenerate*** if

$$\mu(u^{-1}(E)) = 0 \quad \text{for each zero Lebesgue measure set} \quad E \subset \mathbb{R}^d.$$

Theorem 6.3 (Polar Decomposition Theorem) *For each non-degenerate* $u \in L^p(X, \mu, \mathbb{R}^d)$ *there is a unique* $\nabla\psi \in P$ *and a unique measure preserving map* $s : (X, \mu) \to (\Omega, \beta)$ *such that* $u = \nabla\psi \circ s$. *Moreover,* $\nabla\psi$ *is the unique rearrangement of u given in the Symmetric Space Theorem and s is the unique measure preserving map that maximizes* $\int_X u(x) \cdot s(x) d\mu(x)$. *In addition, the map* $u \in L^p(X, \mu, \mathbb{R}^d) \mapsto (\nabla\psi, s) \in L^p(\Omega, \beta, \mathbb{R}^d) \times L^q(X, \mu, \mathbb{R}^d)$ *is continuous for all* $1 \leq q < \infty$.

This is Theorem 1.2 in Brenier [1991]. The usual polar decomposition theorem for invertible complex matrices is obtained from this theorem by taking $(X, \mu) = (\Omega, \beta) = (D, m)$, where D is the unit ball in $\mathbb{C}^d$, and $u(x) = Ax$ for A a $d \times d$ complex matrix. It is easily checked that invertibility of A is equivalent to the non-degeneracy condition in the Polar Decomposition Theorem. Then take $\psi(x) = \frac{1}{2} Rx \cdot x$, for R a positive definite Hermitean matrix, and $s(x) = Ux$, for U a unitary matrix, where $A = RU$ is the usual polar decomposition of A in $GL(N, \mathbb{C})$. The factors guaranteed by Brenier's Polar Decomposition Theorem are then R and U. They happen to be linear for a linear map u when both X and Ω are balls; had we taken the same u and some other domain, this would no longer be true—linear maps have factors that are not linear in general.

Now we return to the annulus and set in all these theorems $X = \Omega = \mathcal{A}$ and consider the monoid of non-degenerate L^2-functions on $\mathcal{A}$ with values in $\mathcal{A}$. On the basis of our convexity theorems and Brenier's work, we propose the following dictionary.

Lie Theory	***Measure Theory/Analysis***
Complex semisimple Lie group G	The monoid $L^2(\mathcal{A},\mathcal{A})_n$ of non–degenerate functions from $\mathcal{A}$ to $\mathcal{A}$
Complex semisimple Lie algebra $\mathfrak{g}$	$L^2(\mathcal{A},\mathbb{C})$ with the usual Poisson bracket whenever it can be performed in the sense of distributions
Compact real form K of G	The semigroup $\overline{\mathrm{SMeas}(\mathcal{A})}$
Compact real form $\mathfrak{k}$ of $\mathfrak{g}$, the $(+1)$–eigenspace of the Cartan involution on $\mathfrak{g}$	Vershik–Saveliev Lie algebra with the continuous roots $n\delta'(t-t')$, thought of as $L^2(\mathcal{A})\cap L^\infty(\mathcal{A})$
Cartan subalgebra $\mathfrak{t}$ of $\mathfrak{k}$	$L^2([0,1])$, real valued functions of $z\in[0,1]$ alone
Maximal torus T of K	Twist maps $\{(z,\theta)\mapsto(z,\theta+\varphi(z))\mid\varphi\,\text{measurable}\}$
Adjoint K–orbit $\mathcal{O}_\mu$ through $\mu\in\mathfrak{k}$	$\mathcal{O}_{\lambda\circ q}=\{x\in L^2(\mathcal{A})\mid\int_{\mathcal{A}}x(z,\theta)^p dz d\theta=\int_0^1\lambda(z)^p dz\ \}$
Weyl group $W=N(T)/T$ of K	Semigroup of measure preserving maps of $[0,1]$
Positive Weyl chamber $\mathfrak{t}^*_+$	Nonincreasing L^2–functions on $[0,1]$
Kostant polytope through an element in $\mathfrak{t}^*_+$	The convex set $\{f\in L^2([0,1])\mid f\prec\lambda\}$ for λ a non–increasing right-continuous function on $[0,1]$
Weyl group orbit $\mathcal{O}_\mu\cap\mathfrak{t}^*_+$	Extreme points of the convex set above
$\mathfrak{p}$, the (-1)–eigenspace of the Cartan involution on $\mathfrak{g}$	$L^2(\mathcal{A},i\mathbb{R})$, the purely imaginary functions on $\mathcal{A}$
$P=\exp(\mathfrak{p})$	$\{\nabla\psi\mid\psi\in H^1(\mathcal{A})\,\text{convex}\}$
Polar decomposition $G_{\mathbb{R}}=PK$	Brenier's polar decomposition $x=\nabla\psi\circ s$, where $x\in L^2(\mathcal{A},\mathcal{A})_n$, $\psi\in H^1(\mathcal{A})$ is convex, and $s\in\overline{\mathrm{SMeas}(\mathcal{A})}$

This table has some major gaps in it, such as the identification of the Borel subgroup and the Iwasawa decomposition. We hope to address these issues in future work.

References

N. Akhiezer [1965] *The Classical Moment Problem*, University Mathematical Monographs, Oliver & Boyd Ltd., Edinburgh and London.

S. Alpern [1978] Approximation to and by measure preserving homeomorphisms, *J. London Math. Soc.* **18**, 305-315.

M.F. Atiyah [1982] Convexity and commuting Hamiltonians, *Bull. Lon. Math. Soc.* **14**, 1-15.

M.F. Atiyah, R. Bott [1982] The Yang-Mills equations over Riemann surfaces, *Phil. Trans. Roy. Soc. Lond. A*, **308**, 523-615.

M. Audin [1991] *The Topology of Torus Actions on Symplectic Manifolds*, Progress in Mathematics **93**, Birkhäuser, Basel.

D. Bao, T. Ratiu [1992] A candidate maximal torus in infinite dimensions, *Contemporary Mathematics* **132**, 117-124.

G. Birkhoff [1946] Three observations on linear algebra, *Universidad Nacional de Tucumán, Revista, Serie A* **5** 147-151.

A. Bloch, H. Flaschka, T. Ratiu [1993] A Schur-Horn-Kostant convexity theorem for the diffeomorphism group of the annulus, *Inv. Math.* **113**, 511-529.

A. Bloch, H. Flaschka, T. Ratiu [1996] The Toda PDE and the geometry of the diffeomorphism group of the annulus, *Fields Institute Communications* **7**, 57-92.

M. Bordemann, E. Meinrenken, M. Schlichenmaier [1994] Toeplitz quantization of Kähler manifolds, *Comm. Math. Phys.* **165**, 281-296.

Y. Brenier [1991] Polar factorization and monotone rearrangement of vector valued functions, *Comm. Pure and Appl. Math.* **44**, 375-417.

R.W. Brockett, A. Bloch [1990] Sorting with the dispersionless limit of the Toda lattice, *Proceedings of the CRM Workshop on Hamiltonian Systems, Transformation Groups and Spectral Transform Methods*, J. Harnad and J. E. Marsden, eds.. Publ. CRM, Montréal, 103-112.

J. R. Brown [1966] Approximation theorems for Markov operators, *Pac. J. Math.* **16**, 13-23.

K.M. Chong and N.M. Rice [1971] *Equimeasurable Rearrangements of Functions*, Queen's Papers in Pure and Applied Mathematics, **28**, Queen's University, Kingston, Ontario, Canada.

D. G. Ebin, J. Marsden [1970] Groups of diffeomorphisms and the motion of an incompressible fluid, *Annals of Math.* **92**, 102-163.

M. El Hadrami [1996] *Poisson Algebras and Convexity*, Ph.D. Dissertation, The University of Arizona, Tucson.

V. Guillemin [1979] Some classical theorems in spectral theory revisited, *Seminar on Singularities of Solutions of Linear Partial Differential Equations* (L. Hörmander ed.), Annals of Mathematics Studies, **91**, Princeton University Press and University of Tokyo Press, 219-259.

V. Guillemin, S. Sternberg [1982] Convexity properties of the momentum mapping I, *Invent. Math.* **67**, 491-513.

P. Halmos [1944] Approximation theorems for measure preserving transformations, *Trans. AMS* **55**, 1-18.

P. Halmos [1950] *Measure Theory*, D. Van Nostrand, Princeton, NJ.

G.H. Hardy, J.E. Littlewood, G. Pólya [1934] *Inequalities*, Cambridge, At the University Press.

A. Horn [1954] Doubly stochastic matrices and the diagonal of a rotation matrix, *Am. J. Math.* **76**, 620-630.

B. Kostant [1973] On convexity, the Weyl group and the Iwasawa decomposition, *Ann. Sci. École Norm. Sup.* **6**, 413-455.

A.W. Marshall, I. Olkin [1979] *Inequalities: Theory of Majorization and its Applications*, Academic Press, New York.

J. Moser [1992] Remark on the smooth approximation of volume-preserving homeomorphisms by symplectic diffeomorphisms, ETH Zürich preprint.

J. C. Oxtoby [1973] Approximation by measure–preserving homeomorphisms, *Recent Advances in Topological Dynamics*, Lecture Notes in Mathematics **318**, Springer–Verlag, Berlin, 206-217.

A. Pressley and G. Segal [1986] *Loop Groups*, Oxford Mathematical Monographs, Clarendon Press, Oxford.

H. Royden [1963] *Real Analysis*, Macmillan, New York.

J. V. Ryff [1963] On the representation of doubly stochastic operators, *Pac. J. Math.* **13**, 1379-1386.

J. V. Ryff [1965] Orbits of L^1 functions under doubly stochastic transformations, *Trans. AMS* **117**, 92-100.

J. V. Ryff [1967] Extreme points of some convex subsets of $L^1(0,1)$, *Proc. AMS* **18**, 1026-1034.

M. V. Saveliev, A. M. Vershik [1989] Continuum analogues of contragredient Lie algebras, *Comm. Math. Phys.* **126**, 367-378.

M. V. Saveliev, A. M. Vershik [1990] New examples of continuum graded Lie algebras, *Phys. Lettr.* **A143**, 121-128.

I. Schur [1923] Über eine Klasse von Mittelbildungen mit Anwendungen auf die Determinantentheorie, *Sitzungsberichte der Berliner Mathematischen Gesellschaft* **22**, 9-20.

ON 2D YANG-MILLS THEORY AND INVARIANTS OF LINKS

MICHAEL POLYAK
IHES,
35 route de Chartres, F-91440 Bures-sur-Yvette, France
polyak@ihes.fr

AND

NICOLAI RESHETIKHIN
Department of Mathematics, University of California,
Berkeley, CA 94720-0001, USA
reshetik@math.berkeley.edu

Abstract. In these notes we discuss some aspects of 2D Yang-Mills theory and its relation to invariants of knots in circle bundle over surfaces. We show how it is related to the quantization of the algebra of functions on the moduli space flat bundles over a surface. An easy proof is given that the partition function of the YM_2 with Wilson loops can be regarded as the integral of a function over the moduli space of flat bundles over surface with respect to the symplectic volume form. It is also shown that there is an interesting relation between YM_2 and invariants of plain curves, in particular with Arnold's J^+-invariant.

1. Introduction

Two dimensional Yang-Mills theory (YM_2) has been extensively studied in physical literature as one of the simplest models of quantum field theory and the simplest model with gauge invariance. Then there was a revival of interest in YM_2, related mostly to its interpretation as a string theory [13, 14, 19].

In dimension 2 Yang-Mills theory is locally trivial, thus the theory on compact surfaces has an almost topological character, depending only on a surface and volume element. Gauge invariant observables in this quantum field theory are known as Wilson loops. They can be regarded as immersed curves equipped with finite dimensional representations of the gauge group. Their correlation functions in the small area limit are topological invariants.

D. Sternheimer et al. (eds.), Deformation Theory and Symplectic Geometry, 223–248.

Mathematically, YM_2 is very interesting because of its relation to moduli spaces of flat connections on surfaces (e.g. [6, 26]). It has also been discovered remarkable relation of this theory with moduli spaces of branched covers of surfaces (e.g. [14]). For this, one has to study the theory based on the gauge group loops in $SU(N)$ in the limit $N \to \infty$.

The relation between $2D$ Yang-Mills theory and Chern-Simons theory is well known [26]. The $2D$ Yang-Mills can be obtained from the Chern-Simons theory in the limit of zero coupling. Here we will elaborate this relation on the level of state sums and establish the relation to Reshetikhin-Turaev link invariants and to invariants of curves on surfaces.

We will also show that the partition function can be regarded as a solution to a certain differential equation, generalizing the results of [16].

An important role for us is played by the notion of shadows of links (for the details see [23]).

For questions well-known in mathematical literature we mostly refer the reader to the literature, giving only an informal overview.

The first three sections contain a review of the 2D Yang-Mills theory. In the second section we will recall path integral phenomenology of two dimensional Yang-Mills theory. Section 3 contains a brief summary of the theory in terms of CW-approximations of a surface. Section 4 is devoted to a combinatorial description of the theory. In Section 5 we describe a Poisson algebra structure on the algebra of graphs on a surface. We then discuss its relation to the functions on the moduli space. In Section 6 we introduce a differential equation, satisfied by the partition function and describe the relation between YM_2 and invariants of curves on surfaces. In the same section we also show that the partition function of YM_2 with zero areas is the integral over the moduli space of the simplectic volume form. Then, in section 7, we give a short overview of quantum link invariants and describe the relation between them and the partition function of YM_2. Finally, in Section 8 we show how the "quantization" of the partition function of YM_2 (described in section 7) is related to the quantization of the moduli space of flat connections in the principal G-bundle over Σ.

Acknowledgements

The authors wish to thank J. Andersen, J. Bernstein, D. Freed, J. Mattes and O. Viro for stimulating discussions.

2. Gauge theories in dimension 2

2.1. EUCLIDEAN YANG-MILLS THEORY

Here we will give a phenomenological description of the $2D$ Yang-Mills theory based on the path integral. It is not rigorously defined mathematically, but describes the fundamental properties of the theory. These axiomatic properties allow, as we will see, to reformulate the theory in terms of finite dimensional integral.

Let Σ be a compact surface, G a compact Lie group, $\mathcal{P} \to \Sigma$ a principle G-bundle (with a fixed trivialization on the boundary, if any) and $\mathfrak{g} = \mathrm{Aut}(\mathcal{P})$ the gauge group. In what follows we will assume that Σ is orientable and G is connected and simply connected simple Lie group. Assume that we have a volume (area) 2-form vol on Σ, normalized by $\int_\Sigma vol = A$. The volume element defines the Hodge operator $* : \Lambda^k(\Sigma) \to \Lambda^{2-k}(\Sigma)$ on the differential forms. Denote by $\mathcal{A}$ the space of G-connections on $\mathcal{P}$ (with fixed conjugacy classes of holonomy around the boundary components, if Σ has a non-empty boundary).

The Yang-Mills action is a $\mathfrak{g}$-invariant function $S : \mathcal{A} \to \mathbb{R}$ given by

$$S(\alpha) = \frac{1}{4g^2} \int_\Sigma \mathrm{Tr}(F_\alpha \wedge *F_\alpha),$$

where $F_\alpha = d\alpha + \alpha^2$ is the curvature of $\alpha \in \mathcal{A}$ and g is the gauge coupling constant of the theory.

One of the main ingredients of 2-dimensional Euclidean quantum Yang-Mills theory – called also $2D$ quantum chromodynamics – is the partition function given by

$$Z(\Sigma, A) = \int_\mathcal{A} e^{-S(\alpha)} \mathcal{D}\alpha. \tag{1}$$

Here the integral is taken over all gauge classes of connections. The exact meaning of such integral was discussed in [10, 22]. In what follow we will not be focusing on the measure aspects of this integral, but rather on its "functorial" properties and will reformulate it as a finite dimensional integral.

Let us start with the discussion of some elementary properties, which we would expect from the integral (1), provided the later makes sense.

2.2. SMALL AREA LIMIT OF YM_2

The partition function $Z(\Sigma, A)$ is a function on $G^{|\partial\Sigma|}$, which also depends on the coupling constant g^2 and on the volume form on Σ. We expect that the integral (1) is invariant with respect to diffeomorphism of Σ. From here one concludes that it may depend only on the total volume (area) A of Σ. Notice that the transformations $g^2 \to g^2\lambda$, $vol \to \lambda^{-1} vol$ leave the action of YM_2 invariant, which suggests that the partition function of *the theory depends only on the value of* $g^2 A$.

So-called small volume limit of YM_2 corresponds to $A \to 0$. Since we expect that the theory should depend only on the value $g^2 A$, the small area limit coincides with the limit $g^2 \to 0$. In this case the path integral (1) concentrates on the extremums, which are flat connections, so can be computed as a Gaussian integral [26]. Since the action of YM_2 is gauge invariant and in (1) we are integrating only over gauge classes of connections, extremums of the action correspond to the points on the moduli space of flat connections.

Thus it is expected that in the small area limit the integral (1) becomes proportional to the volume of the moduli space $\mathcal{M}_\Sigma^G \cong \mathrm{Hom}(\pi_1(\Sigma), G)/G$ of G-flat connections on Σ. It has been shown in [26], that it is indeed proportional to the simplectic volume $\int_{\mathcal{M}_\Sigma^G} \omega^n$ of the moduli space $\mathcal{M}_\Sigma^G$. Here $n = \frac{\dim(\mathcal{M}_\Sigma^G)}{2}$ and ω is a natural symplectic form on $\mathcal{M}_\Sigma^G$.

2.3. INSERTING WILSON LOOPS

Let C_j, $j = 1, \ldots, m$ be a collection of immersed circles on Σ s.t. each C_j is equipped with a representation R_j of G. We assume that $C = \cup C_j$ is in general position, i.e. the only singularities of C are transversal double points.

Given a connection $\alpha \in \mathcal{A}$, let $\mathrm{hol}_{C_j}(\alpha)$ be the holonomy $P\exp \oint_{C_j} \alpha$ of α around C_j. A Wilson loop associated with C_j is the functional $\mathcal{W}(C_j, R_j) = \chi_{R_j}(\mathrm{hol}_{C_j}(\alpha))$.

Let P be the set of connected components of $\Sigma - C$. Denote by $A : P \to \mathbb{R}$ the function, assigning to each component $p \in P$ of $\Sigma - C$ its area A_p. We say that we have a *metrized 2-cobordism* if such a function is fixed for the surface Σ (with immersed circles) representing this cobordism.

The Yang-Mills partition function $Z_M(\Sigma)$ defined by (1) generalizes to a partition function $Z_M(\Sigma, C_j, R_j, A)$, associated to a surface Σ with inserted Wilson loops:

$$Z_M(\Sigma, C, R, A) = \int_{\mathcal{A}} e^{-S(\alpha)} \prod_{j=1}^{m} \mathcal{W}(C_j, R_j) \mathcal{D}\alpha. \tag{2}$$

One can use the same reasoning as in Section 2.1 to argue that the partition function with Wilson loops depends only on the areas A_p via the combinations $g^2 A_p$. Once more repeating the arguments from Section 2.2, one can show that in the small volume limit $A \to 0$, the partition function is given by the integral $\int_{\mathcal{M}_\Sigma^G} \prod_{j=1}^{m} \mathcal{W}(C_j, R_j) \omega^n$ of the corresponding Wilson loops over $\mathcal{M}_\Sigma^G$ with respect to the symplectic volume measure [12].

2.4. FROM PATH INTEGRALS TO FINITE DIMENSIONAL ONES: FACTORIZATION

Regarded as an integral, partition function of YM_2 is expected to have natural "gluing" and "factorization" properties.

Let us start with "factorization". Naively, the measure $\mathcal{D}\alpha$ is the ratio of the measure on the space of all connection, divided over the volume of the gauge group. The measure on the space of all connections can be thought as the "product of standard measures on $\mathfrak{g}$ over all points on Σ". Since Σ is cut by the curves C into the set $P = \{p\}$ of the connected components of $\Sigma - C$, it is natural to assume that

$$\mathcal{D}\alpha = d\alpha_C \prod_{p\in P} \mathcal{D}\alpha_p. \tag{3}$$

Here $d\alpha_C$ is the measure on the gauge classes of connections on C and $\mathcal{D}\alpha_p$ is the measure on the space of gauge classes of connections on p. We also have $S(\alpha) = \sum_p S_p(\alpha)$ and, therefore, we expect that

$$Z(\Sigma, C, R, A) = \int_{\mathcal{A}_C} \prod_p Z(p, A_p) \prod_{j=1}^{m} \mathcal{W}(C_j, R_j) d\alpha_C. \tag{4}$$

Here $Z(p, A_p)$ is the partition function on the connected component p of $\Sigma - C$. The integral is taken over the gauge classes of connections on C. Let dU be the Haar measure on G normalized by $\int_G dU = 1$. Consider C as a graph with vertices in the double points of the immersion. Then $d\alpha_C = dU^{|E|}$, where E is the set of edges in the graph C. Thus, if the "factorizability" holds, we can express the partition function of YM_2 with Wilson loops as a finite dimensional integral of products of partition functions without Wilson loops.

2.5. GLUING DISCS TOGETHER

Now, let us discuss the "gluing" property.

Consider two surfaces Σ_1 and Σ_2, having a common boundary segment $\partial \cong [0,1]$. Let $\Sigma_1 \#_\partial \Sigma_2$ be the result of gluing these surfaces along this arc.

Denote by $C^{(i)}$ and $R^{(i)}$ the data, characterizing Wilson loops in Σ_i, $i = 1, 2$. The partition functions $Z(\Sigma_i, C^{(i)}, R^{(i)})$ are functions of the holonomy along the boundary components. From the path integral description of the partition function for YM_2 we expect the following "gluing" property

$$\int_G Z(\Sigma_1, C^{(1)}, R^{(1)}) Z(\Sigma_2, C^{(2)}, R^{(2)}) dU = Z(\Sigma_1 \#_\partial \Sigma_2, C^{(1)} \cup C^{(2)}, R^{(1)} \cup R^{(2)}) \tag{5}$$

Here the integral is taken over the holonomy U along the common boundary ∂ of surfaces. Apparently, this property together with the factorizability determines the partition function uniquely, when it is known for a disc.

2.6. PARTITION FUNCTION ON A DISC

Let us show, that the gluing property determines area dependence of the partition function. Let $Z_D(U, A)$ be the partition function on a disk with the conjugacy class U of the holonomy around the boundary. The gluing property is equivalent to the following relation

$$\int_G Z_D(VU, A_1) Z_D(U^{-1}W, A_2) dU = Z_D(VW, A_1 + A_2) \tag{6}$$

Let us represent Z as a linear combination of characters:

$$Z_D(U, A) = \sum_R \chi_R(U) \dim(R) F_R(A). \tag{7}$$

Here the sum is taken over all (highest weights of) irreducible finite dimensional representations of G. "Gluing equation" (6) gives the relation

$$F_R(A_1) F_R(A_2) = F_R(A_1 + A_2)$$

which implies $F_R(A) = \exp(f(R)A)$ for some function $f(R)$.

In the derivation of the partition function on a disk we essentially followed the work [18] by Migdal. There he also presents additional arguments, which imply that the partition function of the YM_2 on a disk should also satisfy the heat equation

$$\frac{\partial Z_D(g, A)}{\partial A} = \frac{1}{2} \Delta Z_D(g, A).$$

Here Δ is the Laplacian acting on the Lie group. Together with the gluing property this requirement fixes the partition function uniquely and we have:

$$Z_D(U, A) = \sum_R \dim(R) \cdot \chi_R(U) \cdot e^{-Ac_2(R)/2}, \tag{8}$$

where $c_2(R)$ is the value of the second Casimir element of the irreducible representation R of G (and also a corresponding eigenvalue of $-\Delta$).

2.7. PARTITION FUNCTION DETERMINED BY GLUING AND FACTORIZATION

Gluing property (5) and equation (8) determine the partition function for a surface without Wilson loops [26]. Let Σ be a surface of genus g and area A with b components $E_1, \ldots, E_b$ of the boundary $\partial\Sigma$. Fix a holonomy $U_j = \mathrm{hol}_{E_j}(\alpha)$ around E_j, $j = 1, \ldots, b$. Then

$$Z(\Sigma, A) = \sum_R (\dim R)^{2-2g-b} \prod_{j=1}^{b} \chi_R(U_j) \cdot e^{-Ac_2(R)/2}. \tag{9}$$

Combining this with factorization property (4), we arrive to the following theorem [26].

Theorem 1 (Witten [26]). *Denote by E the set of edges of the Wilson loops C on Σ (C is regarded as a graph, with vertices being the double points of the immersion).*

The partition function of YM_2 is given by the following finite dimensional integral:

$$Z(\Sigma, C, R, A) = \int_{G^{|E|}} \prod_{p \in \Sigma - C} Z(p, A_p) \prod_{j=1}^{m} \mathcal{W}(C_j, R_j) dU^{|E|} \tag{10}$$

The partition functions $Z(p, A_p)$ in (10) depend only on the holonomy around ∂p and are given by integrals (9).

2.8. AREA PRESERVING QUANTUM FIELD THEORY

It is clear that the partition function of the YM_2 determines the quantum field theory in the functorial sense [5, 21]. It gives a functor from the category of metrized 2-cobordisms (possibly with immersed circles, colored by finite dimensional G-modules) to the category of vector spaces. The disjoint union of n circles (which is an object in the category of 2-cobordisms) is mapped to the space of functions on $[G]^{\times n}$. The 2-cobordisms (which are morphisms) are mapped to partition functions, regarded as kernels of integral operators, acting from one space of functions to the other.

This quantum field theory is almost topological, in a sense that the partition function depends on the metric only via the area dependence. Another good name for such theory would be *area preserving quantum field theory.*

Parallel to the Euclidean YM_2 one may consider in a similar manner "Minkovski" version of the theory with the partition function

$$z(\Sigma, C, R, A) = \int_{\mathcal{A}} e^{iS(\alpha)} \prod_{j=1}^{b} \mathcal{W}(C_j, R_j) \mathcal{D}\alpha \tag{11}$$

This version of YM_2 is related naturally to invariants of links in 3-manifolds (see Section 7).

It is clear that in the Minkovski case we might have certain convergence problems. In particular, this made the corresponding functorial quantum field theory ill defined. We will not discuss here the ways to fix such a problem.

3. Lattice 2D Yang-Mills theory

3.1. LATTICE CONNECTIONS AND METRIZED CW-COMPLEXES

By a *lattice* (following the terminology common in the physics literature) we understand here a finite CW-approximation Σ_T of a surface Σ. Denote by V, E and P the sets of vertices, edges and regions, i.e. 0, 1 and 2 cells of Σ_T respectively. In physical literature, regions are usually called plaquettes.

If there are circles immersed in a surface, we assume that they are contained in the 1-skeleton of $V \cup E$ of Σ_T.

Orient the edges in an arbitrary way. One may apply a well-developed theory of connections on oriented graphs for the restriction of bundle $\mathcal{P}$ to the 1-skeleton $V \cup E$ of Σ_T. In this formalism the bundle $\mathcal{P}$ restricts to the bundle $\cup_v \mathcal{P}_v$ over the vertices of 1-skeleton. Therefore, gauge transformations are concentrated on vertices, i.e. are given by functions $g : V \to G$, assigning an element g_v to each vertex v. In a similar way, connections are approximated by 1-cochains, i.e. functions $U : E \to G$, assigning to each edge e with the endpoints x, y an operator $U_e : \mathcal{P}_x \to \mathcal{P}_y$ of parallel transport along e. Gauge transformation g acts on U_e by $U_e \to g_y U_e g_x^{-1}$.

The integration over the space of lattice connections $U : E \to G$ is the integration over $G^{|E|}$ with the natural measure $dU^{|E|}$, where dU is the Haar measure on G.

We call Σ_T a *metrized CW-approximation*, if an additive function $A : P \to \mathbb{R}$ is fixed. We will call such numbers areas of regions. If Σ_T is PL with a metric on it, assigning areas to 2-cells gives an example of a metrized CW-approximation. We will use this example in the rest of this section.

A metrized subdivision of Σ_T is *measure preserving*, if areas of the resulting parts add up to the areas of the initial 2-cells. Two metrized CW-approximations are *stably equivalent*, if there exists a third CW-approximation, which can be obtained from them by a measure preserving subdivision.

3.2. AXIOMS OF YM_2 AND LATTICE PARTITION FUNCTION

The goal of the lattice approach to YM_2 is to find a partition function on metrized CW-approximations, which will satisfy all the fundamental properties we expect from the YM_2 theory:

- the partition function depends on areas A_p of 2-cells, and takes the same values on stably equivalent metrized CW-complexes (i.e. is in-

variant with respect to measure preserving subdivisions of metrized CW approximations of Σ);
- it is a function on $[G]^{|\partial\Sigma|}$, which satisfies gluing (5) and factorization (4) axioms;
- it should have the asymptotic of [18] in the limit $A \to 0$, which is more or less equivalent to the heat equation for Z_D.

We will call them axioms of YM_2.

It has been shown in the previous section, that these axioms determine uniquely the partition function for a disk. The consecutive use of the gluing and factorization axioms gives the following formula for the partition function on a metrized CW-approximation with Wilson loops:

$$Z(\Sigma, C, R, A) = \int_{G^{|E_{int}|}} \prod_{p\in P} Z_D(\operatorname{hol}_p(U), A_p) \prod_{j=1}^{m} \mathcal{W}(C_j, R_j) dU^{|E_{int}|}. \quad (12)$$

Here E_{int} is the set of edges which do not belong to the boundary of Σ. The holonomy $\operatorname{hol}_p(U)$ of U around the boundary ∂p of p is defined as follows. Suppose that ∂p consists of cyclically ordered edges $e_1^{\varepsilon_1}, e_2^{\varepsilon_2}, \dots, e_n^{\varepsilon_n}$, where $\varepsilon_j = +1$ if the orientation of e_j coincides with the one of ∂p and $\varepsilon_j = -1$ otherwise. Define the holonomy $\operatorname{hol}_p(U)$ by

$$\operatorname{hol}_p(U) = U_{e_1}^{\varepsilon_1} U_{e_2}^{\varepsilon_2} \dots, U_{e_n}^{\varepsilon_n}.$$

Final verification of the axioms of YM_2 for the partition function (12) is a simple combinatorics. In particular, the invariance under subdivisions follows immediately from (8), additivity of area under subdivisions and the following orthogonality relations for characters:

$$\int_G \chi_R(VU)\chi_{R'}(U^{-1}W)dU = \frac{\delta_{RR'}}{\dim R}\chi_R(VW) , \quad (13)$$

$$\int_G \chi_R(VUWU^{-1})dU = \frac{1}{\dim R}\chi_R(V)\chi_R(W) . \quad (14)$$

4. Combinatorial description of YM_2

It this section we will show, essentially, how to compute the integral (12) in case when we have Wilson loops.

4.1. LOCAL STATE SUMS ON CW-COMPLEXES

Recall the notion of a local state sum on a CW-complex (or, more generally, on a stratified space). This notion is well-known in statistical mechanics as a partition function of a lattice (in general, non regular) with local interaction. It also appears in combinatorics as a generating function for some configurations on CW-complexes.

For simplicity we will focus on the 2-dimensional case (the generalization to higher dimensions is straightforward). Fix sets S_0, S_1, S_2 of states on vertices, edges and 2-cells respectively. A *state* on a CW-complex Σ_T is a map from the strata of Σ_T to the set of states, which assigns to each i-cell, $i = 0, 1, 2$, an element of S_i. Denote the space of states by $S(\Sigma_T)$. Clearly $S(\Sigma_T) \simeq S_0^V \times S_1^E \times S_2^P$.

A weight of a stratum s is a function $W_s : S(\Sigma_T) \to \mathbb{C}$, which depends only on the state on s and on the states on the adjacent strata. Assume that we have a *local system of weights*, i.e. such a weight W_s is fixed for each stratum s of Σ_T. Then the *local state sum* is the following complex number:

$$\mathcal{Z}_W(\Sigma_T) = \sum_{s \in S(\Sigma_T)} \prod_{v \in V} W_v(s) \prod_{e \in E} W_e(s) \prod_{p \in P} W_p(s) \tag{15}$$

One can consider complex weights as well.

4.2. G-COLORINGS AND CLASSICAL 6J-SYMBOLS

For each triple R_i, R_j, R_k of irreducible representations of G denote by $\mathcal{H}_{ijk}$ the finite-dimensional vector spaces (so-called multiplicity modules or intertwines) $\mathcal{H}_{ijk} = (R_i \otimes R_j \otimes R_k)^{\mathrm{Ad}(G)}$. Pick a basis $\{\epsilon_\mu(ijk)\}$ of $\mathcal{H}_{ijk}$. By a *G-coloring* of Σ with immersed curves C we mean a function, assigning to each connected component $p \in P$ an irreducible representation R_p of G and to each edge $e \subset C_i$ an element ϵ_e of $\{\epsilon_\mu(ijk)\}$, where e is adjacent to regions with representations R_j, R_k. Clearly a coloring can be viewed as a state on Σ_T.

Pick a coloring *col* and let $v \in C_i \cap C_l$ be a double point of C. In a neighborhood of v we have 6 representations of G: representations R_i and R_l of the curves C_i and C_l and representations R_j, R_k, R_m, R_n of four adjacent regions, see Fig. 1. Also, on 4 edges meeting in v we have 4 basis elements $\epsilon_1, \ldots, \epsilon_4$ of the corresponding spaces $\mathcal{H}_{ijk}$, $\mathcal{H}_{lkm}$, $\mathcal{H}_{imn}$ and $\mathcal{H}_{lnj}$ of intertwines. There is a well-known way to extract a number out of such data by taking appropriate pairings of vectors corresponding to edges adjacent to the vertex v. This produces so-called symmetrized classical Wigner-Racah $6j$-symbols $\begin{vmatrix} i & j & k \\ l & m & n \end{vmatrix}_{\epsilon_1\epsilon_2}^{\epsilon_3\epsilon_4} \in \mathbb{C}$. We refer the interested reader to e.g. [17, 23, 24] for exact definitions of $6j$-symbols and more details.

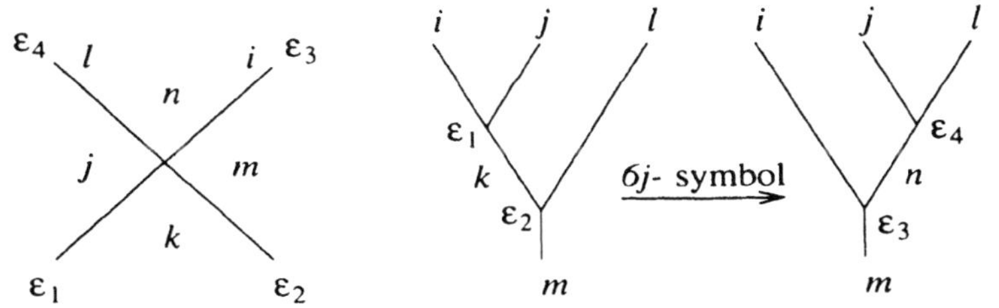

Figure 1. Colorings and $6j$-symbols

Clearly $6j$-symbols determine a system of local weights on Σ with immersed curves.

Algebraically (illustrated symbolically in Fig. 1) $6j$-symbols are matrix elements of associativity isomorphisms (associativity constrains). Indeed, consider a tensor product $R_i \otimes R_j \otimes R_l$. There are two ways to decompose it into the sum of irreducible components. We may tensor R_i with R_j, pick the ϵ_1-th copy of R_k in the decomposition of this product, tensor it with R_l and pick the ϵ_2-th copy of R_m in the decomposition into irreducible factors. Alternatively, we can do it according to the scheme depicted in the right part of Fig. 1. Both of these decompositions are isomorphic. The isomorphism is given by the associativity constraint for the category of finite dimensional G-modules. Its matrix elements are $6j$-symbols.

More generally one can consider any monoidal braided category. Matrix elements of the associativity constraint in such category will give an analog of $6j$-symbols. We will return to this, more general, discussion later.

4.3. COMBINATORIAL FORMULA FOR THE PARTITION FUNCTION OF YM_2

Fix a coloring *col* and denote the classical $6j$-symbol assigned to v in this coloring by $W_v(col) \in \mathbb{C}$. Define also the weight $W_p(col)$ of a region p of Euler characteristic $\chi(p)$, area A_p and assigned representation R_p by

$$W_p(col) = (\dim R_p)^{\chi(p)} \cdot e^{-A_p c_2(R_p)/2}.$$

We are now in a position to give an explicit expression for the partition function $Z_M(\Sigma, C_j, R_j, A)$.

Theorem 2. *Let Σ be a surface with boundary components $E_1, \dots, E_b$ and a collection $C = C_1 \cup \dots \cup C_n$ of immersed circles with representations $R_1, \dots, R_n$ of G assigned to them. Denote by $A : P \to \mathbb{R}$ the function, assigning to each component $p \in P$ of $\Sigma - C$ its area. Then the partition function $Z(\Sigma, C_j, R_j, A)$ of YM_2 with fixed conjugacy classes of holonomies*

$[U_j] = hol_{E_j}(\alpha)$ *around* E_j *is given by the following sum over colorings:*

$$Z(\Sigma, C, R, A) = \sum_{col} \prod_{p \in P} W_p(col) \prod_{v \in V} W_v(col) \prod_{j=1}^{b} \chi_{R_{p_j}}(U_j), \tag{16}$$

where p_j *is the component of* $\Sigma - C$ *adjacent to* E_j.

The proof of this theorem involves a lengthy but elementary combinatorics and is based mostly on the identity

$$\int_G \pi_1(U) \otimes \ldots \otimes \pi_n(U) dU = P_0 \tag{17}$$

where P_0 is the projector on the invariant subspace in the tensor product of representations $R_1 \otimes \ldots \otimes R_n$ and $\pi_k : G \to Aut(R_k)$ are corresponding homomorphisms.

An important case is when there are no Wilson loops. In this particular case we recover equation (9).

5. Poisson algebra of graphs on Σ and functions on the moduli space

In this section we consider the small area limit of YM_2 and treat the following problems. The definition of the partition function of YM_2 with inserted Wilson loops can be generalized to the case of inserted colored graphs in a straightforward manner. In the small areas limit it depends only on the homotopy type of a graph. The space of such colored graphs on a surface has a nice structure of Poisson algebra. The mapping of this algebra into the space of functions on the moduli space $\mathcal{M}_\Sigma^G$, given by YM_2 partition function, turns out to be a surjective Poisson algebra homomorphism.

5.1. GRAPHS AND DIAGRAMS

Let us start from some basic definitions.

By a *graph* Γ we mean further a graph with oriented edges and fixed total order of edges around each vertex. A *diagram* D_Γ of Γ on a surface Σ is a generic immersion of Γ to Σ. Note that besides vertices of Γ, the diagram D_Γ has additional singularities - the double points of the immersion. We will address them as *crossings* of D_Γ. A graph Γ may be reconstructed from its diagram D_Γ in an obvious manner, by ignoring the crossings. Thus, we will usually suppress the subindex Γ in D_Γ, unless a confusion may occur. In figures, we will reserve fat points for vertices, to distinguish them from crossings (see e.g. Fig. 2).

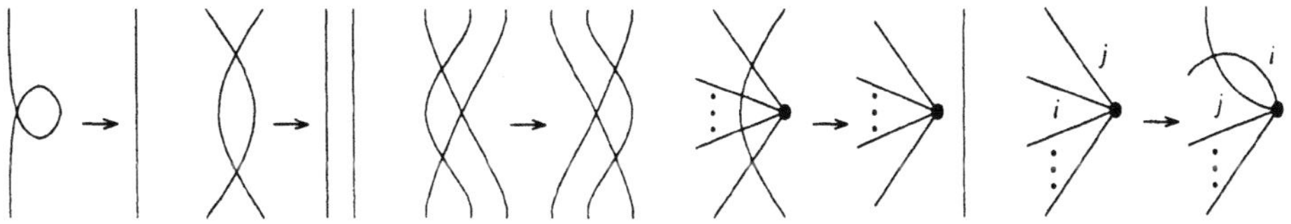

Figure 2. Moves of graph diagrams

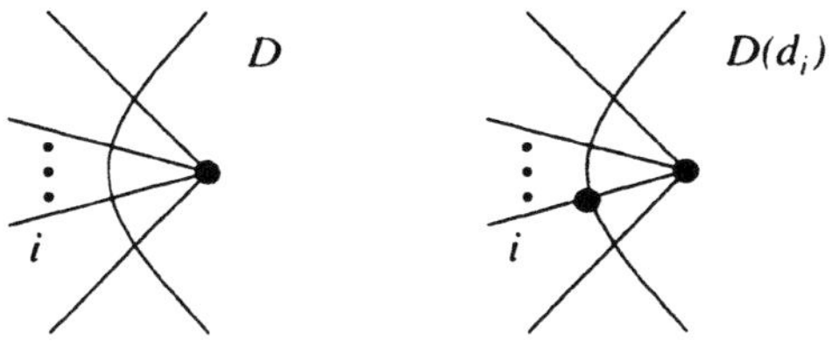

Figure 3. Replacing a crossing by a vertex

A *graph* $[D]$ *on a surface* is the homotopy class of a graph diagram D. It can be thought of as the equivalence class of graph diagrams, with respect to regular homotopy and moves of Fig. 2.

5.2. ALGEBRA OF GRAPHS ON Σ

Let us introduce the Poisson algebra of graphs on surfaces. The formal κ-linear combinations of graphs on Σ provide a vector space which is also a commutative associative algebra. The multiplication is generated by the operation of disjoint union. Consider the quotient algebra of this commutative algebra over the ideal, generated by the following relations.

Let an arc e of some diagram D "go around" a vertex of D, intersecting all the edges e_i, $i = 1, \dots, n$ incident to it, see Fig. 3. Fix $1 \leq i \leq n$ and consider the crossing $d_i = e \cap e_i$. Replacing it by a vertex, as illustrated in Fig. 3, we get a new graph diagram $D(d_i)$.

We impose the relation

$$\sum_{i=1}^{n} \varepsilon_i [D(d_i)] = 0, \tag{18}$$

where ε_i is the local intersection number of e with e_i in d_i. Denote the resulting commutative algebra of graphs on the surface Σ by $g(\Sigma)$.

Remark 5.1. Consider a particular case when the vertex above is 4-valent, with the incoming edges e_1, e_2 and the outgoing edges e_3, e_4. Relation (18) for the diagrams $D(d_i)$, $i = 1, \dots, 4$ becomes in this case the $4T$-relation,

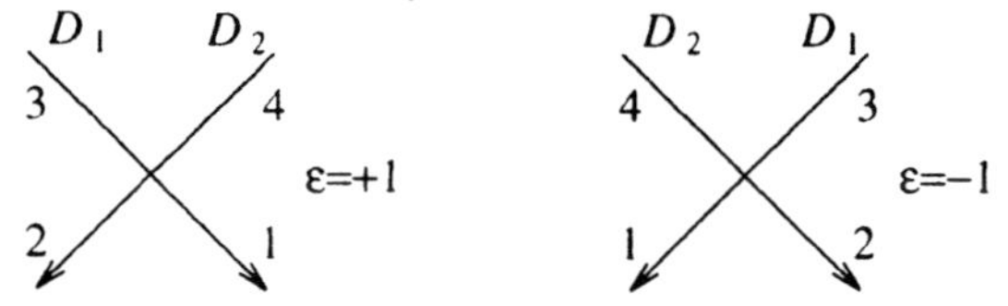

Figure 4. Ordering the edges

well-known in the theory of Vassiliev invariants. Thus equation (18) is a natural generalization of the $4T$-relation to the case of general graphs.

5.3. POISSON BRACKET ON DIAGRAMS AND GRAPHS ON SURFACES

For any two graph diagrams D_1 and D_2 in general position, $D_1 \cup D_2$ is a graph diagram, corresponding to the disjoint union of graphs. Each point $d \in D_1 \cap D_2$ of transversal double intersection is its crossing. For such a crossing d, we define $D_1 \#_d D_2$ as the diagram, obtained from $D_1 \cup D_2$ by replacing d by a vertex. The order of edges in this new vertex is given by the rule shown in Fig. 4.

Denote by $\varepsilon_{12}(d)$ the local intersection number of D_1 and D_2 in d (see Fig. 4).

It is easy to check, that $\{D_1, D_2\} = \sum_d \varepsilon_{12}(d) D_1 \#_d D_2$ is skew-symmetric and satisfies the Jacobi identity. This, actually, determines a Poisson bracket not only on diagrams, but also on the algebra $g(\Sigma)$ of graphs on Σ:

Proposition 5.2. *1) The expression $\sum_d \varepsilon_{12}(d)[D_1 \#_d D_2]$ depends only on the homotopy classes $[D_1]$, $[D_2]$ of diagrams D_1 and D_2;*
2) The bracket

$$\{[D_1], [D_2]\} = \sum_{d \in D_1 \cap D_2} \varepsilon_{12}(d)[D_1 \#_d D_2] \tag{19}$$

determines the structure of the Poisson algebra on $g(\Sigma)$.

Remark 5.3. The Poisson bracket (19) is compatible with the forgetful map, given by suppressing the order of the edges in each vertex. Therefore, it also determines the Poisson algebra structure on the algebra of non-ordered graphs.

5.4. COLORINGS OF DIAGRAMS

Let G be a finite dimensional Lie group over κ, such that its Lie algebra $\mathfrak{g} = Lie(G)$ has a non-degenerate $\mathfrak{g}$-invariant symmetric bilinear form $(.\,,\,.)$. Note that the bilinear form identifies $\mathfrak{g}$ with $\mathfrak{g}^*$, so determines an element

$t \in \mathfrak{g} \otimes \mathfrak{g}$. Denote by rep$(G)$ the set of (irreducible) finite dimensional representations of G and by $\mathbb{1} \in \text{rep}(G)$ its 1-dimensional representation.

Let $E(D)$ and $V(D)$ be the sets of edges and vertices of a graph diagram D. For $v \in V(D)$, denote by $e_1(v), \ldots, e_n(v)$ the (ordered) set of adjacent edges. Also, define $\varepsilon_i(v)$ to be the incidence number of e_i and v, so $\varepsilon_i(v) = +1$ for an incoming edge and $\varepsilon_i(v) = -1$ for an outgoing one.

A *coloring* of a graph $[D]$ on Σ is the following collection of data.

(1) A map $E(D) \to \text{rep}(G)$, $e \mapsto R_e$.
(2) For each $v \in V(D)$, a map $c_v : \mathbb{1} \to R_v$, where $R_v = R_1 \otimes \ldots \otimes R_n$ with $R_i = R_{e_i}$, if $\varepsilon_i(v) = +1$ and $R_i = R^*_{e_i}$, if $\varepsilon_i(v) = -1$.

Equivalently, one can say that a coloring of a vertex v determines a vector $c_v \in R_v = \otimes_{i=1}^{n(v)} R_i$. A *colored graph on* Σ is a pair $([D], c)$, where $[D]$ is a graph on Σ and c is its coloring.

5.5. POISSON BRACKET OF COLORED GRAPHS ON Σ

Let D_1, D_2 be colored graph diagrams in general position and $d \in D_1 \cap D_2$, say, on R_1 and R_2 colored edges of D_1 and D_2 respectively. Consider the diagram $D_1 \#_d D_2$ as in Section 5.3. The coloring of all of its edges and vertices, except for d, is determined by the ones of D_1 and D_2. We assign to d the linear map

$$c_d : \mathbb{1} \to R_1^* \otimes R_2^* \otimes R_1 \otimes R_2,$$

which maps 1 to $1 \otimes t$.

To write this map in an explicit coordinate form, pick a basis $\{I_\mu\}_{\mu=1}^{\dim(\mathfrak{g})}$ in $\mathfrak{g}$, which is orthonormal with respect to the scalar product $(.\,,\,.)$. Pick also a linear basis $\{f_i\}_{i=1}^{\dim(R)}$ for each representation R of G. In these notations $t = \sum_\mu I_\mu \otimes I_m u$ and

$$c_d(1) = \sum_{i=1}^{dim R_1} \sum_{j=1}^{dim R_2} \sum_{\mu=1}^{\dim(\mathfrak{g})} f^i \otimes f^j \otimes I_\mu f_i \otimes I_\mu f_j.$$

Now the coloring of $D_1 \#_d D_2$ is completely defined, so we can use (19) to define the Poisson bracket of colored graphs.

Proposition 5.4. *Colored graphs form a Poisson algebra with the commutative multiplication determined by the disjoint union and with the Poisson bracket (19).*

We will denote the Poisson algebra of colored graphs by $g(\Sigma, G)$.

5.6. WEIGHTS OF COLORED DIAGRAMS

Weights of colored diagrams are examples of local state sums from Section 4 (for 1-dimensional stratified space D).

Let (D, c) be a colored graph diagram. Assume that to each (R_e-colored) edge e is assigned a function $f_e : R_e \to R_e$.

Consider the product $\otimes_{v \in V(D)} R_v$ of spaces assigned to the vertices of D. Each edge e corresponds to a factor R_e in the space R_v assigned to its tail and to R_e^* in the space assigned to its head. Taking the tail of e, we first apply $f_e : R_e \to R_e$ and then consider a usual pairing $R_e \otimes R_e^* \to \kappa$. Doing this for each edge, we notice that all the factors in $\otimes_v R_v$ got paired, giving a function $\otimes_v R_v \to \kappa$.

We define the *weight* $W(D, c, \{f_e\}) \in \kappa$ as the value of this function on the (product of) vectors c_v corresponding to each vertex.

5.7. COLORED GRAPHS AND FUNCTIONS ON MODULI

Fix a flat connection α on the principal G-bundle $\mathcal{P}$ over Σ and denote by $[\alpha]$ its gauge class $[\alpha] \in \mathcal{M}_\Sigma^G$.

For each edge e of a graph diagram D on Σ denote by $\alpha_e : R_e \to R_e$ the parallel transport along e. Considering α_e as the function f_e in the definition of the weight W above, we put

$$F(D, c)(\alpha) = W(D, c, \{\alpha_e\}). \tag{20}$$

Lemma 5.5. *The function (20) depends only on the isotopy class of D and only on the gauge class of α.*

The proof of this lemma is parallel to the proof of Lemma 3 from [2]. Lemma 5.5 implies that the following definition is correct.

Definition 5.6. For each colored graph $([D], c)$ on Σ define the function on the moduli space $\mathcal{M}_\Sigma^G$ by

$$\mathcal{F}([D], c)([\alpha]) \stackrel{\text{def}}{=} F(D, c)(\alpha). \tag{21}$$

It is well known that the moduli space $\mathcal{M}_\Sigma^G$ is a symplectic manifold, thus the space $C(\mathcal{M}_\Sigma^G)$ of functions has a natural Poisson structure.

Theorem 5.7. *The map $\mathcal{F} : g(\Sigma, G) \to C(\mathcal{M}_\Sigma^G)$, $\mathcal{F} : ([D], c) \mapsto \mathcal{F}([D], c)$ is a homomorphism of Poisson algebras.*

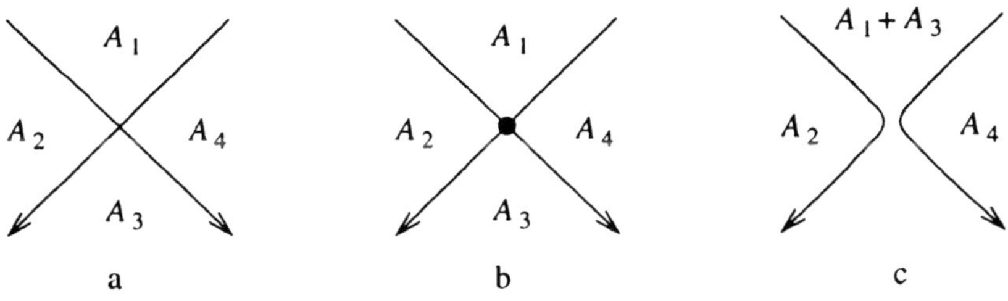

Figure 5. Making vertex from a crossing and smoothing it

The proof of this theorem is completely similar to the proof of Theorem 4 from [2]. Also, it follows from [11] that the map above is surjective.

6. Relating YM_2 to invariants of links and curves

6.1. DIFFERENTIAL EQUATIONS FOR THE PARTITION FUNCTION OF YM_2

First, notice that the definition of the partition function of YM_2 with inserted Wilson loops can be trivially generalized to the case of inserted graphs. Now we will derive important differential equations for the YM_2 partition functions.

Pick a crossing (a singular point of the immersion) d of a graph diagram D and denote by $A_1, \ldots, A_4$ the areas of the regions $p_1 \ldots p_4$ adjacent to v, as shown in Fig. 5a.

Denote by $D(d)$ the diagram obtained from D by replacing a crossing in d by a vertex, see Fig. 5b. A coloring c of D determines a coloring c' of $D(d)$, similarly to Section 5.5 (with the coloring of d described in Section 5.5 for $\varepsilon = +1$ and the coloring c for the rest of the graph).

Theorem 3. *The partition function of YM_2 satisfies the following differential equation at every crossing d*

$$\left(\frac{\partial}{\partial A_1} - \frac{\partial}{\partial A_2} + \frac{\partial}{\partial A_3} - \frac{\partial}{\partial A_4}\right) Z(\Sigma, D, c, A) = Z(\Sigma, D(d), c', A) \qquad (22)$$

The differential equation (22) in the particular case when $G = SU(N)$ and all loops are colored by N-dimensional irreducible representations gives Kazakov equations [16]. In this case the vertex in the diagram $D(v)$ can be replaced by the smoothed (with respect to the orientation) vertex, see Fig. 5c.

Let $E_j \subset \partial\Sigma$ be a connected component of the boundary of Σ. Denote by A_j the area of a connected component of $\Sigma - D$ adjacent to E_j and by

U_j the holonomy around E_j. Then

$$\frac{\partial}{\partial A_j} Z(\Sigma, D, c, A) = \Delta_{U_j} Z(\Sigma, D, c, A) \tag{23}$$

6.2. SMALL AREAS LIMIT OF YM_2 AND FUNCTIONS ON MODULI

From Theorem 3 one can derive the following statement for the small areas limit of YM_2.

Theorem 4. *Let (D_i, c_i) $i = 1, 2$ be colored graphs on Σ and $(\{D_1, D_2\}, c_{12})$ be their Poisson bracket (see Section 5.5). Extend the partition function of YM_2 (with zero areas) on graphs on Σ to the algebra of graphs on Σ by linearity. Then*

$$Z(\Sigma, \{D_1, D_2\}, c_{12}, 0) = 0 \tag{24}$$

The partition function of YM_2 with zero areas determines a non-degenerate linear functional on functions on the moduli space $\mathcal{M}_\Sigma^G$ of flat connections. Non-degeneracy means that the corresponding bilinear form on the commutative algebra of functions is non-degenerate. It follows from the previous theorem and non-degeneracy, that the partition function for YM_2 with zero areas is the integral over the moduli space with respect to the simplectic volume form.

6.3. YM_2 AND ARNOLD'S INVARIANT J^+

We will have to bypass an interesting question about the relations of YM_2 to invariants of curves on surfaces, as it would take us too far away. However, we will make a short digression to mention an elementary example of such a relation, expressing Arnold's invariant $J^+(C)$ of plane curves via coefficients of Z_M. This invariant, introduced recently by Arnold [4], is an elementary invariant of planar curves C without direct self-tangencies. If a curve C is deformed so that it passes a moment of direct self-tangency and, as a result, the number of double points grows by 2, the value of J^+ increases by 2. This property together with additivity under connected sum of curves completely determines J^+. This invariant is, in fact, closely related ([4]) to the Bennequin invariant of Legendrian knots in the manifold of contact elements of $\mathbb{R}^2$.

Let $C \in \mathbb{R}^2$ be a 1-component curve equipped with a basic (N-dimensional) representation of $U(N)$. Consider, as above, the partition function $Z_M(\mathbb{R}^2, C, R_N, A)$ of $U(N)$ theory as a function of areas A_p of regions $p \in P$.

Expanding Z_M in power series in A_p, we get

$$Z_M(\mathbb{R}^2, C, R, A) = N + \frac{N^2}{2} \sum_p c_p A_p + \dots .$$

The first coefficient c_p can, in fact, be explicitly computed from Kazakov's equation (22) and the normalization condition [16] $Z(\mathbb{R}^2, \tilde{C}, R_N, 0) = N^2$ for any 2-component curve $\tilde{C}$ with 0 areas. Let $\mathrm{ind}_C(p)$ be the index of p with respect to the curve C, i.e. the algebraic number of rotations of C around p. Using Kazakov's equation and the normalization condition above (applied to a curve $\widetilde{(C)} = C(d)$ obtained from C by smoothing a crossing d) one can check, that $c_p = \mathrm{ind}_C(p)^2$. The following interesting result easily follows from this computation. Extend the first coefficient c to the edges and vertices of C by averaging its values on the adjacent regions (on 2 regions in a generic point of C and all 4 adjacent regions in a vertex). Denote by c_e and c_v the values of c on $e \in E$ and $v \in V$ respectively.

Theorem 6.1.

$$J^+(C) = \sum_p c_p - \sum_e c_e + \sum_v c_v + \frac{|V|}{2} - 1$$

7. Relating YM_2 to quantum link invariants

The goal of this section is to sketch, rather briefly, the basic facts about quantum invariants of links in 3-manifolds and to show how YM_2 is related to these invariants.

On a phenomenological level this was analyzed in [26]. It has been shown there, that some basic properties of functional integrals for YM_2 on a surface Σ and for the Chern-Simons theory on a 3-manifold $S^1 \times \Sigma$ suggests that the YM_2 with zero areas can be regarded as quasiclassical limit of the Chern-Simons theory.

Here we will show how this argument can be verified on the level of state sums. Also, we will show how to obtain the YM_2 (Minkovski version) with nonzero areas in a similar way.

7.1. SHADOWS OF LINKS

In the partition function of YM_2 with Wilson loops we have the following data: a surface Σ with a collection of immersed curves $C = \cup_{j=1}^n C_j$. In addition, each region p of $\Sigma - C$ is equipped with a real number A_p (its area).

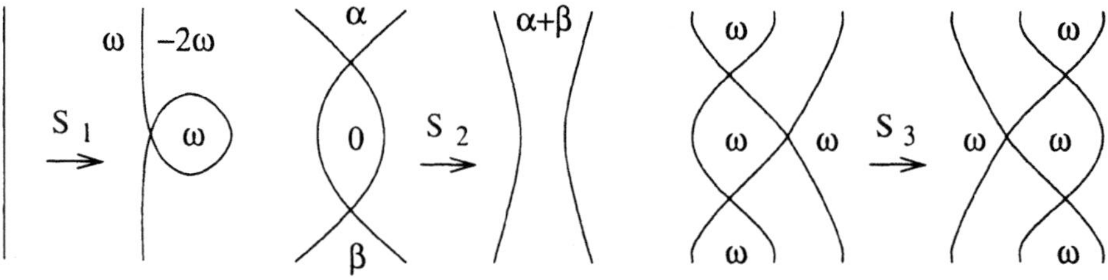

Figure 6. Shadow moves

A very similar data appears in Turaev's theory of link shadows. An n-component link $L = \cup_{j=1}^{n} L_j$ in a circle bundle $\pi : Y \to \Sigma$ over Σ is encoded by its projection $C = \pi(L)$ on Σ with a half-integer a_p assigned to each region p of $\Sigma - C$. Let us briefly review Turaev's construction of shadows (for details and a more general case of links in shadowed 2-polyhedra see [24]).

Let L be a link in a circle bundle $\pi : Y \to \Sigma$ over Σ. A *shadow* of L is a pair (C, a), where $C = \pi(L)$ is the bundle projection of L and a is a function $P \to \frac{1}{2}\mathbb{Z}$, assigning to each region p so-called *gleam* a_p.

The gleam a_p is, roughly speaking, a number of turns in the S^1 fiber made by the preimage in Y of the boundary ∂p of p as we go along the boundary counterclockwise. Unfortunately, the preimage $\pi^{-1}(\partial p) \subset Y$ consists of some non-connected pieces of L. So, over each double point we use one of the two possible pieces of the fiber to close the gap between two preimages of this double point. We end up with a closed curve in π and compute its number of turns in the fiber. The gleam a_p is the average of these numbers over all possible closures as above (2 choices for each double point).

Further we will be interested only in framed links; clearly, a shadow determines the blackboard framing (the one with framing vectors in the fiber).

Notice that the total gleam $\sum_p a_p$ equals to the Euler number of π (thus is 0 for trivial bundles $\Sigma \times S^1$ and ± 1 for the Hopf fibration $S^3 \to S^2$).

For links in bundles over S^2 a shadow determines the link up to isotopy. This remains "almost" true for links in bundles over other surfaces: a shadow determines the link up to isotopy and the action of $H_1(\Sigma)$ on L, see [23]. There is a natural shadow counterpart of framed Reidemeister moves, see Fig. 6, where $\omega = \pm\frac{1}{2}$. These moves do not change the isotopy class of the corresponding framed link and shadows of two isotopic framed links may be connected by a (finite) sequence of these moves. Thus any invariant of shadows, which is preserved under the shadow moves, gives an invariant of framed links.

7.2. TENSOR CATEGORIES AND REPRESENTATIONS OF QUANTUM GROUPS

The Category of finite-dimensional modules over an algebraic group have natural tensor product structure and certain properties which makes it a tensor category. For the details see [9]. In the modern terminology they should be called symmetric tensor categories.

A tensor category is an Abelian category $\mathcal{C}$ with tensor product bifunctor $\otimes : \mathcal{C} \times \mathcal{C} \to \mathcal{C}$, with the associativity constraint, identity object, etc. which satisfy the axioms of monoidal category, with the commutativity constraints that satisfy the hexagon identities and with the appropriate notion of a dual object. For more details about such categories see for example [15].

For each tensor category one can define analogues of dimension, of $6j$-symbol, etc. Category of modules over Hopf algebras are examples of monoidal categories. Categories of modules over quasitriangular Hopf algebras are examples of tensor categories.

The Category of finite dimensional modules over quantized universal enveloping algebra $U_q(\mathfrak{g})$ is an especially interesting example of a tensor category. For generic values of q (and appropriate definition of $U_q(\mathfrak{g})$) one can say that the category of finite-dimensional $U_q(\mathfrak{g})$-modules is a deformation of the category of $U(\mathfrak{g})$-modules in the following sense:

- categories $U(\mathfrak{g})$–mod and $U_q(\mathfrak{g})$–mod are equivalent as semisimple Abelian categories,
- the tensor structure is deformed.

As a result one can introduce deformed $6j$-symbols, which are also known as q-$6j$-symbols. They are matrix elements of the associativity constraint in the category $U_q(\mathfrak{g})$–mod. Because for generic q the categories $U(\mathfrak{g})$–mod and $U_q(\mathfrak{g})$–mod are equivalent as semisimple Abelian categories, the deformed q-$6j$-symbols are parameterized by the same data as the undeformed ones.

The situation is different when q is a root of one. Then the category of modules over $U_q(\mathfrak{g})$ is not semisimple and is not equivalent to the category of $U(\mathfrak{g})$–modules. In this case there exists a finite dimensional quotient Hopf algebra $U_q'(\mathfrak{g})$. There exists an ideal in the category $U_q(\mathfrak{g})$–mod, such that the quotient category is a semisimple tensor category with finite number of simple objects. This category is sometimes called a *truncated category of modules* over $U_q'(\mathfrak{g})$; we will denote it by $\mathcal{C}_q\mathfrak{g}$.

The q-$6j$-symbols for roots of one are the matrix elements of the associativity constraints in the truncated category of modules over $U_q(\mathfrak{g})$. They can be obtained by an analytic continuation of the q-$6j$-symbols for generic q to roots of one.

7.3. QUANTUM LINK INVARIANTS VIA SHADOWS AND YM_2

Tensor categories and, in particular, categories of modules over quantum groups can be used as data for construction of invariants of links. For the details of the construction and references see [24].

In [23] Turaev gave a shadow version of these invariants for links in circle fibers over surfaces using the truncated category of modules over $U_q'(\mathrm{sl}_2)$. This construction can be generalized to arbitrary semisimple finite tensor category [24].

To restate these results, we proceed in a way similar to Section 4.2. The only changes which have to be made are in the definition of $6j$-symbols. Now the coloring is determined by a semisimple finite tensor category $\mathcal{C}$ and the system of weights is given by matrix elements of associativity constraints in this category.

Let $C = \cup_{j=1}^n C_j$ be an n-component curve on Σ, with numbers a_p assigned to the regions p of $\Sigma - C$ and irreducible objects $R_j \in \mathcal{R}$ of $\mathcal{C}$ assigned to the components C_j. By a coloring we mean now a function, assigning to each region $p \in P$ an irreducible object $R_p \in \mathcal{R}$ of the category $\mathcal{C}$ and to each edge $e \subset C_i$ an element ϵ_e. Here, just as in Section 4.2, ϵ_e is a basis vector in the space $\mathcal{H}_{ijk}$ of intertwines (if e is an edge colored by R_i with adjacent regions colored by $R_j, R_k \in \mathcal{R}$).

For coloring *col* of surface Σ with immersed circles denote the quantum $6j$-symbol (appropriate matrix elements of associativity constraints) assigned to a double point $v \in C_i \cap C_l$ by $w_v(col)$. Define also the weight $w_p(col)$ of a region p with assigned number a_p and representation R_p by

$$w_p(col) = (\dim_{\mathcal{C}} R_p)^{\chi(p)} \cdot v(R_p)^{a_p},$$

where $\dim_{\mathcal{C}}(R_p)$ is the dimension of the irreducible object R_p, and $v(R_p)$ is the value of the ribbon element (see the previous subsection) on the object R_p and $\chi(p)$ is the Euler characteristic of p.

Theorem 7.1 ([**23**]). *Let $L = \cup_{j=1}^n L_j$ be a framed link in S^1-bundle $Y \to \Sigma$ over Σ and denote by (C, a) the shadow of L. Assign irreducible objects $R_j \in \mathcal{R}$ of a semisimple finite tensor category $\mathcal{C}$ to the components L_j. Then*

$$z_{\mathcal{C}}(\Sigma, C_j, R_j, a) = \sum_{col} \prod_{p} w_p(col) \prod_{v} w_v(col) \tag{25}$$

is an invariant of the link L.

7.4. YM_2 AS A QUASICLASSICAL LIMIT

Denote by $z_{\mathfrak{g},q}(\Sigma, C_j, R_j, a)$ the invariant of links obtained from the truncated category of modules over $U_q'(\mathfrak{g})$.

As explained in Section 7.2, when $q \to 1$, the truncated category of modules over $U'_q(\mathfrak{g})$ becomes the category of finite dimensional $\mathfrak{g}$-modules. The invariant of links (25) becomes, in this limit, the partition function of the YM_2.

Indeed, consider a generic link L_1 in a S^1-bundle Y_1 over Σ with the Euler number 1. Take now tensor powers of this bundle, factorizing S^1 by r-th roots of unity, to obtain a sequence of S^1-bundles Y_r over Σ with the Euler number r with embedded links $L_r \subset Y_r$ (the images of L_1). The shadows of L_r consist of the same colored curve (C_j, R_j), but with the gleams $a_p(r)$ growing proportionally to r. The set of limit gleams $A_p = \lim_{r\to\infty} \frac{\pi a_p(r)}{2r}$ is well-defined. One can think about the limit of this sequence as about a link in an "infinitely twisted" S^1-bundle over Σ. When $r \to \infty$, the limit of $z(\Sigma, C_j, R_j, a_p(r))$ is easily seen to coincide with $Z_M(\Sigma, C_j, R_j, A_p)$:

Theorem 5. *Define $A_p = \frac{\pi a_p}{2r}$ and let $q \to 1$ so that the values of A_p remain fixed. Then link invariants $z_{\mathfrak{g},q}(\Sigma, C_j, R_j, a)$ tend in the limit to the partition function $Z_{\mathfrak{g}}(\Sigma, C_j, R_j, A)$ of the YM_2 with Wilson loops.*

Note, that the "areas" A_p can remain finite in the limit $q \to 1$ only if the bundle Y_r over Σ becomes "infinitely twisted" (proportionally to r) as $r \to \infty$. This is why the limit does not give an invariant of immersed circles, but only of "equiarea isotopies" of immersed circles on a surface.

However, if the bundle is fixed when $q \to 1$ ($A_p \to 0$), then the limit of $z(\Sigma, C_j, R_j, a)$ coincide with the small area limit $Z_M(\Sigma, C_j, R_j, 0)$ of the Minkovski YM_2 theory. Thus it is an invariant of immersed circles with respect to isotopies.

Thus, one can say that the invariant of links (25) can be regarded as a quantization of the YM_2. In the next section we will clarify how this "quantization" is related to the quantization of $C(\mathcal{M}_\Sigma^G)$.

8. Quantization of the algebra of functions on $\mathcal{M}_\Sigma^G$

8.1. QUANTIZATION OF POISSON ALGEBRAS

Quantization of Poisson algebras has a long history [7] and by now there are several approaches to quantization (see for example [25].

A *weak quantization* of a Poisson algebra A (over κ) consists of the following data.

- A topological space T with the limit point 0 which does not belong to T.
- A collection of associative algebras $\{A_t\}_{t\in T}$.
- A collection of linear maps $\phi_t : A \to A_t$.
- A continuous map $h : T \to \kappa$ with $\lim h = 0$ as $t \to 0$.

- A linear functional $\nu : A \to \mathbb{C}$ on the Poisson algebra A, such that $\nu(\{a, b\}) = \nu(\{b, a\})$.
- A cyclic linear functional ν_t for each algebra A_t: $\nu_t(ab) = \nu_t(ba)$.

This data should satisfy the following axioms.

- $\nu_t(\phi_t(a)) = \nu(a) + O(h(t))$, as $t \to 0$.
- $\nu_t([\phi_t(a), \phi_t(b)]\phi_t(c)) = h(t)\nu(\{a, b\}c) + O(h(t)^2)$, as $t \to 0$.

Usually T is either a set of positive integers, or an interval, or a disk.

A weak quantization is called *non-degenerate* if, for each $t \in T$, the symmetric bilinear form $\nu_t(ab)$ on A_t is non-degenerate and the bilinear form $\nu(ab)$ on A is also non-degenerate. By taking quotients over kernels of bilinear forms one can make any weak quantization non-degenerate.

The linear maps ϕ_t are not defined canonically. But we will not discuss here corresponding equivalences as well as the equivalences of weak quantizations.

Now let us show that the invariants of 3-manifolds constructed in [20] give a weak quantization of the algebra of functions over the moduli space $\mathcal{M}_\Sigma^G$.

8.2. QUANTIZATION OF THE MODULI SPACE AND INVARIANTS OF LINKS IN THE $\Sigma \times S^1$

Consider the algebra $G(\Sigma)$ of graphs in $\Sigma \times I$. The multiplication in this algebra is given by the disjoint union of graphs (see [24, 3]). It is an associative filtered algebra and one can show (similarly to [3], where it has been done for chord diagrams) that its associate graded algebra is a Poisson algebra, which is isomorphic to the Poisson algebra of graphs on Σ.

Invariants of links in $\Sigma \times I$ constructed in [20] can be regarded as linear operators acting in a space associated with the surface. This vector space, known also as Verlinde space, is parameterized by a nonnegative integer k and a simple Lie algebra $\mathfrak{g}$. This data determine both the corresponding truncated category of modules over $U_q(\mathfrak{g})$ and the Chern-Simons-Witten topological field theory. The vector spaces corresponding to surfaces have a simple combinatorial description in terms of truncated categories (see [20, 24]). They also have a nice geometric description (going back to original Verlinde's work) as sections of certain line bundles over the moduli space of complex holomorphic bundles over Σ. In other words, they can be obtained via geometrical quantization of the moduli space.

Denote the vector space corresponding to the surface Σ as $V_k(\mathfrak{g})$. Then the quantum invariant of links and graphs in $\Sigma \times I$ can be regarded as a homomorphism of algebras $\pi_k : G(\Sigma) \to EndV_k(\mathfrak{g})$.

Choose a linear map $\phi : g(\Sigma) \to G(\Sigma)$ in such a way the composition of the map $[\phi]$ from $g(\Sigma)$ to the associate graded of $G(\Sigma)$ with the natural map from the associate graded algebra to $g(\Sigma)$ (see [3]) is the identity map.

Composing linear maps ϕ and π_k we obtain the linear map $\psi_k : g(\Sigma) \to EndV_k(\mathfrak{g})$. Choose any section $f : C(\mathcal{M}_\Sigma^G) \to g(\Sigma)$ of the map $\mathcal{F}$. Denote the composition of f and ψ_k by $\phi_k : C(\mathcal{M}_\Sigma^G) \to EndV_k(\mathfrak{g})$.

Theorem 6. *The data $A_k = EndV_k(\mathfrak{g})$, $\nu(D,c) = Z(\Sigma, D, c, 0)$, $\nu(a) = tr(a)$ and $\phi_k : C(\mathcal{M}_\Sigma^G) \to EndV_k(\mathfrak{g})$ determines a weak quantization of the Poisson algebra $A = C(\mathcal{M}_\Sigma^G)$ with T being the set of nonnegative integers.*

The limit point in this quantization is $k = \infty$.

This theorem follows from the observations that: i) the quantum invariant of graphs in $\Sigma \times S^1$ gives the partition function of the YM_2 in the limit $k \to \infty$, ii) the quantum invariant of graphs in $\Sigma \times S^1$ is proportional to $tr(\pi_k)$, iii) the partition function of YM_2 is the integral over the moduli space with the respect to the simplectic volume form. One has to use also some simple properties of quantum invariants in the limit $k \to \infty$.

Another approach to quantization of the algebra of functions on the moduli space has been developed in [1, 8]. It is based on discrete YM_2.

References

1. A. Alekseev, H. Grosse, V. Schomerus, *Combinatorial quantization of the Hamiltonian Chern-Simons theory I*, Comm. Math. Phys. **172** (1995), 317–358.
2. J. Andersen, J. Mattes, N. Reshetikhin, *The Poisson structure on the moduli space of flat connections and chord diagrams*, Topology **35** (1996), 1069–1083.
3. J. Andersen, J. Mattes, N. Reshetikhin *Quantization of the algebra of chord diagrams*, q-alg/9701018.
4. V.I. Arnold, *Topological invariants of plane curves and caustics*, Univ. Lec. Ser. **5**, Providence, RI, 1994.
5. M. Atiyah, *Topological quantum field theories*, Publ. Math. IHES 68 (1988), 175–186.
6. M. Atiyah, R. Bott, *The Yang-Mills equations over Riemann surfaces*, Phil. Trans. R. Soc. London **A 308** (1983), 523–615.
7. F. Bayen, M. Flato, C. Frønsdal, A. Lichnerovicz and D. Sterheimer, *Deformation theory and quantization, I, II* Ann. Phys. **111** (1978), 61–151.
8. E. Buffenoir, P. Roche, *Link invariants and combinatorial quantization of Hamiltonian Chern Simons theory*, Comm. Math. Phys. **181** (1996), 331–365.
9. P. Deligne, J. Milne, *Categories Tannakiennes*, in: *Hodge Cycles, Motives and Shimura Varieties*, Lec. Notes Math. **900**, Springer, Berlin, 1982, 101–228.
10. D. Fine, *Quantum Yang-Mills on the two-sphere*, Comm. Math. Phys. **134** (1990) 273–292; *Quantum Yang-Mills on a Riemann surface*, Comm. Math. Phys. **140** (1991), 321–338.
11. V. Fock, A. Rosly, *Poisson structure on moduli of flat connections on Riemann surfaces and r-matrices*, Preprint ITEP 72-92 (1992).
12. M. Forman, *On small volume limits od 2-d Yang-Mills*, Comm. Math. Phys. **151** (1993), 39–52.
13. D. Gross, *Two dimensional QCD as a string theory*, Nucl. Phys B 400 (1993), 161–180.

14. D. Gross, A. Taylor, *Twists and loops in the string theory of two dimensional QCD*, Nucl. Phys B **403** (1993), 395–452.
15. C. Kassel, *Quantum groups*, Grad. Texts in Math. Springer-Verlag, New York–Heidelberg–Berlin, 1994.
16. V. Kazakov, *Wilson loop average for an arbitrary contour in two-dimensional $U(N)$ gauge theory*, Nucl. Phys. B **179** (1981), 283–292.
17. A.N. Kirillov, N. Reshetikhin, *Representations of the algebra $U_q(sl(2))$, q-orthogonal polynomials and invariants of links*, in: *Infinite dimensional Lie algebras and groups* (ed V. Kac) Adv. Ser. Math. Phys. **7** (1988), 285–337.
18. A. Migdal, *Recursion equations in gauge field theories*, Zh. Eksp. Teor. Fiz. **69** (1975), 810 (Sov. Phys. JETP **42**, (1975) 413–418).
19. G. Moore, *2D Yang-Mills theory and topological field theories*, in: Proceedings of Int. Cong. Math. 1994.
20. N. Reshetikhin, V. Turaev, *Ribbon graphs and their invariants derived from quantum qroups*, Comm. Math. Phys. **127** (1990) 1–26; *Invariants of 3-manifolds and quantum groups*, Inv. Math. **103** (1991) 547–597.
21. G. Segal, *The definition of conformal field theory*, preprint (1989).
22. A. Sengupta, *Quantum gauge theory on compact sphere*, Ann. Phys. **221** (1993), 17–52.
23. V. Turaev, *Shadow links and face models of statistical mechanics*, J. Diff. Geom. **36** (1992), 35–74.
24. V. Turaev, *Quantum invariants of knots and 3-manifolds*, de Gruyter, Berlin, 1994.
25. A. Weinstein, *Deformation quantization*, Séminaire Bourbaki, Volume 1993/94, Exp. 789, Astérisque **27** (1995), 389–409.
26. E. Witten, *On quantum gauge theories in two dimensions*, Comm. Math. Phys. **141** (1991), 153–209; *Two dimensional gauge theories revisited*, J. Geom. Phys. **9** (1992), 303–368.

SOME APPLICATIONS OF QUANTUM SHUFFLES

MARC ROSSO
I.R.M.A.,
Université Louis Pasteur et Institut Universitaire de France,
7, rue René Descartes, F-67084 Strasbourg Cedex

Abstract. We review recent applications of the "quantum shuffle" approach to quantized enveloping algebras [11, 12].

Introduction

The main motivation for the constructions made in this paper was to find some intrinsic definition of the quantized enveloping algebras, i.e. a definition which does not rely on generators and relations. It turns out that there is a general framework, that of cotensor Hopf algebras on a Hopf bimodule, in which they fit naturally. In simple terms, the data of a Hopf bimodule on a Hopf algebra H is equivalent to that of a module on the double of H, and so they is a braid group action around. The algebra structure on the cotensor Hopf algebra is essentially made by "shuffling with respect to the braid group action". This graded algebra is not generated by its degree 0 and 1 elements, and the sub-Hopf algebra generated by them (a natural quantum version of the symmetric algebra) has very remarkable properties. In the examples of interest, this is exactly the "upper triangular part" of the quantized enveloping algebra. This line of approach suggests that one should be looking for representations of the cotensor Hopf algebra. We give a general method for this in terms of iterated integrals.

The plan of the paper is as follows. In the first section, we give the abstract general construction of the cotensor Hopf algebra built on a Hopf bimodule, and show that its algebra structure has a nice expression in terms of quantum shuffles. In the second section, we explain the main example, the one leading to the "upper triangular" part of quantized enveloping algebras. The third section provides a general method to construct representations of the quantum shuffle algebra via iterated integrals. As an application of this method, we give a simple proof of a result of P. Bouwknegt, J. McCarthy

D. Sternheimer et al. (eds.), Deformation Theory and Symplectic Geometry, 249–258.

and K. Pilch, stating that contour integrals of vertex operators satisfy the quantum Serre relations. Details of the first two sections are in [11], and of the last two sections in [12].

Notation

In all this paper, we denote by k a commutative field.

If H is a k-Hopf algebra, we denote by Δ its coproduct and by S its antipode. We shall use Sweedler's notation: for h in H, $\Delta(h) = \sum h_{(1)} \otimes h_{(2)}$. When we consider H with the opposite algebra structure, we denote it by H^{op}.

As we shall be dealing with tensor products between tensor powers of a given space, we shall often write $(x_1, \ldots, x_n)$ instead of $x_1 \otimes \ldots \otimes x_n$.

We denote by Σ_n the symmetric group of $\{1, 2, \ldots, n\}$ and, for $i = 1, \ldots, n$, by s_i the transposition $(i, i+1)$. The braid group on n strands is denoted by B_n, and for $i = 1, \ldots, n$, the i-th generator by σ_i.

For $l_1 + \cdots + l_r = n$, the set of $(l_1, \ldots l_r)$-shuffles, i.e. the set of permutations w such that $w(1) < w(2) < \cdots < w(l_1)$, $w(l_1+1) < w(l_1+2) < \cdots < w(l_1+l_2) \ldots w(l_1 + \cdots l_{r-1}+1) < \ldots < w(n)$ is denoted by $\Sigma_{(l_1, \ldots l_r)}$.

The length of a permutation w, defined as the length of any reduced expression in terms of the standard generators s_i, $i = 1, \ldots, n$, is denoted by $l(w)$.

If w is in Σ_n, we denote by T_w the corresponding lift in B_n, defined as follows: if $w = s_{i_1} \cdots s_{i_k}$ is any reduced expression of w, then $T_w = \sigma_{i_1} \cdots \sigma_{i_k}$.

We put $\mathcal{B}_{(l_1, \ldots l_r)} = \sum_{w \in \Sigma_{(l_1, \ldots l_r)}} T_w$.

For $q \in k$ different from 1 or -1, or q an indeterminate, and for all $n \in \mathbf{Z}$, we define $[n]_q = \frac{q^n - q^{-n}}{q - q^{-1}}$, and $[n]_q! = [n]_q[n-1]_q \cdots [1]_q$.

1. The quantum shuffle Hopf algebra

We recall general facts, due to W. Nichols [8], about Hopf bimodules and the cotensor algebra. We complete them by clarifying the algebra structure in terms of the natural action of the braid group.

1.1. HOPF BIMODULES

Definition 1. Let H be a k-Hopf algebra. A Hopf bimodule over H is a k-vector space M given with a H-bimodule structure, a H-bicomodule structure (i.e. left and right coactions $\delta_L : M \mapsto H \otimes M$, $\delta_R : M \mapsto M \otimes H$ which commute in the following sense: $(\delta_L \otimes id)\delta_R = (id \otimes \delta_R)\delta_L$), and such that δ_L and δ_R are morphisms of H-bimodules.

This notion has also been considered by S. Woronowicz [13], under the name of bicovariant bimodule, in his work on non-commutative differential calculus on quantum groups.
W. Nichols shows that, taking tensor products over H, Hopf bimodules form a tensor category $\mathcal{E}$.
The structure of Hopf bimodules is clarified if one considers the subspaces of left or right coinvariants: $M^L = \{m \in M; \delta_L(m) = 1 \otimes m\}$ and $M^R = \{m \in M; \delta_R(m) = m \otimes 1\}$. In fact, a classical result of Sweedler says that M is isomorphic, as left module and comodule, to the trivial one $H \otimes M^L$, or, as a right module and comodule, to the trivial one $M^R \otimes H$. Furthermore, M^L is a sub-right comodule of M, and inherits a structure of right H-module given by:

$$m \cdot h = \sum S(h_{(1)}) m h_{(2)}$$

with $m \in M$ and $h \in H$. In the same way, M^R is a sub-left comodule of M, and inherits a left H-module action given by:

$$h \cdot m = \sum h_{(1)} m S(h_{(2)})$$

with $m \in M$ and $h \in H$.

Proposition 2. *With these two structures, M^R is a crossed module over H in the sense of Yetter [14]; and if H and M are finite dimensional, it is a module over the quantum double. Furthermore, a morphism of Hopf bimodules induces on the spaces of right coinvariants a morphism of crossed modules.*

1.2. BRAIDINGS

Woronowicz [13] introduced a remarkable braiding in the category of Hopf bimodules.

Proposition 3. *Let M and N be H-Hopf bimodules. There exists a unique morphism of H-bimodules $\sigma_{M,N} : M \otimes_H N \mapsto N \otimes_H M$ such that, for $\omega \in M^L$ and $\eta \in M^R$ $\sigma_{M,N}(\omega \otimes \eta) = \eta \otimes \omega$. Furthermore, $\sigma_{M,N}$ is invertible, is a morphism of bicomodules and satisfies the following braid equation (where M, N and P are Hopf bimodules):*

$$(I_P \otimes \sigma_{M,N})(\sigma_{M,P} \otimes I_N)(I_M \otimes \sigma_{N,P}) = (\sigma_{N,P} \otimes I_M)(I_N \otimes \sigma_{M,P})(\sigma_{M,N} \otimes I_P)$$

This makes $\mathcal{E}$ a braided tensor category.

One observes that $\sigma_{M,M}$ sends $M^R \otimes M^R$ into itself, and defines a representation T of the braid group B_n in $(M^R)^{\otimes n}$. An easy computation gives:

$\sigma(x \otimes y) = \delta_L(x)(y \otimes 1)$, and so this representation is nothing but the one coming from the crossed bimodule structure. More precisely, we have:

Proposition 4. *The functor sending M to M^R is an equivalence of braided tensor categories between $\mathcal{E}$ and the category of crossed bimodules.*

1.3. THE COTENSOR HOPF ALGEBRA

The following construction generalizes the classical shuffle algebra.

Definition 5. (i) Let M and N be H-Hopf bimodules. Their cotensor coproduct $M \sqcup N$ is the kernel of $\delta_L \otimes I_N - I_M \otimes \delta_R : M \otimes N \mapsto M \otimes H \otimes N$.
(ii) The cotensor coalgebra constructed on M is $T_H^c(M) = H \oplus \bigoplus_{n \geq 1} M^{\sqcup n}$.

In fact, it is easy to see that $M \sqcup N$ is again a Hopf bimodule, and that $T_H^c(M)$ has a graded coalgebra structure given as follows: the component of bidegree (p,q) of the coproduct on $M^{\sqcup n}$ is given by the restriction of the map: $(x_1, \ldots, x_n) \mapsto (x_1, \ldots, x_p) \otimes (x_{p+1}, \ldots, x_n)$, for $p \geq 1$ and $q \geq 1$, and by $\delta_L \otimes id$ for $p = 0$, and $id \otimes \delta_R$ for $q = 0$.

Nichols showed that this coalgebra has a universal property, from which one derives that it is a Hopf algebra. The product is characterized as the unique coalgebra map $T_H^c(M) \otimes T_H^c(M) \to T_H^c(M)$ extending the given product on H and such that on elements of degree 1 in $T_H^c(M) \otimes T_H^c(M)$, i.e. on $H \otimes M + M \otimes H$, it is given by the left or right action of H on M.

When H and M are finite dimensional, it is very easy to understand the cotensor Hopf algebra: H^* is a Hopf algebra, M^* is a Hopf bimodule and the graded dual of the cotensor Hopf algebra $T_H^c(M)$ is the tensor Hopf algebra $T_{H^*}(M^*)$

As $T_H^c(M)$ is a Hopf bimodule, it is a free right -H-module over its subspace of right coinvariants, and this subspace is naturally but not trivially isomorphic to the tensor space $T(V)$ (see below). So, as a right H-module and right H-comodule, $T_H^c(M)$ is isomorphic to $T(V) \otimes H$. We are going to exhibit a Hopf algebra structure on $T(V) \otimes H$, and we shall show that it coincides (via the isomorphism above) with the one coming from the universal property.

Proposition 6.

1. *There is an associative algebra structure on $T(V)$, given by*

$$(x_1 \otimes \ldots \otimes x_p) \cdot (x_{p+1} \otimes \ldots \otimes x_n) = \sum_{w \in \Sigma_{p,n}} T_w(x_1 \otimes \ldots \otimes x_n)$$

for $x_1, \ldots, x_n$ in V, where $\Sigma_{p,n-p}$ is the set of $(p, n-p)$-shuffles.

2. *For the diagonal action of H on each $V^{\otimes n}$, $T(V)$ is an H-module algebra, and $T(V) \otimes H$ inherits the crossed product algebra structure.*

3. The following defines a coalgebra structure on $T(V) \otimes H$:

$$\Delta((v_1, \dots, v_n) \otimes h) = \sum_{k=0}^{n} [(v_1, \dots, v_k) \otimes v_{k+1(-1)} \cdots v_{n(-1)} h_{(1)}] \otimes \otimes [v_{k+1(0)} \cdots v_{n(0)} \otimes h_{(2)}],$$

for $(v_1, \dots, v_n)$ in V and h in H.

4. The algebra structure of 3. and coalgebra structure of 4. are compatible and make $T(V) \otimes H$ a Hopf algebra.

Theorem 7. *The map $\hat{\phi} : T(V) \otimes H \to T^c_H(M)$, given by:*

$$\hat{\phi}[(v^1, \dots, v^n) \otimes h] = \sum v^1 v^2_{(-1)} v^3_{(-2)} \cdots v^n_{(-n+1)} h_{(1)} \otimes \otimes v^2_{(0)} v^3_{(-1)} \cdots v^n_{(-n+2)} h_{(2)} \otimes \cdots \otimes v^n_{(0)} h_{(n)}$$

is a Hopf algebra isomorphism.

1.4. THE QUANTUM SYMMETRIC ALGEBRA.

The sub-Hopf algebra $S_H(M)$ of $T^c_H(M)$ generated by H and M has very remarkable properties. It is again a Hopf bimodule and its subspace of right coinvariants is isomorphic, via ϕ, to the subalgebra of $T(V)$ generated by V. We shall denote this subalgebra by $S_\sigma(V)$. It can be considered as a quantum version of the symmetric algebra and, as an algebra, $S_H(M)$ is nothing but the crossed product of H by $S_\sigma(V)$.

This Hopf algebra was also introduced by W. Nichols under the name bialgebra (or Hopf algebra) of type one.

2. The examples from abelian group algebras

We are going to apply the preceding machinery to a very simple situation, but which leads to interesting examples.

Let $H = k[G]$ be the group algebra of an abelian group G, written as a product $\mathbf{Z}^r \times \mathbf{Z}/l_1 \times \mathbf{Z}/l_2 \times \ldots \times \mathbf{Z}/l_p$. We shall fix generators $K_1, \cdots, K_N$ of G ($N = r+p$). As H is both commutative and cocommutative, the structure of the subspace V of right coinvariants of a Hopf bimodule M is just that of a left module and a left comodule, with no compatibility condition. We shall consider the following family of examples: V is a k-vector space of dimension N, with a left coaction of H reproducing the coproduct of H, and a left action of H which is completely reducible. So, there is a basis $(e_1, \dots, e_N)$ of V and N^2 non zero scalars q_{ij} such that: $\delta(e_i) = K_i \otimes e_i$, $K_i(e_j) = q_{ij} e_j$.

Note that the braiding in $V \otimes V$ is given by: $\sigma(e_i \otimes e_j) = q_{ij}(e_j \otimes e_i)$. The adjoint action in $S_H(M)$ is completely described by: for any x in $S_H(M)$ and any $i = 1, \cdots, N$, $ade_i(x) = e_i x - K_i(x)e_i$; so $ad(e_i)(S_\sigma(V)) \subset (S_\sigma(V))$

In order to study the structure of $S_H(M)$, the following lemma will be very useful.

Lemma 8.

a) for all $i = 1, \cdots, N$ *and* $r \in \mathbf{N}$, $e_i^r = \prod_{k=1}^r (\frac{q_{ii}^k - 1}{q_{ii} - 1}) e_i^{\otimes r}$

b) for $i \neq j$ *and* $r \in \mathbf{N}$,

$$ad(e_i)^r(e_j) = \prod_{k=1}^{r} (\frac{q_{ii}^k - 1}{q_{ii} - 1}) \prod_{k=0}^{r-1} (1 - q_{ii}^k q_{ij} q_{ji})\, e_i^{\otimes r} \otimes e_j.$$

Example. Let $A = (a_{ij})_{1 \leq i,j \leq N}$ be a symmetrizable generalized Cartan matrix, $(d_1, \ldots, d_N)$ positive integers relatively prime such that $(d_i a_{ij})$ is symmetric. Take for G either $\mathbf{Z}^N$ or $(\mathbf{Z}/l)^N$. Let $q \in \mathbf{C}$ and define $q_{ij} = q^{d_i a_{ij}}$.(For $G = (\mathbf{Z}/l)^N$, q has to be an l-th root of unity.

One gets immediately from lemma 14. that:

a) if q is a root of unity of order l, then for all $i = 1, \ldots, N$, $e_i^l = 0$.

b) assuming q generic, for $i \neq j$ there is a smallest integer r such that $ad(e_i)^r(e_j) = 0$, and it is given by: $(r-1)d_i a_{ii} + 2d_i a_{ij} = 0$, i.e. $r = 1 - a_{ij}$.

In fact, we have:

Theorem 9. *In the situation of the example above,*

1. *For* $G = (\mathbf{Z}^N)$, *if* q *is not a root of unity, then the Hopf algebra* $S_H(M)$ *is isomorphic, as a Hopf algebra, to the sub Hopf algebra* U_q^+ *of the quantized universal enveloping algebra associated with* A.
2. *For* $G = (\mathbf{Z}/l)^N$, *if* q *is a primitive* l*-th root of unity,* $S_H(M)$ *is isomorphic to the quotient of the restricted quantized enveloping algebra* u_q^+ *of Lusztig [6] by the two-sided Hopf ideal generated by the elements* $(K_i^l - 1)$, $i = 1, \ldots, N$.

Remark. The computations of Lemma 8 give a very simple way to check that the quantized Serre relations hold. In fact, for our framework, they give a little more: one sees that the coefficients q_{ij} occur only via the symmetrized expression $q_{ij}q_{ji}$. This leaves the freedom to chose the antisymmetric part. If one introduces an antisymmetric matrix (r_{ij}) and put $q_{ij} = q^{d_i a_{ij} + r_{ij}}$, $S_H(M)$ is isomorphic to the dual of the "multiparametric versions" of U_q^+ introduced by Reshetikhin [9] via twistings.

3. Iterated integrals

We now give a general procedure for constructing representations of the quantum shuffle algebra $T(V)$.

Definition 10. Suppose that $V_i, (1 \leq i \leq N)$ are N continuous functions defined on the interval $[0, 2\pi]$ with values in a topological algebra $(A, \circ)$. For each $t \in [0, 2\pi]$, and each n-tuple of indices $(i_1, \dots, i_n)$, one defines the iterated integral $I_{(i_1,\dots,i_n)}(t)$ by induction on n as:

$$I_{(i_1,\dots,i_n)}(t) = \int_0^t I_{(i_1,\dots,i_{n-1})}(s) \circ V_{i_n}(s)\, ds,$$

i.e.

$$I_{(i_1,\dots,i_n)}(t) = \int_{0<t_1<t_2<\cdots<t_n<t} V_{i_1}(t_1) \circ V_{i_2}(t_2) \circ \cdots V_{i_n}(t_n)\, dt_1 dt_2 \cdots dt_n.$$

Theorem 11. *Suppose that $V_i, (1 \leq i \leq N)$ are N continuous functions defined on the interval $[0, 2\pi]$ with values in a topological algebra $(A, \circ)$, and that they satisfy the following "exchange relations":*

for $t < t'$, $V_i(t') \circ V_j(t) = \sum R_{i,j}^{k,l} V_k(t) \circ V_l(t')$,

with $R_{i,j}^{k,l}$ being the matrix coefficients of a braiding σ on a N dimensional vector space W. Fix a basis $(e_1, \dots, e_N)$ of W in which the matrix coefficients of σ are $R_{i,j}^{k,l}$. Then, for each t in $[0, 2\pi]$, the linear map $\rho_t : T_\sigma(W) \to A$, given by

$$\rho_t((e_{i_1}, \dots, e_{i_n})) = I_{(i_1,\dots,i_n)}(t)$$

is an algebra homomorphism. In particular, assigning the integral $\int_0^t V_i(s)\, ds$ to e_i defines a representation of the quantum symmetric algebra $S_\sigma(W)$.

Proof. We first observe that we have the following inductive definition of the quantum shuffle product.

Proposition 12. *Let $x_{i_1}, \cdots, x_{i_p}, x_{j_1}, \cdots, x_{j_q}$ be in V. Then:*

$$\begin{aligned}(x_{i_1}, \dots, x_{i_p}).(x_{j_1}, \dots, x_{j_q}) &= ((x_{i_1}, \dots, x_{i_p}).(x_{j_1}, \dots, x_{j_{q-1}})) \otimes x_{j_q} + \\ &+[((x_{i_1}, \dots, x_{i_{p-1}})) \otimes 1].\tau(x_{i_p}, (x_{j_1}, \dots, x_{j_{q-1}}))\end{aligned}$$

where $\tau = \sigma_{q-1}\sigma_{q-2} \cdots \sigma_1$, and the second term of the right hand side has the following meaning: consider $\tau(x_{i_p}, (x_{j_1}, \dots, x_{j_{q-1}}))$ has an element of $V^{\otimes q} \otimes V$ and make the quantum shuffle product of the part in $V^{\otimes q}$ with $(x_{i_1}, \dots, x_{i_{p-1}})$.

As a consequence, we have a simple way to check whether we have a representation of the quantum shuffle algebra.

Corollary 13. *Suppose that V is finite dimensional, and that we are given a basis $(e_1, \dots, e_n)$ of V. Let $(A, \circ)$ be an associative algebra, and suppose that we have a linear map: $\rho : T(V) \to A$ such that for all indices $i_1, \dots, i_p, j_1, \dots, j_q$ in $1, \dots, n$,*

$$\begin{aligned}\rho(e_{i_1}, \dots, e_{i_p}) \circ \rho(e_{j_1}, \dots, e_{j_q}) &= \rho((x_{i_1}, \dots, x_{i_p}).(x_{j_1}, \dots, x_{j_{q-1}})) \otimes x_{j_q}) \\ &\quad + \rho([((x_{i_1}, \dots, x_{i_{p_1}})) \otimes 1].\tau(x_{i_p}, (x_{j_1}, \dots, x_{j_{q-1}}))\end{aligned}$$

Then ρ is an algebra homomorphism.

We can now come to the proof of the theorem. We use Corollary 13 and proceed by induction on $n = p + q$.

Observe that: $\frac{d}{dt} I_{(i_1, \dots, i_n)}(t) = I_{(i_1, \dots, i_{n_1})}(t) \circ V_{i_n}(t)$. So,
$\frac{d}{dt} (I_{(i_1, \dots, i_p)} \circ I_{(j_1, \dots, j_q)})(t) =$

$$I_{(i_1, \dots, i_{p-1})}(t) \circ V_{i_p}(t) \circ I_{(j_1, \dots, j_q)}(t) + I_{(i_1, \dots, i_p)}(t) \circ I_{(j_1, \dots, j_{q-1})}(t) \circ V_{j_q}(t)$$

We shall rewrite the right hand side, and then integrate between 0 and t to get the formula of Corollary 13.

Using induction hypothesis, one sees that the second term of the right hand side is: $\rho_t((e_{i_1}, \dots, e_{i_p}).(e_{j_1}, \dots, e_{j_{q-1}})) \circ V_{j_q}(t)$, and integrating between 0 and t gives:

$$\rho_t((e_{i_1}, \dots, e_{i_p}).(e_{j_1}, \dots, e_{j_{q-1}}) \otimes e_{j_q}).$$

In order to compute the first term, one pulls $V_{i_p}(t)$ over $I_{(i_1, \dots, i_q)}(t)$ using the exchange relations q times: one gets a linear combination of terms of the form: $I_{(k_1, \dots, k_q)}(t) \circ V_l(t)$, the coefficients of which being the matrix coefficients of $\sigma_q \dots \sigma_1$; one can then apply the induction hypothesis to $I_{(i_1, \dots, i_{p-1})}(t) \circ I_{(k_1, \dots, k_q)}(t)$ and, after integrating between 0 and t, one has a linear combination of terms $\rho_t((e_{i_1}, \dots, e_{i_{p-1}}).(e_{k_1}, \dots, e_{k_{q-1}}) \otimes e_{i_l})$ which is exactly the second term of the right hand side of the formula in Corollary 13.

4. Application to bosonic vertex operators

In order to fix notation, let us recall a little bit of the formalism.
One starts with an Heisenberg algebra, generated by $b^i_n, q_i, n \in \mathbf{Z}, 1 \leq i \leq N$, with the relations:

$$[b^i_n, b^j_m] = n(\alpha_i, \alpha_j)\delta_{n+m,0}\beta^2$$

$$[q_i, b^j_m] = (\alpha_i, \alpha_j)\delta_{m,0}\beta^2$$

where β is a complex parameter, and the data α_i, (,) are those associated with our finite dimensional complex Lie algebra $\mathcal{G}$.

Let $z = exp(it)$, and define:

$$\phi_j^+(z) = \sum_{n>0} -\frac{1}{n} b_n^j z^{-n} \qquad \phi_j^-(z) = \sum_{n<0} -\frac{1}{n} b_n^j z^{-n}.$$

Then, the vertex operators are the normal ordered exponentials:

$$V_j(z) = e^{\phi_j^-(z)}.exp(-q_j).exp(itb_0^j).e^{\phi_j^+(z)}.$$

They are operators sending the direct sum of Fock modules to the direct product of their homogeneous components.

One usually writes down "commutation formulas" for $V_j(z)$, $V_k(z')$, when z and z' are any complex numbers such that $|\frac{z}{z'}| < 1$, using the following formula:

$$[\phi_j^+(z'), \phi_k^-(z)] = (\alpha_j, \alpha_k)\beta^2 log(1 - \frac{z}{z'}).$$

This formula also makes sense for z and z' of modulus 1, say $z = exp(it)$ and $z' = exp(it')$, with $t, t' \in [0, 2\pi]$ and $t' = t + \theta$ with $\theta > 0$ if one takes the analytic continuation of $log(1-u)$ for $u \in \mathbf{C} \setminus [1, +\infty[$.

This leads immediately to:

Proposition 14. *Let $t, t' \in [0, 2\pi]$ with $t' > t$. Then one has:*

$$V_j(exp(it')).V_k(exp(it)) = exp(i\pi\beta^2(\alpha_i, \alpha_k))V_k(exp(it)).V_j(exp(it')).$$

So, we are in position of applying Theorem 11.

Theorem 15. [1, 2] *Set $q = exp(i\pi\beta^2)$; then contour integrals (between 0 and 2π) of the vertex operators $V_j(exp(it))$ satisfy the quantum Serre relations.*

References

1. P. Bouwknegt, J. McCarthy, K. Pilch, Quantum group structure in the Fock space resolutions of $\hat{sl}_n$ representations, *Commun. Math. Phys.* **131** (1990), 125-155.
2. P. Bouwknegt, J. McCarthy, K. Pilch, Free field approach to two-dimensional conformal field theory, *Progr. Theor. Phys. Suppl.* **102**, (1990), 67-135.
3. B. Feigin, E. Frenkel, Integrals of motions and quantum groups, *Lect. Notes in Math.* **1620**, (1995), 349-418.
4. V. Drinfeld: Quantum Groups, *Proceedings of the International Congress of Mathematicians, Berkeley*, (1986), 798-820
5. M. Jimbo: A q-difference analoque of $U(\mathcal{G})$ and the Yang-Baxter equation, *Lett. Math. Phys.* **10** (1986) 63-69

6. G. Lusztig: Quantum groups at a root of 1, *Geom. Dedic.* **35** (1990), 89-114
7. S. Majid: Doubles of quasitriangular Hopf algebras, *Comm. Alg.* **19** (1991) 3061-3073
8. W. Nichols: Bialgebras of type I, *Comm. Alg.* **15** (1978), 1521-1552.
9. N. Reshetikhin: Multiparameter quantum groups and twisted quasitriangular Hopf algebras, *Lett. Math. Phys.* **20** (1990) 331-335.
10. M. Rosso: Certaines formes bilineaires sur les groupes quantiques et une conjecture de Schechtman et Varchenko, *C. R. Acad. Sci. Paris (I)*, **314**, (1992), 5-8.
11. M. Rosso: Groupes quantiques et algèbres de battage quantiques, *C. R. Acad. Sci. Paris (I)*, **320**. Detailed and augmented version: Quantum groups and quantum shuffles, preprint.
12. M. Rosso: Integrals of vertex operators and quantum groups, preprint.
13. S. Woronowicz: Differential calculus on compact matrix pseudogroups (quantum groups, *Commun. Math. Phys.* **122** (1989) 125-170
14. D. Yetter: Quantum groups and representations of monoidal categories, *Math. Proc. Camb. Phil. Soc.* **108** (1990), 261-290.

CHARACTER FORMULAS AND LOCALIZATION OF INTEGRALS

WILFRIED SCHMID*
Department of Mathematics, Harvard University, Cambridge, MA 02138, USA
schmid@math.harvard.edu

Abstract. Two character formulas for representations of semisimple Lie groups are presented. The question of a direct link between the two formulas is raised. In the (known) case of a compact group, the localization formula of Berline–Vergne provides such a link. An argument in the general case might follow from a hypothetical localization formula for Hamiltonian actions of nonabelian, noncompact groups.

In these notes, I shall describe two character formulas for semisimple Lie groups. Both are of interest by themselves, but the potential connections between the two formulas raise some intriguing questions. The formulas represent joint work with Kari Vilonen; full details will appear elsewhere.

The formulas and their relation are well understood in the case of a compact group. For motivation, I shall start out with a discussion of the compact case. Thus I consider K, a connected, compact Lie group, and $T \subset K$, a maximal torus. I write $\mathfrak{k}_{\mathbb{R}}$, $\mathfrak{t}_{\mathbb{R}}$ for the Lie algebras of K, T, and $\mathfrak{k}$, $\mathfrak{t}$ for their complexified Lie algebras.

Let π be an irreducible unitary representation of K. Because of the compactness of K, π must be finite dimensional. The *Weyl character formula* describes the character Θ_π as a function of T, and thus – since the maximal torus meets every conjugacy class – globally as function on all of K. To recall the formula, I use the exponential map to identify the torus T with the quotient of its Lie algebra by the *unit lattice*,

$$T \simeq \mathfrak{t}_{\mathbb{R}}/L\,, \text{ with } L = \text{unit lattice} =_{\text{def}} \{\, x \in \mathfrak{t}_{\mathbb{R}} \mid \exp(x) = \text{id} \,\}\,. \quad \text{(1a)}$$

Dually, the character group $\widehat{T}$ is isomorphic to the *weight lattice*,

$$\widehat{T} \simeq \Lambda = \text{weight lattice} =_{\text{def}} \{\, \lambda \in i\mathfrak{t}^*_{\mathbb{R}} \mid \langle \lambda, L \rangle \subset 2\pi i \mathbb{Z} \,\}\,. \quad \text{(1b)}$$

*Partially supported by an NSF grant

D. Sternheimer et al. (eds.), Deformation Theory and Symplectic Geometry, 259–270.

Under this isomorphism, $\lambda \in \Lambda$ corresponds to $e^\lambda \in \widehat{T}$, where $e^\lambda(\exp x) = e^{\langle \lambda, x\rangle}$ for $x \in \mathfrak{t}_\mathbb{R}$. The *Weyl group* $W = N_K(T)/T$ operates on $i\mathfrak{t}^*_\mathbb{R}$. This action preserves a positive definite inner product $(\cdot\,,\,\cdot)$ on $i\mathfrak{t}^*_\mathbb{R}$. It also preserves the weight lattice and the *root system* Φ, i.e., the set of non-zero $\alpha \in \Lambda$ such that there exist $x \neq 0$ in $\mathfrak{k}$ on which T acts according to the character e^α. The Weyl character formula can be stated as follows:

$$\Theta_\pi|_T \;=\; \sum_{w\in W} \frac{e^{w\lambda}}{\prod_{\alpha\in\Phi,\,(w(\lambda+\rho),\alpha)>0}(1-e^{-\alpha})}\,. \qquad (2)$$

Here $\lambda = \lambda(\pi)$ is a particular element of the weight lattice, the *highest weight* of the representation π, and ρ a certain small vector in $i\mathfrak{t}^*_\mathbb{R}$, independent of π, which will be described presently.

The *root hyperplanes* $\{\,\mu \in i\mathfrak{t}_\mathbb{R} \mid (\alpha,\mu) = 0\,\}$, $\alpha \in \Phi$, divide $i\mathfrak{t}_\mathbb{R}$ into a number of convex cones, the *Weyl chambers*, which are permuted simply transitively by W. Customarily, one singles out one of the Weyl chambers and calls it the *dominant chamber.* When K happens to be simply connected, ρ can be characterized as the shortest element of Λ lying in the interior of the dominant Weyl chamber. In general, the definition of ρ can be reduced to the case of a simply connected group: some finite covering of any connected, compact Lie group splits into the product of a torus and a simply connected compact group. To complete the statement of the Weyl character formula, one only needs to add that an element λ of the weight lattice Λ arises as highest weight of an irreducible representation precisely when it is dominant, i.e., when it lies in the (closed) dominant Weyl chamber.

The Weyl character formula (2) is a beautiful, explicit formula – in some sense, one could hardly ask for more. Still, it has inspired attempts to explain it, or at least interpret it, in geometric terms.

The Atiyah–Bott fixed point formula [1] provides one such geometric interpretation. It depends on the Borel–Weil theorem [4, 17], whose statement I need to recall. I fix a particular dominant weight λ. The splitting $\mathfrak{k}_\mathbb{R} = \mathfrak{t}_\mathbb{R} \oplus [\mathfrak{t}_\mathbb{R}, \mathfrak{k}_\mathbb{R}]$ induces a dual splitting of the vector space $i\mathfrak{k}^*_\mathbb{R}$, which allows me to think of λ and ρ as lying in $i\mathfrak{k}^*_\mathbb{R}$, and to define

$$\Omega_{\lambda+\rho} \;=\; K\text{-orbit of } \lambda+\rho \text{ in } i\mathfrak{k}^*_\mathbb{R}\,. \qquad (3)$$

As homogeneous space, $\Omega_{\lambda+\rho}$ is isomorphic to the quotient space K/T, and thus to the *flag variety* of $\mathfrak{k}$. Since λ exponentiates to a character e^λ of the isotropy subgroup T of K at $\lambda+\rho$, it determines a K-equivariant line bundle

$$\mathcal{L}_\lambda \longrightarrow \Omega_{\lambda+\rho}\,. \qquad (4)$$

Except for the multiplication by $i = \sqrt{-1}$, $\Omega_{\lambda+\rho}$ is a coadjoint orbit of K – an orbit in the dual of the Lie algebra; as such, it carries a canonical symplectic structure. An *complex polarization* for $\Omega_{\lambda+\rho}$ is a K-invariant pseudo-Kähler structure compatible with the intrinsic symplectic structure. Complex polarizations exist; they correspond naturally and bijectively to the Weyl chambers. Once a polarization is chosen, the line bundle $\mathcal{L}_\lambda$ has a unique structure of K-equivariant *holomorphic* line bundle. The polarization corresponding to the dominant Weyl chamber is the unique *positive polarization* – positive in the sense that the pseudo-Kähler structure is positive definite, i.e., a true Kähler structure.

Equip $\Omega_{\lambda+\rho}$ with the positive polarization, and let $\mathcal{O}(\mathcal{L}_\lambda)$ denote the sheaf of holomorphic sections of the line bundle $\mathcal{L}_\lambda$. Since the action of K preserves the complex structures of the space $\Omega_{\lambda+\rho}$ and of the line bundle $\mathcal{L}_\lambda$, it induces an action also on the cohomology groups of the sheaf $\mathcal{O}(\mathcal{L}_\lambda)$. The Borel–Weil theorem describes the resulting representations of K:

$$\begin{aligned} &H^p(\Omega_{\lambda+\rho}, \mathcal{O}(\mathcal{L}_\lambda)) = 0 \ \text{ if } p \neq 0, \\ &H^0(\Omega_{\lambda+\rho}, \mathcal{O}(\mathcal{L}_\lambda)) \text{ is non-zero, irreducible, and has highest weight } \lambda, \end{aligned} \tag{5}$$

provided λ is dominant and the polarization positive, as has been assumed. Bott's generalized Borel–Weil theorem [4, 11] describes the cohomology groups attached to an arbitrary invariant polarization and every $\lambda \in \Lambda$; the precise statement need not concern us here.

The maximal torus T operates on $\Omega_{\lambda+\rho}$ with isolated fixed points, namely the various W-translates of the base point $\lambda+\rho$. The Atiyah–Bott fixed point formula, applied to the T-action on $\Omega_{\lambda+\rho}$ and the holomorphic line bundle $\mathcal{L}_\lambda$, expresses the alternating sum of the T-characters of the cohomology groups $H^p(\Omega_{\lambda+\rho}, \mathcal{O}(\mathcal{L}_\lambda))$ as a sum of terms corresponding to the fixed points; the term corresponding to any one of the fixed points $w(\lambda+\rho)$, $w \in W$, turns out to be the summand in the formula (2) indexed by w. In this way, the Weyl character formula becomes a formal consequence of the fixed point formula, as was observed already by Atiyah and Bott.

Kirillov's formula provides a totally different expression for the irreducible characters of K. By definition,

$$\theta_\pi = (\det \exp_*)^{1/2} \exp^* \Theta_\pi \tag{6}$$

is the *character on the Lie algebra* of the irreducible representation π. Except for the factor $(\det \exp_*)^{1/2}$ – which has the effect of relating invariant differential operators on the group and on the Lie algebra – this is the pull-back of the global character Θ_π to the Lie algebra. It is a smooth, bounded function, but is not square integrable. The exponential map of the connected, compact Lie group K is surjective and generically non-singular, so

θ_π completely determines Θ_π. It will be convenient to define the Fourier transform $\widehat{\phi}$ of a test function $\phi \in C_c^\infty(\mathfrak{k}_\mathbb{R})$ without the customary factor of $i = \sqrt{-1}$ in the exponent, as a function on $i\mathfrak{k}_\mathbb{R}^*$:

$$\widehat{\phi}(\zeta) \;=\; \int_{\mathfrak{k}_\mathbb{R}} \phi(x) e^{\langle \zeta, x\rangle} dx\,. \tag{7}$$

The Fourier transform of θ_π exists as distribution, in the weak sense; with the convention (7), it is given by the relation

$$\widehat{\theta}_\pi \;=\; \text{integration over } \Omega_{\lambda+\rho} \tag{8}$$

(integration relative to the measure induced by the canonical symplectic structure). Kirillov calls this the "universal formula" for irreducible characters [10]; he established it not only in the compact case, but more importantly, in the nilpotent case where he used it to describe the unitary dual.

One may picture the Weyl character formula, its interpretation in terms of fixed points, and Kirillov's formula for $\widehat{\theta}_\pi$ as the vertices of a triangle. One side of the triangle – the connection between the Weyl character formula and its fixed point interpretation – is filled in by the Atiyah–Bott fixed point formula. Kirillov's proof of (8) provides the second side, by direct computation. Berline–Vergne posed, and later answered the question of whether one can fill in the third side of the triangle: an identification of the right hand side of (8) with the sum of terms attached to the fixed points of the action of T on $\Omega_{\lambda+\rho}$, without the "detour" via the Weyl character formula.

The argument of Berline–Vergne involves the K-equivariant cohomology groups $H_K^*(\Omega_{\lambda+\rho}, \mathbb{C})$. These were defined, for any manifold with a differentiable K-action, by Cartan, who also expressed this cohomology in terms of a complex of differentiable forms,

$$H_K^*(\Omega_{\lambda+\rho}, \mathbb{C}) \;\simeq\; H^*(\{\mathcal{A}^{\cdot}(\Omega_{\lambda+\rho}) \otimes S(\mathfrak{k}^*)\}^K)\,; \tag{9a}$$

here $\mathcal{A}^{\cdot}(\Omega_{\lambda+\rho})$ denotes the deRham complex, $S(\mathfrak{k}^*)$ the symmetric algebra of $\mathfrak{k}^*$, and $\{\ldots\}^K$ the space of K-invariants, relative to the translation action on $\mathcal{A}^{\cdot}(\Omega_{\lambda+\rho})$ and the conjugation action on $S(\mathfrak{k}^*)$. The differential

$$d\,:\,\{\mathcal{A}^{\cdot}(\Omega_{\lambda+\rho}) \otimes S(\mathfrak{k}^*)\}^K \;\longrightarrow\; \{\mathcal{A}^{\cdot}(\Omega_{\lambda+\rho}) \otimes S(\mathfrak{k}^*)\}^K \tag{9b}$$

is the sum of the deRham differential and a term involving the infinitesimal action of $\mathfrak{k}$ as a Lie algebra of vector fields and the multiplication map $\mathfrak{k}^* \otimes S(\mathfrak{k}^*) \to S(\mathfrak{k}^*)$. It should be remarked that the equivariant cohomology can be defined whether or not K is compact. The description (9), on the other hand, depends crucially on the compactness of K. In the case of a one point space,

$$H_K^*(\{\mathrm{pt}\},\mathbb{C}) \simeq \{S(\mathfrak{k}^*)\}^K, \tag{10}$$

since the differential reduces to zero. The analogue of (10) for the maximal torus T simplifies because T is abelian: $H_T^*(\{\mathrm{pt}\},\mathbb{C}) \simeq S(\mathfrak{t}^*)$.

Integration of the deRham component over the space $\Omega_{\lambda+\rho}$ and restriction from K to T induce a sequence of morphisms

$$H_K^*(\Omega_{\lambda+\rho},\mathbb{C}) \xrightarrow{\int} H_K^*(\{\mathrm{pt}\},\mathbb{C}) \xrightarrow{\mathrm{res}_T^K} H_T^*(\{\mathrm{pt}\},\mathbb{C}) \simeq S(\mathfrak{t}^*). \tag{11}$$

Just as in the absolute case, it is possible to define a K-equivariant index of the line bundle $\mathcal{L}_\lambda$ and to prove an equivariant Riemann–Roch theorem. However, in the equivariant setting, the index formally involves cohomology classes in all degrees, so its image in the equivariant cohomology of a point lies in the completion of $\{S(\mathfrak{k}^*)\}^K$, i.e., in the space of K-invariant formal power series on $\mathfrak{k}$. Berline–Vergne [2] identify the image of the equivariant index in the completion of $\{S(\mathfrak{k}^*)\}^K$ with θ_π and deduce the Kirillov's formula. The crux of the matter is the process of localization – the expression of the composite of the homomorphisms (11) in terms of contributions of the various fixed points of the action of T.

Berline–Vergne localization applies to any Hamiltonian torus action, with isolated fixed points, on a compact symplectic manifold. Their ideas have been generalized by Witten [19] and Jeffrey–Kirwan [5] to the case of a Hamiltonian action by a compact, not necessarily abelian, group.

Let me turn to the noncompact case. I consider a connected, linear, semisimple Lie group G and a maximal compact subgroup $K \subset G$. Any two maximal compact subgroups are conjugate, so the particular choice of K does not matter. By a *representation* of G, I shall mean a continuous representation on a complete, locally convex Hausdorff topological vector space. A representation π is said to have *finite length* if every increasing chain of closed, invariant subspaces breaks off after finitely many steps. One calls π *admissible* if its restriction to K contains any irreducible representation of K at most finitely often. Admissibility is automatic for irreducible unitary representations. Loosely speaking, the admissible representations of finite length constitute the smallest universe of representations closed under all standard constructions, containing at least the irreducible unitary representations. They are the natural objects of study: on the one hand, admissible representations of finite length provide the most appropriate context for harmonic analysis on G and its quotient spaces; on the other, inadmissible irreducible representations are truly exotic – the first explicit example is relatively recent [18].

An admissible representation of finite length π has a character Θ_π, not in the naive sense since π will not be finite dimensional in general, but in a suitable weak sense: it is a conjugation invariant distribution on G, and the distributions $D\,\Theta_\pi$, with D ranging over the space of bi-invariant linear differential operators, span a finite dimensional space. An irreducible admissible representation π is determined by its character up to *infinitesimal equivalence*, in other words, up to isomorphism of the underlying infinitesimal representation. For irreducible unitary representations, infinitesimal equivalence forces unitary equivalence, so the character of an irreducible unitary representation π completely determines π as unitary representation. These facts are all due to Harish-Chandra, as is the *regularity theorem*. This theorem asserts that the character Θ_π of any admissible π of finite length is (integration against) a locally L^1 function, which is even real analytic on G_{rs}, the set of regular semisimple elements in G. The complement of G_{rs} is a proper real analytic subvariety, so one may think of Θ_π as a "real meromorphic function", with singularities only along the complement of G_{rs}. Every $g \in G_{\text{rs}}$ lies in a (unique) Cartan subgroup, of which there are only finitely many conjugacy classes. In particular, Θ_π is fully determined by its restriction to finitely many Cartan subgroups, one from each conjugacy class.

The Weyl character formula, its interpretation in terms of fixed points, and Kirillov's formula all have analogues in the present setting. It will simplify the discussion to restrict attention to characters Θ_π of representations with *trivial infinitesimal character*, meaning representations whose character is annihilated by all bi-invariant linear differential operators with zero constant term. All interesting phenomena show up already in this special case. Everything that will be said carries over to the case of an arbitrary admissible representation of finite length, though the statements and the proofs become somewhat more complicated.

Thus I consider the character Θ_π of a representation π with trivial infinitesimal character, a Cartan subgroup $H \subset G$, and a connected component C of $H_{\text{rs}} = H \cap G_{\text{rs}}$. Since G was assumed to be linear, it has a complexification $G_{\mathbb{C}}$, and H has a complexification $H_{\mathbb{C}}$ which is a Cartan subgroup of the complex semisimple group $G_{\mathbb{C}}$. The quotient $W_{\mathbb{C}} = N_{G_{\mathbb{C}}}(H)/H_{\mathbb{C}}$ is the *complex Weyl group* associated to H. According to results of Harish-Chandra, there exist integers $n_w = n_w(\pi, C)$ indexed by $w \in W_{\mathbb{C}}$, such that

$$\Theta_\pi|_C \;=\; \sum_{w\in W_{\mathbb{C}}} \frac{n_w}{\prod_{\alpha\in\Phi,\,(w\rho,\alpha)>0}\,(1-e^{-\alpha})}\,; \tag{12}$$

in this formula, Φ and ρ are defined analogously to the corresponding quantities in the compact case. When $G = K$ is compact, "Cartan subgroup" is synonymous with "maximal torus", the complex Weyl group coincides with the usual Weyl group, and the only irreducible representation with trivial infinitesimal character is the trivial representation which has highest weight $\lambda = 0$. Thus, in this very special case, the Harish-Chandra character formula (12) reduces to the Weyl character formula (2), with $n_w = 1$ for all $w \in W_{\mathbb{C}}$ and all connected components $C \subset H_{rs}$. In distinction to the compact case, there exists no simple, completely explicit formula for the n_w in the general.

The description of the n_w in terms of a fixed point formalism depends on a geometric realization of the representation π – in effect, a generalization of the Borel–Weil–Bott theorem. I write $\mathfrak{g}_{\mathbb{R}}$, $\mathfrak{h}_{\mathbb{R}}$, $\mathfrak{k}_{\mathbb{R}}$ for the Lie algebras of G, H, K, and $\mathfrak{g}$, $\mathfrak{h}$, $\mathfrak{k}$ for the complexified Lie algebras. The *flag variety* X of $\mathfrak{g}$ is a complex projective variety with a transitive action of $G_{\mathbb{C}}$, and is universal with respect to these properties: any other complex projective variety Y with transitive $G_{\mathbb{C}}$-action is an equivariant quotient of X. As homogeneous space, the flag variety can be identified with $G_{\mathbb{C}}/B$, the quotient of $G_{\mathbb{C}}$ by a *Borel subgroup* B, i.e., by a maximal solvable subgroup. When B contains $H_{\mathbb{C}}$, as can be arranged, the identity coset is a particular fixed points for the action of H on X, and the complex Weyl group $W_{\mathbb{C}}$ acts simply transitively on the set of all such fixed points. In the following, I shall view X as a complex manifold, rather than complex algebraic variety, and write $\mathcal{O}$ for the sheaf of holomorphic functions on X.

As subgroup of $G_{\mathbb{C}}$, G acts on X; this action has a finite number of orbits. Let $\mathcal{F}$ be a *G-equivariant sheaf* of finite dimensional complex vector spaces, for example, the direct image of a G-equivariant local system on a G-orbit $S \subset X$ via the embedding $S \hookrightarrow X$. More generally, I want to allow any finite extension of direct images of G-equivariant local systems – technically speaking, any element $\mathcal{F}$ of the *G-equivariant derived category* $D_G(X)$ defined by Bernstein and Lunts [3]. Then, for every $p \in \mathbb{Z}$,

$$\text{Ext}^p(\mathcal{F}, \mathcal{O}) \text{ carries a natural Fréchet topology and a natural continuous } G\text{-action; the resulting representation is admissible, of finite length.} \tag{13}$$

This statement was conjectured by Kashiwara [7], and was subsequently proved by Kashiwara and myself [9]. The proof of (13) also shows that every admissible, finite length representation π, with trivial infinitesimal character, is infinitesimally equivalent to the G-module $\text{Ext}^0(\mathcal{F}, \mathcal{O})$ for some $\mathcal{F} \in D_G(X)$, all of whose other Ext-groups vanish. When $G = K$ is compact, with $\mathcal{F}$ equal to the constant sheaf $\mathbb{C}_X$, (13) reduces to the Borel–Weil realization of the trivial representation.

There exists a natural notion of duality in the equivariant derived category, the *Verdier duality* operator $\mathbb{D} : D_G(X) \to D_G(X)$. For $\mathcal{F} \in D_G(X)$,

I set

$$\Theta(\mathcal{F}) \;=\; \sum_p (-1)^p \, \Theta\left(\overset{p}{\operatorname{Ext}}(\mathbb{D}\mathcal{F}, \mathcal{O})\right) ; \tag{14}$$

here $\Theta(\ldots)$ denotes the character of the admissible, finite length G-module $\ldots$; the substitution of $\mathbb{D}\mathcal{F}$ for $\mathcal{F}$ serves the purpose of making the dependence of the Ext-groups on $\mathcal{F}$ covariant. In view of the remarks about (13), every character Θ_π of a representation π with trivial infinitesimal character can be realized as $\Theta(\mathcal{F})$, for some $\mathcal{F} \in D_G(X)$.

In the Harish-Chandra character formula (12), each coefficient n_w corresponds to a fixed point for the action of H on X, as follows. I fix a dominant Weyl chamber for the pair $(\mathfrak{g}, \mathfrak{h})$, consistently with the choice of the quantity ρ in (12). The direct sum of the root space corresponding to the negative roots constitutes a maximal nilpotent subalgebra $\mathfrak{n} \subset \mathfrak{g}$, and $B = N_{G_\mathbb{C}}(\mathfrak{n})$ is a particular Borel subgroup which contains $H_\mathbb{C}$. I identify $X \simeq G_\mathbb{C}/B$ as before, and set $x_w = w$-translate of the identity coset; then $\{\, x_w \mid w \in W_\mathbb{C} \,\}$ is precisely the set of fixed points for H acting on X. As in (12), I let C denote a connected component of H_{rs}. For $h \in C$, the action of h on the tangent space $T_{x_w}X$ at the fixed point x_w splits that space into the direct sum of, respectively, the contracting and the non-contracting directions, and that splitting depends only on the connected component C, not on h itself. Using the structure theory of $G_\mathbb{C}$, one can construct an $H_\mathbb{C}$-invariant, locally closed submanifold $N(x_w, C) \subset X$ passing through the point x_w, whose tangent space at x_w contains precisely the non-contracting directions. The action of H on the triple $(X, N(x_w, C), \{x_w\})$ and the G-equivariant structure of $\mathcal{F}$ induce a linear action of H on $H^*_{\{x_w\}}(N(x_w, C), \mathcal{F})$, the local cohomology of the sheaf $\mathcal{F}$ restricted to the subspace $N(x_w, C)$ relative to the one point subspace $\{x_w\}$.

Kashiwara [6] conjectured that

$$\Theta(\mathcal{F})(h) \;=\; \sum_{w \in W_\mathbb{C}} \frac{L\left(h^{-1}, H^p_{\{x_w\}}(N(x_w, C), \mathcal{F})\right)}{\prod_{\alpha \in \Phi,\, (w\rho, \alpha) > 0} \left(1 - e^{-\alpha}(h)\right)} \tag{15a}$$

for all $h \in C$, where

$$L\left(h^{-1}, H^*(\ldots)\right) \;=\; \textstyle\sum_p (-1)^p \operatorname{tr}\left(h^{-1} : H^p(\ldots) \to H^p(\ldots)\right) \tag{15b}$$

refers to the Lefschetz number of h^{-1} acting on $H^*(\ldots)$. This Lefschetz number is a homotopy invariant, hence constant for h in the connected component C. In particular, (15a) expresses the constants n_w in the formula (12) for the virtual character $\Theta(\mathcal{F})$ as local Lefschetz numbers at the fixed points x_w. A proof of the fixed point formula (15) is sketched in [14], with details to appear in [16].

As illustration of the fixed point formula, let me consider the simplest case, of a compact group $G = K$ and the constant sheaf $\mathbb{C}_X$, which corresponds to the trivial representation π. The Cartan subgroup H is then $H = T$, a maximal torus, whose action at the fixed points cannot have any strictly expanding or contracting directions. In this case, then, the noncontracting manifold $N(x_w, C)$ is all of X. The local cohomology of the constant sheaf relative to a one-point subspace is one dimensional and concentrated in degree equal to the dimension of the ambient space, i.e., in degree $2\dim_{\mathbb{C}} X$. The identity in $H = T$ must act trivially, of course, but the trace of the action is locally constant as function on the torus T, which is connected. The Lefschetz numbers in (15a) are therefore all equal to 1, as they must be for the trivial representation.

Kirillov's "universal formula" fails in the noncompact semisimple case, as can be seen already for $G = SL(2,\mathbb{R})$ – there are simply not enough coadjoint orbits! A remedy for the failure of Kirillov's formula was suggested by Rossmann [13]. For any admissible representation π of finite length, define θ_π, the character on the Lie algebra, by the formula (6) and normalize the Fourier transform $\widehat{\theta}_\pi$ as in (7). I continue with the simplifying assumption that π has trivial infinitesimal character. The complex coadjoint orbit

$$\Omega_{\mathbb{C}} \;=\; G_{\mathbb{C}}\text{-orbit of } \rho \text{ in } \mathfrak{g}^* \tag{16}$$

intersects $i\mathfrak{g}_{\mathbb{R}}^*$ in a finite number of coadjoint orbits for G. Rossmann's idea is to express $\widehat{\theta}_\pi$ as an integral not necessarily over one of these G-coadjoint orbits, but as an integral over a more general type of middle dimensional cycle in the complex coadjoint orbit $\Omega_{\mathbb{C}}$. If – contrary to the standing assumption – π were allowed to have arbitrary infinitesimal character, one would have to work with arbitrary complex coadjoint orbits, rather than the single orbit (16).

To describe the idea concretely, I let $U \subset G_{\mathbb{C}}$ denote the (unique) compact real form which contains K and is invariant under complex conjugation with respect to the real form $G \subset G_{\mathbb{C}}$. Its Lie algebra $\mathfrak{u}_{\mathbb{R}}$ is then a real form in $\mathfrak{g}$. Counting dimensions, one finds that U acts transitively on X; in fact,

$$X \;\simeq\; U\text{-orbit of } \rho \text{ in } i\mathfrak{u}_{\mathbb{R}}^* \;=\; \Omega_{\mathbb{C}} \cap i\mathfrak{u}_{\mathbb{R}}^*, \tag{17}$$

as homogeneous space for U. Thus X can be identified with the coadjoint orbit for U passing through ρ. As such, it carries a distinguished symplectic form, which makes the U-action on X Hamiltonian. The resulting moment map

$$m_U : X \longrightarrow i\mathfrak{u}_{\mathbb{R}}^* \subset \mathfrak{g}^* \tag{18}$$

realizes the isomorphism (17); it is U-equivariant and real algebraic. On the other hand, the complex algebraic action of $G_{\mathbb{C}}$ on the flag variety X

induces a complex algebraic, Hamiltonian action on the cotangent bundle T^*X – Hamiltonian with respect to the intrinsic complex algebraic, non-degenerate 2-form which T^*X carries as cotangent bundle of an algebraic manifold. By construction, its moment map

$$m : T^*X \longrightarrow \mathfrak{g}^* \tag{19}$$

is $G_{\mathbb{C}}$-equivariant and complex algebraic. Let $p : T^*X \to X$ denote the natural projection. Rossmann calls

$$m_\rho =_{\text{def}} m + m_U \circ p \; : \; T^*X \longrightarrow \mathfrak{g}^* \tag{20a}$$

the *twisted moment map* and shows that

$$m_\rho : T^*X \xrightarrow{\sim} \Omega_{\mathbb{C}}, \tag{20b}$$

real algebraically and U-equivariantly.

Recall that G acts on X with only finitely many orbits, which stratify X real algebraically. Thus

$$T^*_G X =_{\text{def}} \text{union of the conormal bundles of the } G\text{-orbits} \tag{21}$$

is a real algebraic, Lagrangian subvariety of T^*X; in particular, it has dimension $2n$, where $n = \dim_{\mathbb{C}} X$. By a "$2n$-cycle in $T^*_G X$ with possibly infinite support", I mean an element of $H^{\text{inf}}_{2n}(T^*_G X, \mathbb{C})$, the *Borel–Moore homology* of $T^*_G X$ in degree $2n$; this is the homology theory built on cycles that are allowed to run off to infinity. Since $2n$ is the dimension of $T^*_G X$, any $c \in H^{\text{inf}}_{2n}(T^*_G X, \mathbb{C})$ can be regarded as a true geometric cycle, rather than a homology class. I can now state Rossmann's result: for each admissible π of finite length, subject to our standing assumption of trivial infinitesimal character, there exists a unique $2n$-cycle $c \in H^{\text{inf}}_{2n}(T^*_G X, \mathbb{C})$, such that

$$\int_{\mathfrak{g}_{\mathbb{R}}} \theta_\pi \phi \, dx = \int_{m_\rho(c)} \widehat{\phi}\, \sigma^n \qquad \text{for any } \phi \in C^\infty_c(\mathfrak{g}_{\mathbb{R}}) \, ; \tag{22}$$

here σ refers to the intrinsic holomorphic symplectic form on the complex coadjoint orbit $\Omega_{\mathbb{C}}$, and the Fourier transform $\widehat{\phi}$ of the test function ϕ, normalized as in (7), is regarded as a holomorphic function on $\mathfrak{g}^*$. According to the Paley–Wiener theorem, $\widehat{\phi}$ decays rapidly in the imaginary directions, and that makes the integral on the right in (22) converge. Loosely paraphrased, (22) expresses the Fourier transform $\widehat{\theta}_\pi$ of the distribution θ_π as integration over the cycle $m_\rho(c)$ in $\Omega_{\mathbb{C}}$.

Rossmann's argument does not specify the cycle c, except in the special case of a complex group. The linear span of the various θ_π lies in the space of *invariant eigendistributions* with trivial infinitesimal character, i.e., conjugation invariant distributions on $\mathfrak{g}_{\mathbb{R}}$ that are annihilated by all conjugation invariant, constant coefficient linear differential operators with zero constant term. Whatever the cycle c, the right hand side of (22) defines such an invariant eigendistribution, and the resulting map from $H_{2n}^{\inf}(T_G^*X, \mathbb{C})$ to the space of invariant eigendistributions is injective. The crux of Rossmann's proof of (22) is the equality of the dimensions of the two vector spaces, which he establishes by calculating each and comparing them.

To make the character formula (22) explicit, it is necessary to describe the cycle c in terms of the representation π. This was done in the announcement [15], with details to appear in [16]. The description of the cycle c involves first of all the geometric realization of the character as $\theta_\pi = \theta(\mathcal{F})$ for some $\mathcal{F} \in D_G(X)$, as in (13-14), and secondly, Kashiwara's notion of the *characteristic cycle* of a constructible sheaf: to each such sheaf $\mathcal{F}$ on a real analytic manifold, Kashiwara associates $\mathrm{CC}(\mathcal{F})$, a Lagrangian, $\mathbb{R}^+$-conical cycle in the cotangent bundle; details of the construction can be found in [8]. In the case at hand, CC defines a $\mathbb{Z}$-linear map

$$\mathrm{CC} : D_G(X) \longrightarrow H_{2n}^{\inf}(T_G^*X, \mathbb{Z}). \tag{23}$$

With these ingredients, the Fourier transform of $\theta(\mathcal{F})$ is given by the formula

$$\int_{\mathfrak{g}_{\mathbb{R}}} \theta(\mathcal{F})\phi\, dx \;=\; \frac{1}{(2\pi)^n\, n!}\int_{m_\rho(\mathrm{CC}(\mathcal{F}))} \widehat{\phi}\;\sigma^n \qquad (\,\phi \in C_c^\infty(\mathfrak{g}_{\mathbb{R}})\,). \tag{24}$$

In what sense can (24) be regarded as an extension of Kirillov's formula to the noncompact semisimple case? The integrand $\widehat{\phi}\;\sigma^n$ is closed; thus, if we were dealing with ordinary (finite) cycles, the integral would remain unchanged if the cycle $m_\rho(\mathrm{CC}(\mathcal{F}))$ were replaced by any homologous cycle in $\Omega_{\mathbb{C}}$. Even without the finiteness condition, one can still deform the cycle without changing the value of the integral, provided the notion of homology is taken in a restricted sense, taking into account the growth of chains at infinity. The characteristic cycles associated to so-called *tempered representations* turn out to be homologous, in this restricted sense, to G-coadjoint orbits. Thus (24) reduces to Kirillov's formula for those representations of G for which it is known to hold [12], namely the tempered representations.

Via the process of going to the Lie algebra and taking the Fourier transform, the integral formula (24) describes the same object as the fixed point formula. The equivalence of the two expressions can be stated in terms of the sheaves $\mathcal{F}$ alone, without any reference to their representation-theoretic significance. The existing proof, on the other hand, depends totally on representation theory. It is natural to ask whether this equivalence can be seen directly, just as in the compact case. A hypothetical argument would most likely involve a version of localization of integrals, in the manner of Berline–Vergne, Witten, and Jeffrey–Kirwan, but for actions of noncompact groups. The technical problems of such an approach appear formidable!

References

1. Atiyah, M. F. and R. Bott, A Lefschetz fixed point formula for elliptic complexes I, II. *Ann. of Math.* **86** (1967), 374-407; **88** (1968), 451–491.
2. Berline, N. and M. Vergne, Classes characteristiques equivariantes. Formule de localisation en cohomologie equivariante. *C. R. Acad. Sci. Paris* **295** (1982), 539–541.
3. Bernstein, J. and V. Lunts *Equivariant sheaves and functors.* Lecture Notes in Mathematics, **1578**. Springer, Berlin-Heidelberg-New York, 1994.
4. Bott, R., Homogeneous vector bundles. *Ann. of Math.* **66** (1957), 203–248.
5. Jeffrey, L. C. and F. C. Kirwan, Localization for nonabelian group actions. *Topology* **34** (1995), 291–327.
6. Kashiwara, M., Character, character cycle, fixed point theorem, and group representations. *Advanced Studies in Pure Mathematics*, vol. 14. Kinokuniya, Tokyo, 1988, pp. 369–378.
7. Kashiwara, M., *Open problems in group representation theory.* Proceedings of Taniguchi symposium held in 1986, RIMS preprint 569, Kyoto University, 1987.
8. Kashiwara, M. and P. Schapira *Sheaves on manifolds* Springer, Berlin-Heidelberg-New York, 1990.
9. Kashiwara, M., and W. Schmid, Quasi-equivariant $\mathcal{D}$-modules, equivariant derived category, and representations of reductive Lie groups. *Lie Theory and Geometry, in Honor of Bertram Kostant.* Progress in Mathematics, vol. 123. Birkhäuser, Boston, 1994, pp. 457–488.
10. Kirillov, A. A. *Elements of the Theory of Representations.* Springer, Berlin-Heidelberg-New York, 1976.
11. Kostant, B., Lie algebra cohomology and the generalized Borel-Weil theorem. *Ann. of Math.* **74** (1961), 329–387.
12. Rossmann, W., Kirillov's Character Formula for Reductive Lie Groups. *Inventiones Math.* **48** (1978), 207–220.
13. Rossmann, W., Invariant Eigendistributions on a Semisimple Lie Algebra and Homology Classes on the Conormal Variety I, II. *Journal of Functional Analysis* **96** (1991), 130–193.
14. Schmid, W. and K. Vilonen, Characters, Fixed Points and Osborne's Conjecture. *Contemp. Math.* **145** (1993), 287–303.
15. Schmid, W. and K. Vilonen, Weyl Group Actions on Lagrangian Cycles and Rossmann's formula. *Noncompact Lie Groups and Some of Their Applications.* Kluwer, Dordrecht-Boston-London,1994, pp. 243–250.
16. Schmid, W. and K.Vilonen, Two geometric character formulas for semisimple Lie groups. To appear.
17. Serre, J.-P., Représentations linéaires et espaces homogènes Kählériennes des groupes de Lie compacts. *Séminaire Bourbaki, 6ᵉ année, 1953/54, Exposé 100.* Inst. Henri Poincaré, Paris, 1954. Reprinted with corrections, 1965.
18. Soergel, W., An irreducible not admissible Banach representation of $SL(2,\mathbf{R})$. *Proc. Amer. Math. Soc.* **104** (1988), 1322–1324.
19. Witten, E., Two-dimensional gauge theories revisited. *Jour. Geom. Phys.* **9** (1992), 303–368.

DEFORMATION THEORY AND THE BATALIN–VILKOVISKY MASTER EQUATION

JIM STASHEFF*
Department of Mathematics, University of North Carolina
Chapel Hill, NC 27599-3250, USA
jds@math.unc.edu

Abstract. The Batalin-Vilkovisky master equations, both classical and quantum, are precisely the integrability equations for deformations of algebras and differential algebras respectively. This is not a coincidence; the Batalin-Vilkovisky approach is here translated into the language of deformation theory.

The following exposition is based in large part on work by Marc Henneaux (Bruxelles) especially and with Glenn Barnich (Penn State and Bruxelles) and Tom Lada and Ron Fulp of NCSU (The Non-Commutative State University). The first statement of the relevance of deformation theory to the construction of interactive Lagrangians, that I am aware of, is due to Barnich and Henneaux [3]:

> We point out that this problem can be economically reformulated as a deformation problem in the sense of deformation theory [13], namely that of deforming consistently the master equation.

The 'ghosts' introduced by Fade'ev and Popov [12] were soon incorporated into the BRST-cohomology approach [7] to a variety of problems in mathematical physics. There they were reinterpreted by Stora [25] and others in terms of the Maurer-Cartan forms in the case of a finite dimensional Lie group and more generally as generators of the Chevalley-Eilenberg complex [10] for Lie algebra cohomology. This led eventually to the Batalin-Vilkovisky approach [4, 5, 6] to quantizing particle Lagrangians and then to string field theory, both classical and quantum [30]. With hindsight, the Batalin-Vilkovisky machinery can be recognized as that of homological algebra [17]. The 'quantum' Batalin-Vilkovisky master equation has the form

* Research supported in part by NSF grant DMS-9504871.

D. Sternheimer et al. (eds.), Deformation Theory and Symplectic Geometry, 271–284.

of the Maurer-Cartan equation for a flat connection, while the 'classical' version has the form of the integrability equation of deformation theory. In the context of the present conference, my goal is to show that these are more than analogies; the master equations are indeed the integrability equations of the deformation theory of, respectively, differential graded commutative algebras and graded commutative algebras.

First I will review the jet bundle approach to Lagrangian field theory. Here we could already encounter cohomological physics in the form of the variational bicomplex of differential forms on the jet bundle, but I will omit that today and proceed instead to the anti-field, anti-bracket formalism, the rubric under which physicists reinvented homological algebra. Here the 'standard construction' is the Batalin-Vilkovisky complex.

Just as the Maurer-Cartan equation makes sense in the context of Lie algebra cohomology, so the Batalin-Vilkovisky master equation has an interpretation in terms of strong homotopy Lie algebras (L_∞-algebras), as I explain next. After a brief recollection of deformation theory for differential graded algebras, I will look at various physical examples where free Lagrangians are deformed to interactive Lagrangians. Particularly interesting examples are provided by Zwiebach's closed string field theory [30] and higher spin particles [8].

1. The jet bundle setting for lagrangian field theory

Let us begin with a space Φ of **fields** regarded as the space of sections of some bundle $\pi : E \to M$. For expository and coordinate computational purposes, I will assume E is a trivial vector bundle and will write a typical field as $\phi = (\phi^1, \ldots, \phi^k) : M \to \mathbb{R}^k$. In terms of local coordinates, we start with a trivial vector bundle $E = F \times M \to M$ with base manifold M, locally R^n, with coordinates $x^i, i = 1, \ldots, n$ and fibre R^k with coordinates $u^a, a = 1, \ldots, k$. We 'prolong' this bundle to create the associated jet bundle $J = J^\infty E \to E \to M$ which is an infinite dimensional vector bundle with coordinates u^a_I where $I = i_1 \ldots i_r$ is a symmetric multi-index (including, for $r = 0$, the empty set of indices, meaning just u^a). The notation is chosen to bring to mind the mixed partial derivatives of order r. Indeed, a section of J is the (infinite) jet $j^\infty\phi$ of a section ϕ of E if, for all r, we have $\partial_{i_1}\partial_{i_2}...\partial_{i_r}\phi^a = u^a_I \circ j^\infty\phi$ where $\phi^a = u^a \circ \phi$ and $\partial_i = \partial/\partial x^i$.

Definition 1.1 A **local function** $L(x, u^{(p)})$ is a smooth function in the coordinates x^i and the coordinates u^a_I, where the order $|I| = r$ of the multi-index I is less than or equal to some integer p.

Thus a local function is in fact the pullback of a smooth function on some finite jet bundle J^pE, i.e. a composite $J \to J^pE \to R$.

The **space of local functions** will be denoted $\mathcal{C}(\mathcal{J})$.

Definition 1.2 A **local functional**

$$\mathcal{L}[\phi] = \int_M L(x, \phi^{(p)}(x)) dvol_M = \int_M (j^\infty \phi)^* L(x, u^{(p)}) dvol_M \tag{1.1}$$

is the integral over M of a local function evaluated for sections ϕ of E. (Of course, we must restrict M and ϕ or both for this to make sense.)

The variational approach is to seek the critical points of such a local functional. More precisely, we seek sections ϕ such that $\delta\mathcal{L}[\phi] = 0$ where δ denotes the variational derivative corresponding to an 'infinitesimal' variation: $\phi \mapsto \phi + \delta\phi$. The condition $\delta\mathcal{L}[\phi] = 0$ is equivalent to the Euler-Lagrange equations on the corresponding local function L as follows: Let

$$D_i = \frac{\partial}{\partial x^i} + u^a_{Ii} \frac{\partial}{\partial u^a_I} \tag{1.2}$$

be the total derivative acting on local functions and

$$E_a = (-D)_I \frac{\partial}{\partial u^a_I} \tag{1.3}$$

the Euler-Lagrange derivative. The notation $(-D)_I$ means $(-1)^r D_{i_1} \cdots D_{i_r}$. The Euler-Lagrange equations are then

$$E_a(L) = 0. \tag{1.4}$$

Two local functionals $\mathcal{L}$ and $\mathcal{K}$ are equivalent if and only if the Euler-Lagrange derivatives of their integrands agree, $E_a(L) = E_a(K)$, for all $a = 1, \ldots, k$. The kernel of the Euler-Lagrange derivatives is given by total divergences,

$$E_a(L) = 0, \text{ for all } a = 1, \ldots, k \iff L = (-1)^k D_i j^i \tag{1.5}$$

for some local functions j^i. Equivalently,

$$E_a(L) = 0, \text{ for all } a \iff L(x, u^{(p)}) dvol_M = (-1)^{i-1} dj^i dvol_M / dx^i \tag{1.6}$$

where if $dvol_M = dx^1 \cdots dx^n$, then $dvol_M/dx^i = dx^1 \cdots dx^{i-1} dx^{i+1} \cdots dx^n$.

Since $\mathcal{L}$ is the integral of an n-form on J, it is not surprising that this all makes sense in the deRham complex $\Omega^*(J)$, which remarkably splits as a bicomplex (though the finite level complexes $\Omega^*(J^pE)$ do not). The appropriate 1-forms in the fibre directions are not the du^a_I but rather the **contact forms** $\theta^a_I = du^a_I - u^a_{Ii} dx^i$. The total differential d splits as $d = d_H + d_V$ where $d_H = dx^i \partial/\partial x^i$ is the usual exterior differential on M pulled up to J. We will henceforth restrict the coefficients of our forms to be local functions, although we will not decorate $\Omega^*(J)$ to show this.

The Euler-Lagrange operators assemble into an operator on forms:

$$E(Ldvol_M) = E_a(L)\theta^a dvol_M. \tag{1.7}$$

A Lagrangian $\mathcal{L}$ determines a **stationary surface** or **solution surface** or **shell** $\Sigma \subset J^\infty$ such that ϕ is a solution of the variational problem (equivalently, the Euler-Lagrange equations) if $j^\infty\phi$ has its image in Σ. The corresponding algebra is the **stationary ideal** $\mathcal{I}$ of local functions which vanish 'on shell', i.e. when restricted to the solution surface Σ.

The Euler-Lagrange equations generate $\mathcal{I}$ as a differential ideal, but this means we may have not only **Noether identities**

$$r^a_\alpha E_a(L) = 0 \tag{1.8}$$

but also

$$r^{aI}_\alpha D_I E_a(L) = 0. \tag{1.9}$$

Of course we have 'trivial' identities of the form

$$D_J E_b(L)\mu^{bJaI}_\alpha D_I E_a(L) = 0, \tag{1.10}$$

since we are dealing with a commutative algebra of functions. One can show [17] that all Noether identities, where the coefficients r^a_α vanish on shell, are of the above form. We now assume we have a set of indices $\{\alpha\}$ such that the above identities generate all the non-trivial relations in $\mathcal{I}$. According to Noether [22], each such identity corresponds to an **infinitesimal gauge symmetry**, i.e. an infinitesimal variation that preserves the space of solutions or, equivalently, a vector field tangent to Σ. For each Noether identity indexed by α, we denote the corresponding vector field by δ_α. We denote by Ξ, the **space of gauge symmetries**, considered as a vector space but also as a module over $C(\mathcal{J})$. We can regard δ_α as a (constant) vector field on the space of fields Φ and hence δ as a linear map

$$\delta : \Xi \to Vect\Phi.$$

Since the bracket of two such vector fields $[\delta_\alpha, \delta_\beta]$ is again a gauge symmetry, it agrees with something in the image of δ when acting on solutions. If we denote that something as $[\alpha, \beta]$, one says this bracket 'closes on shell'. It is not in general a Lie bracket, since the Jacobi identity may hold only 'on shell'. (Later I will address the issue that $[\delta_\alpha, \delta_\beta]$ may not be constant on Φ.)

To make this more explicit, write

$$[\delta_\alpha, \delta_\beta] = \delta_{[\alpha,\beta]} + \nu^a_{\alpha\beta}\frac{\delta L}{\delta u^a}. \tag{1.11}$$

The possible failure of the Jacobi identity results from those last terms which vanish only on shell and the fact that we are working in a module over $\mathcal{C}(\mathcal{J})$. (For example, we have structure *functions* rather than structure constants in terms of our generators.)

All of this, including these latter subtleties, are incorporated into a remarkable complex by Batalin-Vilkovisky , which we shall describe below.

2. Deformation theory

A deformation theoretic approach to producing interactive Lagrangians is to start with a 'free' Lagrangian L_0 on a space of fields Φ with an abelian Lie algebra Ξ represented as gauge symmetries by a Lie map

$$\delta_0 : \Xi \to Vect\Phi,$$

and then study formal deformations

$$L = \Sigma t^i L_i, \quad \delta = \Sigma t^i \delta_i,$$

keeping Φ and Ξ fixed as vector spaces, such that $\delta_\xi \int L dvol_M = 0$ for all $\xi \in \Xi$.

Classical deformation theory emphasizes the following two problems:

• 1. Consider a candidate **infinitesimal** L_1 and see if there exists a full formal deformation L with corresponding δ. Such candidate terms L_1 are often suggested by physical descriptions of elementary interactions.

• 2. Classify all formal deformations up to appropriate equivalence.

A third subsequent problem, that of convergence of the power series involved, is a problem in analysis; cohomological techniques apply to the formal algebraic theory - I don't do estimates.

The cohomological approach to deformation theory, as initiated by Gerstenhaber [13], situates the problem in an appropriate complex, the Hochschild cochain complex $C^*(A, A)$ in the case of an associative algebra A. For the Lagrangian problem, the complex is due to Batalin and Vilkovisky [4, 5, 6] using anti-field and ghost technology and the anti-bracket of Zinn-Justin [29].

3. Anti-fields, (anti-)ghosts and the anti-bracket

Let me take you 'through the looking glass' and present a 'bi-lingual' (math and physics) dictionary.

From here on, we will talk in terms of algebra extensions of $C^\infty(E)$ and $\mathcal{C}(\mathcal{J})$, but the extensions will all be free graded commutative. We could instead talk in terms of an extension of E or J as a super-manifold, the new generators being thought of as (super)-coordinates.

We first extend $C^\infty(E)$ by adjoining generators of various degrees to form a free graded commutative algebra $\mathcal{A}_\infty$ over $C^\infty(E)$, that is, even graded generators give rise to a polynomial algebra and odd graded generators give rise to a Grassmann (= exterior) algebra. The generators (and their products) are, in fact, bigraded (p,q); the graded commutativity is with respect to the total degree $p-q$.

For each variable u^a, adjoin an anti-field u_a^* and for each r_α, adjoin a corresponding ghost C^α and a corresponding anti-ghost C_α^*. Here is a table showing the corresponding math terms and the bidegrees.

Physics Term	Math Term	Ghost Degree	Anti-ghost Degree	Total Degree
field	section	0	0	0
anti-field	Koszul generator	0	1	-1
ghost	Cartan-Eilenberg generator	1	0	1
anti-ghost	Tate generator	0	2	-2

Note that the anti-field coordinates depend on E but the ghosts and anti-ghosts depend also on the specific Lagrangian.

This algebra in turn can be given an **anti-bracket** (,) of degree -1 which, remarkably, combines with the product we began with to produce precisely an 'up-to-homotopy' analog of a Gerstenhaber algebra [21, 18], though this was not recognized until quite recently.

Definition 3.1 A **Gerstenhaber algebra** is a graded commutative and associative algebra A together with a bracket $[\cdot,\cdot] : A \otimes A \to A$ of degree -1, such that for all homogeneous elements x, y, and z in A,

$$[x,y] := -(-1)^{(|x|-1)(|y|-1)}[y,x],$$

$$[x,[y,z]] = [[x,y],z] + (-1)^{(|x|-1)(|y|-1)}[y,[x,z]],$$

and

$$[x,yz] = [x,y]z + (-1)^{(|x|-1)|y|}y[x,z].$$

In the field-anti-field formalism (without jet coordinates), the anti-bracket looks like

$$(\phi^a(x), \phi_b^*(y)) = \delta_b^a \delta(x-y) \tag{3.1}$$

where $\delta(x-y)$ is the Dirac delta 'function' (distribution). In the corresponding jet bundle formalism, the anti-bracket is defined on generators and then extended to polynomials by applying the graded Leibnitz identity so that $(\psi,\ )$ is a graded derivation for any ψ in this algebra. The only non-zero anti-brackets of generators are

$$(u^a, u_b^*) = \delta_b^a \quad \text{and} \quad (C^\alpha, C_\beta^*) = \delta_\beta^\alpha.$$

Now we further extend $\mathcal{C}(\mathcal{J})$ with corresponding jet coordinates u_a^{I*}, C_I^α and C_α^{I*} with the corresponding pairings giving the extended anti-bracket. The resulting Batalin-Vilkovisky algebra we denote $\mathcal{A}$.

Define an operator s_0 of degree -1 on $\mathcal{A}$ as $(L,\ )$.

We call the antifields Koszul generators because

$$s_0 u_a^* = \frac{\delta L}{\delta u^a} \tag{3.2}$$

as in the Koszul complex for the ideal, so that $H^{0,0} \subset \Phi$ is given by $\frac{\delta L}{\delta u^a} = 0$, but

$$s_0(r_\alpha^a u_a^*) = r_\alpha^a \frac{\delta L}{\delta u^a} = 0, \tag{3.3}$$

so that $H^{0,1} \neq 0$.

Now consider the extended Lagrangian

$$L_1 = L_0 + u_a^* r_\alpha^a C^\alpha \tag{3.4}$$

and $s_1 = (L_1,\ )$, so that

$$s_1 C_\alpha^* = u_a^* r_\alpha^a \tag{3.5}$$

and $H^{0,1}$ is now 0, as in Tate's extension of the Koszul complex of the ideal to produce a resolution [27]. That is why we refer to the anti-ghosts as Tate generators. (If needed, Tate tells us to add further generators in bidegree $(0, q)$ for $q > 2$ so that $H^{0,q} = 0$ for $q > 0$.)

Further extend L_1 to

$$L_2 = L_1 + C_*^\alpha c_{\beta\gamma}^\alpha C^\beta C^\gamma, \tag{3.6}$$

so that

$$s_2 C^\alpha = c_{\beta\gamma}^\alpha C^\beta C^\gamma$$
$$s_2 u^a = r_\alpha^a C^\alpha$$

which is how the Chevalley-Eilenberg coboundary looks in terms of bases for a Lie algebra and a module and corresponding structure constants. However, we may not have $(s_2)^2 = 0$ since r_α^a and $c_{\beta\gamma}^\alpha$ are functions. Batalin and Vilkovisky prove that all is not lost. First, they add to L_2 a term involving the functions $\nu_{\alpha\beta}^a$.

Theorem 3.1 *L_2 can be further extended by terms of higher degree in the anti-ghosts to L_∞ so that $(L_\infty, L_\infty) = 0$ and hence the corresponding s_∞ will have square zero.*

With hindsight, we can see that the existence of these terms of higher order is guaranteed because the antifields and antighosts provide a resolution of the stationary ideal.

We refer to this complex $(\mathcal{A}, \int_\infty)$ as the **Batalin-Vilkovisky complex**.

What is the significance of $(s_\infty)^2 = 0$ in our Lagrangian context, or, equivalently, of the **Master Equation** $(L_\infty,\ L) = 0$? There are three answers: in higher homotopy algebra, in deformation theory and in mathematical physics. It is the deformation theory that provides the transition between the other two.

4. The Master Equation and higher homotopy algebra

If we expand $s = s_0 + s_1 + \ldots$ where the subscript indicates the change in the ghost degree p, the individual s_i do not correspond to $(x,\ )$ for any term x in L_∞ but do have the following graphical description:

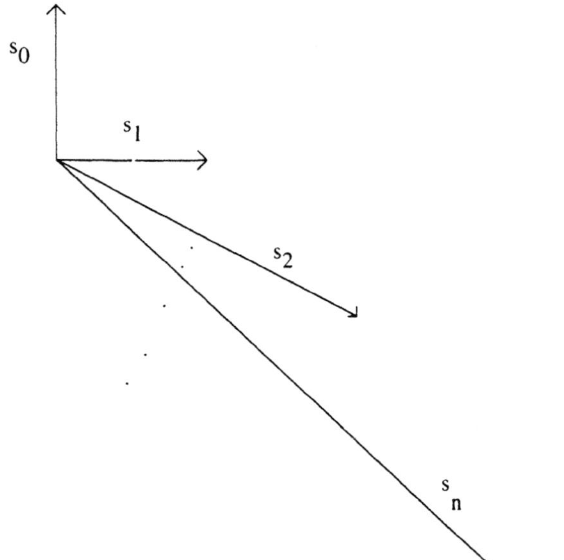

so that we see the B-V-complex as a multi-complex. The differential s_1 gives us the Koszul-Tate differential d_{KT} and part of s_2 looks like that of Chevalley-Eilenberg. That is, $C^*_\alpha c^\alpha_{\beta\gamma} C^\beta C^\gamma$ describes the (not-quite-Lie) bracket on Ξ. Further terms with one anti-ghost C^*_α and three ghosts $C^\beta C^\gamma C^\delta$ describe a tri-linear $[\ ,\ ,\ ]$ and so on for multi-brackets of possibly arbitrary length. Moreover, the graded commutativity of the underlying algebra of the B-V-complex implies appropriate symmetry of these multi-brackets. The condition that $s_\infty^2 = 0$ translates to the following identities, which are the defining identities for a **strong homotopy Lie algebra** or L_∞ **algebra**.

$$d[v_1, \dots, v_n] + \sum_{i=1}^{n} \epsilon(i)[v_1, \dots, dv_i, \dots, v_n]$$
$$= \sum_{k+l=n+1} \sum_{\text{unshuffles } \sigma} \epsilon(\sigma)[[v_{i_1}, \dots, v_{i_k}], v_{j_1}, \dots, v_{j_l}], \tag{4.1}$$

where $\epsilon(i) = (-1)^{|v_1|+\dots+|v_{i-1}|}$ is the sign picked up by taking d through $v_1, \dots, v_{i-1}$ and, for the unshuffle $\sigma : \{1, 2, \dots, n\} \mapsto \{i_1, \dots, i_k, j_1, \dots, j_l\}$, the sign $\epsilon(\sigma)$ is the sign picked up by the elements v_i passing through the v_j's during the unshuffle of $v_1, \dots, v_n$, as usual in superalgebra.

Remark 4.1 Here we follow the physics grading and sign conventions in our definition of a strong homotopy Lie algebra [28, 30]. These are equivalent to but different from those in the existing mathematics literature, c.f. Lada and Stasheff [20], in which the n-ary bracket has degree $2-n$. (The correspondence and additional insights are presented in full detail in Kjeseth's dissertation [19].) With those mathematical conventions, L_∞−algebras occur naturally as deformations of Lie algebras. If L is a Lie algebra and V is a complex with a homotopy equivalence to the trivial complex $0 \to L \to 0$, then V is naturally a homotopy Lie algebra, see Schlessinger and Stasheff [23], Barnich, Fulp, Lada and Stasheff [2] and Getzler and Jones [14].

Realized in the Batalin-Vilkovisky complex, these defining identities tell us, for small values of n, that d_{KT} is a graded derivation of the bracket, that the bracket may not satisfy the graded Jacobi identity but that we do have (with the appropriate signs)

$$[[v_1, v_2], v_3] \pm [[v_1, v_3], v_2] \pm [[v_2, v_3], v_2] =$$
$$-d[v_1, v_2, v_3] \pm [dv_1, v_2, v_3] \pm [v_1, dv_2, v_3] \pm [v_1, v_2, dv_3]. \tag{4.2}$$

i.e. the Jacobi identity holds *up to homotopy* or, for closed forms, the Jacobi identity holds modulo an exact term - the tri-linear bracket.

Note that the identity is the Jacobi identity if $d_{KT} = 0$ and all the other brackets vanish and that the identity has content even if only one n-linear bracket is non-zero and all the others vanish. Precisely that situation has recent been studied quite independently of my work and of each other by Hanlon and Wachs [16] (combinatorial algebraists), by Gnedbaye [15] (of Loday's school) and by Azcarraga and Bueno [11] (physicists). At the

Ascona conference, Flato brought to my attention that Takhtajan's identity [26] for his trilinear 'bracket' is not the one above, but suitably symmetrized does agree with it. (This identity was also known to Flato and Frønsdal in 1992, though unpublished.)

5. The Master Equation and deformation theory

The Master Equation $(L_\infty, L_\infty) = 0$ has precisely the form of Gerstenhaber's condition for L_∞ to be a deformation of L_0. Classical (formal) algebraic deformation theory uses a differential graded Lie algebra (dgla) $\mathbb{L}$ (e.g. the Hochschild cochain complex with the Gerstenhaber bracket) to study the problem of 'integrating' an infinitesimal deformation θ to a full formal deformation θ_t. The primary obstruction, regarding θ as a class in H^* of this dgla $\mathbb{L}$, is $[\theta, \theta]$ and further obstructions can be described in terms of multi-brackets on this cohomology. Alternatively, the formal deformation θ_t itself as a cochain must satisfy $[\theta_t, \theta_t] = 0$. The analogy with the Master Equation is manifest.

Following the historical pattern in algebraic deformation theory, we could hope to calculate this homology to be 0 in the relevant dimensions in certain cases, thus obtaining results of unobstructedness for the integrability question or of rigidity for the classification problem. Such calculations are highly non-trivial, however, and to my knowledge have been carried out only in the case of electricity and magnetism (Maxwell's equations), Yang-Mills and gravity.

6. The Master Equation in field theory

In the Lagrangian setting, we wish to deform not just the local functional, but rather the underlying local function L. In the case of electricity and magnetism (Maxwell's equations), Yang-Mills and gravity, the relevant algebra of gauge symmetries is described by a finite dimensional Lie algebra which, moreover, holds off shell. In terms of an appropriate basis and in the notation of section 1, we have

$$[\delta_\alpha, \delta_\beta] = c^\gamma_{\alpha\beta}\delta_\gamma \tag{6.1}$$

for structure *constants* $c^\gamma_{\alpha\beta}$ and acting on all fields, not just on solutions. This allows the extended Lagrangian to be no more than quadratic in the ghosts.

As field theories, the photon can be described by a field of spin 1, as can a Yang-Mills particle, while the graviton can be described by a field of spin 2. Somehow this is related to the strict Lie algebra structures just described. For higher spin particles, however, we have quite a different story, which first caught my attention in the work of Burgers, Behrends and van Dam [8, 9], though I have since learned there was quite a history before that and major questions still remain open. By higher spin particle Lagrangians, I mean that the fields are symmetric s-tensors (sections of the symmetric s-fold tensor product of the tangent bundle). If the power is s, the field is said to be of spin s and represents a particle of spin s. Burgers, Behrends and van Dam start with a free theory with abelian gauge symmetries and calculate all possible infinitesimal interaction terms up to the appropriate equivalence (effectively calculating the appropriate homology group). They then sketch the problem of finding higher order terms for the Lagrangian, but do not carry out the full calculation. In fact, according to the folklore in the subject, a consistent theory for $s \geq 3$ will require additional fields of arbitrarily high spin s. For $s = 3$, the conjecture is that all higher integral spins are needed. From the deformation theory point of view, this suggests the following attack: Compute the primary obstructions and discover that all infinitesimals are obstructed. Add additional fields to kill the obstructions and calculate that indeed additional fields of arbitrarily high spin s are needed. In one memorable phrase, this would be 'doing string field theory the hard way".

Zwiebach [30] does indeed have a consistent closed string field theory (CSFT), but produced in an entirely different way. Recall one of the earliest examples of deformation quantization, the Moyal bracket. Moyal was able to produce a non-trivial deformation of a commutative algebra $C^\infty(M)$ on a symplectic manifold, with infinitesimal given by the Poisson bracket, by writing down the entire formal power series. Similarly, Zwiebach is able to describe the entire CSFT Lagrangian (at tree level) by giving it in terms of the differential geometry of the moduli space of punctured Riemann spheres (tree level = genus 0). In fact, Zwiebach has the following structure: a differential graded Hilbert space $(\mathcal{H}, < \, , \, >, \mathcal{Q})$ related to the geometry of the moduli spaces from which he deduces n-ary operations $[\, , \dots , \,]$ which give an L_∞ structure:

$$d[\phi_1, \dots, \phi_N] + \textstyle\sum_{i=1}^N \pm [\phi_1, \dots, d\phi_i, \dots, \phi_N] = \textstyle\sum_{k=2}^{N-1} \pm [[\phi_{i_1}, \dots, \phi_{i_k}], \phi_{i_{k+1}}, \dots, \phi_{i_N}]$$

with d being Q and $[\phi] = Q\phi$.

The deformed Lagrangian (still classical) and hence the Master Equation is satisfied for

$$S(\Psi) = \frac{1}{2}\langle \Psi, Q\Psi\rangle + \sum_{n=3}^{\infty} \frac{\kappa^{n-2}}{n!}\{\Psi \dots \Psi\}.$$

The expression $\{\Psi \dots \Psi\}$ contains n-terms and will be abbreviated $\{\Psi^n\}$; it is given in terms of the brackets by $\{\Psi \dots \Psi\} = \langle \Psi, [\Psi, \dots, \Psi]\rangle$.

The field equations follow from the classical action by simple variation:

$$\delta S = \sum_{n=2}^{\infty} \frac{\kappa^{n-2}}{n!} \{\delta\Psi, \Psi^{n-1}\}$$

with gauge symmetries given by

$$\delta_\Lambda \Psi = \sum_{n=0}^{\infty} \frac{\kappa^n}{n!} [\Psi^n, \Lambda].$$

7. Quantization

So far our description of the anti-field, anti-bracket formalism has been in the context of deformations of 'classical' Lagrangians. Batalin and Vilkovisky (as well as much of the work on BRST cohomology) were motivated by problems in quantum theory. The quantum version of the anti-field, anti-bracket formalism involves a further 'second order' differential operator Δ of square 0 on the B-V complex relating the graded commutative product and the bracket - namely, the bracket is the deviation of the operator Δ from being a derivation of the product. This has led to the abstract definition of a BV-algebra.

Definition 7.1 A **BV-algebra** is a Gerstenhaber algebra with an operator (necessarily of degree -1 for a bracket of degree -1) such that

$$[A, B] = \Delta(AB) - \Delta(A)B + (-1)^A A\Delta(B).$$

Alternatively, a definition can be given in terms of a graded commutative algebra with an appropriate operator Δ [1, 24].

The quantization of Zwiebach's CSFT involves further expansion of the Lagrangian in terms of (the moduli space of) Riemann surfaces of genus $g \geq 0$. Here the operator Δ is determined by the self-sewing of a pair of pants (a Riemann sphere with 3 punctures). Now Zwiebach's CSFT provides a solution of the 'quantum Master Equation' which in the context of a BV-algebra, is

$$(S, S) = \Delta S.$$

Again we see an analog of the Maurer-Cartan equation or of a flat connection, but why?

References

1. F. Akman, *On some generalizations of Batalin-Vilkovisky algebras*, preprint, Cornell University, 1995, to appear in JPAA, q-alg/9506027.

2. G. Barnich, R. Fulp, T. Lada, and J. Stasheff, *The sh Lie structure of Poisson brackets in field theory*, pre-preprint, ULB,NCSU,UNC-CH, 1996.
3. G. Barnich and M. Henneaux, *Consistent couplings between fields with a guage freedom and deformations of the master equation*, Phys. Lett. **B 311** (1993), 123–129, hep-th/9702176.
4. I.A. Batalin and G.S. Vilkovisky, *Quantization of gauge theories with linearly dependent generators*, Phys. Rev. D **28** (1983), 2567–2582.
5. ———, *Closure of the gauge algebra, generalized Lie equations and Feynman rules*, Nucl. Phys. B **234** (1984), 106–124.
6. ———, *Existence theorem for gauge algebra*, J. Math. Phys. **26** (1985), 172–184.
7. C. Becchi, A. Rouet, and R. Stora, *Renormalization of the abelian Higgs-Kibble model*, Commun. Math. Phys. **42** (1975), 127–162.
8. F.A. Berends, G.J.H. Burgers, and H. van Dam, *On the theoretical problems in constructing interactions involving higher spin massless particles*, Tech. report, UNC-CH, 1984, preprint IFP234-UNC.
9. G.J.H. Burgers, *On the construction of field theories for higher spin massless particles*, Ph.D. thesis, Rijksuniversiteit te Leiden, 1985.
10. C. Chevalley and S. Eilenberg, *Cohomology theory of Lie groups and Lie algebras*, Trans. Amer. Math. Soc. **63** (1948), 85–124.
11. J. A. de Azcarraga and J. C. Perez Bueno, *Higher-order simple Lie algebras*, preprint, Univ. de Valencia, 1996, q-alg/9605213.
12. L. D. Fadde'ev and V. N. Popov, *Feynman diagrams for the Yang-Mills field*, Phys. Lett. **25B** (1967), 29–30.
13. M. Gerstenhaber, *On the deformation of rings and algebras*, Ann. of Math. **79** (1964), 59–103.
14. E. Getzler and J.D.S. Jones, *n-algebras and Batalin-Vilkovisky algebras*, preprint, 1993.
15. V. Gnedbaye, *Operads of k-ary algebras*, Operads: Proceedings of Renaissance Conferences (J.-L. Loday, J. Stasheff, and A. A. Voronov, eds.), Amer. Math. Soc., 1996, pp. 83–114.
16. P. Hanlon and M.L. Wachs, *On Lie k-algebras*, Adv. in Math. **113** (1995), 206–236.
17. M. Henneaux and C. Teitelboim, *Quantization of gauge systems*, Princeton Univ. Press, 1992.
18. T. Kimura, A.A. Voronov, and G. Zuckerman, *Homotopy Gerstenhaber algebras and topological field theory*, Operads: Proceedings of Renaissance Conferences (J.-L. Loday, J. Stasheff, and A. A. Voronov, eds.), Contemporary Math. **202**, Amer. Math. Soc., 1996, pp.305–333.
19. L. Kjeseth, *BRST cohomology and homotopy Lie-Rinehart pairs*, Dissertation, UNC-CH, 1996.
20. T. Lada and J.D. Stasheff, *Introduction to sh Lie algebras for physicists*, Intern'l J. Theor. Phys. **32** (1993), 1087–1103.
21. B.H. Lian and G.J. Zuckerman, *New perspectives on the BRST-algebraic structure of string theory*, Commun. Math. Phys. **154** (1993), 613–646, hep-th/9211072.
22. E. Noether, *Invariante variationsprobleme*, Nachr. könig. Gesell. Wissen. Göttingen, Math,-Phys. Kl. (1918), 235–257, in English: Transport Theory and Stat. Phys. 1 (1971),186–207.
23. M. Schlessinger and J.D. Stasheff, *The Lie algebra structure of tangent cohomology and deformation theory*, J. of Pure and Appl. Alg. **38** (1985), 313–322.
24. A. Schwarz, *Geometry of Batalin-Vilkovisky quantization*, CMP **155** (1993), 249–260.
25. R. Stora, *Continuum gauge theories*, New Developments in Quantum Field Theory (M.Levy and P.Mitter, eds.), Plenum, 1977, pp. 201–224.
26. L. Takhtajan, *On foundation of the generalized Nambu mechanics*, CMP **160** (1994), 295–315.
27. J. Tate, *Homology of Noetherian rings and local rings*, Ill. J. Math. **1** (1957), 14–27.

28. E. Witten and B. Zwiebach, *Algebraic structures and differential geometry in two-dimensional string theory*, Nucl. Phys. B **377** (1992), 55–112.
29. J. Zinn-Justin, *Renormalization of gauge theories*, Trends in Elementary Particle Theory (H. Rollnick and K. Dietz, eds.), vol. 37, Springer, 1975, pp. 2–39.
30. B. Zwiebach, *Closed string field theory: Quantum action and the Batalin-Vilkovisky master equation*, Nucl. Phys. B **390** (1993), 33–152.

THE FUKAYA TYPE CATEGORIES FOR ASSOCIATIVE ALGEBRAS

RYSZARD NEST
University of Copenhagen
2100 Copenhagen, Denmark

AND

BORIS TSYGAN
Penn State University
University Park, PA, 16803, USA

1. Introduction

In his study of Lagrangian intersections Fukaya discovered a new algebraic structure on Morse-Floer complexes of loop (path) spaces of symplectic manifolds [7]. He constructed a "category" whose objects are symplectomorphisms and whose space of morphisms are Morse-Floer complexes. The composition in this "category" yields multiplication on Floer cohomology. At the level of cochains this composition is "strongly associative up to homotopy"; in other words, it defines a structure of an A_∞ category.

It turns out that a very similar algebraic structure can be defined in terms of noncommutative differential geometry for any associative ring. When the ring in question is a result of deformation quantization of a symplectic manifold then the resemblance with the Fukaya construction becomes especially strong.

It is well known that most geometric properties of a manifold can be recovered from the algebra A of functions on this manifold. By (affine) noncommutative geometry one usually means extending corresponding constructions to the case when A is not necessarily commutative. For example, consider the standard cochain complex $C^*(A, A)$ computing the Hochschild cohomology

$$H^*(A, A) = Ext^*_{A\otimes A^o}(A, A)$$

and the standard chain complex computing the Hochschild homology

$$HH_*(A) = H_*(A, A) = Tor_*^{A\otimes A^o}(A, A).$$

D. Sternheimer et al. (eds.), Deformation Theory and Symplectic Geometry, 285–300.

It has been shown in [14] that, if A is the algebra of regular functions on an affine nonsingular algebraic variety X over a field k of characteristic zero, then

$$HH_i(A) \simeq \Omega^i_{X/k}$$

$$H^i(A, A) \simeq \Gamma(X, \wedge^i TX)$$

The standard cochain complex $C^*(A, A)$ is a differential graded associative algebra (with the cup product which can also be interpreted as the Yoneda product). On the other hand, the standard chain complex $C_*(A, A)$ is not a differential graded algebra unless A is commutative. It turns out, however, that if one considers all differential operators on $\Omega^*(X)$ (not just zero order multiplication operators) one gets a ring which has a noncommutative generalization. The construction we are going to present rests upon an idea of Yu.I. Manin [17].

Let R be a monoidal category. For an object A of R one calls an object $\underline{End}\ A$ of R the inner object of endomorphisms of A if there are natural morphisms

$$\underline{End}\ A \times \underline{End}\ A \to \underline{End}\ A \quad ; \quad \underline{End}\ A \times A \to A \tag{1}$$

which are universal and associative in a natural way. In the category of associative algebras inner objects of endomorphisms do not exist (for example, the set of endomorphisms of an algebra is not a linear space). In this paper, however, we will show that, if one takes for $\underline{End}\ A$ the differential graded algebra $C^*(A, A)$ of Hochschild cochains, the maps (1), in a sense, still exist. More precisely, they exist if one passes from the category of algebras to the category of complexes by means of some well known homological functors.

Let A be an associative unital algebra over a commutative unital ground ring k. Consider A as a bimodule over itself; by $\mathcal{E}^*_A$ we denote the differential graded algebra $(C^*(A, A), \delta, \smile)$ which is the standard complex for computing $Ext^*_{A\otimes A^o}(A, A) = H^*(A, A)$ (the Hochschild cohomology) equipped with the Yoneda product (Sect. 2). We construct the map of complexes

$$\bullet : C_*(A) \otimes C_*(\mathcal{E}^*_A) \to C_*(A) \tag{2}$$

$$\bullet : C_*(\mathcal{E}^*_A) \otimes C_*(\mathcal{E}^*_A) \to C_*(\mathcal{E}^*_A) \tag{3}$$

Here C_* stands for the Hochschild complex computing $HH_*(A)$ or for the periodic cyclic complex.

For the case of bar complex an analogous operation was discovered by Getzler–Jones and by Gerstenhaber–Voronov [11, 9]. Therefore one can say informally that in the homotopy category of (differential graded) algebras the inner objects of endomorphisms always exist.

Next we construct construct "the Fukaya category" of an associative algebra A; the objects of this category are automorphisms of A and, for two endomorphisms α and β, the complex $Hom(\alpha, \beta)$ is the twisted cochain complex $C_*(\mathcal{E}^*_{A},_\alpha \mathcal{E}^*_\beta)$. We also construct a functor putting in correspondence to α the twisted chain complex $C_*(A, A_\alpha)$. Conjecturally, these are an A_∞ category and an A_∞ functor.

In other words, we construct the products

$$\bullet : C_*(A, A_\alpha) \otimes C_*(\mathcal{E}^*_{A},_\alpha \mathcal{E}^*_\beta) \to C_*(A, A_\beta) \tag{4}$$

$$\bullet : C_*(\mathcal{E}^*_A,\ {}_\alpha\mathcal{E}^*_\beta) \otimes C_*(\mathcal{E}^*_A,\ {}_\beta\mathcal{E}^*_\gamma) \to C_*(\mathcal{E}^*_A,\ {}_\alpha\mathcal{E}^*_\gamma) \tag{5}$$

associative up to homotopy.

Depending on whether one uses finite or infinite cochains, one can define two versions of the homology category $\mathcal{H}^*_A$ or $\mathcal{H}^*_{A,\infty}$.

Next we consider some examples. First, in case when $A = k[V]$ where V is affine nonsingular, one has

$$\mathcal{H}^*_A(1, 1) = \mathcal{D}(\Omega^*(V))^{op}$$

where $\mathcal{D}$ is the ring of differential operators and ^{op} stands for the opposite ring. We consider a partial case in which we prove that

$$\mathcal{H}^*_A(\alpha, \alpha) = \mathcal{D}(\Omega_*(V^\alpha))^{op}$$

where V^α is the fixed point set; $\mathcal{H}^*_A(\alpha, \beta)$ can be obtained from a standard D-module construction. On the other hand,

$$\mathcal{H}^*_{A,\infty}(1, 1) = End(\Omega^*(V))$$

and it seems that in a rather big generality (whether in the commutative case or not)

$$\mathcal{H}^*_{A,\infty}(\alpha, \beta) = Hom(H_*(A, A_\alpha), H_*(A, A_\beta))$$

Next we consider the case when A is a deformed ring of functions on a symplectic manifold M over the field of Laurent series $\mathbb{C}[\hbar, \hbar^{-1}]]$ [1]. We show that

$$\mathcal{H}^*_A(1, 1) = H^*(M^{S^1}, \mathbb{C}[\hbar, \hbar^{-1}]]). \tag{6}$$

After this we study the subcategory of the category $\mathcal{H}_A$ in a partial case when A is a deformed ring of functions on S^2 and the automorphisms are rotations. We show how to compute $\mathcal{H}^*_A(\alpha, \beta)$ in terms of the cohomology of the space of loops based on points x for which $\alpha x = \beta x$.

Note the resemblance with the Fukaya construction from [7]. Recall that deformation quantization is an algebra indexed by a formal parameter from

$H^2(M)$. Its construction, but not its isomorphism class, depends on an additional geometric structure, like a Kähler structure on A. It looks like there is, together with the categories $\mathcal{H}^*_A$ and $\mathcal{H}^*_{A,\infty}$, some third, semi-infinite version which is more closely related to the Fukaya category and the quantum cohomology. It also seems plausible, in light of the formality conjecture of Kontsevich [15], that there exists a quantum cohomology theory for any Poisson manifold.

Acknowledgements

The idea that the cohomology of deformed algebras may be related to Lagrangian intersections is due to Boris Feigin. We thank M. Kontsevich, A. Radul and A. Voronov for helpful discussions.

The authors are very grateful to the organizers of the Ascona meeting.

2. The Hochschild cochain complex

Let A be a graded algebra with unit over a commutative unital ring k. Let M be a graded bimodule over A. A Hochschild d-cochain is a linear map $A^{\otimes d} \to M$. Put, for $d \geq 0$,

$$C^d(A, M) = Hom_k(\overline{A}^{\otimes d}, M)$$

where $\overline{A} = A/k \cdot 1$. Put

$$deg\, D = (\textit{degree of the linear map } D) + d$$

$$|D| = deg\, D - 1; \quad |a| = deg\, a - 1$$

Given a tensor $a_1 \otimes \cdots \otimes a_N$ in $A^{\otimes N}$, we will denote it by $(a_1, \ldots, a_n)$. We will write $\eta_j = \sum_{i=1}^{j} |a_i|$ (as in [10]). Put for cochains D and E from $C^*(A, A)$

$$(D \smile E)(a_1, \ldots, a_{d+e}) = (-1)^{deg\, E \cdot \eta_d} D(a_1, \ldots, a_d) \times$$

$$\times E(a_{d+1}, \ldots, a_{d+e});$$

$$(D \circ E)(a_1, \ldots, a_{d+e-1}) = \sum_{j \geq 0} (-1)^{|E|\eta_j} D(a_1, \ldots, a_j,$$

$$E(a_{j+1}, \ldots, a_{j+e}), \ldots);$$

$$[D,\ E] = D \circ E - (-1)^{|D||E|} E \circ D$$

These operations define the graded associative algebra $(C^*(A, A), deg\,, \smile)$ and the graded Lie algebra $(C^*(A, A), |\cdot|, [\ ,\ ])$ (cf. [2, 8]). Let

$$m(a_1, a_2) = (-1)^{deg\, a_1}\, a_1 a_2;$$

this is a 2-cochain of A (not in C^2). Put

$$\delta D = [m, D];$$

$$(\delta D)(a_1, \ldots, a_{d+1}) = (-1)^{|a_1||D|+|a_1|+1} \times$$

$$\times a_1 D(a_2, \ldots, a_{d+1}) +$$

$$+ \sum_{j=1}^{d} (-1)^{|D|+\eta_j} D(a_1, \ldots, a_j\, a_{j+1}, \ldots, a_{d+1})$$

$$+ (-1)^{|D|+\eta_d+1} D(a_1, \ldots, a_d) a_{d+1}$$

The last formula defines a cochain differential on $C^*(A, M)$ for any M. For an element x of A, let $\underline{x}$ be the corresponding zero-cochain in $C^*(A, A)$. By definition

$$(\delta \underline{x})(a) = (-1)^{deg\, x}[x, a];$$

a one-cochain D is a cocycle iff it is a derivation. One has

$$\delta^2 = 0; \quad \delta(D \smile E) = \delta D \smile E + (-1)^{deg\, D} D \smile \delta E$$

$$\delta[D, E] = [\delta D, E] + (-1)^{|D|}\, [D, \delta E]$$

($\delta^2 = 0$ follows from $[m, m] = 0$).

Thus $C^*(A, A)$ becomes a complex; the cohomology of this complex is $H^*(A, A)$ or the Hochschild cohomology. The $\smile$ product induces the Yoneda product on $H^*(A, A) = Ext^*_{A \otimes A^0}(A, A)$. The operation $[\ ,\ ]$ is the Gerstenhaber bracket [8].

If $(A,\ \partial)$ is a differential graded algebra then one can define the differential ∂ acting on A by

$$\partial D \ = \ [\partial, D]$$

Definition 1 *Define the differential graded algebra* $\mathcal{E}_A^*$ *as the complex* $C^*(A, A)$ *with the differential* δ *(or* $\delta + \partial$ *if* (A, ∂) *is a differential graded algebra), the grading deg and the product* $\smile$.

In Sections 2–4 the only cochains we will be considering will be those from $C^*(A, A)$. For Hochschild cochains D and D_i define a new Hochschild cochain by the following formula of Gerstenhaber [8] and Getzler [10]:

$$D_0\{D_1, \ldots, D_m\}(a_1, \ldots, a_n) =$$

$$= \sum (-1)^{\sum_{p=1}^{m} \eta_{i_p} |D_p|} D_0(a_1, \ldots, a_{i_1}, D_1(a_{i_1+1}, \ldots), \ldots, D_m(a_{i_m+1}, \ldots), \ldots)$$

Proposition 2 *One has*

$$(D\{E_1,\ldots,E_k\})\{F_1,\ldots,F_l\} = \sum(-1)^{\sum_{q\leq i_p}|E_p||F_q|}\times$$

$$\times D\{F_1,\ldots,E_1\{F_{i_1+1},\ldots,\},\ldots,E_k\{F_{i_k+1},\ldots,\},\ldots,\}$$

nibProof Direct computation □

For a cochain D let $D^{(k)}$ be the following k-cochain of $\mathcal{E}_A^*$:

$$D^{(k)}(D_1,\ldots,D_k) = D\{D_1,\ldots,D_k\}$$

Proposition 3 *The map*

$$D \mapsto \sum_{k\geq 0} D^{(k)}$$

is a morphism of differential graded algebras $\mathcal{E}_A^* \to \mathcal{E}_{\mathcal{E}_A^*}^*$.

3. The Hochschild chain complex

Let M be a bimodule over A. Recall that the Hochschild homological complex $(C_*(A,M),b)$ is the following:

$$C_n(A,M) = M\otimes \overline{A}^{\otimes n};$$

We will always write, as in Section 1,

$$(a_0,\ldots,a_n) = a_0\otimes\ldots\otimes a_n$$

For a_0 in M and a_i in A, $i>0$, define

$$C_n(A,A) = A\otimes \overline{A}^{\otimes n};$$

$$b(a_0,\ldots,a_n) = \sum_{j=0}^{n-1}(-1)^{\eta_{j+1}+1}(a_0,\ldots,a_j\, a_{j+1},\ldots,a_n)$$

$$+(-1)^{(|a_n|+1)(\eta_n+1)+1}(a_na_0,a_1,\ldots,a_{n-1}).$$

We shall denote $C_*(A,A)$ simply by $C_*(A)$. We introduce a grading on $C_*(A,A)$ by the formula

$$deg\,(a_0,\ldots,a_n) = \sum deg\; a_i + n$$

(A rule for remembering the signs: let $|\epsilon| = 1$, $\epsilon^2 = 0$; map $C_n(A)$ to $A[\epsilon]/[A[\epsilon],A[\epsilon]]$, $a_0\otimes\cdots\otimes a_n \mapsto a_0\epsilon a_1\epsilon\cdots a_n\epsilon$; then b becomes $\frac{\partial}{\partial\epsilon}$).

The homology of $C_*(A, M)$ is the Hochschild homology $HH_*(A, M) = Tor_*^{A\otimes A^\circ}(A, M)$. We will denote $HH_*(A, A)$ simply by $HH_*(A)$. If A is a differential graded algebra and ∂ the differential in A, one extends ∂ to $C_*(A)$:

$$\partial(a_0, \ldots, a_n) = \sum_{j=0}^{n}(-1)^{\eta_j}(a_0, \ldots, \partial a_j, \ldots, a_n)$$

For a in $C_*(A, A)$ and x in $C_*(\mathcal{E}_A^*, \mathcal{E}_A^*)$ define

$$a \bullet x = a \bullet_1 x + a \bullet_2 x \tag{7}$$

where

$$(a_0, \ldots, a_n) \bullet_1 (D_0, \ldots, D_m) = \sum(-1)^{(\eta_{n+1}-\eta_1)\deg D_0 + \sum_{p=1}^{m}(\eta_{n+1}-\eta_{i_p+1})|D_p|} \times$$

$$\times\, (a_0 D_0(a_1, a_2, \ldots), \ldots, a_{i_1}, D_1(a_{i_1+1}, \ldots), \ldots, D_m(a_{i_m+1}, \ldots), \ldots) \tag{8}$$

$$(a_0, \ldots, a_n) \bullet_2 (D_0, \ldots, D_m) = \tag{9}$$

$$= \sum_{q\leq n+1}(-1)^{(\eta_{n+1}-\eta_q)\eta_q + (\eta_q - \eta_{i_0+1})\deg D_0 + \sum_{p=1}^{m-1}(\eta_q-\eta_{i_p+1})|D_p| + \eta_{n+1}|D_m|}$$

$$\times(-1)^{|D_m|(\sum_{p\geq 0}|D_p|+1)}\times$$

$$\times(D_m(a_q, \ldots, a_n, a_0, \ldots, a_{i_0})D_0(a_{i_0+1}, \ldots), \ldots, a_{i_1},$$

$$, D_1(a_{i_1+1}, \ldots), \ldots, D_{m-1}(a_{i_{m-1}+1}, \ldots), \ldots)$$

The sum in (8) is taken over all q, $i_0, \ldots,$ i_{m-1} for which a_0 is inside D_m.

Theorem 4 *The map*

$$\bullet : C_*(A, A) \otimes C_*(\mathcal{E}_A^*, \mathcal{E}_A^*) \to C_*(A, A)$$

is a morphism of complexes, i.e.

$$b(a \bullet x) = (ba) \bullet x + (-1)^{\deg a} a \bullet (b + \delta)x$$

Proof. Direct computation. □

Consider some examples of the product $\bullet$. When all D_i are zero -cochains then one gets the shuffle product from [2]. On the other hand, put, for a d-cochain D and for a in $C_*(A, A)$, $i_D(a) = (-1)^{\deg a\, \deg D}(a \bullet D)$;

$$i_D(a_0, \ldots, a_n) = (-1)^{\deg a_0 \cdot \deg D_0}(a_0 D(a_1, \ldots, a_d), \quad a_{d+1}, \ldots, a_n)$$

One gets the cap product from [2]:

$$C^d(A, A) \otimes C_n(A, A) \to C_{n-d}(A, A)$$

Now, for a d-cochain D and for a in $C_*(A,A)$, let $L_D(a) = (-1)^{|D|\cdot deg\, a} a \bullet (1, D)$; one has

$$L_D(a_0, \ldots, a_n) = \sum_{q=0}^{n-d+2} (-1)^{(\eta_{n+1}-\eta_q)\eta_q+|D|}(D(a_q, \ldots, a_0, \ldots), \ldots, a_{q-1}) +$$

$$+\sum_{k=1}^{n-d}(-1)^{(\eta_{k+1}+1)|D|}(a_0, \ldots, a_k, D(a_{k+1}, \ldots), \ldots,)$$

One has

$$[b, L_D] = -L_{\delta D}$$

Now construct the product

$$\bullet : C_*(\mathcal{E}^*_A, \mathcal{E}^*_A) \otimes C_*(\mathcal{E}^*_A, \mathcal{E}^*_A) \to C_*(\mathcal{E}^*_A, \mathcal{E}^*_A) \tag{10}$$

by composing the one for the algebra $\mathcal{E}^*_A$ with the map $\mathcal{E}^*_A \to \mathcal{E}^*_{\mathcal{E}^*_A}$ from Proposition 3.

The product (10) is given by formulas (8), (9) where a_i are now viewed as Hochschild cochains and $D(a_1, \ldots, a_d)$ is replaced by $D\{a_1, \ldots, a_d\}$.

Proposition 5 *The product (10) is homotopically associative.*

We omit the proof.

4. The periodic cyclic complex

In the situation of Sect. 3 define

$$B(a_0, \ldots, a_n) = \sum_{j=0}^{n}(-1)^{(\eta_{n+1}-\eta_j)\eta_j}(1, a_j, \ldots, a_n, a_0, \ldots, a_{j-1})$$

Then $B^2 = b^2 = Bb + bB = 0$; denote

$$CC_n^{per}(A) = \prod_{i \equiv n (mod 2)} A \otimes \overline{A}^{\otimes n}$$

with the differential $b + B + \partial$ (for the differential graded algebra (A, ∂)). Put

$$(a_0, \ldots, a_n) \bullet_3 (D_1, \ldots, D_m) = \tag{11}$$

$$= \sum_{0 \le p \le m} (-1)^{\sum_{r<p}(|D_r|+1)\sum_{r\ge p}(|D_r|+1)+\sum |D_r|} \times$$

$$\times \sum_{0 \le j \le n}(-1)^{(\eta_{n+1}-\eta_j)\eta_j}\lambda'(D_p, \ldots, D_m, D_0, \ldots, D_{p-1})(1, a_j, \ldots, a_n,\ a_0, \ldots, a_{j-1})$$

Here

$$\lambda(D_1,\ldots,D_m)(a_0,\ldots,a_n) =$$
$$= \sum_{j_1>0}(-1)^{|D_1|(\eta_{n+1}-\eta_{j_1})+\ldots+|D_m|(\eta_{n+1}-\eta_{j_m})}\times$$
$$\times(a_0,\ldots,D_1(a_{j_1}\ldots),\ldots,D_m(a_{j_m},\ldots),\ldots);$$

λ' is the sum of all those terms in λ which contain $D_0(\ldots)$ after a_0.

Theorem 6 *The map*

$$\bullet : CC_*^{per}(A)\otimes CC_*^{per}(\mathcal{E}_A^*)\to CC_*^{per}(A)$$

is a morphism of complexes.

Proof. Direct computation. □

For a d-cochain D and for a in $CC_*^{per}(A)$, let $I_D(a) = (-1)^{deg\,D\cdot deg\,a}(a\bullet D)$; $L_D(a) = (-1)^{|D|\cdot deg\,a}a\bullet(1,D)$; one has

$$[B+b,\ I_D]-I_{\delta D} = L_D.\square$$

This is the homotopy formula of Rinehart [21].

One can define the product

$$\bullet : CC_*^{per}(\mathcal{E}_A^*)\otimes CC_*^{per}(\mathcal{E}_A^*)\to CC_*^{per}(\mathcal{E}_A^*)$$

exactly in the same way as in the end of Section 2.

Remark 7 *Considering $CC_*^{per}(A) = CC_*^{per}(\mathcal{E}_A^0)$ as a part of $CC_*^{per}(\mathcal{E}_A^*)$ and restricting the product $\bullet$ to it, we get the Hood-Jones product on $CC_*^{per}(A)$. When A is commutative then $CC_*^{per}(A)$ is a subcomplex of $CC_*^{per}(\mathcal{E}_A^*)$ and one gets a product on the complex $CC_*^{per}(A)$.*

5. The Fukaya category

$$\mathcal{C}^n(A) = \bigoplus_{i+j=n} C_{-i}(\mathcal{E}^*,\mathcal{E}^*)^j \tag{12}$$

$$\mathcal{C}^n_\infty(A) = \prod_{i+j=n} C_{-i}(\mathcal{E}^*,\mathcal{E}^*)^j \tag{13}$$

In this Section we extend the results of Section 3 by "moving away from the diagonal". For any automorphism α of an algebra A we define the twisted Hochschild complex $C_*(A,A_\alpha)$ (related to noncommutative geometry of fixed points of α) and construct in these terms a homotopically associative category for which $C^*(A)$ or $C^*_\infty(A)$ is the ring of endomorphisms of the identity.

Let α, β be two automorphisms of A. Define the new bimodule ${}_\alpha A_\beta$ over A as follows: ${}_\alpha A_\beta = A$ as k-modules and $a \cdot m \cdot b = \alpha(a)m\beta(b)$ for a, m, b in A. Consider the chain complex $C_*(A, {}_\alpha A_\beta)$ as in Section 2 and the cochain complex ${}_\alpha\mathcal{E}^*_\beta = C^*(A, {}_\alpha A_\beta)$ as in Section 1. We shall write $A_\alpha =_{id} A_\alpha$. The cup product on Hochschild cochains is well defined as a morphism ${}_\alpha\mathcal{E}^*_\beta \otimes_\beta \mathcal{E}^*_\gamma \to_\alpha \mathcal{E}^*_\gamma$.

Given an automorphism α of A, one can define its action on a Hochschild cochain from $\mathcal{E}^*_A$ in two ways:

$$(\alpha D)(a_1, \ldots, a_n) = \alpha(D(a_1, \ldots, a_n));$$

$$(D\alpha)(a_1, \ldots, a_n) = D(\alpha a_1, \ldots, \alpha a_n)$$

Both yield morphisms of complexes $\mathcal{E}^*_A \to_\alpha \mathcal{E}_\alpha$. We define an $\mathcal{E}^*_A$-bimodule structure on ${}_\alpha\mathcal{E}^*_\beta$ as follows:

$$D \cdot M \cdot E = D\alpha \smile M \smile \beta E$$

for $D, E \in \mathcal{E}^*_A$ and $M \in_\alpha \mathcal{E}^*_\beta$.

Theorem 8 *The pairings (2, 3) can be extended to the homotopically associative natural morphisms of complexes*

$$\bullet : C_*(A, A_\alpha) \otimes C_*(\mathcal{E}^*_A, {}_\alpha \mathcal{E}^*_\beta) \to C_*(A, A_\beta)$$

$$\bullet : C_*(\mathcal{E}^*_A, {}_\alpha\mathcal{E}^*_\beta) \otimes C_*(\mathcal{E}^*_A, {}_\beta\mathcal{E}^*_\gamma) \to C_*(\mathcal{E}^*_A, {}_\alpha\mathcal{E}^*_\gamma)$$

Proof. As in Section 2, put for $a \in C_*(A, A_\alpha)$ and $x \in C_*(\mathcal{E}^*_A, {}_\alpha\mathcal{E}_\beta)$

$$a \bullet x = a \bullet_1 x + a \bullet_2 x$$

where

$$(a_0, \ldots, a_n) \bullet_1 (D_0, \ldots, D_m) =$$

$$\sum \pm(a_0 \cdot D_0(a_1, \ldots), \ldots, a_{i_1}, D_1(a_{i_1+1}, \ldots), \ldots, D_m(a_{i_m+1}, \ldots), \ldots)$$

$$(a_0, \ldots, a_n) \bullet_2 (D_0, \ldots, D_m) =$$

$$= \sum_{q \leq n+1} \pm(D_m(a_q, \ldots, a_n, a_0, \alpha a_1, \ldots, \alpha a_{i_0}) \cdot D_0(a_{i_0+1}, \ldots), \ldots, a_{i_1},$$

$$, D_1(a_{i_1+1}, \ldots), \ldots, D_{m-1}(a_{i_{m-1}+1}, \ldots), \ldots)$$

For $A \in C_*(\mathcal{E}^*_A, {}_\alpha\mathcal{E}_\beta)$ and $x \in C_*(\mathcal{E}^*_A, {}_\beta\mathcal{E}_\gamma)$

$$A \bullet x = A \bullet_1 x + A \bullet_2 x$$

where

$$(A_0, \ldots, A_n) \bullet_1 (D_0, \ldots, D_m) =$$

$$\sum \pm (A_0 \smile D_0\{A_1, A_2, \ldots\}, \ldots, A_{i_1}, D_1\{A_{i_1+1},$$

$$\ldots\}, \ldots, D_m\{A_{i_m+1}, \ldots\}, \ldots)$$

$$(A_0, \ldots, A_n) \bullet_2 (D_0, \ldots, D_m) =$$

$$= \sum_{q \leq n+1} \pm (D_m\{\{A_q, \ldots, A_n, A_0, A_1, \ldots, A_{i_0}\}\} \smile D_0\{A_{i_0+1}, \ldots\}, \ldots, A_{i_1},$$

$$, D_1\{A_{i_1+1}, \ldots\}, \ldots, D_{m-1}\{A_{i_{m-1}+1}, \ldots\}, \ldots)$$

and

$$D_m\{\{A_q, \ldots, A_n, A_0, A_1, \ldots, A_{i_0}\}\}(a_1, \ldots, a_n) =$$

$$= \sum \pm D_m(\alpha a_1, \ldots, A_q(\alpha a_j, \ldots,), \ldots,$$

$$, \alpha a_t, A_0(a_{t+1}, \ldots,), \beta a_p, \ldots, \beta A_1(a_r, \ldots), \beta a_s, \ldots)$$

One checks that these maps are homotopically associative morphisms of complexes. □

One gets the cohomology groups $\mathcal{H}^*(\alpha, \beta)$ and $\mathcal{H}^*_\infty(\alpha, \beta)$ which form a category $\mathcal{H}^*_A$ or $\mathcal{H}^*_{A,\infty}$.

6. Examples

Theorem 9 *Let $A = k[X]$ be the algebra of regular functions on an affine nonsingular algebraic variety X over a field k of characteristic zero. Then*

$$\mathcal{H}^*(A) \rightrightarrows \mathcal{D}(\Omega^*_{X/k})^{op}$$

(the ring opposite to the ring of all differential operators on Ω);

$$\mathcal{H}^*_\infty(A) \rightrightarrows End(\Omega^*_{X/k})^{op}$$

Note that $HH_*(A) \simeq \Omega^*_{X/k}$; the $\bullet$ action of $\mathcal{H}^*$ on HH^* is the obvious one.

Proof of 9.

Lemma 10 *Let $\mathcal{E}^*$, $\mathcal{F}^*$ be two differential graded algebras together with a quasi-isomorphism $f : \mathcal{E}^* \to \mathcal{F}^*$. Then f induces an isomorphism*

$$\bigoplus_{i+j=n} C_{-i}(\mathcal{E}^*, \mathcal{E}^*)^j \rightrightarrows \bigoplus_{i+j=n} C_{-i}(\mathcal{F}^*, \mathcal{F}^*)^j$$

Proof. The corresponding spectral sequence converges (this is not true if one replaces $\bigoplus$ by $\prod$). □

Note that by [14] $\mathcal{E}_A^*$ is quasi-isomorphic to $\mathcal{F}^* = \Gamma(X, \bigwedge^* TX)$ as differential graded algebras. Locally, $\mathcal{F}_U^* = \mathcal{O}_U \otimes \bigwedge^*(\partial_{x_1}, \ldots, \partial_{x_n})$; $HH_*(\mathcal{F}_U^*) = \Omega_{U/k}^* \otimes \bigwedge^*(\partial_{x_1}, \ldots, \partial_{x_n}) \otimes S^*(d\partial_{x_1}, \ldots, d\partial_{x_n})$ (classes $d\partial_{x_i}$ are represented by cycles $(1, d\partial_{x_i})$). Now, looking at the $\bullet$ action on $HH^*(\mathcal{O}_U) \simeq \Omega_{U/k}^*$, one sees that $\Omega_{U/k}^*$ acts by multiplication, ∂_{x_i} acts by contraction $i_{\partial_{x_i}}$ and $d\partial_{x_i}$ acts by Lie derivative $L_{\partial_{x_i}}$. Whence the local isomorphism

$$\mathcal{H}^*(\mathcal{O}_U) \rightrightarrows \mathcal{D}(\Omega_{U/k}^*)^{op};$$

this statement may be globalized exactly as in [14]. Now let us compute $\mathcal{H}_\infty^*(A)$. We have to use the other spectral sequence whose E_1 term is $HH_*(\mathcal{E}_A^*)$ if one forgets about the differential δ. But, as a graded algebra, $\mathcal{E}_A^*$ is just $T^*(A')\widehat{\otimes}A$ by which we mean the space of all multilinear maps from A to A. One can show that

$$E_2^{-p,n} = HC^{n-1}(A)\widehat{\otimes}\Omega^p + HC^{n-2}(A)\widehat{\otimes}\Omega^{p+1}$$

$$E_3^{-p,n} = (KerS)^{n-1}\widehat{\otimes}\Omega^p + (CokerS)^{n-2}\widehat{\otimes}\Omega^{p+1}$$

where $S : HC^n \to HC^{n+2}$ is the Bott operator [16]. Recall the Gysin exact sequence [3, 22, 16]:

$$0 \longrightarrow (CokerS)^n \longrightarrow HH^n(A) \longrightarrow (KerS)^{n-1} \longrightarrow 0$$

It is not hard to construct for any element of $HH^*(A)\widehat{\otimes}\Omega_{X/k}^*$ a corresponding cycle of the complex $\mathcal{C}^*$. This shows that the spectral sequence degenerates at E_3 term and that $\mathcal{H}_\infty^*(A) \rightrightarrows HH^*(A)\widehat{\otimes}\Omega_{X/k}^*$. □

Let A be a one-dimensional affine space and $A = k[V] = k[x]$. We will compute the full subcategory of $\mathcal{H}_A^*$ whose objects are dilations $x \mapsto \alpha x$. We view $\mathcal{D} = \mathcal{D}(\Omega^*(V))$ as a graded algebra $k[x, \zeta, \frac{\partial}{\partial x}, \frac{\partial}{\partial \zeta}]$ where $\zeta = dx$.

Proposition 11

$$H^*(1,1) \rightrightarrows \mathcal{D}^{op};$$

$$H^*(\alpha, \beta) \rightrightarrows k, \quad \alpha \neq 1, \quad \beta \neq 1 \ ;$$

$$H^*(\alpha, 1) \rightrightarrows k \otimes_{k[x, \partial_\zeta]} \mathcal{D}^{op};$$

$$H^*(1, \alpha) \rightrightarrows \mathcal{D}^{op} \otimes_{k[x, \partial_\zeta]} k.$$

We denote by V^α the set of fixed points of α and consider $\mathcal{D}(\Omega^*(V^\alpha))$ as the algebra of functions on the supermanifold TV^α. Let us identify TV with T^*V; consider the maps

$$TV^\alpha \xleftarrow{j_\alpha} T_{V^\alpha \cap V^\beta} V \xrightarrow{j_\beta} TV^\beta$$

where j_α is the composition

$$T_{V^\alpha \cap V^\beta} V \xrightarrow{i} T_{V^\alpha} V \xrightarrow{\pi} TV.$$

Then

$$H^*(\alpha, \alpha) = \mathcal{D}(\Omega^*(V^\alpha))^{op};$$

$$H^*(\alpha, \beta) = j_{\alpha *} j_\beta^* \mathcal{D}(\Omega^*(V^\beta))$$

which is a left $H(\alpha, \alpha)$-module and a right $H(\alpha, \alpha)$-module.

Denote by $e_{\alpha\beta}$ the generators of $H^0(\alpha, \beta)$ for $\alpha \neq 1$ and by $D_{\alpha\beta}$ the generators of $H^1(\alpha, \beta)$ for $\alpha = 1$ or $\beta = 1$.

Proposition 12 *One has*

$$e_{\alpha\beta} \bullet e_{\beta\gamma} = e_{\alpha\gamma}; \quad e_{\alpha\beta} \bullet D_{\beta\gamma} = D_{\alpha\gamma};$$

$$D_{\alpha\beta} \bullet e_{\beta\gamma} = D_{\alpha\gamma}; \quad D_{\alpha\beta} \bullet D_{\beta\gamma} = 0;$$

the right, resp.left $\mathcal{D}^{op}$-module structure on $H^*(1, \alpha)$, resp. $H^*(\alpha, 1)$, is the obvious one.

Proposition 13 *One has*

$$H_*(A, A) = \Omega^*(V); \quad H_*(A, A_\alpha) = k, \quad \alpha \neq 1.$$

Let 1_α be the generator of $H_0(A, A_\alpha)$. Then for $\omega \in \Omega^*(V)$ $\omega \bullet D_{1\alpha} = (i_{\frac{\partial}{\partial x}} \omega)(0)$; $1_\alpha \bullet D_{\alpha\beta} = 1_\beta$ if $\alpha, \beta \neq 1$; $1_\alpha \bullet D_{\alpha 1} = 0$.

In other words, $H_*(A, A_\alpha) = \Omega^*(V^\alpha)$ and one has

$$\omega \bullet D_{\alpha\beta} = j_{\beta *} j_\alpha^* \omega.$$

Here j_α^* is the usual restriction; $j_{\beta *}$ is either zero when j_β is a closed embedding or the Berezin integral when j_β is a projection along the odd fiber.

Proofs of the Propositions. The Hochschild homology ${}_\alpha H_\beta^* \stackrel{def}{=} H^*(A,{}_\alpha A_\beta)$ is computed by the complex $k[x] \stackrel{\alpha-\beta}{\rightarrow} k[x]$. So for $\alpha \neq \beta$ ${}_\alpha H_\beta^1 \simeq k$ (we denote the generator by $D_{\alpha\beta}$)and ${}_\alpha H_\beta^i = 0$ if $i \neq 1$. Consider the spectral sequence with the E_2 term $H_*(H^*,{}_\alpha H_\beta^*)$ where $H^* ={}_1 H_1^*$. When $\alpha = \beta$, we get the ring $H^* = k[x, \partial_x]$ acting on itself: $x \cdot a = xa$, $a \cdot x = \alpha a x$, $\partial_x \cdot a = \alpha \partial_x a$, $a \cdot \partial_x = a \partial_x$. So, for $\alpha = \beta \neq 1$, $E_2 = k$ and the spectral sequence collapses.

When $\alpha \neq \beta$ then the E_2 term is $k[dx, d\partial_x] \cdot D_{\alpha\beta}$ where $dx, d\partial_x$ are the generators from $H_1(H^*, k)$. We claim that the spectral sequence degenerates at E_2 if $\alpha = 1$ or $\beta = 1$ and that $E_3 = k$ if $\alpha,\ \beta \neq 1$. Indeed, if $\alpha,\ \beta \neq 1$ then

$$d_2(D_{\alpha\beta} \cdot dx \cdot (d\partial_x)^n = (\alpha - \beta)(\alpha - 1)(\beta - 1)(D_{\alpha\beta} \cdot (d\partial_x)^n;$$

so all terms vanish in E_3 except $D_{\alpha\beta} \cdot dx$. This terms always survives in E_∞ if $\alpha \neq \beta$. It is represented by the cycle

$$(D_{\alpha\beta}, x) + \frac{\beta - 1}{\alpha - \beta} \cdot 1.$$

In case when $\alpha = 1$, resp. $\beta = 1$, the element above is equal to $x \bullet D_{\alpha\beta}$, resp. $D_{\alpha\beta} \bullet x$. Therefore $D_{\alpha\beta} \neq 0$; it generates the $\mathcal{D}^{op}$-module $\mathcal{H}^*(\alpha, \beta)$ and is annihilated by x and $\frac{\partial}{\partial\zeta}$. This determines this module uniquely. The rest of the statements of the Propositions are obtained by direct computations (note that $H_*(A, A_\alpha)$ is the homology of the complex $k[x] \overset{\alpha}{\to} k[x]$). □

Now let us turn to the case when A is a deformed algebra on a symplectic manifold. Let $*$ be a star product on a symplectic manifold M. Denote by $\mathbb{A}^\hbar(M)$ the algebra $(C^\infty(M), *)$.

Theorem 14 *One has*

$$H^*(\mathbb{A}^\hbar(M)[\hbar^{-1}]) \rightrightarrows H^*(M^{S^1}, \mathbb{C}[[\hbar, \hbar^{-1}])$$

Proof.

Lemma 15 *The differential graded algebra $\mathcal{E}^*_{\mathbb{A}^\hbar(M)[\hbar^{-1}]}$ is equivalent to the algebra $(\Omega^*(M)[[\hbar, \hbar^{-1}], d)$.*

By them being equivalent we mean that they are connected by a chain of quasi-isomorphisms of differential graded algebras. The Theorem follows from the Lemma because of Lemma 10. To prove the Lemma, consider the Weyl bundle $\mathbb{W}$ with a Fedosov connection D^F [6]. Consider the double complex $\Omega^*(M, C^*(\mathbb{W}))$ with the differential D^F. This is a differential graded algebra which, after inverting $\hbar$, is quasi-isomorphic to both algebras in the statement of the Lemma. Indeed, on the one hand, for every p the complex $\Omega^*(M, C^p(\mathbb{W}))[\hbar^{-1}]$ has only zero-dimensional cohomology group which is $C^p(\mathbb{A}^\hbar(M)[\hbar^{-1}])$; on the other hand, the embedding $\mathbb{C}[[\hbar, \hbar^{-1}] \to C^*(\mathbb{A}^\hbar(M)[\hbar^{-1}])$ is a quasi-isomorphism invariant under all derivations of $\mathbb{A}^\hbar(M)$ (which can be seen by filtering the complex by powers of $\hbar$ and considering the corresponding spectral sequence). □

Next we will compute "the Fukaya subcategory" of a deformed algebra of functions on $M = S^2$. We will consider the deformation of the algebra of algebraic functions. Let $A = U(\mathfrak{sl}_2)/(\Delta - c^2)$ where Δ is the Casimir operator. Consider the subgroup S^1 of SU_2 acting on A;

$$X_\pm \mapsto \alpha^{\pm 1} X_\pm; H \mapsto H$$

where $\alpha \in S^1$.

Let ${}_\alpha\Omega_\beta(M) = \{l : S^1 \to M : \alpha l(1) = \beta l(1)\}$.

Proposition 16 *One has*

$$\mathcal{H}^*_A(\alpha, \beta) = H^{*-d_{\alpha\beta}}({}_\alpha\Omega_\beta(M))$$

where $d_{\alpha\beta} = \mathrm{codim}\{x|\alpha x = \beta x\}$;

$$H_*(A, A_\alpha) = H^{2-*}(M), \quad \alpha = 1; \tag{14}$$

$$H_*(A, A_\alpha) = \bigoplus_{\alpha x = x} \mathbb{C}, \quad \alpha \neq 1. \tag{15}$$

Proof. Recall that if V is a $U(\mathfrak{g})$-bimodule then $H^*(U(\mathfrak{g}), V)$ is computed by the Koszul complex $(C_*(\mathfrak{g}, V), \partial)$ where the $\mathfrak{g}$-module structure on V is given by $X \cdot v = Xv - vX$. Consider the operator $D : C_* \to C_{*+1}$,

$$D(v \otimes (X_1 \wedge \ldots \wedge X_n)) = \sum_i (e_i v + v e_i) \otimes (e^i \wedge X_1 \ldots \wedge X_n)$$

where (e_i, e^i) are dual bases with respect to the Killing form. The mixed complex

$$(R_n = \bigoplus C_{n-2i}(\mathfrak{g}, A \otimes A); \ \partial + D)$$

is a free bimodule resolution of A (the subcomplex of the standard bar resolution). Computing the spectral sequences for $R_* \otimes_{A \otimes A^o} A_\alpha$ or for $Hom_{A \otimes A^o}(R_*, {}_\alpha A_\beta)$, we see that they degenerate at E_2. One sees that the homology H_* is given by (14, 15) and the cohomology is given by

$$H^*(A, {}_\alpha A_\beta) = H^*(S^2), \quad \alpha = \beta;$$

$$H^*(A, {}_\alpha A_\beta) = \bigoplus_{\{x \mid \alpha x = \beta x\}} \mathbb{C}[-2], \quad \alpha \neq \beta.$$

The spectral sequence with the E_2 term $H_*(H^*, {}_\alpha H^*_\beta)$ degenerates at E_2 (there is no place for a nonzero differential). □

References

1. F. Bayen, M. Flato, C. Frønsdal, A. Lichnerowicz, D. Sternheimer(1978) Deformation and quantization, I, II, *Annals of Physics*, **Vol. 111**, 61–151.
2. A. Cartan and S. Eilenberg (1956) *Homological Algebra*. Princeton University Press.
3. A. Connes (1994) *Non Commutative Geometry*. Academic Press.
4. A. Connes, M. Flato, D. Sternheimer (1992) Closed star products and cyclic homology, *Letters in Math. Phys.* **Vol. 24**, 1-12.
5. P. Deligne (1995) Déformations d'algèbre des fonctions d'une varieté symplectique: comparaison entre Fedosov et DeWilde, Lecomte, *Selecta Math.*, **Vol. 1**, 667–697.

6. B.V. Fedosov (1996) *Deformation Quantization and Index Theory*. Akademie Verlag, Berlin.
7. K. Fukaya (1994) Morse homotopy, A_∞ categories and Floer homologies, *MSRI Preprint* No. 020-94.
8. M. Gerstenhaber (1963) The cohomology structure of an associative ring, *Ann. of Math.*, **Vol. 78**, 267-288.
9. M. Gerstenhaber, A. Voronov (1995) Homotopy G-algebras and moduli space operad, *Int. Math. Res. Notices*, 141–153.
10. E. Getzler (1993) Cartan homotopy formulas and the Gauss-Manin connection in cyclic homology, *Israel J. of Math.*, **Vol. 7**, 65–78.
11. E. Getzler and J. Jones (1994) Operads, homotopy algebra and iterated integrals for double loop spaces, *hep-th/9403055*.
12. E. Getzler and J. Jones (1990) A_∞ algebras and cyclic complexes, *Illinois J. of Math.*, **Vol. 34 no. 2**, 256–278.
13. C.E. Hood and J. Jones (1987) Some algebraic properties of cyclic homology groups, *K-theory* **Vol. 1**, 361–384.
14. G. Hochschild, B. Kostant and A. Rosenberg (1962) Differential forms on regular affine algebras, *Trans. Amer. Math. Soc.* **Vol. 102**, 383–408.
15. M. Kontsevich (1996) Lectures at IHES.
16. J.L. Loday (1993) *Cyclic homology*. Springer Verlag, New York.
17. Yu.I. Manin (1991) *Topics in non-commutative geometry*. Princeton University Press.
18. D. MacDuff and D. Salamon (1994) *Quantum Comology and J-holomorphic Curves* . AMS University Lecture Series, Vol. 6.
19. R. Nest and B. Tsygan (1995) Algebraic index theorem for families, *Advances in Math.*, **Vol. 113 no. 2**, 151–205.
20. R. Nest and B. Tsygan (1996) On the cohomology ring of an algebra, *Manin Festschrift*, to appear.
21. G. Rinehart (1963) Differential forms on commutative algebras, *Trans. Amer. Math. Soc.*, **Vol. 108**, 139–174.
22. B. Tsygan (1983) Homology of matrix Lie algebras over rings and Hochschild homology, *Uspekhi Mat. Nauk*, **Vol. 38 no. 2**, 217–218.

TANGENTIAL DEFORMATION QUANTIZATION AND POLARIZED SYMPLECTIC GROUPOIDS

ALAN WEINSTEIN*
Department of Mathematics, University of California
Berkeley, CA 94720, USA
alanw@math.berkeley.edu

Abstract. We derive geometric analogues of theorems of Cahen–Gutt–Rawnsley and Asin Lares on deformation quantizations compatible with coadjoint orbits.

1. Introduction

The symmetrization isomorphism between the universal enveloping algebra $U(\mathfrak{g})$ and the symmetric algebra $S(\mathfrak{g})$ can be used [5, 14] to produce a "standard" deformation quantization [4] for the Lie–Poisson structure on the dual $\mathfrak{g}^*$ of the Lie algebra $\mathfrak{g}$. Since the symplectic leaves of $\mathfrak{g}^*$ are the coadjoint orbits, it is interesting to know whether this, or any, deformation quantization of $\mathfrak{g}^*$ restricts to give deformation quantizations of the orbits. Cahen, Gutt, and Rawnsley [6] have shown that a deformation quantization of $\mathfrak{g}^*$ by bidifferential operators compatible with the coadjoint orbit decomposition can exist only if $\mathfrak{g}$ satisfies a very strong algebraic condition. No semisimple Lie algebras satisfy this condition. In addition, Asin Lares [2] has shown that the *standard* deformation quantization of $\mathfrak{g}^*$ restricts to a given coadjoint orbit $\mathcal{O}$ if and only if $\mathcal{O}$ is an open subset of an affine subspace in $\mathfrak{g}^*$.

Cahen–Gutt–Rawnsley (CGR) and Asin Lares give algebraic proofs which provide no geometric connections between the hypotheses and conclusions of their theorems. The purpose of this note is to establish such connections by using the idea of quantization via symplectic groupoids. Although our results do not (yet?) lead to new proofs of the theorems on deformation quantization, they do establish a new analogy between defor-

*Research partially supported by NSF Grant DMS-96-25122.

D. Sternheimer et al. (eds.), Deformation Theory and Symplectic Geometry, 301–314.

mation quantization and quantization by symplectic groupoids, adding to the evidence in [24] for a close relation between these two kinds of quantization. In fact, we now propose to call quantization by symplectic groupoids **geometric deformation quantization**; the similarity to the standard term "geometric quantization" is entirely intentional, as we shall explain below.

The main idea of our paper is to introduce a structure called a **strict geometric deformation quantization** and to show (Theorems 6.1 and 6.2 below) that the precise conclusions of the Asin Lares and CGR theorems hold when algebraic deformation quantizations are replaced by their geometric analogues.

In the CGR case, our conclusion is even stronger, and our work suggests the following possible sharpening of the CGR result, as well as a possible route to a more conceptual proof.

Conjecture 1.1 *Let G be a Lie group whose Lie algebra is $\mathfrak{g}$. The Poisson manifold $\mathfrak{g}^*$ admits a deformation quantization by bidifferential operators compatible with the coadjoint orbit decomposition if and only if a neighborhood of the identity in G admits a flat torsion-free affine connection for which the induced connection on the vector bundle T^*G is compatible with the decomposition into the left (or right) translates of coadjoint orbits.*

Flat torsion free connections have appeared before in deformation quantization (e.g. in quantizations of Moyal and Fedosov type), but they seem to occur here for quite different reasons.

We end this introduction with two questions.

1. It is known [1] that when $\mathfrak{g}$ is semisimple of compact type, there *is* an algebraic deformation quantization of the polynomials on $\mathfrak{g}^*$ which is compatible with the coadjoint orbit decomposition on the regular orbits (and on all orbits for some special cases, such as $\mathfrak{su}(2)$). The operators defining this quantization are *not* bidifferential. We also note that the idea was raised already by Flato and Frønsdal and reported in [20] that an orbit-compatible deformation quantization of $\mathfrak{g}^*$ might involve nonbidifferential, possibly *pseudo*-differential operators. Is there something in our geometric notion of deformation quantization which corresponds to bidifferentiability?

2. Which Lie algebras admit a connection of the type described in Conjecture 1.1?

Acknowledgements

This work was stimulated by conversations with Simone Gutt, Santos Asin Lares, Michel Cahen, and John Rawnsley at the conference on Deformation Theory and Mathematical Physics, Ascona, June 1996 and by lectures of Gutt and Asin Lares at the CIMPA Summer School on Quantization, Nice,

July 1996. I would like to thank the sponsors and organizers of both of those conferences for providing an excellent working atmosphere.

2. Geometric deformation quantization

The cotangent bundle T^*P is a "geometric model" for the space of functions (or half-densities) on a manifold P. In this model, multiplication of functions is represented by addition in the fibres of T^*P. (For instance, the multiplication of exponential functions corresponds to the addition of their frequency covectors.) Independently, Karasev [15], Zakrzewski [26], and the author [8, 24] have developed a geometric model for deformed multiplications in which T^*P with its fibrewise addition operation is replaced by a general symplectic groupoid whose base is P. Such a groupoid determines a Poisson structure on P, just as a deformed multiplication does. Even though it is still unclear how to pass back and forth between deformed multiplications and these geometric objects, they have quite a rich structure in their own right. For precision, we make two definitions.

The first definition corresponds to the case where we have a 1-parameter family of associative algebras deforming $C^\infty(P)$, but where the algebras are not necessarily isomorphic as vector spaces to $C^\infty(P)$. The deformation parameter will appear in Section 3.

Definition 2.1 *A* **geometric deformation quantization** *of a Poisson manifold P is a symplectic groupoid Γ over P.*

The second definition corresponds to the case where all the deformed algebras are identified with $C^\infty(P)$ as vector spaces. It may be related to Rieffel's analytic notion of strict deformation quantization [18]. (See Section 7 of [24].)

Definition 2.2 *A* **strict geometric deformation quantization** *of a Poisson manifold P is a symplectic groupoid Γ over P together with a symplectic isomorphism from Γ to T^*P which maps the identity section to the zero section and which conjugates inversion in Γ to multiplication by -1 in T^*P.*

We call two manifolds **globally transverse** if they are transverse and intersect in a single point. The following proposition is a simple globalization of Theorem 7.1 in [22]. We omit the proof.

Proposition 2.3 *There is a one-to-one correspondence between strict geometric deformation quantizations and pairs consisting of a geometric deformation quantization Γ and a real polarization of the symplectic manifold Γ with complete (in the natural flat affine connection) simply connected leaves, each of which is globally transverse to the zero section and is invariant under inversion.*

For many purposes, local quantizations are sufficient. Recall that a **local groupoid** is essentially an object having the properties of a neighborhood of the identity in a groupoid. See Section 4.2.2 of [10] for a definition.

Definition 2.4 *A [strict]* **local geometric deformation quantization** *of a Poisson manifold P is a local symplectic groupoid over P [together with a local polarization which is transverse to the identity section].*

A strict local geometric deformation quantization exists for any Poisson manifold. Without the strictness condition, it is unique up to isomorphism, but there are in general many ways in which the polarization can relate to the groupoid structure. It is not clear how these facts are connected to the existence and uniqueness of algebraic deformation quantizations.

3. Explosion of manifolds

To see the geometric objects defined above as deformations, we recall the notion of "explosion of manifolds". (This term was introduced in [23], but the idea appears already in algebraic form in a 1966 paper by Gerstenhaber [12]. See [23] for other references.)

Definition 3.1 *The* **explosion** $E(X,Y)$ *of a differentiable manifold* X *along a submanifold* Y *is, as a set, the product* $X \times \mathbb{R}$ *with the slice* $X \times \{0\}$ *replaced by the normal bundle* $N(X,Y) = T_Y X/TY$. *A precise definition of the differentiable structure on* $E(X,Y)$ *is given in [23].*

If (y,z) are local coordinates on X, with Y defined by $z=0$, then there is a corresponding set of local coordinates (y,z',ϵ) on $E(X,Y)$, where ϵ is the projection to $\mathbb{R}$. Over the set where $\epsilon \neq 0$, the projection from $E(X,Y)$ to X is given in these coordinates by $(y,z',\epsilon) \mapsto (y,\epsilon z)$. This projection extends over $\epsilon = 0$ to a smooth map $p : E(X,Y) \to X$ when we map $N(X,Y)$ to $Y \subset X$ by the bundle projection.

If X carries a symplectic structure ω, we can pull ω back by the projection p to a closed 2-form on $E(X,Y)$. We consider this pullback instead as a family $p^{-1}\omega$ of 2-forms on the fibres of the projection ϵ. If Y is an isotropic submanifold of X, the ratio $p^{-1}\omega/\epsilon$ is still a smooth family of closed forms. In particular, if Y is a lagrangian submanifold of X, then by using local Darboux coordinates on X which are adapted to Y we can see that $p^{-1}\omega/\epsilon$ is symplectic on every fibre of ϵ, and that the symplectic structure on the fibre $\epsilon = 0$ is exactly that given by the usual identification of the normal bundle of a lagrangian submanifold with its cotangent bundle. To summarize:

Proposition 3.2 *If* Y *is a lagrangian submanifold of the symplectic manifold* (X,ω), *then the exploded manifold* $E(X,Y)$ *is a smooth family of* *symplectic manifolds obtained from the product* $X \times \mathbb{R}$ *carrying the form* ω/ϵ *on each* $X \times \{\epsilon\}$ *by replacing the fibre* $X \times \{0\}$ *by the cotangent bundle* T^*Y *with its canonical symplectic structure.*[1]

When X is a (possibly local) geometric deformation quantization of a Poisson manifold P and Y is the identity section, the fibre $\epsilon = 0$ in $E(X,Y)$ is just T^*P with its canonical symplectic structure. Now suppose that the deformation is strict. By choosing Darboux coordinates (y, z) on X in which Y is defined by $z = 0$ and the leaves of the polarization are defined by $y =$ constant, we get a family of polarizations on the fibres of ϵ in $E(X,Y)$ which for $\epsilon = 0$ is the polarization of T^*P by cotangent spaces.

We continue with a discussion of the effect of explosion on some other geometric structures.

If Y is a leaf of a foliation of X, then the foliations on the fibres for $\epsilon \neq 0$ extend over $\epsilon = 0$ to give a foliation on all of $E(X,Y)$. (One sees this easily by using coordinates (y, z) in which the foliation is given by $z =$ constant.) The leaves of the induced foliation on the slice $\epsilon = 0 = N(X,Y)$ are the parallel sections for a flat linear ("Bott") connection on the normal bundle to the leaf. When X is symplectic and the foliation is a polarization, then the induced foliation on T^*Y is lagrangian, so that the connection is not only flat but torsion free as well. This is the flat affine structure on a leaf of a polarization discovered by Libermann [16] (also see [22]).

Finally we investigate when happens to a Lie groupoid structure on X when we explode along its identity section Y. When X is a group, $E(X,Y)$ is a smooth 1-parameter family of Lie groups, with the group $\epsilon = 0$ being the Lie algebra $N(X,Y)$ with the addition operation. For a general Lie groupoid, it follows from our discussion of foliations transverse to Y that the source and target foliations of X *both* have as their limits the foliation by fibres in $N(X,Y)$; the limit of the groupoid multiplication is addition in these fibres. In other words, $E(X,Y)$ is a 1-parameter family of groupoids over Y which are isomorphic for $\epsilon \neq 0$ to X, and for $\epsilon = 0$ to the vector bundle $N(X,Y)$, which is just the Lie algebroid of X. It can also be considered as a single groupoid over $X \times \mathbb{R}$.

If $X = Y \times Y$ is the pair groupoid, then $N(X,Y)$ is the tangent bundle TY, and $E(X,Y)$ is called the **tangent groupoid** of Y. We refer to [9] for a discussion of the central role played by this groupoid in index theory.

If X is a symplectic groupoid over a Poisson manifold Y, $E(X,Y)$is a 1-parameter family of symplectic groupoids over Y (or a single Poisson groupoid over $Y \times \mathbb{R}$). The fibres for $\epsilon \neq 0$ are isomorphic to X as groupoids, but with symplectic structure multiplied by $1/\epsilon$, while the fibre $\epsilon = 0$ is

[1]The reader might want to compare this result with Corollary 5.3 in [23], where a manifold exploded along a legendrian submanifold yields a Poisson manifold in which the open dense subset $\epsilon \neq 0$ is a single symplectic leaf.

the vector bundle T^*Y with its canonical symplectic structure. If X is strict as a geometric deformation quantization of Y, then the limit in T^*Y of the polarization on X is always the polarization by cotangent spaces. Thus, geometric deformation quantizations are deformations of T^*Y as a symplectic groupoid.

4. From geometric to algebraic deformations

In this section we shall review from [24] the use of geometric quantization and microlocal analysis to construct noncommutative algebras from symplectic groupoids. In the next sections, we will apply these ideas to obtain geometric analogues of certain conditions on algebraic deformations.

We appeal to the quantization "dictionary" (see, for example [3]) connecting symplectic geometry and linear algebra (or analysis): thus, a symplectic manifold X should correspond to a complex vector space $V(X)$, while a lagrangian submanifold L of X corresponds to an element of $V(X)$. Taking the product of symplectic manifolds produces the tensor product of their vector spaces, and multiplying a symplectic structure by -1 (we denote the resulting symplectic manifold by $\overline{X}$) produces the dual vector space. A lagrangian submanifold R of $\overline{X} \times Y$, called a canonical relation from X to Y, then corresponds to a linear mapping $V(R)$ from $V(X)$ to $V(Y)$. Quantization should be functorial in the sense that a composition of canonical relations should go over to the composition of the corresponding linear mappings.

Continuing with our hypothetical mode of argument, we note that the multiplication in a symplectic groupoid X is a canonical relation from $X \times X$ to X, so it corresponds to a bilinear multiplication on $V(X)$, whose associativity follows from that of the groupoid if quantization is functorial. The identity section Y of X corresponds to the unit element of the algebra $V(X)$, and inversion in X to an anti-automorphism. The explosion $E(X, Y)$ of X along Y corresponds to a 1-parameter deformation of unital involutive algebras starting with the quantization of T^*Y, which is just $C^\infty(Y)$ with pointwise multiplication, unit element 1, and complex conjugation.

To pass from the hypothetical to a constructive mode, we use ideas from geometric quantization and microlocal analysis. According to geometric quantization, we can construct $V(X)$ for a symplectic manifold X by introducing a complex line bundle F with a connection whose curvature is the symplectic form on X, and then taking the space of sections which are parallel along a chosen polarization. To quantize a lagrangian submanifold L in X, we lift L to a parallel section s of F (which exists when L satisfies a quantization condition) and then take the parallel section of F which agrees with s along L. Of course, this procedure works nicely only when

L is globally transverse to each leaf of the polarization, or more generally when L has clean intersection with the leaves of the polarization, in which case the support of the quantization of L should be the set of leaves which intersect L.

The quantization of lagrangian submanifolds can be carried out via microlocal analysis in the special case where the polarization is a fibration by manifolds which are simply connected and are complete with respect to their flat affine connection. If we assume the existence of a "base" lagrangian submanifold Y which is globally transverse to each leaf of the polarization, then X is isomorphic to the cotangent bundle of Y with the polarization by fibres. Lagrangian submanifolds of T^*Y can then be quantized as asymptotic distributions on Y, i.e. distributions depending on a parameter ϵ. (See [3] or, for more details, the last chapter of [13].) The asymptotic support of such a distribution (i.e. the set where it is not rapidly decreasing as $\epsilon \to 0$) is contained in the projection of L into Y.

When X is a symplectic groupoid, applying the construction just described to the graph of multiplication leads to an asymptotic distribution on $Y \times Y \times Y$ which is a candidate for the Schwartz kernel of a multiplication on $C^\infty(Y)$. We discuss this further in the next section.

5. Deformations compatible with a submanifold

In this section, we will define and analyze a property of geometric deformation quantizations which is analogous to the compatibility of algebraic deformations with a subset of the underlying manifold. We begin with some algebraic and heuristic arguments, which will lead us to a precise geometric condition.

If we have an associative multiplication $*$ on the space $C^\infty(P)$ (which we may think of as a deformation of pointwise multiplication, although this is not essential), it restricts to a multiplication on the functions on a closed subset $M \subset P$ when the functions vanishing on P form a two-sided ideal in the algebra determined by $*$. If $*_\epsilon$ depends on a parameter ϵ, then we should impose this condition for all ϵ, or the consequent condition on the coefficients of all powers of ϵ in case the deformation is formal. When the condition is satisfied, we call the deformation **compatible with** M.

Now let Γ be a (possibly local) strict geometric deformation quantization of the Poisson manifold P, and identify P with the lagrangian submanifold of identity elements in Γ. We will denote by π the projection from Γ to P along the leaves of the polarization.

Quantizing the canonical relation of groupoid multiplication in Γ produces a product operation on $C^\infty(P)$ which is defined by a kernel distribution of the form $K(u', u'', u)\, du'\, du''$ on $P \times P \times P$; i.e. the new multiplication of two functions f and g on P is given by an integral

$$(f * g)(u) = \int_{P \times P} f(u')g(u'')K(u', u'', u)\, du'\, du''.$$

The asymptotic support of K is contained in the set of triples

$$S_\Gamma = \{(\pi(\gamma'), \pi(\gamma''), \pi(\gamma'\gamma''))|(\gamma', \gamma'') \text{ is a composable pair in } \Gamma \times \Gamma\}.$$

Unless there are special cancellations in the quantization construction, the asymptotic support should be equal to S_Γ; we will take this condition as a working hypothesis.

To relate the support of K to compatibility with a submanifold, we observe the following.

Lemma 5.1 *Let M be a subset of P, If the set J_M of smooth functions on P which vanish on some neighborhood of M forms an ideal with respect to the multiplication $*$ defined by the kernel K, then if (u', u'', u) is in the support of K and u is in M, u' and u'' must also be in M.*

Proof. Suppose that there is a triple (u', u'', u) in the support of K which does not satisfy the condition of the Lemma. Then we can find bump functions f and g supported near u' and u'' respectively such that $(f * g)(u)$ is not zero.

Q.E.D.

Actually, we would like to replace J_M by all the functions which vanish on M; nevertheless, the discussion above makes the following definition appear to be a reasonable one.

Definition 5.2 *Let Γ be a (possibly local) strict geometric deformation quantization of a Poisson manifold P, and let M be a subset of P. Let $\pi : \Gamma \to P$ be the projection along the leaves of the polarization We say that the deformation is* **compatible with** *M if for any two composible groupoid elements γ' and γ'', $\pi(\gamma'\gamma'') \in M$ implies that $\pi(\gamma') \in M$ and $\pi(\gamma'') \in M$.*

Notice that the compatibility condition on M in the preceding definition involves only the subset $\pi^{-1}(M)$ of the groupoid Γ: a product must lie in $\pi^{-1}(M)$ if and only if both of its factors do. We will call such a subset of a groupoid **saturated**. (Equivalently, the complement of the set is absorbing under multiplication.)

Saturated subsets prove to be very special. We recall some terminology. Let Γ be a groupoid over P with source and target maps β and α, so that a product $\gamma'\gamma''$ is defined whenever $\beta(\gamma') = \alpha(\gamma'')$. Two elements of P are called equivalent if some element of Γ has them as its source and target. This is indeed an equivalence relation, and the equivalence classes are called **orbits** of Γ in P.

As usual, we identify P with the set of identity elements in Γ, so that α and β are retractions.

Proposition 5.3 *Let Γ be a groupoid over P with source and target mappings β and α. A subset C of Γ is saturated if and only if it is of the form $\alpha^{-1}(U)$, where U is a union of orbits in P (in which case $C = \beta^{-1}(U)$ as well).*

Proof. Let $C = \alpha^{-1}(U) = \beta^{-1}(U)$, where U is a union of orbits in P. If $\gamma'\gamma'' \in C$, then $\alpha(\gamma') = \alpha(\gamma'\gamma'')$ and $\beta(\gamma'') = \beta(\gamma'\gamma'')$ belong to U. This implies that γ' and γ'' belong to C, so C must be saturated.

Conversely, suppose that C is saturated. For any $\gamma \in C$, the identity $\alpha(\gamma)\gamma = \gamma = \gamma\beta(\gamma)$ implies that C is closed under the source and target retractions. Let U be the image of C under either of these mappings, which also must equal the intersection of C with P. The identities $\alpha(\gamma) = \gamma\gamma^{-1}$ and $\beta(\gamma) = \gamma^{-1}\gamma$ imply that C is contained in, and hence equal to, $\alpha^{-1}(U)$ and $\beta^{-1}(U)$. Finally, the fact that $C = \alpha^{-1}(U)$ is closed under β implies that U is a union of orbits.

Q.E.D.

The following Corollary follows immediately from Definition 5.2, Proposition 5.3, and the fact that the mapping π associated with a polarization is a retraction.

Corollary 5.4 *A (possibly local) strict geometric deformation quantization of the Poisson manifold P is compatible with a subset M of P if and only if M is a union of orbits in P and $\pi^{-1}(M) = \alpha^{-1}(M) = \beta^{-1}(M)$.*

Remarks. If Γ is α-connected, the orbits in P are just the symplectic leaves. Thus, M must be a union of symplectic leaves. The inverse image under α of a symplectic leaf is coisotropic; hence, so is $C = \alpha^{-1}(M) = \beta^{-1}(M)$. Since, for u in M, $\pi^{-1}(u)$ is lagrangian and contained in the coisotropic submanifold C, $\pi^{-1}(u)$ must contain the characteristic submanifold of C which passes through u. If C is a single orbit, this characteristic submanifold is just the isotropy subgroup Γ_u of u.

To close this section, we consider the case where a strict geometric deformation quantization is compatible with the decomposition of the Poisson manifold P into its symplectic leaves, i.e. the quantization is compatible with each leaf separately. By analogy with the algebraic case, we will such a quantization **tangential**. For simplicity, we will assume that Γ is α-connected, so that the leaves are the orbits of Γ. The inverse images of the leaves under α or β form a decomposition of Γ into coisotropic subgroupoids which we will call the **fundamental decomposition** of the groupoid Γ.

Corollary 5.5 *A (possibly local) strict deformation quantization Γ of the Poisson manifold P is tangential if and only if the polarization of Γ is a refinement of the fundamental decomposition.*

6. The Lie–Poisson case

By applying the results of the previous section when $P = \mathfrak{g}^*$ is the dual of the Lie algebra of the Lie group G, we will now obtain geometric analogues of the theorems of Asin Lares and Cahen–Gutt–Rawnsley on algebraic deformation quantizations compatible with coadjoint orbits.

The cotangent bundle $\Gamma = T^*G$ is the underlying symplectic manifold of a geometric deformation quantization of $\mathfrak{g}^*$, with a groupoid structure for which the source and target maps are the left and right translations to the cotangent space T_e^*G at the group identity, this cotangent space being the manifold of identities in Γ. (See [8] or the exposition in [21].) The cotangent bundle of course carries a natural polarization by its fibres, but this does not constitute a strict geometric deformation quantization because the leaves are not transverse to the groupoid identity manifold. In fact, when the method of geometric quantization is applied to this polarization, the resulting algebra is the convolution algebra $A(G)$ of densities on the group G.

Berezin [5] connected the group convolution algebra with the functions on $\mathfrak{g}^*$ by the composition of two operations, pullback by the exponential map from $\mathfrak{g}$ to G and Fourier transform from densities on the Lie algebra $\mathfrak{g}$ to functions on $\mathfrak{g}^*$. Applying these operations to the subalgebra of $A(G)$ consisting of distribution densities supported at e produces the standard "symmetrization" identification between the universal enveloping algebra $U(\mathfrak{g})$ and the symmetric algebra $S(\mathfrak{g})$ of polynomial functions on $\mathfrak{g}^*$. He expressed the Schwartz kernel of the new multiplication on $\mathfrak{g}^*$ in terms of the Campbell–Hausdorff mapping $\mathfrak{g} \times \mathfrak{g} \to \mathfrak{g}$, and he observed that the commutator in $U(\mathfrak{g})$ is a perturbation of the Lie–Poisson bracket on $\mathfrak{g}^*$. (No deformation parameter appears in Berezin's work.) The symmetrization correspondence was later shown by Gutt [14] to give a formal algebraic deformation quantization, while Rieffel [19] used the Fourier transform picture to show that one had a strict deformation quantization in the sense of C^*-algebras.

It is well known in geometric quantization (see, for example [25]) that the Fourier transform between functions (or densities) on a vector space V and its dual V^* is represented geometrically by the pairing between the spaces obtained from the "horizontal" and "vertical" polarizations of the symplectic manifold $V \times V^*$. This leads us to make the following construction. By the exponential mapping (restricted if necessary to an open neighborhood of 0 in $\mathfrak{g}$ on which it is an embedding, although we will seldom mention this restriction explicitly from now on), we identify T^*G with $T^*\mathfrak{g}$ and hence with $\mathfrak{g} \times \mathfrak{g}^*$. The latter symplectic space has a "horizontal" polarization whose leaves are globally transverse to $\{0\} \times \mathfrak{g}^*$. Under our exponential identification, this gives a polarization of T^*G whose leaves are globally transverse to T_e^*G, which we will call the **exponential polarization**. The symplectic groupoid T^*G with this polarization constitutes a (possibly local) strict geometric deformation quantization of $\mathfrak{g}^*$ which will may call the **exponential quantization**.

Our version of the algebraic theorem of Asin Lares [2] now goes as follows.

Theorem 6.1 *Let $\mathcal{O}$ be a coadjoint orbit in $\mathfrak{g}^*$. Then the following three conditions are equivalent: (i) the exponential quantization is compatible with $\mathcal{O}$; (ii) $\mathcal{O}$ is flat–i.e. it is an open subset of an affine subspace of $\mathfrak{g}^*$; (iii) the isotropy algebra $\mathfrak{g}_u$ of any $u \in \mathcal{O}$ is an ideal in $\mathfrak{g}$.*

Proof.

We begin with a proof of the equivalence of (ii) and (iii), close to the one for the nilpotent case which can be found in [7]. (See Theorem 3.2.3 in this book, which also exhibits the importance of flat orbits in representation theory.) Recall that the tangent space to a coadjoint orbit $\mathcal{O}$ at a point u, when considered as a linear subspace of $\mathfrak{g}^*$, is the annihilator of the isotropy algebra $\mathfrak{g}_u$. The orbit is flat if and only if its tangent spaces are all equal, i.e. when the isotropy is independent of u. The isotropy subgroups of the points of $\mathcal{O}$ are the conjugates of that of u, so the isotropy subalgebras are all the same when the isotropy subgroup is normal, i.e. when $\mathfrak{g}_u$ is an ideal.

The rest of the proof will based on the formula for the derivative of the exponential mapping of a Lie group:

$$T_a \exp = T_e \ell_{\exp a} \circ \frac{1 - e^{\mathrm{ad}_a}}{\mathrm{ad}_a},$$

which, following a suggestion of Duistermaat, we write in the integral form

$$T_a \exp = T_e \ell_{\exp a} \circ \int_0^1 e^{-s\mathrm{ad}_a}\, ds.$$

In both of these formulas, ℓ_g denotes left translation by the group element g.

We will use this formula to compute the mapping $\pi : T^*G \to \mathfrak{g}^*$ associated with the exponential polarization, using the identification of T^*G with $G \times \mathfrak{g}^*$ given by left translations. Under this identification, the mappings α and β from $G \times \mathfrak{g}^*$ to $\mathfrak{g}^*$ are given by the projection of $G \times \mathfrak{g}^*$ and the coadjoint action respectively.

For $g = \exp a \in G$ and $\mu \in \mathfrak{g}^*$, we have

$$\begin{aligned}\pi(g,\mu) &= (T_g\ell_g^{-1})^*\mu \circ T_a \exp \\ &= \mu \circ T_g\ell_g^{-1} \circ T_e\ell_g \circ \int_0^1 e^{-s\mathrm{ad}_a}\, ds \\ &= \mu \circ \int_0^1 \mathrm{Ad}_{\exp(-sa)}\, ds \\ &= \int_0^1 \mathrm{Ad}^*_{\exp(-sa)}\mu\, ds. \qquad (1)\end{aligned}$$

Suppose now that $\mathcal{O}$ is flat and that $(g,\mu) \in \alpha^{-1}(\mathcal{O})$, i.e. $\mu \in \mathcal{O}$. For g near e, $\pi(g,\mu)$ is a convex combination of elements of $\mathcal{O}$ near μ and hence must belong to the flat orbit $\mathcal{O}$; thus we have shown that $\alpha^{-1}(\mathcal{O}) \subseteq \pi^{-1}(\mathcal{O})$. Since $\alpha^{-1}(\mathcal{O})$ and $\pi^{-1}(\mathcal{O})$ are manifolds of the same dimension, both containing $\mathcal{O}$, they must be equal near $\mathcal{O}$, i.e. near the identity section of T^*G, so the quantization is compatible with $\mathcal{O}$.

Conversely, suppose that the quantization is compatible with $\mathcal{O}$. Fix $\mu \in \mathcal{O}$ for the moment and consider the map $\phi : g \mapsto \pi(g,\mu)$. Since $(g,\mu) \in \alpha^{-1}(\mathcal{O})$, by compatibility we have $(g,\mu) \in \pi^{-1}(\mathcal{O})$, so $\phi(g,\mu)$ is an element of $\mathcal{O}$ which is also in the convex hull of $\mathcal{O}$. On the other hand, it is easy to check (note that $\pi(\exp a,\mu)$ is an average of points on a path from μ to $\mathrm{Ad}^*_{\exp(-a)}\mu$) that $T_e\phi : \mathfrak{g} \to \mathfrak{g}^*$ is the map $b \mapsto -\frac{1}{2}\mathrm{Ad}^*_b\mu$, whose image is $T_\mu\mathcal{O}$. Thus ϕ maps a neighborhood of e in G to a neighborhood of μ in $\mathcal{O}$. It follows that μ is locally convex near μ, and since μ was arbitrary, $\mathcal{O}$ must be flat.

Q.E.D.

We turn now to the theorem of Cahen–Gutt–Rawnsley [6]. Our geometric version is Corollary 6.3 below, which will follow from the more general theorem which follows, and which inspired Conjecture 1.1.

Theorem 6.2 *Let G be a Lie group whose Lie algebra is $\mathfrak{g}$. Then $\mathfrak{g}^*$ admits a local tangential strict geometric deformation quantization if and only if a neighborhood of the identity in G admits a flat torsion-free affine connection for which the associated connection on T^*G is compatible with the decomposition into left (or right) translates of coadjoint orbits.*

Proof. Any local symplectic groupoid for $\mathfrak{g}^*$ is a neighborhood of the identity section in T^*G. By Corollary 5.5, its polarization as a tangential strict deformation quantization refines the fundamental decomposition by the left or right translates of coadjoint orbits. In particular, the translates of the coadjoint orbit $\{0\}$ form the zero section of T^*G, which must therefore be a leaf of the polarization. If we explode T^*G around this leaf, as in Section 2, we obtain a flat torsion free affine connection on a neighborhood of the identity in G for which the associated flat connection on T^*G refines the exploded fundamental decomposition. But since multiplication by constants on T^*G leaves the fundamental decomposition invariant, the exploded decomposition is the same as the original one. Thus we have a connection of the type described in the statement of the theorem.

For the converse, we simply take the parallel sections for the given connection as a polarization of the fundamental groupoid.

Q.E.D.

The conclusion of the following Corollary is precisely that of the main theorem in [6].

Corollary 6.3 *If $\mathfrak{g}^*$ admits a local tangential strict geometric deformation quantization and B is an ad^*-invariant bilinear form on $\mathfrak{g}^*$, then the image of the associated linear map $\tilde{B} : \mathfrak{g}^* \to \mathfrak{g}$ is a 2-step nilpotent ideal in $\mathfrak{g}$.*

Proof. The fact that $\tilde{B}(\mathfrak{g})$ is an ideal is just a consequence of ad*-invariance and has nothing to do with quantization. (We give a proof below.) To prove that it is 2-step nilpotent, we begin with the case where B is nondegenerate. The pullback $\overline{B}$ of B to T^*G by left or right translations is the kinetic energy function of a bi-invariant pseudoriemannian metric on G. Since B is invariant, it is constant on coadjoint orbits; hence $\overline{B}$ is constant on each of the submanifolds in the fundamental decomposition. Since the connection produced in Theorem 6.2 leaves the fundamental decomposition invariant, it leaves $\overline{B}$ invariant. Being torsion free, it must be the Levi-Civita connection for this metric. But we also know that this connection is flat. Now the riemannian curvature of a bi-invariant metric on G can be expressed in terms of brackets on $\mathfrak{g}$ by the following formula. (See for example [17], p. 113; positive definiteness of the metric is unnecessary.)

$$R(X,Y)Z = \frac{1}{4}[[X,Y],Z]$$

for left invariant vector fields X, Y, and Z. The vanishing of R means that all triple brackets are zero, which is just the condition that $\mathfrak{g}$ be 2-step nilpotent.

In the nondegenerate case, $\mathfrak{g} = \tilde{B}(\mathfrak{g}^*)$, and our proof is complete. For the general case, we note first that, since $\tilde{B}$ is ad*-invariant, so is its kernel, which is the annihilator of $\mathfrak{h} = \tilde{B}(\mathfrak{g})$. It follows that $\mathfrak{h}$ is ad-invariant, and hence an ideal. Since the connection preserves $\overline{B}$, it must preserve the left (which are the same as the right) translates of $\mathfrak{h}$, and therefore the corresponding subgroup $H \subset G$ is totally geodesic. It follows that the connection restricts to a flat, torsion free metric connection on H, from which it follows as above that $\mathfrak{h}$ is 2-step nilpotent.

Q.E.D.

References

1. Arnal, D., Cahen, M., and Gutt, S., Deformations on coadjoint orbits, *J. Geom. Phys.* **3** (1986), 327-351.
2. Asin Lares, S., On tangential properties of the Gutt $*$-product, University of Warwick preprint 1996.
3. Bates, S., and Weinstein, A., Lectures on the geometry of quantization, *Berkeley Math. Lecture Notes*, U. of Calif. Berkeley, (1995).
4. Bayen, F., Flato, M., Frønsdal, C., Lichnerowicz, A., and Sternheimer, D., Deformation theory and quantization, I and II, *Ann. Phys.* **111**, (1977), 61–151.
5. Berezin, F.A., Some remarks about the associated envelope of a Lie algebra, *Funct. Anal. Appl.* **1**, (1967), 91–102.
6. Cahen, M., Gutt, S., and Rawnsley, J., Tangential star products for the coadjoint Poisson structure, *Comm. Math. Phys.* **180**, (1996), 99–108.
7. Corwin, L.J., and Greenleaf, F.P., *Representations of Nilpotent Lie Groups and their Applications*, Cambridge U. Press, Cambridge, 1990.
8. Coste, A., Dazord, P., et Weinstein, A., Groupoïdes symplectiques, *Publications du Département de Mathématiques, Université Claude Bernard-Lyon I* **2A** (1987), 1–62.
9. Connes, A., *Noncommutative Geometry*, Academic Press, San Diego, 1994.
10. Est, E.T. van, Rapport sur les S-atlas, in *Structure Transverse des Feuilletages, Toulouse 17-19 fév 1982, Astérisque* **116** (1982), 235–292.
11. Fedosov, B., A simple geometrical construction of deformation quantization, *J. Diff. Geom.* **40** (1994), 213–238.
12. Gerstenhaber, M., On the deformation of rings and algebras:II, *Annals of Math.*, **84** (1966), 1–19.
13. Guillemin, V., and Sternberg, S., *Geometric Asymptotics*, Math. Surveys **14**, Amer. Math. Soc., Providence, 1977.
14. Gutt, S., An explicit $*$-product on the cotangent bundle of a Lie group, *Lett. Math. Phys.* **7** (1983), 249–258.
15. Karasev, M.V., Analogues of objects of Lie group theory for nonlinear Poisson brackets, *Math. USSR Izvestiya* **28**, (1987), 497–527.
16. Libermann, P., Sur le problème d'équivalence de certaines structures infinitesimales, *Annali di Matematica Pura e Aplicata* **36** (1954), 27–120.
17. Milnor, J., *Morse Theory*, Annals of Mathematics Studies **51**, Princeton U. Press, Princeton, 1963.
18. Rieffel, M.A., Deformation quantization of Heisenberg manifolds, *Commun. Math. Phys.* **122** (1989), 531–562.
19. Rieffel, M.A., Lie group convolution algebras as deformation quantizations of linear Poisson structures, *Amer. J. Math.* **112** (1990), 657–686.
20. Sternheimer, D., Phase-space representations, in *Applications of Group Theory in Mathematical Physics*, M. Flato, P. Sally, G. Zuckerman eds., *Lect. Appl. Math.* **21**, Amer. Math. Soc., Providence, 1985, pp. 255–267.
21. Vaisman, I., *Lectures on the geometry of Poisson manifolds*, Birkhäuser, Basel, 1994.
22. Weinstein, A., Symplectic manifolds and their lagrangian submanifolds, *Advances in Math.* **6** (1971), 329–346.
23. Weinstein, A., Blowing up realizations of Heisenberg–Poisson manifolds, *Bull. Sc. Math.* **113** (1989), 381–406.
24. Weinstein, A., Noncommutative geometry and geometric quantization, *Symplectic Geometry and Mathematical Physics: actes du colloque en l'honneur de Jean-Marie Souriau*, P. Donato et al eds., Birkhäuser (1991), 446–461.
25. Woodhouse, N.M.J., *Geometric Quantization*, 2nd Ed., Oxford U. Press, Oxford, 1992.
26. Zakrzewski, S., Quantum and classical pseudogroups, I and II, *Comm. Math. Phys.* **134** (1990), 347–395.

FORMAL GNS CONSTRUCTION AND WKB EXPANSION IN DEFORMATION QUANTIZATION

MARTIN BORDEMANN AND STEFAN WALDMANN
Fakultät für Physik, Universität Freiburg
Hermann-Herder-Str. 3, D-79104 Freiburg i. Br.

1. The concept of deformation quantization has been defined and exemplified in [2]. The existence of star-products on every symplectic manifolds has been established in [5] and in [6]. This gives a reasonable physical picture of the noncommutative algebra of quantum observables with built-in classical limit. However, the discussion of a formal analogue of representations of the deformed algebra in some 'Hilbert space' seems to have been restricted to examples in the literature up to now. In [4] the authors have proposed how to construct formal pre-Hilbert spaces for the deformed algebra in the same category by means of a generalized Gel'fand-Naimark-Segal (GNS) construction.

We shall now briefly review this construction and shall apply this in the next section to the Weyl star-product on the cotangent bundle of $\mathbb{R}^n$ (which will give back the usual Schrödinger representation together with the Weyl symmetrization rule; details thereof can be found in [4]). The third section contains new material: We shall describe how to incorporate the usual WKB expansion into the framework of star-products and GNS representations by means of a certain positive linear functional on the deformed algebra having support on a projectable Lagrangean submanifold graph(dS) of $T^*\mathbb{R}^n$. The main trick is to use a suitable form of the star-exponential $e^{*\frac{it}{\lambda}S}$.

Let $\mathbb{C}((\lambda))$ (resp. $\mathbb{R}((\lambda))$) be the field of complex (resp. real) Laurent series in λ with finite principal part (i.e. a finite number of negative powers in λ) and for any complex or real vector space V let $V((\lambda))$ denote the vector space of all formal Laurent series (with finite principal part) in λ with coefficients in V. Let (M,ω) be a symplectic manifold equipped with a differential star-product $*$ satisfying $\overline{f * g} = \bar{g} * \bar{f}$. Let $C^\infty(M)$ be the space of all smooth complex-valued functions on M. Then $(C^\infty(M)((\lambda)), *)$ becomes an associative algebra over the Laurent field $\mathbb{C}((\lambda))$. Since the

D. Sternheimer et al. (eds.), Deformation Theory and Symplectic Geometry, 315–319.

field $\mathbb{R}((\lambda))$ is an ordered field by using the real order of the lowest non-vanishing coefficient, a $\mathbb{C}((\lambda))$-linear functional ω of $C^\infty(M)((\lambda))$ can be called *positive* iff $\omega(\bar{f} * f)$ is positive in the Laurent field $\mathbb{R}((\lambda))$ for all $f \in C^\infty(M)((\lambda))$. As in the case of C^*-algebras it can be shown that the so-called *Gel'fand ideal* $J_\omega := \{f \in C^\infty(M)((\lambda)) | \omega(\bar{f} * f) = 0\}$ is a left ideal of the algebra $C^\infty(M)((\lambda))$, that the quotient $\mathcal{H}_\omega := C^\infty(M)((\lambda))/J_\omega$ (with canonical projection denoted by $f \mapsto \Psi_f$) is a left module for $C^\infty(M)((\lambda))$, equipped with a positive definite $\mathbb{C}((\lambda))$-valued inner product defined by $\langle \Psi_f, \Psi_g \rangle := \omega(\bar{f} * g)$, and that the GNS representation $\pi_\omega(f)\Psi_g := \Psi_{f*g}$ is a star-representation in the sense that $\pi_\omega(f)$ and $\pi_\omega(\bar{f})$ are formal adjoints. In [4] it was shown that on any Kähler manifold equipped with the star-product of Wick type (see e.g. [3]) every delta-functional (evaluation at a point) is a positive functional whose GNS-representation gives rise to a formal Bargmann representation.

2. Consider the cotangent bundle $\pi : T^*\mathbb{R}^n \to \mathbb{R}^n$ of $\mathbb{R}^n$ with its canonical symplectic structure, the standard Weyl product $*$, and the standard volume form $d^n q := dq^1 \cdots dq^n$ on $\mathbb{R}^n$. We define

$$C^\infty(T^*\mathbb{R}^n)_{\mathbb{R}^n} := \{f \in C^\infty(T^*\mathbb{R}^n) | \ \mathrm{supp}(f) \cap i(\mathbb{R}^n) \ \text{ is compact.}\} \tag{1}$$

where $i : \mathbb{R}^n \to T^*\mathbb{R}^n$ is the canonical embedding of $\mathbb{R}^n$ as the zero section. Then we define the $\mathbb{C}((\lambda))$-linear functional $\omega_0 : C^\infty(T^*\mathbb{R}^n) \to \mathbb{C}((\lambda))$ by

$$\omega_0(f) := \int_{\mathbb{R}^n} d^n q \; i^* f, \tag{2}$$

and let $\mathcal{S} : C^\infty(T^*\mathbb{R}^n)((\lambda)) \to C^\infty(T^*\mathbb{R}^n)((\lambda))$ be defined as $\mathcal{S} := \exp(-\frac{i\lambda}{2}\Delta)$ where $\Delta := \frac{\partial^2}{\partial q^k \partial p_k}$ then the following Lemma is found by integration by parts:

Lemma 1 *For any two* $f, g \in C^\infty(T^*\mathbb{R}^n)_{\mathbb{R}^n}((\lambda))$ *we have*

$$\omega_0(f * g) = \int_{\mathbb{R}^n} d^n q (i^* \bar{\mathcal{S}} f)(i^* \mathcal{S} g)$$

whence ω_0 *is a positive linear functional. The Gel'fand ideal of* ω_0 *is given by*

$$J_0 := \{f \in C^\infty(T^*\mathbb{R}^n)_{\mathbb{R}^n}((\lambda)) \mid i^* \mathcal{S} f = 0\}.$$

In [4] we have shown that J_0 is the left ideal generated by the n momentum functions $p_1, \ldots, p_n$. Moreover we can determine the GNS representation induced by ω_0 explicitly using the map $\mathcal{S}$ and the fact that $\mathcal{S} \circ \pi^* = \pi^*$:

Theorem 2 *Let* $f \in C^\infty(T^*\mathbb{R}^n)((\lambda))$, $g \in C^\infty(T^*\mathbb{R}^n)_{\mathbb{R}^n}((\lambda))$, *and* $\varphi, \psi \in C_0^\infty(\mathbb{R}^n)((\lambda))$.

1. *There is a canonical isomorphism* $\Phi : C^\infty(T^*\mathbb{R}^n)_{\mathbb{R}^n}((\lambda))/J_0 \to \mathcal{H}_0 := C_0^\infty(\mathbb{R}^n)((\lambda))$ *given by* $\Phi(\Psi_g) := i^*Sg$ *whose inverse is simply given by* $\Phi^{-1}(\varphi) = \Psi_{\pi^*\varphi}$. *The hermitean product on* $\mathcal{H}_0$ *induced by the pull-back of the GNS inner product with respect to* Φ *is the standard* L^2*-inner product* $\langle\varphi,\psi\rangle = \int_{\mathbb{R}^n} d^nq\ \bar{\varphi}\psi$.
2. J_0 *is even a left ideal of* $C^\infty(T^*\mathbb{R}^n)((\lambda))$ *whence the GNS representation is defined for all its elements* f *and given by the following formula on elements of* $\mathcal{H}_0$:

$$\pi_0(f)\varphi = i^*S(f * (\pi^*\varphi)) = \sum_{r=0}^{\infty} \frac{1}{r!}\left(\frac{\lambda}{i}\right)^r \frac{\partial^r(Sf)}{\partial p_{i_1}\cdots p_{i_r}} \frac{\partial^r\varphi}{\partial q^{i_1}\cdots q^{i_r}}. \tag{3}$$

We had also shown in [4] that the representation (3) exactly corresponds to the Weyl symmetrization rule of polynomials in $q^1,\ldots,q^n,p_1,\ldots,p_n$.

3. We shall now use these results to formulate the WKB expansion for $T^*\mathbb{R}^n$ in the framework of deformation quantization. Let $H \in C^\infty(T^*\mathbb{R}^n)$ be a classical real-valued Hamiltonian and assume that there is a smooth real-valued function $S : \mathbb{R}^n \to \mathbb{R}$ satisfying the following (in general nonlinear) first order partial differential equation, the so-called Hamilton-Jacobi equation, for some chosen real number E:

$$H(dS(q)) = E \quad \text{for all } q \in \mathbb{R}^n. \tag{4}$$

Let $L := \operatorname{graph}(dS) \subset T^*\mathbb{R}^n$ be the graph or dS which is known to be a projectable Lagrangean submanifold of $T^*\mathbb{R}^n$ (see e.g. [1, p.28-31]). Equation (4) implies $(H-E)|_L = 0$, and the local flow of the Hamiltonian vector field of H preserves L.

We are now going to construct a suitable GNS functional having support in L. First, consider the 'Heisenberg equation' with respect to the Hamiltonian π^*S:

$$\frac{df}{dt}(t) = \frac{i}{\lambda}((\pi^*S) * f(t) - f(t) * (\pi^*S)) \tag{5}$$

which can be shown to have a unique solution $f(t) = A_t(f_0)$ for each initial function $f_0 \in C^\infty(T^*\mathbb{R}^n)((\lambda))$ where $t \mapsto A_t$ is a one-parameter group of $\mathbb{C}((\lambda))$-linear automorphisms of the $C^\infty(T^*\mathbb{R}^n)((\lambda))$ commuting with complex conjugation. It allows for the factorization

$$A_t = \Phi_t^* \circ T_t \tag{6}$$

where Φ_t is the Hamiltonian flow of π^*S which consists of translations by dS in the direction of the cotangent spaces, and $T_t = 1 + \sum_{r=1}^{\infty} \lambda^r T_t^{(r)}$

is a formal power series in λ whose coefficients are differential operators on $T^*\mathbb{R}^n$. Note that Φ_1 is a diffeomorphism of the configuration space $\mathbb{R}^n$ onto L. Let $C^\infty(T^*\mathbb{R}^n)_L$ be the space of all smooth complex-valued functions on $T^*\mathbb{R}^n$ whose support has compact intersection with L, and let $\omega_1 : C^\infty(T^*\mathbb{R}^n)_L((\lambda)) \to \mathbb{C}((\lambda))$ be the following $\mathbb{C}((\lambda))$-linear functional (where i_L denotes the canonical injection $L \subset T^*\mathbb{R}^n$):

$$\omega_1(f) := \int (\Phi_1^* d^n q)\, i_L^*(T_{-1} f) \;=\; \omega_0 \circ A_{-1}(f) \tag{7}$$

which obviously has support in L and is easily seen to be positive since A_{-1} is a real automorphism. Moreover, the Gel'fand ideal of ω_1 is simply given by $J_1 = A_1 J_0$, and the GNS representation π_1 induced by ω_1 can be constructed as follows: since $C^\infty(T^*\mathbb{R}^n)_L((\lambda)) = A_1(C^\infty(T^*\mathbb{R}^n)_{\mathbb{R}^n}((\lambda)))$ there is a well-defined unitary map $U : \mathcal{H}_0 \to \mathcal{H}_1$ between the two formal GNS pre-Hilbert spaces given by $U(\Psi_f^{(0)}) := \Psi_{A_1 f}^{(1)}$ whence the vector space $\mathcal{H}_1$ is isomorphic to $C_0^\infty(L)((\lambda))$. Observing that the Gel'fand ideal of ω_1 is again a left ideal in the algebra $C^\infty(T^*\mathbb{R}^n)((\lambda))$ we get the following formula for the GNS representation of $f \in C^\infty(T^*\mathbb{R}^n)((\lambda))$:

$$\pi_1(f) = U\pi_0(A_{-1}f)U^{-1}. \tag{8}$$

We consider the time-independent Schrödinger equation $\pi_1(H)\phi^{(1)} = E\phi^{(1)}$ for $\phi^{(1)}$ in the formal distribution space $\mathcal{H}_1' := (C_0^\infty(L))'((\lambda))$. This is well-defined since $\pi_1(H)$ is a series of differential operators. Via the map U, this problem is related to the time-independent Schrödinger equation

$$\pi_0(A_{-1}H)\phi = E\phi \tag{9}$$

where $\phi = \sum_{r=0}^\infty \lambda^r \phi_r$ and all the ϕ_r are distributions in $C_0^\infty(\mathbb{R}^n)'$. The above equation leads to a system of partial differential equations for the ϕ_r given in the following

Theorem 3 (Formal WKB Expansion) *With the above notations we find that the system of equations (9) is equivalent to the following system of partial differential equations for ϕ:*

$$\begin{aligned}(i^*\Phi_{-1}^* T_{-1}^{(1)} H)\phi_r - \frac{i}{2} i^*\Delta((\Phi_{-1}^* H)(\pi^*\phi_r)) + \frac{i}{2} i^*\{\Phi_{-1}^* H, \pi^*\phi_r\} = \\ - \sum_{\substack{a+b+c+d=r+1,\\ a,b,c,d>0,\ r>d}} \frac{(-i/2)^a}{a!}\frac{(i/2)^b}{b!}\Delta^a(M_b(\Phi_{-1}^* T_{-1}^{(c)} H, \pi^*\phi_d))\end{aligned} \tag{10}$$

where M_b are the bidifferential operators of the standard Weyl star product in $T^*\mathbb{R}^n$, i.e. $M_b(f,g) := (\partial_{q^k}\partial_{p'_k} - \partial_{p_k}\partial_{q'^k})^b(f(q,p)g(q',p'))|_{q=q',p=p'}$.

In the particular case of a Hamiltonian of the physical form $H(q,p) = \sum_{i=1}^{n}(p_i)^2 + V(q)$ for a smooth real-valued function V on $\mathbb{R}^n$ the system of equations (10) reduces to the well-known WKB transport equations (see e.g. [1, 13-14]) (which is mainly due to the fact that $A_t H = \Phi_t^* H$ for H at most quadratic in the momenta) where $\phi_r := 0$ for negative r:

$$\frac{\partial^2 S}{\partial q^k \partial q^k}\phi_r + 2\frac{\partial S}{\partial q^k}\frac{\partial \phi_r}{\partial q^k} = i\frac{\partial^2 \phi_{r-1}}{\partial q^k \partial q^k}. \tag{11}$$

Remarks:

1. Since we assumed that the energy surface $H^{-1}(E)$ contains a graph of a one-form, namely dS, the physical situation in the above Theorem is in so far simplified as there are no classically forbidden regions or turning points. It is an interesting problem to perform the same GNS analysis for general cotangent bundles and more general Lagrangean submanifolds incorporating Maslov corrections.
2. The usual WKB phase function $e^{iS/\hbar}$ is recovered in the following way: assume first that H is polynomial in the momenta. Then e. g. $\exp(*\frac{it}{\lambda}(\pi^* S)) * H$ is a well-defined formal Laurent series in $1/\lambda$ converging for all $\lambda = \hbar \in \mathbb{R} \setminus \{0\}$. Hence $A_t(H) = \exp(*\frac{it}{\lambda}(\pi^* S)) * H * \exp(-*\frac{it}{\lambda}(\pi^* S)) = (\pi^* \exp(\frac{it}{\lambda}S)) * H * (\pi^* \exp(-\frac{it}{\lambda}S))$ converges for all $\lambda = \hbar \in \mathbb{R} \setminus \{0\}$. Assuming that the power series ϕ converges for some $\hbar \in \mathbb{R} \setminus \{0\}$ to a distribution, we readily see that eqn (9) is equivalent to $\pi_0(H)\exp(\frac{it}{\hbar}S)\phi = \exp(\frac{it}{\hbar}S)\phi$ since π_0 is a representation.

Acknowledgements

We would like to thank M. Flato, K. Fredenhagen, J. Huebschmann, H. Rehren, and A. Weinstein for useful discussions.

References

1. S. Bates, A. Weinstein: *Lectures on the Geometry of Quantization.* Berkeley Mathematics Lecture Notes, Volume 8.
2. F. Bayen, M. Flato, C. Frønsdal, A. Lichnerowicz, D. Sternheimer: *Deformation Theory and Quantization.* Annals of Physics **111** (1978), part I: 61-110, part II: 111-151.
3. M. Bordemann, S. Waldmann: *A Fedosov Star Product of Wick Type for Kähler Manifolds.* preprint Uni Freiburg FR-THEP-96/9, and q-alg/9605012 May 1996, to appear in Lett. Math. Phys.
4. M. Bordemann, S. Waldmann: *Formal GNS Construction and States in Deformation Quantization.* Preprint Uni Freiburg, FR-THEP-96/12, July 1996, and q-alg/9607019.
5. M. DeWilde, P.B.A. Lecomte: *Existence of star-products and of formal deformations of the Poisson Lie Algebra of arbitrary symplectic manifolds.* Lett. Math. Phys. **7** (1983), 487-496.
6. B. Fedosov: *A Simple Geometrical Construction of Deformation Quantization.* J. of Diff. Geom. **40** (1994), 213-238.

ON THE DEFORMATION OF TIME HARMONIC FLOWS

JENS HOPPE
Institut für Theoretische Physik ETH
Hönggerberg, CH-8093 Zürich, Switzerland

As pointed out some time ago [9], surfaces moving through $\mathbf{R}^3$ in such a way that their (normal) velocity is always equal to the local surface-area density, $\sqrt{g}$ (divided by some non-dynamical 'reference-density', ρ), have the property that the time, at which the surface Σ_t reaches a point in space, is a harmonic function. These motions are related to certain reductions of self-dual $SU(N)$ Yang-Mills theories (which play a central role in the construction of monopoles, cf. [8]), and the Lax-pair formulation of the latter can be taken over identically [7, 17]. One can also show [2] that time-harmonic flows are, what remains when projecting certain diffeomorphism invariant Hamiltonian field theories on to the (integrable) Diff-singlet sector.

In this note, I would like to show that time-harmonic flows may be deformed from the 'w_∞'-invariant parametrized form to 'W_∞'-invariant motions of parametrized surfaces, while keeping integrability in the form of infinitely many constants of motion. Again, the Lax-pair can be taken from the self-dual Yang-Mills theory, or rather: the Nahm equations (i.e. reduced self-dual $SU(N)$ Yang-Mills equations) are just a special case of the most natural first order (in time), quadratically non-linear, evolution equation for a set of (3) operators, X_i. The specification (of the space of operators) that will correspond to deformed time-harmonic flows of surfaces of spherical (toroidal) topology, resp. *-products on $S^2(T^2)$, is discussed in detail.

Let $\mathcal{A}$ be a non-commutative, associative Algebra ('of operators') and $X_i(t)_{i=1}^N$ a set of timedependent operators satisfying the non-linear evolution equation(s)

$$\dot{X}_i = \epsilon_{ii_1 \ldots i_M} X_{i_1}, X_{i_2} \ldots X_{i_M} \tag{1}$$

$i = 1, 2, \ldots, N = M + 1$.

Using [10, 11], one observes that (1) can be written in the form

$$\dot{L} = \Big[L, M_1, \ldots, M_{N-2} \Big], \tag{2}$$

D. Sternheimer et al. (eds.), Deformation Theory and Symplectic Geometry, 321–326.

where L and the M's ($\epsilon\mathcal{A}$) are linear combinations of the X_i, depending on $[\frac{N}{2}]$ spectral parameters, and $[\cdot,\ldots,\cdot]$ (a fully antisymmetric map from $\mathcal{A}\times\ldots\times\mathcal{A}$ to $\mathcal{A}$) denotes the natural M-commutator,

$$[A_1,\ldots,A_M] := \epsilon^{r_1\ldots r_M} A_{r_1}\cdot\ldots\cdot A_{r_M}. \tag{3}$$

In particular, one may take

$$\begin{aligned} L &= \mu(X_1+iX_2) - \Big(\frac{X_1-iX_2}{\mu}\Big) + 2X_3 \\ M_1 &= i(\mu(X_1+iX_2)+X_3) \end{aligned} \tag{4}$$

(just as for the reduced self-dual Yang-Mills equations, see e.g. [8]) for $N=3$, and (just as in [10, 11])

$$\begin{aligned} L &= \mu(X_1+iX_2)+\tilde{\mu}(X_3+iX_4) \\ &- \frac{1}{\mu}(X_1-iX_2)-\frac{1}{\tilde{\mu}}(X_3-iX_4)+\sqrt{8}X_5 \\ M_1 &= \frac{i}{\sqrt{2}}\Big(\mu(X_1+iX_2)+\frac{X_5}{\sqrt{2}}\Big) \\ M_2 &= \frac{i}{\sqrt{2}}\Big(\tilde{\mu}(X_3+iX_4)+\frac{X_5}{\sqrt{2}}\Big) \\ M_3 &= \frac{-1}{\sqrt{2}}\Big(\frac{(X_1-iX_2)}{\mu}-\frac{(X_3-iX_4)}{\tilde{\mu}}\Big) \end{aligned} \tag{5}$$

for $N=5$.

If there exists a trace on $\mathcal{A}$, satisfying

$$TrAB = TrBA, \tag{6}$$

$$Q_n := TrL^n, \quad n\epsilon\mathbf{N} \tag{7}$$

will be automatically time-independent only for $N=3$; for odd $N>3$, at least Q_1 and Q_2 are conserved, while for even N not even the basic M-commutator is traceless.

In the following, I will restrict myself to the case $N=3$, i.e.

$$\dot{X}_i = \frac{1}{2}\epsilon_{ijk}[X_j,X_k] \tag{8}$$

which, if the X_i were finite-dimensional matrices, are just 'Nahm's equations'. They still trivially 'are', for infinite matrices with finitely many non-zero coefficients, but 'all' other infinite dimensional choices for $\mathcal{A}$ (or rather: an infinite-dimensional Lie-algebra, $\mathcal{L}$) are of quite different nature, and it seems that only the time-harmonic [2, 9], w_∞-invariant case (area-preserving limit of $SU(N)$ [4, 7, 12, 17]), where $\mathcal{L}$ is the Lie algebra of (non-constant) symplectic diffeomorphisms of S^2 or $T^2 \ldots$, (8) becoming the following set of first order partial differential equations for time-dependent functions on a two-dimensional manifold $\sum$,

$$\dot{x}_i = \frac{1}{2}\epsilon_{ijk}\frac{\epsilon^{rs}}{\rho(\phi)}\frac{\partial x^j}{\partial \phi^r}\frac{\partial x^k}{\partial \phi^s}, \tag{9}$$

has previously been considered in the literature [2, 7, 10, 11, 17]. Here, I would like to consider *-product deformations of (9), which amounts to (for S^2, [3]) choosing $\mathcal{A}$ to be the enveloping algebra of $SO(3)$, divided by the 'Casimir-ideal', or (for T^2) specific subclasses of infinite-dimensional matrices with only finitely many non-zero off-diagonals (cf. [14]). Both series of infinite dimensional 'W_∞'-algebras admit an invariant trace, making (7) time-independent for all n.

Let me first discuss the 'spherical type' W^∞-algebras (cf. [3, 5, 15]). Let $\mathcal{G}$ be a semi-simple Lie-algebra, $\{T_a\}_{a=1}^{d=dim\mathcal{G}}$ a basis of $\mathcal{G}$,

$$[T_a, T_b] = f_{ab}^c T_c \qquad abc = 1 \ldots d \tag{10}$$

and $U(\mathcal{G})$ be the associative algebra (over $\mathbf{C}$) of polynomials

$$T = c_T \cdot \mathbf{1} + \sum c^{a_1 \ldots a_2} T_{a_1} \ldots T_{a_l}, \tag{11}$$

modulo (10) (i.e. the universal enveloping algebra). The center of U is generated by $r = \text{rank}\mathcal{G}$ 'Casimirs' $C_1, \ldots, C_r$, and U may be divided by the sum of the r two-sided ideals

$$I_j = (C_j - \lambda_j \mathbf{1})U, \qquad \lambda_j \epsilon \mathbf{C}; \tag{12}$$

resulting in $U_{\lambda=(\lambda_1,\ldots,\lambda_r)}(\mathcal{G})$, the algebra of polynomials

$$T^{(\lambda)} = c_T \mathbf{1} + \sum c_T^{a_1 \ldots a_l} T_{a_1}^{(\lambda)} \cdot \ldots \cdot T_{a_l}^{(\lambda)}, \tag{13}$$

where the $T_a^{(\lambda)}$ are irreducible representations of (10), having the property that certain polynomials, like

$$C_1^{(\lambda)} := T_a T_b g^{ab} = \lambda_1 \cdot \mathbf{1} \tag{14}$$

(g^{ab} being the inverse of $g_{ab} := \frac{1}{2} f_{ac}^d f_{bd}^c$, the metric tensor on $\mathcal{G}$) are proportional to $\mathbf{1}$, and the coefficients $c^{a_1 \ldots a_l}$ apart from (for definiteness) being totally antisymmetric, will consequently be taken to satisfy additional requirements like $g_{ab}c^{aba_3\ldots a_l} \equiv 0, \ldots$ (in accordance with the r Casimir relations). It is known (see e.g. [6]) that $U_\lambda(\mathcal{G})$ decomposes, under the action of $\mathcal{G}$, into a direct sum of finite dimensional irreducible $\mathcal{G}$-modules,

$$U_\lambda = \oplus \sum_{(t)_j} U_\lambda^{(t)_j}, \tag{15}$$

$j = 1 \ldots m(t)$, where each (tensor) representation (t) occurs finitely many $(m_{(t)})$ times; the 1-dimensional representation occurs only once and, as $[U_\lambda, U_\lambda] = [\mathcal{G}_\lambda, U_\lambda]$ (see [16]), it is easy to see that U_λ is also the direct sum

$$U_\lambda = \mathbf{C} \cdot \mathbf{1} \oplus [U_\lambda, U_\lambda], \tag{16}$$

implying that

$$TrT := c_T \tag{17}$$

defines an invariant trace on U_λ, $Tr[A,B] = 0$ – which is all one needs to conclude that (8), with $\mathcal{A} = U_\lambda(\mathcal{G})$, will have infinitely many conserved quantities, (7). I am referring to the series of algebras $U_\lambda(\mathcal{G})$ as the 'spherical series' as in the simplest case, $\mathcal{G} = SO(3)$, the (to be extended) map $\phi_{\hbar := \frac{1}{\lambda}}$,

$$\begin{aligned} Y_{lm}(\theta, \phi) &= (\sum c^{(lm)}_{a_1 \ldots a_l} x_{a_1} \cdot \ldots \cdot x_{a_l})_{\vec{x} = (\sin\theta\cos\phi, \sin\theta\sin\phi, \cos\theta)} \\ &\longrightarrow \phi_\hbar(Y_{lm}) = \bar{T}^{(\lambda)}_{lm} := c^{(\lambda)}_{(l)} \sum c^{(lm)}_{a_1 \ldots a_l} \bar{T}^{(\lambda)}_{a_1} \ldots \bar{T}^{(\lambda)}_{a_l}, \end{aligned} \tag{18}$$

where $[\bar{T}^{(\lambda)}_a, \bar{T}^{(\lambda)}_b] = \frac{\epsilon_{abc}}{\lambda} \bar{T}^{(\lambda)}_c, \sum_{a=1}^{3} \bar{T}^{(\lambda)}_a \bar{T}^{(\lambda)}_a = \mathbf{1}$, provides a one to one correspondence between $U_\lambda(SO(3))$ and the (commutative, resp. Poisson-) algebra of complex-valued functions on S^2 (cf. [4, 12, 13]). Moreover, the overall normalization may be chosen such that when $\hbar = \frac{1}{\lambda}$ is appropriately taken to 0,

$$\begin{aligned} \frac{1}{i\hbar}[\phi_\hbar(Y_{lm}), \phi_\hbar(Y_{l'm'})] &\rightarrow \{Y_{lm}, Y_{l'm'}\} \\ &:= \frac{1}{sin\theta}(\frac{\partial Y_{lm}}{\partial\theta}\frac{\partial Y_{l'm'}}{\partial\phi} - (lm \longleftrightarrow l'm')) \end{aligned} \tag{19}$$

which together with suitable properties under complex (hermitean) conjugation, … (see [1] for the definition of a star-product) allows to call the associative multiplication in $U_\lambda(SO(3))$ a *-product on S^2:

$$Y_{lm} * Y_{l'm'} := \phi_\hbar^{-1}(\phi_\hbar(Y_{lm})\phi_\hbar(Y_{l'm'})). \tag{20}$$

For the 'Torus-case', where a *-product may easily be written directly in terms of functions on T^2,

$$f * g := f \cdot g + \sum_{n=1}^{\infty} \frac{(\lambda/2)^n}{n!} \epsilon^{r_1 s_1} \ldots \epsilon^{r_n s_n} \frac{\partial^n f}{\partial\phi^{r_1} \ldots \partial\phi^{r_n}} \frac{\partial^n g}{\partial\phi^{s_1} \ldots \partial\phi^{s_n}}, \tag{21}$$

the equation

$$\lambda \dot{x}_i = \epsilon_{ijk} x_j * x_k, \tag{22}$$

viewed as an evolution-equation for a hypersurface in $\mathbf{R}^3$, may then 'at any stage' (s.b.) be compared with the time-harmonic flow (9) (which is equivalent to (22), $\lambda = 0$). In view of (9) being equivalent to (cf. [9])

$$\vec{\nabla}^2 t(x^1, x^2, x^3) = 0, \tag{23}$$

it is tempting to interchange dependent and independent variables also in (22): making this 'hodograph' transformation,

$$\phi^o = t, \phi^1, \phi^2 \to x^i := x^i(t, \phi^1, \phi^2), \tag{24}$$

one first notes the purely 'kinematical' consequences,

$$\begin{aligned}
\dot{\vec{x}} &\hat{=} J(\vec{\nabla}\phi^1 \times \vec{\nabla}\phi^2) \\
J : &= |(\frac{\partial x^i}{\partial \phi^\alpha})| = \dot{\vec{x}}(\partial_1 \vec{x} \times \partial_2 \vec{x}) = (\vec{\nabla} t \cdot (\vec{\nabla}\phi^1 \times \vec{\nabla}\phi^2))^{-1} \\
\partial_t &\hat{=} \dot{\vec{x}} \cdot \vec{\nabla} = J(\vec{\nabla}\phi^1 \times \vec{\nabla}\phi^2) \cdot \vec{\nabla} =: D_o \\
\frac{\partial}{\partial \phi^r} &\hat{=} J(\vec{\nabla} t \times \vec{\nabla}\phi^r) \cdot \vec{\nabla} =: D^r \\
D^r \phi^s &= \epsilon^{rs}, \quad [D^\alpha, D^\beta] = 0,
\end{aligned} \tag{25}$$

$\alpha, \beta = 0, 1, 2$, while (22) becomes

$$\begin{aligned}
(\vec{\nabla}\phi^1 \times \vec{\nabla}\phi^2)_i &= \frac{\epsilon_{ijk}}{\lambda J} \quad e^{\frac{\lambda}{2}\epsilon_{rs} D^r \otimes D^s} x_j \otimes x_k \Big|_{\text{Diag.}} \\
&= (\vec{\nabla} t)_i \\
&+ \frac{1}{2J}\epsilon_{ijk} \sum_{l=1}^{\infty} \frac{(\lambda^2/4)^l}{(2l+1)!} (\epsilon_{rs} D^r \otimes D^s)^{2l+1} x_j \otimes x_k \Big|_{\text{Diag.}} \\
&=: (\vec{\nabla} t)_i + \sum_{l=1}^{\infty} (\lambda^2)^l (\vec{F}_l(\vec{\nabla} t, \vec{\nabla}\phi^1, \vec{\nabla}\phi^2))_i.
\end{aligned} \tag{26}$$

Note that just as $\sum_i [x_i, \dot{x}_i]_* = 0$ is a consequence of (22), solutions of (26) will satisfy

$$\begin{aligned}
&\sum_i e^{\frac{\lambda}{2}\epsilon_{rs} D^r \otimes D^s} (x_i \otimes J(\vec{\nabla}\phi^1 \times \vec{\nabla}\phi^2)_i - J(\vec{\nabla}\phi^1 \times \vec{\nabla}\phi^2)_i \otimes x_i) \Big|_{\text{Diag.}} \\
&= 0
\end{aligned} \tag{27}$$

Acknowledgements

I would like to thank M. Bordemann and M. Seifert for valuable discussions, the organizers and participants of 'Deformation theory, symplectic geometry and applications' for the stimulating atmosphere at the Monte Verita, and the Eidgenössische Technische Hochschule Zürich for its hospitality.

References

1. F. Bayen, M. Flato, C. Frønsdal, A Lichnerowics and D. Sternheimer, 'Deformation theory and quantization' *Ann. of Physics* **111** (1978), part I: 61-110, part II: 111-151.
2. M. Bordemann, J. Hoppe, 'Diffeomorphism Invariant Integrable Field Theories and Hypersurface Motions in Riemannian Manifolds', ETH-TH/95-31, FR-THEP-95-26.
3. M. Bordemann, J. Hoppe, P. Schaller, *Phys. Lett.* **B232** (1989) 199.
4. M. Cahen, S. Gutt, "Produits * sur les orbites des groupes semi simple de rang 1", *C. R. Acad. Sc. Paris*, 296 (1983) 821-823.
5. M. Cahen, S. Gutt; "An Algebraic Construction of * Product on the Regular Orbits of Semi Simple Lie Groups", in : Gravitation and Geometrie (eds. W. Rindler, A. Trautman) Bibliopolis, Naples 1987.
6. J. Dixmier, 'Enveloping Algebras', North-Holland 1977; J.C. Jantzen, 'Enveloping Algebras of Semi Simple Lie Algebras', Springer 1983.
7. E.G. Floratos, G.K. Leontaris, *Phys. Rev.* **D43** (1991) 2687.
8. N. Hitchin, *Comm. Math. Phys.* **89** (1989) 145.
9. J. Hoppe, *Phys. Lett.* **B335** (1994) 41.
10. J. Hoppe, 'On M-Algebras, the Quantisation of Nambu-Mechanics, and Volume Preserving Diffeomorphism' (Contribution to the Hepp-Hunziker Volume of Helvetia Physica Acta) ETH-TH/95-33.
11. J. Hoppe, 'Higher-Dimensional Integrable Systems from Multilinear Evolution Equations', J. of Physics A 29 (1996) L381; 'Multilinear Evolution Equations for Time-Harmonic Flows in Conformally Flat Manifolds', *Class. Quant. Gravity* 13 (1996) L87.
12. J. Hoppe, 'Quantum Theory of a Massless Relativistic Surface and...', MIT Ph.D. thesis 1982 and Elem.Part.Res.J. (Kyoto) **80** (1989) 145.
13. J. Hoppe, *Int. J. Mod. Physics* **A4** (1989) 5235.
14. E.G. Floratos, *Phys. Lett.* **B232** (1989) 467.
15. J. Hoppe, '2n+1 dimensional Integrable Field Theories', in: Nonperturbative Methods in Low Dimensional Quantum Field Theories (John Hopkins Workshop) Debrecen 1990, World Scientific 1991, eds. Domokos et al.
16. G. Warner, 'Harmonic Analysis on Semi-simple Lie Groups I', Springer 1972.
17. R.S. Ward, *Phys. Lett.* **B234** (1990) 81.

DISTINGUISHED ESSENTIAL EXTENSIONS OF INFINITE DIMENSIONAL CLASSICAL LIE ALGEBRAS

ANDRÉ LICHNEROWICZ
Membre de l'Institut
6 avenue Paul Appell, F-75014 Paris, France

Abstract. In deformation-quantization a fundamental ingredient is a Chevalley 2-cocycle S^3_Γ of the Poisson Lie algebra which is never a coboundary. We show that this cocycle can be traced to the cohomology of the Lie algebra L of all vector fields on a differentiable manifold (of dimension $m \geq 2$), with values in the space ϕ_2 of the 2-forms on the manifold. It is the same for many infinite-dimensional subalgebras of L, given by geometrical structures. The existence of these 2-cocycles can be interpreted as the existence of a distinguished essential extension of these algebras by the abelian Lie algebra ϕ_2.

1. Introduction

Deformation-quantization on a phase space was introduced by us twenty years ago ([4, 1]). It is well known that, in this quantization, a fundamental ingredient is a Chevalley 2-cocycle (classically denoted by S^3_Γ) of the Poisson Lie algebra of functions, cocycle which is never a coboundary (see e.g. [7]).

This cocycle finds its origin in a 2-cocycle of the Lie algebra L of all the vector fields on a differentiable manifold W (of dimension $m \geq 2$), with values in the space ϕ_2 of the 2-forms of W. The existence of this 2-cocycle is strictly connected with the existence on W of linear connexions without torsion. It is the same [8] for many classical infinite dimensional Lie subalgebras of L.

The existence of these 2-cocycles can be interpreted as the existence of a distinguished essential extension of these algebras by ϕ_2 considered as an abelian Lie algebra.

D. Sternheimer et al. (eds.), Deformation Theory and Symplectic Geometry, 327–332.

2. The Lie algebra of all vector fields

a) Let W be a paracompact, connected differentiable manifold of class C^∞ and dimension $m \geq 2$. All the elements are supposed C^∞. We suppose that L acts on ϕ_2 by the Lie derivative $\mathcal{L}$ and we denote by $H^2(L, \mathcal{L}, \phi_2)$ the second space of differentiable Chevalley cohomology of L with values in ϕ_2 associated with the representation $\mathcal{L}$. Flato and myself [3] have showed that this space admits a distinguished generator, defined by an algorithm corresponding to the Chern-Weil homomorphism. Following our suggestion, this point of view has been extended by De Wilde and Lecomte [2] to the other cohomology spaces $H^k(L, \mathcal{L}, \phi_2)$.

b) Let Γ be a linear connection without torsion of $W, \mathcal{L}(X)\Gamma$ the Lie derivative of Γ by the vector field $X \in L$. If $\{x^i\}(i, j, k, \ldots = 1, \ldots m)$ is an arbitrary chart of W of domain U, we have on U

$$(\mathcal{L}(X)\Gamma)^k_{ij} = \partial_{ij}X^k + X^r\partial_r\Gamma^k_{ij} - \partial_r X^k\Gamma^r_{ij} + \partial_i X^r\Gamma^k_{rj} + \partial_j X^r\Gamma^k_{ir}.$$

Define a 2-cochain S^Γ of L, with values in ϕ_2, by (on each domain U)

$$S^\Gamma(X,Y)_{ij} = (\mathcal{L}(X)\Gamma)^k_{li}(\mathcal{L}(Y)\Gamma)^l_{kj} - (\mathcal{L}(Y)\Gamma)^k_{li}(\mathcal{L}(X)\Gamma)^k_{kj} \tag{2.1}$$

Its coboundary is $\partial S^\Gamma(X,Y,Z) = \mathcal{S}\mathcal{L}(X)S^\Gamma(Y,Z) - \mathcal{S}S^\Gamma([X,Y],Z)$ where $X, Y, Z \in L$ and $\mathcal{S}$ is the sum after circular permutation. It follows from the properties of the Lie derivative that $\partial S^\Gamma = 0$ and S^Γ is a 2-cocycle with values in ϕ_2. Substitute to the connection Γ another linear connection Γ' without torsion with $\Gamma' = \Gamma + T$ where T is a tensor. We set

$$\begin{aligned} P(X)|_U &= T^k_{li}(\mathcal{L}(X)\Gamma)^l_{kj} - T^k_{lj}(\mathcal{L}(X)\Gamma)^l_{ki} \\ Q(X)|_U &= T^k_{li}(\mathcal{L}(X)T)^l_{kj} - T^k_{lj}(\mathcal{L}(X)T)^l_{ki} \end{aligned}$$

We have $S^{\Gamma'} - S^\Gamma = \partial(P + \frac{1}{2}Q)$. Therefore the cohomology class β of $H^2(L, \mathcal{L}, \phi_2)$ defined by S^Γ is independent from the choice of the connection without torsion Γ. We have proved in [3] that S^Γ is never exact and β is always $\neq 0$.

c) Considering ϕ_2 as an abelian Lie algebra we can define the extensions of the Lie algebra $(L, [,])$ by ϕ_2 associated with the representation defined by $\mathcal{L}$. We consider the equivalences between extensions as differentiable equivalences. There exists thus an essential extension $(A, \{,\})$ of $(L, [,])$ by ϕ_2 associated with $\mathcal{L}$ and the cohomology 2-class β. An element of A is a couple (X, λ) of a vector $X \in L$ and of a 2-form $\lambda \in \phi_2$. The bracket in A of (X, λ), $(Y, \mu) \in A$ is given in terms of a 2-cocycle $S^\Gamma \in \beta$ by

$$\{(X, \lambda), (Y, \mu)\} = ([X,Y], S^\Gamma(X,Y) + \mathcal{L}(X)\mu - \mathcal{L}(Y)\lambda) \tag{2.2}$$

THEOREM 1. *Let $(L, [\ ,\ ])$ be the Lie algebra of all the vector fields on W of dimension $m \geq 2, \phi_2$ the space of the 2-forms on W considered as an abelian Lie algebra, L acting on ϕ_2 by $\mathcal{L}$. There exists a distinguished essential extension $(A, \{,\})$ of L by ϕ_2 associated with $\mathcal{L}$ and β with bracket given by (2.2).*

This extension is called the universal extension of L by ϕ_2.

3. Lie algebras associated with a generalized foliation

a) On W consider a (generalized) field of planes P generated by a set E of vector fields; P is invariant under a vector field on W if it is invariant under the flow defined by this vector field. We have [9]

THEOREM 2. *If P is invariant under each vector X of E, we have for each point x of W a unique maximal integral manifold $S(x)$ of P passing by $x; S(x)$ is weakly embedded in W. For each $y \in S(x)$, we have $T_yS(x) = P_y$ and the dimension of the elements of P is constant along $S(x)$. Conversely, if such is the case P is invariant under each vector field belonging to P and P defines a generalized foliation of W in the sense of Sussman [9].*

$S(x)$ is called a leaf corresponding to x. A Reeb ordinary foliation is called here a regular foliation. In the general case, a point x of (W, P) is called a regular point if there is an open neighborhood U of x such that the restriction P_U of P to U is a regular foliation. If x is regular, the points of $S(x)$ are all regular. The set of regular points of (W, P) is an everywhere dense open set. If W_ρ is a connected component of this open set, P_{W_ρ} defines a regular foliation of W_ρ. Let L_P^T be the set of vector fields tangent to P; L_P^T is a Lie subalgebra of L. The normalizer L_P of L_P^T in L is the Lie algebra of vector fields preserving (globally) the generalized foliation.

Let Γ be a linear connection without torsion on W, S^Γ the 2-cocycle with values in ϕ_2 defined by (2.1). The restriction of S^Γ to L_P^T defines a cohomology 2-class $\beta_P^T \in H^2(L_P^T, \mathcal{L}, \phi_2)$ independent from the choice of the connection.

b) Suppose that P admits leaves of dimension ≥ 2. Let V be a contractible domain of W such that the foliation P_V induced by P on V is regular; we suppose that V is a domain of a chart adapted to the foliation P_V, $\{x^i\} = \{x^\alpha, x^a\}$ $(\alpha, \beta, \ldots = 1, \ldots p, p \geq 2; a = j + i, \ldots m)$. The leaves are defined on V by x^a constant. Let U be a contractible domain such that $\bar{U} \subset V$, $P_{\bar{U}}$ the foliation induced on $\bar{U}$. If X is a vector field on U, tangent to $P_{\bar{U}}$, X admits an extension to a vector field on W, element of L_P^T. We denote by x_0 a point of U admitting coordinates such that at x_0 we have $x^1 = x^2 = 0$.

Consider two couples (X', Y') and (X'', Y'') of elements of L_P^T for which the restrictions to $\bar{U}$ have only as nonzero components in the chart

$$X'^1 = x^1x^2;\ Y'^2 = x^1x^2,\ X''^2 = \tfrac{1}{2}(x^1)^2,\ Y''^1 = \tfrac{1}{2}(x^2)^2.$$

It is easy to see that on $\bar{U}$ we have $[X', Y']|_{\bar{U}} = -2[X'', Y'']|_{\bar{U}}$. The 1-jets at x_0 of X', Y', X'', Y'' are null. If X, Y are elements of L_P^T admitting null 1-jets at x_0, we have at this point

$$S^\Gamma(X,Y)(x_0) = (\partial_{i\beta}X^\alpha \partial_{j\alpha}Y^\beta - \partial_{j\beta}X^\alpha \partial_{i\alpha}Y^\beta)(x_0) \tag{3.1}$$

For $i = 1, j = 2$ we have

$$S^\Gamma(X', Y')_{12}(x_0) = 1 \qquad S^\Gamma(X'', Y'')_{12}(x_0) = 1 \tag{3.2}$$

But if S^Γ was the coboundary of a differentiable 1-cochain T with values in ϕ_2:

$$S^\Gamma(X,Y) = \mathcal{L}(X)T(Y) - \mathcal{L}(Y)T(Y) - T([X,Y])$$

we would have for the couples (X', Y') and (X'', Y''):

$$S^\Gamma(X', Y')(x_0) = -2S^\Gamma(X'', Y'')(x_0)$$

in contradiction with (3.2). Thus S^Γ is never exact and β_P^Γ is $\neq 0$ [8].

THEOREM 3. *Let P be a generalized foliation of W admitting leaves of dimension ≥ 2, L_P^T the Lie algebra of the vector fields tangent to P. If ϕ_2 is the space of the 2 forms of W considered as an abelian Lie algebra, the restriction to L_P^T of the universal extension of L by ϕ_2 defines a distinguished essential extension of L_P^T by ϕ_2 associated with $\mathcal{L}$ and with the 2-class $\beta_P^T \in H^2(L_P^T, \mathcal{L}, \phi_2)$. The same result is true for the Lie algebra L_p.*

The same proof shows that for $m \geq 2$, the 2-class β of L is everywhere $\neq 0$. A similar proof shows that if W admits a unimodular structure (that is a m-form element of volume η), the restriction of the universal extension of L by ϕ_2 to the Lie algebra L_u of the unimodular vector fields (preserving η) is a distinguished essential extension of L_u by ϕ_2 associated with $\mathcal{L}$ and with a 2-class $\beta_u \in H^2(L_u, \mathcal{L}, \phi_2)$.

4. Poisson manifolds and associated Lie algebras

a) Let W be a differentiable manifold of dimension $m \geq 2$. We now write $N = C^\infty(W; \mathbb{R})$. A Poisson structure is defined on W by a skewsymmetric contravariant 2-tensor $\Lambda \not\equiv 0$ satisfying in the sense of the Schouten brackets $[\Lambda, \Lambda] = 0$. The 2-tensor Λ defines on N a Poisson bracket

$$\{u, v\} = i(\Lambda)(du \wedge dv) \quad (u, v \in N) \tag{4.1}$$

where $i()$ is the inner product; $(N, \{,\})$ is the Poisson Lie algebra of (W, Λ). If m is even, $m = 2n$, and if Λ has everywhere the rank $2n$, (W, Λ) is a symplectic manifold for which the 2-form F is given by the inversion of the 2-tensor Λ.

In the general case we associate with each element $u \in N$ the vector field $X_u = [\Lambda, u]$ called the Hamiltonian vector field associated with u. The Hamiltonian vector fields define a Lie algebra L^H_Λ the elements of which preserve Λ. The map $\sigma : u \in N \longrightarrow X_u \in L^H_\Lambda$ is a homomorphism of $(N, \{,\})$ onto $(L^H_\Lambda, [,])$. If $x \in N$, the values at x of the Hamiltonian vector fields generate a subspace P_x of T_xW. We obtain thus on W a field P of planes, the characteristic field of (W, Λ). It is well-known that P defines a generalized foliation. The restriction of Λ to a leaf $S(x)$ of P defines on $S(x)$ a symplectic structure.

b) Let Γ be a linear connection without torsion on W. The restriction of S^Γ to L^H_Λ defines an element $\beta_\Lambda \in H^2(L^H_\Lambda, \mathcal{L}, \phi_2)$. We associate with every differentiable k-cochain C of L^H_Λ with values in ϕ_2 the k-cochain $C_\Lambda = i(\Lambda)C$; Λ being invariant under the Hamiltonian vector fields, we have $\partial C_\Lambda = i(\Lambda)\partial C$ and a homomorphism

$$i(\Lambda) : H^k(L^H_\Lambda, \mathcal{L}, \phi_2) \longrightarrow H^k(L^H_\Lambda, \mathcal{L}, N) \tag{4.2}$$

Since $\{u, v\} = \mathcal{L}(X_u)v$, we can interpret the cohomology of $(L^H_\Lambda, [\,])$ with values in N in the following way. Introduce for the cochains of N null on the constants on the leaves, the differentiable Chevalley cohomology of $(N, \{,\})$ with values in N corresponding to the adjoint representation; for this cohomology $H^k(N, \{,\}, N)$ is isomorphic to $H^k(L^H_\Lambda, \mathcal{L}, N)$. In particular the 2-cocycle of $(N, \{,\})$ with values in N, given by:

$$S^\Gamma_\Lambda(u, v) = i(\Lambda)S^\Gamma(X_u, X_v) \quad (u, v \in N) \tag{4.3}$$

defines $\beta^1_\Lambda \in H^2(N, \{,\}, N)$; β_Λ and β^1_Λ are independent from the choice of the connection Γ. Choose a domain U of W such that the restriction $\Lambda_{\bar{U}}$ of Λ to $\bar{U}$ defines $(\bar{U}, \Lambda_{\bar{U}})$ as a regular Poisson manifold. The functions elements of $C^\infty(\bar{U}; \mathbb{R})$ can be extended to W. We suppose that $\bar{U} \subset V$ where V is the domain of a chart $\{x^i\} = \{x^\lambda, x^{\bar{\lambda}}, x^a\}(\lambda = 1, \dots p,\ \bar{\lambda} = \lambda + p, a = 2p + 1, \dots, m)$ adapted to Λ and such that Λ admits as only non null components $\Lambda^{\lambda\bar{\lambda}} = -\Lambda^{\bar{\lambda}\lambda} = 1$. We denote by x_0 a point of U admitting coordinates such that $x^1 = x^{\bar{1}} = 0$. Consider two couples (u', v') and (u'', v'') of elements of N for which the restrictions to $\bar{U}$ are

$$u'|_{\bar{U}} = x^1(x^{\bar{1}})^2, v'|_{\bar{U}} = (x^1)^2 x^{\bar{1}}; u''|_{\bar{U}} = (x^{\bar{1}})^3, v''|_{\bar{U}} = (x^1)^3$$

It is easy to see that $\{u'', v''\}|_{\bar{U}} = 3\{u', v'\}|_{\bar{U}}$. Set $S^3_\Gamma(u, v) = -S^\Gamma_\Lambda(u; v)$. For $\alpha, \beta = 1, \ldots, 2p$, we have at x_0

$$S^3_\Gamma(u, v)(x_0) = (\Lambda^{\alpha_1\beta_1}\Lambda^{\alpha_2\beta_2}\Lambda^{\alpha_3\beta_3}\partial_{\alpha_1\alpha_2\alpha_3}u\partial_{\beta_1\beta_2\beta_3}v)(x_0) \tag{4.4}$$

The same argument as in Section 3 gives a contradiction if S^3_Γ is supposed exact. Therefore $\beta^1_\Lambda \neq 0$.

THEOREM 4. *Let L^H_Λ be the Lie algebra of the Hamiltonian fields on a Poisson manifold (W, Λ). If ϕ_2 is the space of the 2-forms on W considered as an abelian Lie algebra, the restriction to L^H_Λ of the universal extension of L by ϕ_2 defines a distinguished essential extension A_Λ of L^H_Λ by ϕ_2 associated with $\mathcal{L}$ and $\beta_\Lambda \in H^2(L^H_\Lambda, \mathcal{L}, \phi_2)$. We obtain by $i(\Lambda)$ an essential extension of L^H_Λ by N considered as an abelian Lie algebra and so an essential extension A^1_Λ of $(N, \{,\})$ by N associated with $\{,\}$ and $\beta^1_\Lambda \in H^2(N, \{,\}, N)$.*

An element of A^1_Λ is a couple (u, u_1) of two elements of N. The bracket in A^1_Λ of $(u, u_1), (v, v_1)$ can be written

$$\{(u, u_1), (v, v_1)\} = (\{u, v\}, S^\Gamma_\Lambda(u, v) + \{u, v_1\} - \{v, u_1\}) \tag{4.5}$$

$S^3_\Gamma = -S^\Gamma_\Lambda$ defines an infinitesimal deformation of the Poisson Lie algebra which is never trivial and is strictly connected with the quantization.

References

1. F. Bayen, M. Flato, C. Frønsdal, A. Lichnerowicz and D. Sternheimer: Deformation theory and quantization I and II. *Annals of Physics* (NY) **111**, 61-110 and 111-151 (1978).
2. M. De Wilde and P. Lecomte: Cohomology of the Lie algebra of smooth vector fields associated to the Lie derivatives of smooth forms. *J. Math Pures et Appl.*, **62**, 197 (1983).
3. M. Flato and A. Lichnerowicz: Cohomologie des représentations définies par la dérivation de Lie et à valeurs dans les formes de l'algèbre de Lie des champs de vecteurs d'une variété différentiable. Premiers espaces de cohomologie. Applications. *C.R. Acad. Sci. Paris*, **291**, A 331-335 (1980).
4. M. Flato, A. Lichnerowicz and D. Sternheimer: Crochet de Moyal-Vey et quantification. *C.R. Acad. Sci. Paris* **283**, A 19-24 (1976).
5. F. Guedira and A. Lichnerowicz: Géométrie des algèbres de Lie de Kirillov. *J. Math Pures et Appl.*, **63**, 407 (1984).
6. A. Lichnerowicz: Variétés de Poisson et feuilletages. *Ann. Fac. Sc. Toulouse*, **4**, 195 (1982).
7. A. Lichnerowicz: Déformations d'algèbres associées à une variété symplectique (les star-produits). *Ann. Inst. Fourier*, **32**, 157 (1982).
8. A. Lichnerowicz and J.A. Pereira Da Silva: Extensions essentielles d'algèbres de Lie classiques de dimensions infinies. *J. Math. Pures et Appl.*, **66**, 159 (1987).
9. H.J. Sussmann: Orbits of families of vector fields and integrability of distributions. *Trans. Amer. Math. Soc.*, **180**, 171 (1973).

NONCOMMUTATIVE CONTACT ALGEBRAS

HIDEKI OMORI
Department of Mathematics,
Faculty of Science and Technology,
Science University of Tokyo, Noda, 278, Japan

YOSHIAKI MAEDA
Department of Mathematics,
Faculty of Science and Technology,
Keio University, Yokohama, 223, Japan

NAOYA MIYAZAKI
Department of Mathematics,
Faculty of Science and Technology,
Science University of Tokyo, Noda, 278, Japan

AND

AKIRA YOSHIOKA
Department of Mathematics,
Faculty of Technology,
Science University of Tokyo, Tokyo, 162, Japan

1. Introduction

In [5], Gerstenhaber proposed a definition of the deformation of an algebra which associates to an associative algebra $\mathcal{A}$ an associative structure on the space $\mathcal{A}[[\hbar]]$ of formal power series in $\hbar$ with coefficients in $\mathcal{A}$. Here $\hbar$ is a central element in $\mathcal{A}[[\hbar]]$.

Stimulated by this work, several authors have studied such deformations from a geometrical point of view. In particular, Bayen et al. [1] proposed a notion of deformation quantization, which is a deformation of Poisson algebras. In this context, the question naturally arises of how to generalize these deformations for various geometric structures.

In this paper, we will introduce the notion of a noncommutative contact algebra. On a contact manifold M, the *Lagrange bracket* $\{\, ,\, \}_L$ defines a Lie

D. Sternheimer et al. (eds.), Deformation Theory and Symplectic Geometry, 333–338.

bracket on the space $C^\infty(M)$ of all smooth functions on M, under which $C^\infty(M)$ forms a Lie algebra. For a deformation of contact structures, one might attempt first to deform the Lie algebra $(C^\infty(M), \{\ ,\ \}_L)$. However, there exists no non-trivial deformation of the Lie algebra $(C^\infty(M), \{\ ,\ \}_L)$ in general [4].

In our process of "noncommutatizing" contact algebras, the deformation parameter will no longer be a central element. A distinguishing feature of contact algebras, as opposed to Poisson algebras, is the existence of the characteristic vector field. We will noncommutatize the characteristic vector field together with the algebra $C^\infty(M)$. Thus, the algebra we present here is a slight generalization of deformation quantization.

Our main purpose in this paper is also to show the existence of the noncommutative contact algebra on an arbitrary contact manifold.

2. Contact algebra

Let M be a contact manifold of dimension $(2n+1)$ and ω its contact 1-form; $(d\omega)^n \wedge \omega \neq 0$. We denote by $C^\infty(M)$ the algebra of all smooth functions on M. There exist local coordinates $(x_1, \cdots, x_n, y_1, \cdots, y_n, z)$ such that ω can be written locally as

$$\omega = dz + \frac{1}{2}\sum_{j=1}^{n}(x_j dy_j - y_j dx_j). \tag{1}$$

The contact 1-form ω gives a Jacobi structure on M (cf. [6]): a pair (Λ, D) consisting of a vector field $D \in \Gamma(TM)$ and a skew-symmetric bivector field Λ on TM satisfying the following;

(i) $[D, \Lambda] = 0$,

(ii) $[\Lambda, \Lambda] = 2D \wedge \Lambda$,

(iii) $\operatorname{rank} \Lambda = 2n$,

where $[\ ,\ ]$ stands for the Schouten bracket. In terms of the local coordinates where the 1-form ω has the form (1), we have

$$\Lambda = \sum_{j=1}^{n} \partial_{x_j} \wedge \partial_{y_j} - \partial_z \wedge \frac{1}{2}\sum_{j=1}^{n}(x_j\partial_{x_j} + y_j\partial_{y_j}) \text{ and } D = \partial_z.$$

For $f, g \in C^\infty(M)$, set

$$\{f, g\} = \Lambda(df, dg) : C^\infty(M) \times C^\infty(M) \to C^\infty(M). \tag{2}$$

The bracket $\{\ ,\ \}$ is skew-symmetric and satisfies

$$\{f, gh\} = \{f, g\}h + g\{f, h\}, \quad D\{f, g\} = \{Df, g\} + \{f, Dg\}. \tag{3}$$

We call the triple $(C^\infty(M), \{\ ,\ \}, D)$ the *contact algebra* on M.

We remark that Lichnerowicz [6] introduced a Lie algebra structure on $C^\infty(M)$ by setting

$$\{f, g\}_L = \{f, g\} + fD(g) - D(f)g. \tag{4}$$

The bracket $\{\ ,\ \}_L$ given by (4) is called the *Lagrange bracket* on M.

3. Noncommutative contact manifold

We first generalize the notion of deformation quantization (cf. [9]): let $\mathcal{A}$ be an associative, complete topological algebra. We denote by $a*b$ the product of $a, b \in \mathcal{A}$. We set $[a, b] = a * b - b * a$ and $\mu^k = \underbrace{\mu * \cdots \mu}_{k}$.

Definition 1 *$\mathcal{A}$ is called a formal μ-regulated algebra if there exists an element $\mu \in \mathcal{A}$ with $\bar{\mu} = \mu$ and a closed linear subspace B such that*

(A.1) $[\mu, \mathcal{A}] \subset \mu*\mathcal{A}*\mu$.
(A.2) $[\mathcal{A}, \mathcal{A}] \subset \mu*\mathcal{A}$.
(A.3) $\mathcal{A} = \prod_{k=0}^{\infty} \mu^k * B$ *(topological direct sum).*
(A.4) *The mappings $\mu* : \mathcal{A} \to \mu*\mathcal{A}$, $*\mu : \mathcal{A} \to \mathcal{A}*\mu$ defined by $a \to \mu * a$, $a \to a * \mu$ respectively are linear isomorphisms.*

μ is called the *regulator*. By (A.2), the product $*$ defines a commutative associative product on the factor space $\mathcal{A}/\mu*\mathcal{A}$ identified with B. We denote the commutative product on B by $a \cdot b$.

Formally setting

$$[\mu^{-1}, a] = -\mu^{-1} * [\mu, a] * \mu^{-1}, \tag{5}$$

we see by (A.1) and (A.4) that $[\mu^{-1}, a]$ is well-defined. Since $[\mu^{-1}, \mathcal{A}] \subset \mathcal{A}$, $\mathrm{ad}(\mu^{-1})$ is a derivation of $(\mathcal{A}, *)$, which also induces a derivation of $(B, \cdot)$.

According to the decomposition (A.3), we can write

$$\begin{aligned} a * b &= a \cdot b + \mu * \pi_1(a, b) + \cdots + \mu^k * \pi_k(a, b) + \cdots, \\ ad(\mu^{-1})a &= \xi_0(a) + \mu * \xi_1(a) + \cdots + \mu^k * \xi_k(a) + \cdots, \end{aligned} \tag{6}$$

uniquely for any $a, b \in B$.

Remark 1 *Since $ad(\mu^{-1})$ is a derivation of $(\mathcal{A}, *)$, ξ_0 is a derivation of $(B, \cdot)$. Note that the commutator bracket $[a, b]$ is a biderivation of $\mathcal{A} \times \mathcal{A}$ to $\mathcal{A}$. Therefore, the skew part π_1^- of π_1 is a skew-biderivation of $B \times B$ to B.*

Definition 2 *Let M be a contact manifold and $(C^\infty(M), \{\ ,\ \}, D)$ the contact algebra on M. A formal μ-regulated algebra $\mathcal{A} = B[[\mu]]$ of formal power series in μ with coefficients in B is called the noncommutative contact algebra on M if $B = C^\infty(M)$, $\pi_1 = \{\ ,\ \}$, and $\xi_0 = D$ in (6).*

Example 1 *A typical example of a noncommutative contact algebra is the noncommutative 3-sphere given in [9], which is obtained by quantizing the Hopf fibration* $\pi : S^3 \to S^2$.

Example 2 *Another example of a noncommutative contact algebra is an algebra* $\Sigma^0(T^*N)$ *of symbols of order 0 on the cotangent bundle* T^*N *over a compact manifold* N *modulo the algebra* $\Sigma^{-\infty}(T^*N)$ *of symbols of order* $-\infty$ *(cf. [10]). This gives a noncommutative contact algebra defined on the unit cosphere bundle* S^*N.

Theorem 1 *For the contact algebra* $(C^\infty(M), \{\,,\,\}, D)$ *on a contact manifold* M, *there exists a noncommutative contact algebra* $C^\infty(M)[[\mu]]$.

Theorem 1 shows that every contact algebra extends to a formal μ-regulated algebra, i.e. every contact algebra is deformation quantizable.

4. Proof of Theorem 1

Let M be a contact manifold. Consider the direct product manifold $R_+ \times M$, with R_+ the set of positive real numbers. Define a closed 2-form $\Omega = d(\rho\omega)$ on $R_+ \times M$, where ρ is the coordinate function on R_+. It is easy to see that Ω is a symplectic 2-form on $R_+ \times M$. Thus, it defines a Poisson bracket $\{\,,\,\}_s$ on $C^\infty(R_+ \times M)$. We see that

$$\{f, g\}_s = \rho^{-1}\{f, g\} \tag{7}$$

for $f, g \in C^\infty(M)$, where $\{\,,\,\}$ is given by (2). Let $\mathcal{B} = C^\infty(M)[[\rho^{-1}]]$ be the set of all formal power series $\tilde{f}$ in ρ^{-1} with coefficients in $C^\infty(M)$:

$$\tilde{f}(\rho, p) = f_0(p) + \rho^{-1} f_1(p) + \cdots + \rho^{-k} f_k(p) + \cdots. \tag{8}$$

It is obvious that the bracket $\{\,,\,\}_s$ in (7) gives a Poisson bracket on $\mathcal{B}$ such that

$$\{\mathcal{B}, \mathcal{B}\}_s \subset \rho^{-1}\mathcal{B}. \tag{9}$$

We now consider a deformation quantization of $\mathcal{B}$: an associative algebra $\mathcal{B}[[\hbar]]$ with a star product $*$ such that $\hbar$ is in the center and for $\tilde{f}, \tilde{g} \in \mathcal{B}$

$$\tilde{f} * \tilde{g} = \tilde{\pi}_0(\tilde{f}, \tilde{g}) + \hbar\tilde{\pi}_1(\tilde{f}, \tilde{g}) + \cdots + \hbar^k \tilde{\pi}_k(\tilde{f}, \tilde{g}) + \cdots, \tag{10}$$

where

$$\tilde{\pi}_0(\tilde{f}, \tilde{g}) = \tilde{f} \cdot \tilde{g}, \quad \tilde{\pi}_1(\tilde{f}, \tilde{g}) = \frac{1}{2}\{\tilde{f}, \tilde{g}\}_s. \tag{11}$$

We define the following 1-parameter conformal symplectomorphism on $(R_+ \times M, \Omega)$;

$$\phi_t(\rho, x) = (e^t \rho, x), \quad t \in R. \tag{12}$$

Setting $\phi_t^*(\hbar) = e^t \hbar$, we extend ϕ_t^* on $\mathcal{B}$ to $\mathcal{B}[[\hbar]]$:

Lemma 1 *Let $\mathcal{B}[[\hbar]]$ be a deformation quantization of $(\mathcal{B}, \cdot, \{\ ,\ \}_s)$. Then, the map ϕ_t^* is an automorphism of $(\mathcal{B}[[\hbar]], *)$ if and only if $\tilde{\pi}_k$ satisfies*

$$\tilde{\pi}_k(\rho^{-l} f, \rho^{-m} g) \in \rho^{-(k+l+m)} C^\infty(M) \tag{13}$$

for all $f, g \in C^\infty(M)$ and for all $k, l, m \geq 0$.

We note the following:

Proposition 1 *Let $\mathcal{B}[[\hbar]]$ be a deformation quantization of $(\mathcal{B}, \cdot, \{\ ,\ \}_s)$. If ϕ_t^* is an automorphism of $(\mathcal{B}[[\hbar]], *)$, then the subalgebra of all ϕ_t^*-invariant elements in $\mathcal{B}[[\hbar]]$ is a noncommutative contact algebra on M.*

Proof. Set $\mu = \hbar\rho^{-1}$. Denote by $\tilde{\mathcal{B}}[[\hbar]]$ the space of ϕ_t^*-invariant elements of $\mathcal{B}[[\hbar]]$. We have $\tilde{\mathcal{B}}[[\hbar]] = C^\infty(M)[[\mu]]$ For $f, g \in C^\infty(M)$, we have

$$\begin{aligned} f * g &= f \cdot g + \frac{\mu}{2}\{f, g\} \pmod{\mu^2}, \\ [\mu^{-1}, f] &= \xi_0(f) \pmod{\mu}, \end{aligned} \tag{14}$$

which gives Proposition 1.

To obtain Theorem 1, it suffices to show the following:

Lemma 2 *There exists a deformation quantization $\mathcal{B}[[\hbar]]$ of $(\mathcal{B}, \cdot, \{\ ,\ \}_s)$ with the property (13).*

Proof. Since Ω is a symplectic form on $R_+ \times M$, the Poisson algebra $(C^\infty(R_+ \times M), \cdot, \{\ ,\ \}_s)$ is deformation quantizable [2, 3, 8]. We recall the stepwise construction of the bilinear map $\tilde{\pi}_k$ in (9). As in [8], given $\{\tilde{\pi}_j\}_{j=0}^{k-1}$, $\tilde{\pi}_k$ is determined by the following equations:

$$\begin{aligned} \tilde{\pi}_k^-(f, gh) &- g\tilde{\pi}_k^-(f, h) - \tilde{\pi}_k^-(f, g)h \\ &= \langle\langle f, g\rangle^-, h\rangle_k^+ + \langle\langle f, h\rangle^-, g\rangle_k^+ + \langle\langle g, h\rangle^+, f\rangle_k^- \end{aligned}$$

$$\begin{aligned} \tilde{\pi}_k^+(f, gh) &- \tilde{\pi}_k^+(h, gf) - \tilde{\pi}_k^+(f, g)h + \tilde{\pi}_k^+(h, g)f, \\ &= \langle\langle f, g\rangle^+, h\rangle_k^+ - \langle\langle h, g\rangle^+, f\rangle_k^+ + \langle\langle h, f\rangle^-, g\rangle_k^-, \end{aligned}$$

where $\tilde{\pi}_i^\pm(f, g) = \frac{1}{2}\{\tilde{\pi}_i(f, g) \pm \tilde{\pi}_i(g, f)\}$, and

$$\langle\langle f, g\rangle^\pm, h\rangle_m^\pm = \sum_{\substack{i+j=m, \\ i,j\geq 1}} \tilde{\pi}_i^\pm(\tilde{\pi}_j^\pm(f, g), h) \qquad (m \geq 2). \tag{15}$$

On each coordinate neighborhood, we can construct $\tilde{\pi}_k$ inductively as a bidifferential operator satisfying (14) and (16) (cf. [8]). A partition of unity argument yields Lemma 2.

Since there always exists a contact one-form on every orientable, compact 3-manifold (cf. [7]), we have

Corollary 1 *Let M be a orientable compact 3-manifold. Then there exists a noncommutative contact algebra on M.*

References

1. F. Bayen, M. Flato, C. Frønsdal, A. Lichnerowicz and D. Sternheimer, *Deformation theory and quantization*, Ann. Phys. **111** (1978), 61-110
2. M. De Wilde, P. Lecomte, *Existence of star-products and formal deformations of the Poisson Lie algebra of arbitrary symplectic manifolds*, Lett. Math. Phys. **7** (1983), 487-496.
3. B. Fedosov, *A simple geometric construction of deformation quantization*, J. Diff. Geom., **40** (1994), 213-238.
4. M. Flato, A. Lichnerowicz and D. Sternheimer, *Déformations 1-différentiables des algèbres de Lie attachées à une variété symplectique ou de contact*, Compositio Math. **31** (1975) 47-82.
5. M. Gerstenhaber, *On the deformation of rings and algebras*, Ann. Math. **79** (1964), 59-103.
6. A. Lichnerowicz, *Les variétés de Jacobi et leurs algèbres de Lie associées*, J. Math. Pures et Appl. **57** (1978), 453-488
7. J. Martinet, *Formes de contact sur les variétés de dimension 3*, Proc. Liverpool Singularities Sympos. II, Lecture Notes in Math., vol. 209, Springer-Verlag, Berlin and New York, 1971, 142-163
8. H. Omori, Y. Maeda, A. Yoshioka, *Deformation quantization of Poisson algebras*, Contemp. Math. **179** (1994), 213-240.
9. H. Omori, Y. Maeda, N. Miyazaki and A. Yoshioka, *Noncommutative 3-sphere*, to appear.
10. H. Omori, Y. Maeda, A. Yoshioka and O. Kobayashi, *On regular Fréchet Lie groups VI*, Tokyo J. Math. **6** (1983), 217-246.

BRAIDED GEOMETRY AND THE INDUCTIVE CONSTRUCTION OF LIE ALGEBRAS AND QUANTUM GROUPS

SHAHN MAJID
Math Department, Harvard, Cambridge MA 02138, USA[†]
DAMTP, University of Cambridge, Cambridge CB3 9EW, UK

Abstract. Double-bosonisation associates to a braided group in the category of modules of a quantum group, a new quantum group. We announce the semiclassical version of this inductive construction.

1. Introduction

A question usually overlooked in deformation theory is that of uniformity: we can quantise this or that Poisson manifold, but do our individual quantisations fit together into a coherent 'quantum geometry'? In classical geometry the co-ordinate rings are assumed *uniformly* to be commutative. When we relax this, each object has many 'directions' in which to become non-commutative and we need to know how to pick these in a coherent way.

This problem is addressed by braided geometry, introduced by the author through about 60 papers since 1989. Rather than deforming one algebra at a time, we deform the tensor product itself; we do group theory and geometry in a braided category in place of Vec. Then all mathematical concepts founded in linear algebra are q-deformed uniformly as we switch on the braiding. Braided geometry has its own method of proofs in which algebraic information 'flows' along braid and tangle diagrams like information in a computer, except that under and over crossings of wires are nontrivial braiding operators Ψ [4, 5, 6, 6, 7, 9, 10]. In physical terms, braided geometry is a generalisation of supergeometry with -1 in Bose-Fermi statistics replaced by braid statistics (e.g. by q). This is conceptually quite different from the usual quantisation picture where $q = e^{\hbar/2}$. But braided-commutative with respect to some $\otimes_q$ still means non-commutative with respect to the usual $\otimes$, so we generate noncommutative algebras, which we can then 'semiclassicalise' via such an expansion; we do not start with Poisson brackets but rather we generate them, i.e. this is a deeper point of view.

[†]Visiting 1995 & 1996. Royal Society University Research Fellow

D. Sternheimer et al. (eds.), Deformation Theory and Symplectic Geometry, 339–344.

The starting point is the concept of *braided group* [4] or Hopf algebra in a braided category. This means an algebra and coalgebra B in the category for which the coproduct $\underline{\Delta} : B \to B\underline{\otimes}B$ is an algebra homomorphism, where $\underline{\otimes}$ is the *braided tensor product* of algebras [4] in a braided category. In concrete terms, $B\underline{\otimes}B$ has product $(a\otimes b)(c\otimes d) = a\Psi(b\otimes c)d$.

The simplest example [7] is the tensor algebra TV on a finite-dimensional vector space V equipped with a braiding $\Psi : V\otimes V \to V\otimes V$. Write $TV = \mathbb{C}\langle x_i\rangle$ and $\Psi(x_i\otimes x_j) = x_b\otimes x_a R^a{}_i{}^b{}_j$ (with summation of indices), where R obeys the Yang-Baxter equation. Then the coproduct has the form [7]:

$$\underline{\Delta} x_{i_1}x_{i_2}\cdots x_{i_m} = \sum_{r=0}^{m} x_{j_1}\cdots x_{j_r}\otimes x_{j_{r+1}}\cdots x_{j_m}\begin{bmatrix} m \\ r \end{bmatrix}; R\Big]^{j_1\cdots j_m}_{i_1\cdots i_m}.$$

In his talk, Rosso [13] mentioned the 'quantum shuffle algebra' but this is just the graded dual of TV. Its product has just the structure of the coproduct of the latter. Writing $y^{i_m\cdots i_1}$ for the dual basis to $x_{i_1}\cdots x_{i_m}$, clearly

$$y^{i_m\cdots i_{r+1}}\cdot y^{i_r\cdots i_1} = \left[\begin{matrix} m \\ r \end{matrix}; R\right]^{i_1\cdots i_m}_{j_1\cdots j_m} y^{j_m\cdots j_1}, \quad \underline{\Delta} y^{i_m\cdots i_1} = \sum_{r=0}^{m} y^{i_m\cdots i_{r+1}}\otimes y^{i_r\cdots i_1}.$$

From standard properties [7] of these *braided binomial* matrices $\left[\begin{smallmatrix} m \\ r \end{smallmatrix}; R\right]$,

$$\pi : T(V^*) \to (TV)^*, \quad \pi(y^{i_m}y^{i_{m-1}}\cdots y^{i_1}) = [m;R]!^{i_1\cdots i_m}_{j_1\cdots j_m} y^{j_m\cdots j_1}$$

is a homomorphism of braided groups, where $[m;R]!$ are the *braided factorial* matrices and $T(V^*) = \mathbb{C}\langle y^i\rangle$. Hence $\mathrm{ev} : T(V^*)\otimes TV \to \mathbb{C}$,

$$\mathrm{ev}(f(y), g(x)) = \pi(f(y))(g(x)) = f(\partial)g(x)|_{x=0} = f(y)g(\overleftarrow{\partial})|_{y=0} \tag{1}$$

is a duality pairing of braided groups. Here ∂ denotes *braided differentiation*

$$\partial^i x_{i_1}\cdots x_{i_m} = x_{j_2}\cdots x_{j_m}[m;R]^{ij_2\cdots j_m}_{i_1\cdots i_m}$$

where $[m;R]$ is the *braided integer* matrix. Similarly for $\overleftarrow{\partial}$. These are some rudiments of braided geometry on free algebras [7].

The kernels of ev may be non-zero; quotienting by them gives new braided groups such as the quantum planes $\mathbb{C}_q^m$. Another choice [11] is $R^i{}_j{}^k{}_l = \delta^i{}_j\delta^k{}_l q^{\beta_{jl}}$, where β is a bilinear form. This is the case considered in [13] and Frønsdal's talk [2]. When β comes from a Cartan matrix, Lusztig [3] computed $\ker\pi$ as the q-Serre relations, i.e. $T(V^*)/\ker\pi = \mathrm{image}(\pi) = U_q(n_+)$.

2. Transmutation and bosonisation; induction principle

Some theorems about braided groups are the following. We use Sweedler notation $\Delta h = h_{(1)} \otimes h_{(2)}$ for coproducts, S for antipodes and $\mathcal{R} = \mathcal{R}^{(1)} \otimes \mathcal{R}^{(2)}$ for Drinfeld's quasitriangular structure [1] (summations understood).

1. Transmutation (SM 1990). Let H be a quantum group (quasitriangular Hopf algebra). Its *transmutation* is the braided group $\underline{H} \in {}_H\mathcal{M}$, the braided category of modules. $\underline{H}$ is H as a module-algebra by Ad, and

$$\underline{\Delta} h = h_{(1)} S\mathcal{R}^{(2)} \otimes \mathrm{Ad}_{\mathcal{R}^{(1)}}(h_{(2)}), \quad \Psi(h \otimes g) = \mathrm{Ad}_{\mathcal{R}^{(2)}}(g) \otimes \mathrm{Ad}_{\mathcal{R}^{(1)}}(h).$$

2. Bosonisation (SM 1991). Let $B \in {}_H\mathcal{M}$ be a braided group with $\underline{\Delta} b = b_{\underline{(1)}} \otimes b_{\underline{(2)}}$ and action $\triangleright$ of H. Its *bosonisation* is the Hopf algebra $B{>\!\!\!\triangleleft} H$ generated by H as a Hopf algebra, B as an algebra, and

$$hb = (h_{(1)} \triangleright b) h_{(2)}, \quad \Delta b = b_{\underline{(1)}} \mathcal{R}^{(2)} \otimes \mathcal{R}^{(1)} \triangleright b_{\underline{(2)}}. \tag{2}$$

3. Biproducts (cf. Radford 1985, SM 1992). Let $B \in {}_F^F\mathcal{M}$, the crossed modules over a Hopf algebra F with bijective S. There is a *biproduct* Hopf algebra $B{>\!\!\!\triangleleft} F$ projecting to F. Every projection to F is of this form.

4. Double-bosonisation (SM 1995). Let $B^{\check{}}$ be dually paired to $B \in {}_H\mathcal{M}$ via $\mathrm{ev} : B \otimes B^{\check{}} \to \mathbb{C}$. There is a quantum group $B{>\!\!\!\triangleleft} H {\triangleright\!\!\!<} B^{\check{}\mathrm{op}}$ containing $B{>\!\!\!\triangleleft} H$ and $H{\triangleright\!\!\!<} B^{\check{}\mathrm{op}}$ as subHopf algebras, defined by (2) and

$$b_{\underline{(1)}} \mathcal{R}^{(2)} c_{\underline{(1)}} \mathrm{ev}(\mathcal{R}^{(1)} \triangleright b_{\underline{(2)}}, c_{\underline{(2)}}) = \mathrm{ev}(b_{\underline{(1)}}, \mathcal{R}^{(2)} \triangleright c_{\underline{(1)}}) c_{\underline{(2)}} \mathcal{R}^{(1)} b_{\underline{(2)}}$$

$$hc = (h_{(2)} \triangleright c) h_{(1)}, \quad \Delta c = \mathcal{R}^{(2)} \triangleright c_{\underline{(1)}} \otimes c_{\underline{(2)}} \mathcal{R}^{(1)}, \quad \mathcal{R}^{\mathrm{new}} = \mathcal{R} \exp^{-1},$$

where $\mathcal{R}^{\mathrm{new}}$ needs a canonical element (coevaluation) $\exp \in B^{\check{}} \otimes B$ for ev.

5. Double-biproducts (SM 1995). Let $B^{\check{}} \in {}_F^F\mathcal{M}$ be dually paired to $B \in {}_F^F\mathcal{M}$ in 3. as in [9]. There is a Hopf algebra $B{>\!\!\!\triangleleft} F {\triangleright\!\!\!<} B^{\check{}\mathrm{op}}$. A functor ${}_H\mathcal{M} \to {}_H^H\mathcal{M}$ allows 2. & 4. to be viewed as special cases of 3. & 5.

Bosonisation has been used to construct inhomogeneous Hopf algebras $\mathbb{C}_q^n {>\!\!\!\triangleleft} \widetilde{U_q(su_n)}$. The $\tilde{}$ denotes a central extension. On the other hand, double-bosonisation can be iterated to provide a graph of quantum groups, including the standard families of $U_q(g)$ as well as new quantum groups without classical limit. At each node H, the branches are the inequivalent $B \in {}_H\mathcal{M}$. The new node is $B{>\!\!\!\triangleleft} H {\triangleright\!\!\!<} B^{\check{}\mathrm{op}}$. The initial node is the quantum group $\mathbb{C}$. Its central extension is the quantum line $U_q(1)$. Adjoining the braided line $\mathbb{C}_q$ to this yields $U_q(su_2)$. There are several braided groups in the category of $\widetilde{U_q(su_2)}$-modules, each yielding a new quantum group. The quantum-braided plane $\mathbb{C}_q^2$ gives us $U_q(su_3)$, etc. Induction up the exceptional and C series is via extended quantum planes. Some technicalities are in [9].

Also, as there is surely *some* braided group B in the category of $\widetilde{U_q(e_8)}$-modules, we obtain at least one quantum group $B{>\!\!\!\triangleleft} \widetilde{U_q(e_8)} {\triangleright\!\!\!<} B^{\check{}\mathrm{op}}$ which could be called $U_q(e_9)$! Presumably it does not survive as $q \to 1$. Also, building up $U_q(g)$ inductively by a series of triple products yields automatically a natural *inductive block basis* for it, which becomes a basis when we fix bases for the braided planes B which are adjoined at each stage. E.g.,

$$U_q(su_n) = \mathbb{C}_q^{n-1}{>\!\!\!\triangleleft}\mathbb{C}_q^{n-2}{>\!\!\!\triangleleft}\cdots{>\!\!\!\triangleleft}\mathbb{C}_q{>\!\!\!\triangleleft}U_q(\beta){\triangleright\!\!\!<}\mathbb{C}_q{\triangleright\!\!\!<}\cdots{\triangleright\!\!\!<}\mathbb{C}_q^{n-2}{\triangleright\!\!\!<}\mathbb{C}_q^{n-1} \quad (3)$$

where the central extensions are collected together as $U_q(\beta) = U(1)^{\otimes n}$ generated by H_i with a quasitriangular structure $\mathcal{R}_\beta = q^{\sum \beta_{ij}^{-1} H_i \otimes H_j}$. Here β is the symmetrised Cartan matrix. This *proves* the PBW theorem for $U_q(g)$ and explicitly constructs $U_q(n_+) = \mathbb{C}_q^{n-1}{>\!\!\!\triangleleft}\mathbb{C}_q^{n-2}\cdots{>\!\!\!\triangleleft}\mathbb{C}_q$. Choosing bases for the $\mathbb{C}_q^i$ gives us a basis for $U_q(n_+)$, as well as all the relations between them (including the q-Serre relations when expressed in terms of the simple roots). The inductive basis is coherent across the graph of quantum groups. Moreover, its restriction to any substring of factors gives a sub-braided or quantum group. In (3), $\mathbb{C}_q^{n-1}{>\!\!\!\triangleleft}\cdots{>\!\!\!\triangleleft}\mathbb{C}_q{>\!\!\!\triangleleft}U_q(\beta) = U_q(b_+)$, $\mathbb{C}_q^2{>\!\!\!\triangleleft}\mathbb{C}_q{>\!\!\!\triangleleft}U_q(\beta){\triangleright\!\!\!<}\mathbb{C}_q = \mathbb{C}_q^2{>\!\!\!\triangleleft}\widetilde{U_q(su_2)}$, etc. If one is interested in only half the story, i.e. only in constructing $U_q(b_+)$, one can also do it by iterated biproducts. Thus, $\mathbb{C}_q^n{>\!\!\!\triangleleft}\widetilde{U_q(b_+)}$ gives the q-Borel of $U_q(su_{n+1})$.

Double-bosonisation also generalises Lusztig's construction. Any β defines a quantum group $U_q(\beta)$ with generators h_i and $\mathcal{R}_\beta = q^{\sum \beta_{ij} h_i \otimes h_j}$. $B = \mathbb{C}\langle y^i\rangle$, $B^\star = \mathbb{C}\langle x_i\rangle$, paired by (1), live in the category of $U_q(\beta)$-modules by $h_i{\triangleright}y^j = \delta_{ij}y^j$, $h_i{\triangleright}x_j = -\delta_{ij}x_j$. So $\mathbb{C}\langle y^i\rangle{>\!\!\!\triangleleft}U_q(\beta){\triangleright\!\!\!<}\mathbb{C}\langle x_i\rangle$ is a Hopf algebra. Quotienting by the kernels of ev we obtain a quantum group $U_q(n_+){>\!\!\!\triangleleft}U_q(\beta){\triangleright\!\!\!<}U_q(n_-)$ with $\mathcal{R} = \mathcal{R}_\beta \exp^{-1}$. For generic β (or generic R-matrix in Section 1) $[m;R]$ are invertible and the coevaluation for (1) is

$$\exp = \textstyle\sum_{m=0}^{\infty} x_{i_m}\cdots x_{i_1}([m;R]!^{-1})^{i_1\cdots i_m}_{j_1\cdots j_m} y^{j_m}\cdots y^{j_1} \in B^{\star\mathrm{op}} \otimes B.$$

Otherwise, quotienting by the kernels is nontrivial but we still have ∂, $\overleftarrow{\partial}$ and the *braided exponential* exp is characterised as their eigenfunction [10].

Note that Frønsdal in his talk and [2] considered recursion relations for an ansatz of the form $\mathcal{R}_\beta f(x,y)$ to obey the Yang-Baxter equation, with resulting Hopf algebra being coboundary. By contrast, double-bosonisation already provides a closed expression for $\mathcal{R}$ via a braided-exponential, proves that quotienting by kernels of ev yields a Hopf algebra and proves that it is quasitriangular. [9] has been circulated in October 1995.

3. Braided-Lie bialgebras and Lie induction

We now announce a semiclassical concept of braided groups. Let g, $\delta : g \to g\otimes g$, $r \in g\otimes g$ be a quasitriangular Lie bialgebra as per Drinfeld [1]. Let $2r_+ = r + \tau(r)$ where τ is transposition. Let $\triangleright$ denote an action of g.

0. A *braided-Lie bialgebra* $b \in {}_g\mathcal{M}$ is a g-covariant Lie algebra and g-covariant Lie coalgebra with cobracket $\underline{\delta} : b \to b \otimes b$ obeying $\forall x, y \in b$,

$$\underline{\delta}([x,y]) = \mathrm{ad}_x \delta y - \mathrm{ad}_y \delta x - \psi(x \otimes y); \quad \psi = 2r_+(\triangleright \otimes \triangleright) \circ (\mathrm{id} - \tau),$$

i.e., $\mathrm{d}\underline{\delta} = \psi$ where d is the Lie coboundary on $\underline{\delta} \in C^1_{\mathrm{ad}}(b, b \otimes b)$ and $\mathrm{d}\psi \equiv 0$.

1. Let $i : g \to f$ be a map of Lie bialgebras. The *transmutation* of f is a braided-Lie bialgebra $\underline{f} \in {}_g\mathcal{M}$ with Lie algebra f and for all $x \in f$,

$$\underline{\delta} x = \delta x + r^{(1)} \triangleright x \otimes i(r^{(2)}) - i(r^{(2)}) \otimes r^{(1)} \triangleright x, \quad \triangleright = \mathrm{ad} \circ i.$$

In particular, g has a braided version $\underline{g} \in {}_g\mathcal{M}$ by ad, the same bracket, and

$$\underline{\delta} x = 2r_+{}^{(1)} \otimes [x, r_+{}^{(2)}]. \tag{4}$$

2. Let $b \in {}_g\mathcal{M}$ be a braided-Lie bialgebra. Its *bosonisation* is the Lie bialgebra $b \rtimes\!\!\!\cdot\, g$ with g as sub-Lie bialgebra, b as sub-Lie algebra and

$$[\xi, x] = \xi \triangleright x, \quad \delta x = \underline{\delta} x + r^{(2)} \otimes r^{(1)} \triangleright x - r^{(1)} \triangleright x \otimes r^{(2)}, \quad \forall \xi \in g,\ x \in b. \tag{5}$$

3. Let f be a Lie bialgebra and ${}^f_f\mathcal{M}$ its category of Lie crossed modules (=modules of the Drinfeld double $D(f)$.) Objects b are simultaneously f-modules $\triangleright$ and f-comodules $\beta : b \to f \otimes b$ obeying $\forall \xi \in f, x \in b$,

$$\beta(\xi \triangleright x) = ([\xi, \,] \otimes \mathrm{id} + \mathrm{id} \otimes \xi \triangleright)\beta(x) + (\delta \xi) \triangleright x.$$

Writing $\beta(x) = x^{(\bar 1)} \otimes x^{(\bar 2)}$, the infinitesimal braiding in this category is

$$\psi(x \otimes y) = y^{(\bar 1)} \triangleright x \otimes y^{(\bar 2)} - x^{(\bar 1)} \triangleright y \otimes x^{(\bar 2)} - y^{(\bar 2)} \otimes y^{(\bar 1)} \triangleright x + x^{(\bar 2)} \otimes x^{(\bar 1)} \triangleright y.$$

Let $b \in {}^f_f\mathcal{M}$ be a braided-Lie bialgebra. The *bisum* Lie bialgebra $b \rtimes\!\!\!\cdot\, g$ has semidirect Lie bracket/cobracket and projects onto f. Any Lie bialgebra projecting onto f is of this form. A functor ${}_g\mathcal{M} \to {}^g_g\mathcal{M}$ relates 2. & 3.

4. Let $b^{\check{}} \in {}_g\mathcal{M}$ be a braided-Lie bialgebra dually paired with b by invariant $\mathrm{ev} : b \otimes b^{\check{}} \to \mathbb{C}$. Its *double-bosonisation* is the Lie bialgebra $b \rtimes\!\!\!\cdot\, g \cdot\!\!\!\ltimes b^{\check{}\mathrm{op}}$ with g as sub-Lie bialgebra, $b, b^{\check{}\mathrm{op}}$ sub-Lie algebras, (5) and

$$[\xi, \phi] = \xi \triangleright \phi, \ [x, \phi] = \mathrm{ev}(x_{\underline{(1)}}, \phi) x_{\underline{(2)}} + \mathrm{ev}(x, \phi_{\underline{(1)}}) \phi_{\underline{(2)}} + 2r_+{}^{(1)} \mathrm{ev}(x, r_+{}^{(2)} \triangleright \phi)$$
$$\delta \phi = \underline{\delta} \phi + r^{(2)} \triangleright \phi \otimes r^{(1)} - r^{(1)} \otimes r^{(2)} \triangleright \phi, \quad r^{\mathrm{new}} = r - \textstyle\sum_a f^a \otimes e_a,$$

$\forall x \in b, \xi \in g$ and $\phi \in b^{\check{}}$. Here $\underline{\delta} x = x_{\underline{(1)}} \otimes x_{\underline{(2)}}$, etc., and r^{new} assumes that ev has a coevaluation, i.e. if $\{e_a\}$ is a basis of b then $\{f^a\}$ is dual w.r.t. ev.

Double bosonisation provides an inductive construction for quasitriangular Lie bialgebras, preserving factorisability (nondegeneracy of r_+). It is a co-ordinate free version of the idea of adjoining a node to a Dynkin diagram (adjoining a simple root vector in the Cartan-Weyl basis). Moreover, building up g iteratively like this also builds up the quasitriangular structure r. The braided-Lie bialgebra used in the induction could be trivial:

Proposition. Let g be a semisimple factorisable (s.s.f) Lie bialgebra and b a faithful isotypical representation such that $\Lambda^2 b$ is isotypical. Then b with zero bracket and zero cobracket is a braided-Lie bialgebra in ${}_{\tilde{g}}\mathcal{M}$, and $b{\bowtie}\tilde{g}{\ltimes} b^{*\mathrm{op}}$ is another s.s.f. Lie bialgebra. Here $\tilde{g}$ is a central extension.

The induction also works at the simple strictly quasitriangular level (with b irreducible). For example, the 2-dimensional and 3-dimensional representations of su_2 have the required property (ensuring $\psi \propto (\mathrm{id} - \tau)$ in ${}_g\mathcal{M}$). They give su_3 and so_5, taking us up the A and B series respectively.

Finally, just as Lie bialgebras extend to Poisson-Lie groups, so braided-Lie bialgebra structures generally extend to the associated Lie group of g. The resulting Poisson bracket does not, however, respect the group product in the usual way but rather up to a 'braiding' obtained from ψ.

Example. The transmutation (4) of the Drinfeld-Sklyanin (or other factorisable) Lie cobracket on semisimple g is the Kirillov-Kostant Lie cobracket. Moreover, this *Kirillov-Kostant braided-Lie bialgebra* extends, in principle, to a braided-Poisson Lie group. Details are to appear elsewhere.

References

1. V.G. Drinfeld. Quantum groups. In *Proc. ICM*, pages 798–820. AMS, 1987.
2. C. Frønsdal. Generalisation and exact deformations of quantum groups. Preprint q-alg/9606020, 1996.
3. G. Lusztig. *Introduction to Quantum groups.* Birkhauser, 1993.
4. S. Majid. Braided groups and algebraic quantum field theories. *Lett. Math. Phys.*, 22:167–176, 1991.
5. S. Majid. Transmutation theory and rank for quantum braided groups. *Math. Proc. Camb. Phil. Soc.*, 113:45–70, 1993.
6. S. Majid. Cross products by braided groups and bosonization. *J. Algebra*, 163:165–190, 1994.
7. S. Majid. Free braided differential calculus, braided binomial theorem and the braided exponential map. *J. Math. Phys.*, 34:4843–4856, 1993.
8. S. Majid. Algebras and Hopf algebras in braided categories. volume 158 of *Lec. Notes in Pure and Appl. Math*, pages 55–105. Marcel Dekker, 1994.
9. S. Majid. Double bosonisation and the construction of $U_q(g)$. Preprint q-alg/9511001, 1995. To appear *Math. Proc. Camb. Phil. Soc.*
10. S. Majid. *Foundations of Quantum Group Theory.* Cambridge Univ. Press, 1995.
11. S. Majid. $\mathbb{C}$-statistical quantum groups and Weyl algebras. *J. Math. Phys.*, 33:3431–3444, 1992.
12. D. Radford. The structure of Hopf algebras with a projection. *J. Algebra*, 92:322–347, 1985.
13. M. Rosso. Groupes quantiques et algebres de battage quantiques. *C. R. Acad. Sci.*, 320:145–148, 1995.

STAR PRODUCTS IN THE TRIANGULAR CASE

CARLOS MORENO
Lab. Phys. Math. C.N.R.S., Univ. Bourgogne
B.P. 138, F-21004 Dijon.
Dpto. de Física Teórica, Univ. Complutense
E-28040 Madrid.
cmo@moka.ccr.jussieu.fr

AND

LUIS VALERO
I.F.P. "Giner de los Rios"
E-28100, Madrid.
Dpto. de Física Teórica, Univ. Complutense
E-28040 Madrid.
lvalero@eucmos.sim.ucm.es

Abstract.

All the invariant star products (ISPs) on a simply connected Lie group $\mathbf{G}$ endowed with a left-invariant symplectic structure β_1 can be obtained from the Moyal star product. We prove a theorem stated by Drinfeld that constructs a bijection between the set of equivalence classes of ISPs and the set $\beta_1 + h \cdot H^2(\mathfrak{g})[[h]]$.

1. The Moyal star product

a) Let $\mathbf{G} \equiv \mathbb{R}^{2n}$ be the commutative Lie group with Lie algebra $\mathfrak{g} \equiv \mathbb{R}^{2n}$. Suppose that $\beta_1 \in \mathfrak{g}^* \wedge \mathfrak{g}^*$ is not degenerate. The manifold $(\mathbb{R}^{2n}, \beta_1)$ is then a Lie group with an invariant symplectic structure. We write $\{x^i, i = 1, ..., 2n\}$ for a generic point in the group $\mathbb{R}^{2n}$ or in its Lie algebra. Nevertheless, we will sometimes write $\exp x$ for the corresponding element in the group when this seems clearer.

The Moyal star product (SP) on the manifold in question is defined by means of the following unitary irreducible projective representation (UIPR)

D. Sternheimer et al. (eds.), Deformation Theory and Symplectic Geometry, 345–352.

of the group $\mathbb{R}^{2n}$ itself: $W(x)\cdot W(y) = e^{\frac{ih}{4\pi}\beta_1(x,y)}\,W(x+y)$. If ψ_1 is a function on the Lie group $\mathbb{R}^{2n}$, its Weyl symbol is by definition the operator,

$$W(\psi_1) = \int_{\mathbb{R}^{2n}} (\mathcal{F}\psi_1)(x)\, W(x)\, dx,$$

where $\mathcal{F}$ is for the Fourier transform [23].

The integral form of the Moyal SP is the Weyl symbol of the product $W(\psi_1)\cdot W(\psi_2)$, i.e.,

$$(\psi_1 * \psi_2)(z) = \left(\frac{2}{h}\right)^{2n} (det\,\beta_1) \int_{\mathbb{R}^{2n}\times\mathbb{R}^{2n}} e^{\frac{4\pi i}{h}\beta_1(z-x;z-y)}\psi_1(x)\psi_2(y)\,dxdy. \tag{1}$$

The bijection between the set of equivalence classes of UIPRs of the group $\mathbb{R}^{2n}$ and the set of equivalence classes of irreducible unitary representations of the Heisenberg group $\mathbf{H}_n$ allows to consider this group as well. Its Lie algebra $\bar{\mathfrak{g}}_{\beta_1}$ is the central extension of $\mathfrak{g}$ by $\mathbb{R}^{2n}$ defined by the cocycle $\beta_1 : \bar{x} = x + aE$, $[\bar{x};\bar{y}] = \beta_1(x,y)E$.

b) Every irreducible (unitary) representation of $\mathbf{H}_n$ is obtained from a (coadjoint) orbit in $\bar{\mathfrak{g}}^*_{\beta_1}$. The coadjoint orbit of the point $\xi + uE^* \in \mathbb{R}^{2n} + \mathbb{R}E^*$ is the set

$$\overline{Ad}^*(\overline{exp}\,\bar{x})\cdot(\xi + uE^*) = \xi + uE^* - u\tilde{\beta}(x), \quad \forall x \in \mathfrak{g},$$

where $\xi \in \mathfrak{g} \equiv \mathbb{R}^{2n}$, $u \in \mathbb{R}$. It is thus the manifold $uE^* + \mathbb{R}^{2n}$.

The following two obvious properties of the coadjoint action are basic for the *commutative* case considered here. Similar properties apply, though only *locally*, in the general case to be considered in the next section.

Proposition 1 *a) The mapping* $\mathcal{K} : \mathbf{G} \equiv \mathbb{R}^{2n} \longrightarrow uE^* + \mathbb{R}^{2n} \equiv uE^* + \mathfrak{g}^*$, *defined as:* $\mathcal{K}(\exp x) = \overline{Ad^*}(\overline{\exp \bar{x}})\cdot(uE^*) = uE^* - u\tilde{\beta}(x)$, *is a global diffeomorphism. b)* $\mathcal{K}\circ\lambda_{\exp y} = \overline{Ad^*}(\overline{\exp y})\circ\mathcal{K}$, $\forall y \in \mathfrak{g}$, *where* $\lambda_{\exp y}$ *stands for the (left!) action of the group on itself.*

The value $h = 2\pi u$ determines the above representation of $\mathbf{H}_n$. The multiplier of the UIPR of $\mathbb{R}^{2n}$ corresponding to it is $e^{-\frac{ih}{4\pi}\beta_1(x,y)}$. The mapping $\frac{h}{2\pi}\mathcal{K}$ is an isomorphism of symplectic manifolds.

By the diffeomorphism $\mathcal{K}$ the star product (1) can be pulled back to the orbit $\mathcal{O}_{\frac{h}{2\pi}E^*} \equiv \frac{h}{2\pi}E^* + \mathfrak{g}^*$.

Proposition 2 *Let* φ_1 *and* φ_2 *be two functions on the orbit* $\mathcal{O}_{\frac{h}{2\pi}E^*}$. *Let*

$$(\varphi_1\hat{*}\varphi_2)(\xi) = ((\varphi_1\circ\mathcal{K}) * (\varphi_2\circ\mathcal{K})\left(\mathcal{K}^{-1}(\xi)\right)$$

by definition. Then

$$(\varphi_1 \hat{*} \varphi_2)\ (\xi) = \left(\frac{2}{h}\right)^{2n} \left(\frac{1}{det\,\beta_1}\right) \int_{\mathbb{R}^{2n}\times\mathbb{R}^{2n}} e^{\frac{4\pi i}{h}\Lambda_1(\xi-\eta;\xi-\zeta)}\,\varphi_1(\eta)\varphi_2(\zeta)\,d\zeta d\eta,$$

where Λ_1 *is the skewsymmetric tensor defined by the relation* $\tilde{\Lambda}_1 \cdot \tilde{\beta}_1 = 1$, *and* $\tilde{\beta}_1$ *is the isomorphism in* $\mathbb{R}^{2n}$ *defined by* β_1.

We can also write this formula in terms of the Fourier transform:

Proposition 3 *The last integral form of the star product on the orbit* $\mathcal{O}_{\frac{h}{2\pi}E^*}$ *can also be written as*

$$(\varphi_1 \hat{*} \varphi_2)(\xi) = \int_{\mathbb{R}^{2n}\times\mathbb{R}^{2n}} e^{-2\pi i(\xi(x+y)+\frac{1}{2}h\beta_1(x;y))}\,(\overline{\mathcal{F}}\varphi_1)(x)(\overline{\mathcal{F}}\varphi_2)(y)\,dxdy.$$

d) The Campbell-Hausdorff formula for the Lie algebra $\overline{\mathfrak{g}}_{\beta_1}$ is $\overline{\gamma}_{\beta_1}(\overline{x};\overline{y}) = x + y + (a + b + \frac{1}{2}\beta_1(x;y))E$. The expression in Prop. 3 thus becomes,

$$(\varphi_1 \hat{*} \varphi_2)(\xi) = \int_{\mathbb{R}^{2n}\times\mathbb{R}^{2n}} e^{-2\pi i(\xi+E^*);h^{-1}\overline{\gamma}_{\beta_1}(hx;hy))}\,(\overline{\mathcal{F}}\varphi_1)(x)(\overline{\mathcal{F}}\varphi_2)(y)\,dxdy. \tag{2}$$

The property of associativity of the $\hat{*}$ -product defined by this integral, like the similar property of its series expansion in powers of h, i.e., the Moyal star product, is a consequence of the associativity of the law group $\overline{\gamma}_{\beta_1}$.

e) We refer to [20, 21] for deformations of multiplier algebras and their corresponding quantum groups.

2. Invariant star products on $(\mathbf{G}, \beta_1)$

a) In the general case of the invariant symplectic manifold $(\mathbf{G};\beta_1)$ we can proceed as in the commutative case. In particular, the starting point for obtaining all the ISPs is the expression corresponding to (2). Let $t \in \mathbb{R}$ be small and $\beta_t = \beta_1 + \beta_2\, t + \cdots + \beta_R\, t^{R-1}$, where R is finite but *arbitrary*, and $\beta_i \in Z^2(\mathfrak{g})$. $\tilde{\beta}_t$ is an isomorphism and $(\mathbf{G};\beta_t)$ is then a symplectic manifold. Let $\overline{\mathfrak{g}}_{\beta_t}$ be the central extension of $\mathfrak{g}$ defined by the cocycle β_t : $\overline{\mathfrak{g}}_{\beta_t} = \mathfrak{g} \oplus \mathbb{R}\,E$; $[\overline{x}\,;\overline{y}] = [x\,;y] + \beta_t(x;y)\,E$. Let $\overline{\mathbf{G}}_{\beta_t}$ be the simply connected Lie group with the Lie algebra $\overline{\mathfrak{g}}_{\beta_t}$ and let $\overline{\gamma}_{\beta_t}(\overline{x};\overline{y})$ be the Campbell-Hausdorff formula for the Lie algebra $\overline{\mathfrak{g}}_{\beta_t}$. The coadjoint orbit of the point $\xi + u\,E^* \in \overline{\mathfrak{g}}^*_{\beta_t}$ is generated by the actions

$$\overline{Ad}^*(\overline{exp}\,\overline{x}) \cdot (\xi + uE^*) = Ad^*(\exp\,x) \cdot \xi + u \cdot f_{\beta_t}(-x) + uE^*,\ \forall x \in \mathfrak{g},$$

where $f_{\beta_t}(x) = \tilde{\beta}_t(x) \cdot A(x) \in \mathfrak{g}^*$ and $A(x) = (\exp(adx) - 1)(adx)^{-1}$. It is an open set in the affine space $\mathfrak{g}^* \oplus uE^*$. For reasons like the ones applying in the commutative case, only the orbit of the point E^* is relevant.

b) Even if the Campbell-Hausdorff formula is generally not convergent everywhere, we may nevertheless consider the following formal integral:

$$(\varphi_1 \hat{*} \varphi_2)(\xi) = \int_{\mathfrak{g}\times\mathfrak{g}} e^{-2\pi i(\xi+E^*;h^{-1}\overline{\gamma}_{\beta_t}(hx;hy))} \, (\overline{\mathcal{F}}\varphi_1)(x) \, (\overline{\mathcal{F}}\varphi_2)(y) \, dxdy,$$

where φ_1, φ_2 are functions on $\mathfrak{g}^* \oplus E^*$. The equality $\varphi_1 \hat{*} 1 = \varphi_1$ follows from the equality $\overline{\gamma}_{\beta_t}(x;0) = x$, and the condition of associativity of $\hat{*}$ on suitable spaces of functions on the orbit follows from the associativity of the law group $\overline{\gamma}_{\beta_t}$. In particular, the space of polynomials in the components of ξ is an algebra for the product $\hat{*}$. The operator $\overline{Ad}(\overline{g})$ is an isomorphism of the Lie algebra $\overline{\mathfrak{g}}_{\beta_t}$. Consequently the product $\hat{*}$ is invariant under $\overline{Ad}^*(\overline{\mathbf{G}}_{\beta_t})$ as can be seen from the exponential term. These four properties apply also in case of the expansion of the integral in powers of h, i.e., for the star product on the desired orbit. $h^{-1}\overline{\gamma}_{\beta_t}(hx;hy)$ is a power series in h, in the components of x, y and ξ. Consequently [1] the cochains of this SP are:

$$P_S(\varphi_1;\varphi_2)(\xi) = B_S\left(\frac{-1}{2\pi i}\frac{\partial}{\partial\xi}; \frac{-1}{2\pi i}\frac{\partial}{\partial\xi}; \xi\right) (\varphi_1 \otimes \varphi_2)(\xi), \; S \in \mathbb{N},$$

where B_S is a bidifferential operator with polynomial coefficients. It can be seen that the degree of its term of *minimal* degree is $S+1$, and every term is of a degree greater than or equal to 1 in both its arguments [16]. We thereby get the following SP

$$(\varphi_1 * \varphi_2)(\xi) = (\varphi_1 \cdot \varphi_2)(\xi) + \sum_{S\geq 1} P_S(\varphi_1;\varphi_2)(\xi)h^S. \tag{3}$$

c) In the general case we are considering, the two basic properties [8] of the coadjoint action $\overline{Ad}^*$ are similar to those in Prop. 1,

Proposition 4 *a) The mapping* $\mathcal{K}_{\beta_t} : \mathbf{G} \to E^* + \mathfrak{g}^*$, $\mathcal{K}_{\beta_t}(\exp x) = \overline{Ad^*}(\overline{\exp x}) \cdot E^* = Ad^*(\exp x) \cdot \xi + f_{\beta_t}(-x) + E^*$, *is a local diffeomorphism at the point* $e \in \mathbf{G}$. *b) In the neighborhood of* $e \in \mathbf{G}$, $\mathcal{K}_{\beta_t} \circ \lambda_{\exp y} = \overline{Ad^*}(\overline{\exp y}) \circ \mathcal{K}_{\beta_t}$, $\forall \exp y$ *in a neighborhood of* $e \in \mathbf{G}$, *where* $\lambda_{\exp y}$ *represents the left action of the group* $\mathbf{G}$ *on itself.*

By property a), the star product (3) can be pulled back to a neighborhood of $e \in \mathbf{G}$. We thereby get a SP on this neighborhood: $(\psi_1 *_t \psi_2)(\exp x) = (\psi_1 \circ \mathcal{K}_t^{-1}) \hat{*} (\psi_2 \circ \mathcal{K}_t^{-1})(\xi)$. Its 2-cochains are of a degree greater than or equal to 1 in both arguments. This is so, first, because the same is true of the product (3) and second, because the equations, in normal coordinate and dual basis, of the diffeomorphism $\mathcal{K}_t^{-1}$ are:

$$x^b = \xi_a \Lambda_t^{ab} + x^i \Lambda_t^{ab} (\beta_t)_{ik} \sum_{r\geq 1} M^k_{i_1,\ldots,i_r,a} x^{i_1} \ldots x^{i_r},$$

where $M^k_{i_1,\ldots,i_r,a}$ are polynomials in the structure constants of $\mathfrak{g}$. By property b), it can be proved that this SP *on the symplectic manifold* $(\mathbf{G}; \beta_t)$ is invariant under $\mathbf{G}$ [15, 16]. It is therefore defined by an element

$$F^t(x;y) = 1 + \sum_{i=1}^{\infty} F_i^t(x;y)\, h^i$$

in the formal power series algebra $\mathfrak{U}^{\otimes 2}\mathfrak{g}[[h]]$, where $\mathfrak{U}\mathfrak{g}$ is the enveloping algebra of $\mathfrak{g}$. The associativity property of this SP on $\mathbf{G}$ and the condition on their cochains can be written respectively [8, 9] as

$$F^t(x{+}y;z)\cdot F^t(x;y) = F^t(x;y{+}z)\cdot F^t(y;z), \quad F^t(x;0) = F^t(0;y) = 1, \tag{4}$$

where + stands for the action of the coproduct in the enveloping algebra. The term h^n in the power series expansion of this expression is

$$dF_n^t = \alpha_n(F_1^t; F_2^t; \cdots; F_{n-1}^t),$$

where d is the coboundary operator of the complex $(\mathfrak{U}\mathfrak{g}^{\otimes *}; d)$ defined by the coalgebra structure of $\mathfrak{U}\mathfrak{g}$ [3]. It is the condition of associativity to the order n [14] of the deformation which is induced on the algebra $\mathfrak{U}\mathfrak{g}$ by the invariant product $*_t$ on $\mathbf{G}$.

We can now prove [16]

Proposition 5 *Take the following Taylor expansion in* t, $F_R^t(x;y) = \sum_{L\geq 0} F_{RL}(x;y)\, t^L$. *The formal series* $F(x;y) = \sum_{R\geq 0} F_R(x;y) h^R$, *where* $F_R(x;y) = \sum_{S+T=R} F_{ST}(x;y)$, *and* $t = h$ *in (4), is an ISP on the symplectic manifold* $(\mathbf{G}, \beta_1)$. *It will be said to be determined by the formal cocycle* $\beta_h = \beta_1 + \beta_2\, h + \beta_3\, h^2 + \cdots + \beta_R\, h^{R-1} + \cdots$

Definition 1 *Let* $F'(x;y), F(x;y)$ *be two ISPs on* $(\mathbf{G};\beta_1)$. *We say they are equivalent if there exists* $E(x) = 1 + \sum_{i\geq 1} E_i(x)\, h^i \in \mathfrak{U}(\mathfrak{g})[[h]]$ *such that* $F'(x;y) = E^{-1}(x+y)\cdot F(x;y)\cdot E(x)\cdot E(y)$.

Proposition 6 *Any star product* F *on* $(\mathbf{G};\beta_1)$ *determines a formal 2-cocycle,* $\beta_h = \beta_1 + \beta_2\, h + \beta_3\, h^2 + \ldots + \beta_R\, h^{R-1} + \cdots$, *on the Lie algebra* $\mathfrak{g}$ *such that* F *is equivalent to the star product determined by* β_h *as in Proposition 5.*

Proposition 7 *Let $F^{\omega_h} = 1 + \sum_{R\geq 1} F_R^{\omega_h} h^R$, $F^{\beta_h} = 1 + \sum_{R\geq 1} F_R^{\beta_h} h^R$ be star products on $(\mathbf{G};\beta_1)$ determined respectively by cocycles $\omega_h = \beta_h + \tilde{\delta}\alpha_h$ and β_h. There must then be an element $E(x) = 1 + \sum_{R\geq 1} E_R h^R \in \mathfrak{U}\mathfrak{g}[[h]]$, such that: $F^{\omega_h}(x;y) = E^{-1}(x+y)\cdot F^{\beta_h}(x;y)\cdot E(x)\cdot E(y)$. That is to say, F^{ω_h} and F^{β_h} are equivalent by $E(x)$.*

We can now also prove [16, 17]

Proposition 8 *Let F, F' be ISP on $(\mathbf{G};\beta_1)$ determined respectively by the formal cocycles β_h and β'_h. Assuming these products are equivalent, there must be $\alpha_2,\cdots,\alpha_R\cdots \in \mathfrak{g}$ such that $\beta'_i - \beta_i = \tilde{\delta}\alpha_i$, for $i = 2, 3.\cdots, R, \cdots$*

The results in the three preceding propositions can be summarized in the following classification theorem:

Theorem 1 *(Drinfeld, V. G. [8]) Choose a vector subspace $\mathbf{V}$ in the space $Z^2(\mathfrak{g})$ of invariant de Rham 2-cocycles on $\mathfrak{g}$, which is a supplementary space of de Rham 2-exact cocycles $B^2(\mathfrak{g})$, i.e., $Z^2(\mathfrak{g}) = \mathbf{V}\oplus B^2(\mathfrak{g})$. Let $F'(x;y)$ be any ISP on $(\mathbf{G};\beta_1)$. Then, $F'(x;y)$ is equivalent to the ISP obtained in Prop. 5 from a cocycle: $\beta_h = \beta_1 + \beta_2 h + \cdots + \beta_R h^{R-1} + \cdots$, such that $\beta_k \in \mathbf{V}$ if $k > 1$. Moreover, $\{\beta_k\}$ is uniquely determined by $F'(x;y)$.*

Clearly, we can identify the set $\beta_1 + h\cdot\mathbf{V}[[h]]$ with the set $\beta_1 + h\cdot H^2(\mathfrak{g})[[h]]$ through the isomorphism: $[\beta]\in H^2(\mathfrak{g}) \to \beta - \tilde{\delta}\alpha \in \mathbf{V}$.

d) This theorem classifies also [9] the equivalence classes of triangular quantum groups on $(\mathbf{G};\beta_1)$ [22].

e) One of the stages in the proof of this theorem is Prop. 5. Suppose that F and F^{β_h}, where $\beta_h = \beta_1 + \beta_2 h + \cdots + \beta_k h^{k-1}$, are equivalent to the order k. The obstruction to being equivalent to the order $k+1$ is a Hochschild 2-cocycle. It can be proved that it is also a Poisson 2-cocycle. It therefore determines an element $\beta_{k+1} \in Z^2(\mathfrak{g})$, since β_1 is not degenerate. SPs F and $F^{\beta'_h}$, where ${\beta'}_h = \beta_h + \beta_{k+1}h^k$, are then equivalent to the order $k+1$. This is the way to remove the obstruction.

f) De Wilde, Lecomte [4, 5] and Fedosov [10, 12] have proved that on any symplectic manifold $(\mathbf{M};\omega_1)$, there exists a star product. We refer to [2, 6, 7, 10, 11, 18, 19] for the proof of the theorem stating that the set $\omega_1 + h\cdot H^2(\mathbf{M})[[h]]$ classifies the equivalence classes of the star products on the manifold $(\mathbf{M};\omega_1)$.

g) For the classification of the equivalence classes of 1-differential *infinitesimal* deformations of the *dynamical* Lie algebra $(C^\infty(\mathbf{M});\{;\}_1)$, the reader is referred to [13]. A bijection has been constructed between the equivalence classes of these deformations and the set: $P^1(\mathbf{M};\omega_1)\oplus H^2(\mathbf{M})/Q^2(\mathbf{M};\omega_1)$ where $P^1(\mathbf{M};\omega_1) = Im(\omega_1\wedge:\ H^1(\mathbf{M})\to H^3(\mathbf{M}))$ and

$Q^2(\mathbf{M};\omega_1) = Ker(\omega_1\wedge : H^2(\mathbf{M}) \to H^4(\mathbf{M}))$. As with the classification of ISPs, it can be introduced the deformations of the geometry, i.e., the symplectic structures defined by the closed 2-forms on $\mathbf{M}$, $\omega_h = \omega_1 + h\cdot\omega_2$. The deformation $(\mathbf{C}^\infty(\mathbf{M})[[h]];\{;\}_h)$, may then be defined. A bijection can be constructed between the set of *infinitesimal* deformations which *are not* equivalent to one defined by some ω_h, and the set $P^1(\mathbf{M};\omega_1)$. It is also possible to prove that the equivalence classes of *pure* [13] 1-differential *infinitesimal* deformations are classified by the set $\omega_1 + h\cdot H^2(\mathbf{M})$. In this case, as with ISPs, every equivalence class of *infinitesimal* deformations is determined by one and only class of equivalent deformations of the geometry, and conversely.

References

1. Bayen, F., Flato,M., Frønsdal, C., Lichnerowicz, A and Sternheimer, D. (1978) Deformation Theory and Quantization. I and II, *Ann. of Phys.*, **111**, pp. 61-151
2. Bertelson, M., Cahen, M. and Gutt S. (1996) Equivalence of star products. *Université Libre de Bruxelles. Travaux de Mathématiques, 1-15*
3. Cartier, P. (1955) Hyperalgèbres et groupes de Lie formels,*Séminaire Sophus Lie, 2ème année 1955/56, Faculté des Sciences de Paris.*
4. De Wilde, M. and Lecomte, P. (1983) Existence of star-products and of formal deformations in Poisson Lie algebra of arbitrary symplectic manifolds,*Lett. Math. Phys.*, **7**, pp. 487-496
5. De Wilde, M. and Lecomte, P. (1990) Existence of star-products revisited, *Suppl. n. 1, Note di Mathematica*, **X**, pp. 205-216
6. De Wilde, M. and Lecomte, P.(1996) Existence et classification des star-produits sur les variétés symplectiques. *Notes du cours dans l'Ecole d'Eté du C.I.M.P.A., "Quantification-Quantification par déformation", 25 juin-12 juillet 1996, Nice, (France)*
7. Deligne, P. (1995) Déformations de l'Algèbre des Fonctions d'une Variété Symplectique: Comparaison entre Fedosov et De Wilde, Lecomte. *Selecta Mathematica, New Series.* **1**, pp. 667-697
8. Drinfeld, V. G. (1983) On constant, quasiclassical solutions of the Yang-Baxter quantum equation, *Sov. Math. Dokl.* **28**, pp. 667-671
9. Drinfeld, V. G. (1990), Quasi-Hopf algebras,*Leningrad Math. J.*, **1**, pp. 1419-1457
10. Fedosov, B. (1985) Formal quantization. in: *Some topics of Modern Mathematics and their applications to problems of mathematical physics*, pp. 129-136, Moscow
11. Fedosov, B. (1991) Index theorems, *Itogi Nauki i Tekhniki*, **65**
12. Fedosov, B. (1994) A simple geometrical constuction of deformation quantization, *J. Diff. Geom.*, **40**, pp. 213-238
13. Flato, M., Lichnerowicz, A. et Sternheimer, D. (1975) Déformations 1-différentiables d'algèbres de Lie attachées à une variété symplectique ou de contact,*Compositio. Math.*, **31**, pp. 47-92
14. Gerstenhaber, M. (1964) Deformation theory of algebraic structures, *Ann. of Math.*, **79**, pp. 59-90
15. Moreno, C. et Valero, L. (1990) Produits star invariants et équation de Yang-Baxter quantique constante, in *Actes des Journées Relativistes, 24-29 avril 1990, Aussois, France.*
16. Moreno, C. et Valero, L. (1992) Star products and quantum groups, *Physics and Manifolds. Proceedings of the International Colloquium in honour of Yvonne Choquet-Bruhat, Paris, June 3-5, 1992. Kluwer Academic Publishers, 1994.*

17. Moreno, C. Produits star et groups quantiques, *Notes du cours dans l'Ecole d'Eté du C.I.M.P.A., "Quantification-Quantification par déformation", 25 juin-12 juillet 1996, Nice, (France)*
18. Nest, R. and Tsygan, B. (1995) Algebraic index theorem. *Commun. Math. Phys.*, **172**, pp. 223-262
19. Nest,R. and Tsygan, B. (1995) Algebraic index theorem for families, *Advances in Mathematics*, **1133**, pp. 151-205
20. Rieffel, M. (1993) Deformation quantization for actions of $\mathbb{R}^d$,*Memoirs A.M.S.* **506**, *Providence, RI: Am. Math. Soc.*
21. Rieffel, M. (1995) Non-Compact Quantum Groups Associated with Abelian Subgroups, *Commun. Math. Phys.* **171**, pp. 181-201
22. Takhtajan, L.A. (1990) Lectures on quantum groups, in: *Introduction to Quantum Groups and Integrable Massive Models of Quantum Field Theory.* Nankai Lectures on Mathematical Physics. World Scientific, London
23. Trèves, F. (1970) *Topological vector spaces, distributions and kernels*, Academic Press, New York, London

ASPETTI FISICI DELLA TEORIA DELLE DEFORMAZIONI

GIUSEPPE DITO
Research Institute for Mathematical Sciences
Kyoto University
Sakyo-ku, Kyoto 606-01 Japan

Riassunto: Presenteremo brevemente e in un modo immaginoso, la nozione matematica di deformazione delle strutture algebriche (gruppi e algebre) ponendo l'accente sulla loro presenza in molti campi della fisica contemporanea. Questa teoria spiega il passaggio fra diversi livelli di descrizione della realtà fisica (permette anche le loro predizioni). L'algebra di Galilei è quella dell'invarianza della meccanica di Newton e dalla deformazione di questa algebra ne risulta che si trova l'algebra di Poincaré, che è niente meno che l'algebra di invarianza della teoria della relatività speciale. Un'altra tappa di deformazione permette il passaggio alla teoria della relatività generale. In un modo simile vedremo in che senso la meccanica quantistica è una deformazione della meccanica di Newton e di Hamilton, e questa sarà il tema centrale della conferenza di questa settimana.

Abstract: We will briefly introduce in an illustrative way the mathematical notion of deformations of algebraic structures (groups and algebras) by stressing their presence in many fields of contemporary physics. This theory explains the transition between different levels of description of the physical reality (it also permits their predictions). The Galilei algebra is the algebra of invariance of Newtonian Mechanics, and from its deformation one finds out the Poincaré algebra which is nothing else than the algebra of invariance of special relativity. One further step of deformations allows the transition to the theory of general relativity. In a similar way, we will see in what sense Quantum Mechanics appears to be a deformation of Newtonian and Hamiltonian Mechanics. This will be the main subject of this week's conference.

References

1. M. Flato. Deformation view of physical theories, Czec. J. Phys. **B32** (1982), 472-475.

D. Sternheimer et al. (eds.), Deformation Theory and Symplectic Geometry, 353.

DEFORMATION THEORY, SYMPLECTIC GEOMETRY AND APPLICATIONS

Organizers: JÜRG FRÖHLICH, MICHEL CAHEN, MOSHÉ FLATO, YVETTE KOSMANN-SCHWARZBACH, MURRAY GERSTENHABER AND DIDIER ARNAL

Centro Stefano Franscini of ETH Zürich
Monte Verità, CH-6612 Ascona, June 17–21, 1996

SCIENTIFIC PROGRAMME

Monday June 17

9h - 10h45 : Getting together and informal discussions.

10h50 : Opening by Michel CAHEN

Chair: André LICHNEROWICZ

11h - 12h : Ludwig FADDEEV (St. Petersburg and Helsinki)
Shift operator for lattice $SL(2)$ current algebra and Sugawara construction

LUNCH BREAK

14h15 - 15h45: [Discussions and Informal Seminars organized by Didier ARNAL]
Jens HOPPE: *Multidimensional Integrable Systems and M-Algebras*
Shahn MAJID: *Braided approach to Lusztig's construction of $U_q(g)$*

Chair: Daniel STERNHEIMER

16h - 17h : Stanislaw L. WORONOWICZ (University of Warsaw)
Manageable multiplicative unitaries

17h30 - 18h30 : Alan WEINSTEIN (U.C. Berkeley)
The modular automorphism group of a Poisson manifold

Tuesday June 18

Chair: Joseph WOLF

9h30 - 10h30 : Bertram KOSTANT (M.I.T.)
Flag manifold, quantum cohomology, the Toda lattice and the irreducible representation with highest weight

11h - 12h : Olivier MATHIEU (C.N.R.S. Strasbourg)
Canonical operations arising from symplectic structures

LUNCH BREAK

14h15 - 15h45 : [Discussions and Informal Seminars]
André LICHNEROWICZ: *On a 2-cocycle which is never exact*
Anthony GIAQUINTO: *Boundary solutions to the classical Yang-Baxter equation*

Chair: Michael SEMENOV-TIAN-SHANSKY

16h - 17h : Murray GERSTENHABER (University of Pennsylvania)
Hecke algebras, $U_q(sl_n)$, and the Donald-Flanigan conjecture for S_n

17h30 - 18h30 : Nicolai RESHETIKHIN (U.C. Berkeley)
On quantization of the moduli space of flat bundle over surfaces

Wednesday June 19

Chair: Yvette KOSMANN-SCHWARZBACH

9h30 - 10h30 : Alain CONNES (Collège de France and I.H.E.S.)
Gravity coupled with matter and spectral action principle

11h - 12h : Jürg FRÖHLICH (E.T.H., Zürich)
Supersymmetric quantum theory and geometry

LUNCH BREAK

14h15 - 15h45 : [Discussions and Informal Seminars]
Pierre CARTIER: *Lie algebras at root of unity*

Chair: Alberto GALINDO

15h - 16h : Tetsuji MIWA (R.I.M.S., Kyoto)
Symmetric approach to solvable lattice models

16h30 - 17h30 : Michio JIMBO (Kyoto University)
Infinite dimensional symmetries in solvable lattice models

17h45 - 18h45 : Christian FRØNSDAL (U.C.L.A.)
Deformations of quantum groups with some results on the elliptic case

19H30: DINNER
followed by piano evening with HOROWITZ (Joël) and BIELIAVSKY (Pierre)

Thursday June 20

Chair: Alexei REYMAN

9h30 - 10h30 : Moshé FLATO (Université de Bourgogne)
Deformation quantization and Nambu mechanics

11h - 12h : Maxim KONTSEVICH (I.H.E.S.)
Associativity equation via antibrackets

LUNCH BREAK

14h15 - 15h15 : [Discussions and Informal Seminars]

Johannes HUEBSCHMANN: *Extended moduli spaces and the Kan construction*

Chair: John RAWNSLEY

15h30 - 16h30 : Simone GUTT (U.L.B. and Metz)
Convergent star-products and Weyl transforms

17h - 18h : Boris TSYGAN (Pennsylvania State University)
On the cohomology ring of an algebra

18h15 - 19h15 : Wilfried SCHMID (Harvard)
Localization of integrals and character formulas for non-compact groups

Evening Popular Lecture by Giuseppe DITO: *Aspetti fisici della teoria delle deformazioni*

Friday June 21

Chair: Yoshiaki MAEDA

9h - 10h : Marc ROSSO (Université Louis Pasteur, Strasbourg)
Some applications of quantum shuffles

10h15 - 11h15 : Tudor RATIU (U.C. Santa Cruz)
Gradient systems, integrability and convexity in finite and infinite dimension

11h45 - 12h45 : James STASHEFF (University of North Carolina)
Deformation theory and the Batalin-Vilkovisky master equation

LUNCH BREAK

14h15 - 17h: [Discussions and Informal Seminars]

14h15. Martin BORDEMANN:
A formal GNS construction and states in deformation quantization
15h15. Carlos MORENO: *Star products in the triangular case*
16h15. Joseph WOLF: *Cycle space on a flag domain*

PARTICIPANTS

Lecturers

Alain Connes, I.H.E.S., Bures sur Yvette, France
connes@ihes.fr

Ludwig Faddeev, Steklov Institute of Mathematics, St. Petersburg and University of Helsinki, Finland
faddeev@pdmi.ras.ru & faddeev@pcu.helsinki.fi

Moshé Flato, Université de Bourgogne, France
flato@u-bourgogne.fr

Jürg Fröhlich, E.T.H. Honggerberg, Zürich, Switzerland
juerg@itp.phys.ethz.ch

Christian Frønsdal, U.C.L.A., USA
fronsdal@physics.ucla.edu

Murray Gerstenhaber, University of Pennsylvania, USA
murray@math.upenn.edu

Simone Gutt, Université Libre de Bruxelles, Belgium and Université de Metz, France
sgutt@ulb.ac.be & gutt@poncelet.univ-metz.fr

Michio Jimbo, Mathematics, Kyoto University, Japan
jimbo@ginkaku.kusm.kyoto-u.ac.jp

Maxim Kontsevich, I.H.E.S., Bures sur Yvette, France
maxim@ihes.fr

Bertram Kostant, M.I.T., USA
kostant@math.mit.edu

Olivier Mathieu, Université Louis Pasteur, Strasbourg, France
mathieu@math.u-strasbg.fr

Tetsuji Miwa, R.I.M.S., Kyoto University, Japan
miwa@kurims.kyoto-u.ac.jp

Tudor Ratiu, Mathematics, U.C. Santa Cruz, USA
ratiu@cats.ucsc.edu

Nicolai Reshetikhin, U.C. Berkeley, USA
reshetik@math.berkeley.edu

Marc Rosso, Université Louis Pasteur, Strasbourg, France
rosso@math.u-strasbg.fr

Wilfried Schmid, Harvard University, USA
schmid@math.harvard.edu

James Stasheff, University of North Carolina, USA
jds@math.unc.edu

Boris Tsygan, Pennsylvania State University, USA
tsygan@math.psu.edu

Alan Weinstein, U.C. Berkeley, USA
alanw@math.berkeley.edu

Stanislaw L. Woronowicz, University of Warsaw, Poland
Stanislaw.Woronowicz@fuw.edu.pl

Other Participants

Anton Alekseev, E.T.H. Honggerberg, Zürich, Switzerland
alekseev@itp.phys.ethz.ch

Didier Arnal, Université de Metz, France
arnal@poncelet.univ-metz.fr

Santos Asin Lares, University of Warwick, United Kingdom
asin@maths.warwick.ac.uk

Fréderic Bidegain, Université Libre de Bruxelles, Belgium
bidegain@ulb.ac.be [currently bidegain@poncelet.univ-metz.fr]

Pierre Bieliavsky, U.C. Berkeley, USA
and Université Libre de Bruxelles, Belgium
piebie@math.Berkeley.edu & pbiel@ulb.ac.be

Philippe Bonneau, Université de Bourgogne, France
bonneau@u-bourgogne.fr

Martin Bordemann, Universität Freiburg, Germany
mbor@phyq1.physik.uni-freiburg.de

Michel Cahen, Université Libre de Bruxelles, Belgium
mcahen@ulb.ac.be

Pierre Cartier, Ecole Normale Supérieure, Paris, France
cartier@ihes.fr

Véronique Chloup-Arnould, Université de Metz, France
chloup@poncelet.univ-metz.fr

Jean-Claude Cortet, Université de Bourgogne, France
cortet@u-bourgogne.fr

Pierre Dazord, Université de Lyon I, France
dazord@geometrie.univ-lyon1.fr

Vivian De Smedt, Université Libre de Bruxelles, Belgium
vdesmedt@ulb.ac.be
Giuseppe Dito, R.I.M.S., Kyoto University, Japan
and Université de Bourgogne, France
dito@kurims.kyoto-u.ac.jp & ditog@u-bourgogne.fr

Benjamin Enriquez, Ecole Polytechnique, France
enriquez@orphee.polytechnique.fr

Alberto Galindo Tixaire, Universidad Complutense de Madrid, Spain
galindo@eucmos.sim.ucm.es

Tony Giaquinto, Mississipi State University, USA
tonyg@math.MsState.Edu

Jens Hoppe, E.T.H., Switzerland
hoppe@itp.phys.ethz.ch

Joel Horowitz, Université Libre de Bruxelles, Belgium
horowitz@ulb.ac.be

Johannes Huebschmann, Université de Lille I, France
huebschm@gat.univ-lille1.fr

Christophe Kapoudjian, Université de Lyon I, France
ckapoudj@desargues.univ-lyon1.fr

Yvette Kosmann-Schwarzbach, Ecole Polytechnique, France
yks@orphee.polytechnique.fr

André Lichnerowicz, Membre de l'Institut de France, Paris

Maria Lledo, Physics, U.C.L.A., USA
lledo@physics.ucla.edu

Yoshiaki Maeda, Mathematics, Keio University, Japan
maeda@math.keio.ac.jp

Shahn Majid, D.A.M.T.P., University of Cambridge, United Kingdom
Shahn.Majid@damtp.cam.ac.uk

Martin Markl, Mathematical Institute of the Academy, Czech Republic
mmarkl@math.unc.edu & markl@mbox.cesnet.cz

Danilo Marlini, CERFIM, Locarno, Switzerland

Christiane Martin, Université de Bourgogne, France
cmartin@u-bourgogne.fr

Vladimir Matveev, Université de Bourgogne, France
matveev@u-bourgogne.fr

Sébastien Michea, Université de Bourgogne, France
michea@u-bourgogne.fr

Carlos Moreno Gonzalez, Universidad Complutense de Madrid, Spain
cmoreno@eucmos.sim.ucm.es

Ryszard Nest, University of Copenhagen, Denmark
rnest@math.ku.dk

Georges Pinczon, Université de Bourgogne, France
physmath@u-bourgogne.fr (pour Pinczon)

John Rawnsley, University of Warwick, United Kingdom
jhr@maths.warwick.ac.uk

Alexei Reyman, Université de Lyon I, France
reyman@geometrie.univ-lyon1.fr

Olivier Schiffmann, Ecole Normale Supérieure, France
schiffma@clipper.ens.fr

Michael Semenov-Tian-Shansky, Université de Bourgogne, France
semenov@u-bourgogne.fr

Alexei Sevostyanov, Uppsala University, Sweden
seva@rhea.teorfys.uu.se

Pierre Sleewaegen, Université Libre de Bruxelles, Belgium
ulbmath@ulb.ac.be (pour Sleewaegen)

Daniel Sternheimer, Université de Bourgogne, France
dastern@u-bourgogne.fr

Stefan Waldmann, Universität Freiburg, Germany
waldman@phyq2.physik.uni-freiburg.de

Joseph A. Wolf, U.C. Berkeley, USA
jawolf@math.Berkeley.edu

This photograph appears by courtesy of Frau Katia Bastianelli, Centro Stefano Franscini, Monte Verità, Ascona, Switzerland.

MONTE VERITÀ
ASCONA

Centro Stefano Franscini
Politecnico federale di Zurigo

Workshop sulla

Teoria di deformazione, geometria simplettica e applicazioni

16-22 giugno 1996

In questo convegno verranno presentati i più recenti sviluppi attorno alla teoria della deformazione e la geometria simplettica. Verranno sottolineati specialmenti i collegamenti tra quantizzazione star (star-quantization), la geometria simplettica e Poisson, i gruppi di quanti (affini e non) e la geometria non comunicativa. Verranno discusse le applicazioni alle varie aree della fisica matematica incluso le teorie di campo e la fisica della materia condensata.

Il programma consiste in relazioni plenarie e seminari dedicati a problemi di carattere più tecnico. Parte del tempo sarà messo a disposizione per degli incontri informali tra i partecipanti.

Organizzatori: Prof. J. Fröhlich (Theo. Physics, ETH Zurigo), Prof. M. Cahen (Dépt. Math. ULB, Bruxelles, Belgio), Prof. M. Flato (Dépt. Math., Université de Bourgogne, Dijon, Francia), Prof. Y. Kosmann-Schwarzbach (Centre de Mathématique, Ecole Polytechnique, Palaiseau, Francia)

La conferenza è sponsorizzato dal Fondo Nazionale Svizzero per la ricerca scientifica (Berna), dal Centro Stefano Franscini (ETH Zurigo) e dall' UE.

La conferenza pubblica verrà annunciata in seguito

This photograph appears by courtesy of Frau Katia Bastianelli, Centro Stefano Franscini,

MONTE VERITÀ
ASCONA

Centro Stefano Franscini
ETH Zurich

Workshop on

Deformation Theory, Symplectic Geometry and Applications

June 16-22, 1996

Recent developments around deformation theory and symplectic geometry will be presented in this meeting. There will be emphasis on links between star-quantization, symplectic and Poisson geometry, quantum groups (affine or not) and noncommutative geometry. Applications to various areas of mathematical physics including field theories and condensed matter physics will be discussed.

The programme consists of plenary lectures and seminars devoted to more technical questions. Some time will be left free for informal meetings between participants.

Organizers: Prof. J. Fröhlich (Theo. Physics, ETH Zurich), Prof. M. Cahen (Dépt. Math. ULB, Bruxelles, Belgian), Prof. M. Flato (Dépt. Math., Université de Bourgogne, Dijon, French), Prof. Y. Kosmann-Schwarzbach (Centre de Mathématique, Ecole Polytechnique, Palaiseau, French)

The conference is sponsored by the Swiss National Science Foundation (Berne), the CSF (ETH Zurich) and by the EU.

Public lecture: to be announced

Mathematical Physics Studies

1. F.A.E. Pirani, D.C. Robinson and W.F. Shadwick: *Local Jet Bundle Formulation of Bäcklund Transformations.* 1979 ISBN 90-277-1036-8
2. W.O. Amrein: *Non-Relativistic Quantum Dynamics.* 1981 ISBN 90-277-1324-3
3. M. Cahen, M. de Wilde, L. Lemaire and L. Vanhecke (eds.): *Differential Geometry and Mathematical Physics.* 1983 Pb ISBN 90-277-1508-4
4. A.O. Barut (ed.): *Quantum Theory, Groups, Fields and Particles.* 1983 ISBN 90-277-1552-1
5. G. Lindblad: *Non-Equilibrium Entropy and Irreversibility.* 1983 ISBN 90-277-1640-4
6. S. Sternberg (ed.): *Differential Geometric Methods in Mathematical Physics.* 1984 ISBN 90-277-1781-8
7. J.P. Jurzak: *Unbounded Non-Commutative Integration.* 1985 ISBN 90-277-1815-6
8. C. Fronsdal (ed.): *Essays on Supersymmetry.* 1986 ISBN 90-277-2207-2
9. V.N. Popov and V.S. Yarunin: *Collective Effects in Quantum Statistics of Radiation and Matter.* 1988 ISBN 90-277-2735-X
10. M. Cahen and M. Flato (eds.): *Quantum Theories and Geometry.* 1988 ISBN 90-277-2803-8
11. Bernard Prum and Jean Claude Fort: *Processes on a Lattice and Gibbs Measures.* 1991 ISBN 0-7923-1069-1
12. A. Boutet de Monvel, Petre Dita, Gheorghe Nenciu and Radu Purice (eds.): *Recent Developments in Quantum Mechanics.* 1991 ISBN 0-7923-1148-5
13. R. Gielerak, J. Lukierski and Z. Popowicz (eds.): *Quantum Groups and Related Topics.* Proceedings of the First Max Born Symposium. 1992 ISBN 0-7923-1924-9
14. A. Lichnerowicz, Magnetohydrodynamics: Waves and Shock Waves in Curved Space-Time. 1994 ISBN 0-7923-2805-1
15. M. Flato, R. Kerner and A. Lichnerowicz (eds.): *Physics on Manifolds.* 1993 ISBN 0-7923-2500-1
16. H. Araki, K.R. Ito, A. Kishimoto and I. Ojima (eds.): *Quantum and Non-Commutative Analysis.* Past, Present and Future Perspectives. 1993 ISBN 0-7923-2532-X
17. D. Ya. Petrina: *Mathematical Foundations of Quantum Statistical Mechanics.* Continuous Systems. 1995 ISBN 0-7923-3258-X
18. J. Bertrand, M. Flato, J.-P. Gazeau, M. Irac-Astaud and D. Sternheimer (eds.): *Modern Group Theoretical Methods in Physics.* Proceedings of the Conference in honour of Guy Rideau. 1995 ISBN 0-7923-3645-3
19. A. Boutet de Monvel and V. Marchenko (eds.): *Algebraic and Geometric Methods in Mathematical Physics.* Proceedings of the Kaciveli Summer School, Crimea, Ukraine, 1993. 1996 ISBN 0-7923-3909-6

Mathematical Physics Studies

20. D. Sternheimer, J. Rawnsley and S. Gutt (eds.): *Deformation Theory and Symplectic Geometry*. Proceedings of the Ascona Meeting, June 1996. 1997
ISBN 0-7923-4525-8

Kluwer Academic Publishers – Dordrecht / Boston / London

GPSR Compliance

The European Union's (EU) General Product Safety Regulation (GPSR) is a set of rules that requires consumer products to be safe and our obligations to ensure this.

If you have any concerns about our products, you can contact us on ProductSafety@springernature.com

In case Publisher is established outside the EU, the EU authorized representative is:

Springer Nature Customer Service Center GmbH
Europaplatz 3
69115 Heidelberg, Germany

Batch number: 10614741

Printed by Printforce, the Netherlands